AF835487

TRANSLATIONAL RESEARCH IN TRAUMATIC BRAIN INJURY

FRONTIERS IN NEUROSCIENCE

Series Editor

Sidney A. Simon, PhD

Published Titles

Apoptosis in Neurobiology

Yusuf A. Hannun, MD, Professor of Biomedical Research and Chairman, Department of Biochemistry and Molecular Biology, Medical University of South Carolina, Charleston, South Carolina

Rose-Mary Boustany, MD, tenured Associate Professor of Pediatrics and Neurobiology, Duke University Medical Center, Durham, North Carolina

Neural Prostheses for Restoration of Sensory and Motor Function

John K. Chapin, PhD, Professor of Physiology and Pharmacology, State University of New York Health Science Center, Brooklyn, New York

Karen A. Moxon, PhD, Assistant Professor, School of Biomedical Engineering, Science, and Health Systems, Drexel University, Philadelphia, Pennsylvania

Computational Neuroscience: Realistic Modeling for Experimentalists

Eric DeSchutter, MD, PhD, Professor, Department of Medicine, University of Antwerp, Antwerp, Belgium

Methods in Pain Research

Lawrence Kruger, PhD, Professor of Neurobiology (Emeritus), UCLA School of Medicine and Brain Research Institute, Los Angeles, California

Motor Neurobiology of the Spinal Cord

Timothy C. Cope, PhD, Professor of Physiology, Wright State University, Dayton, Ohio

Nicotinic Receptors in the Nervous System

Edward D. Levin, PhD, Associate Professor, Department of Psychiatry and Pharmacology and Molecular Cancer Biology and Department of Psychiatry and Behavioral Sciences, Duke University School of Medicine, Durham, North Carolina

Methods in Genomic Neuroscience

Helmin R. Chin, PhD, Genetics Research Branch, NIMH, NIH, Bethesda, Maryland

Steven O. Moldin, PhD, University of Southern California, Washington, D.C.

Methods in Chemosensory Research

Sidney A. Simon, PhD, Professor of Neurobiology, Biomedical Engineering, and Anesthesiology, Duke University, Durham, North Carolina

Miguel A.L. Nicolelis, MD, PhD, Professor of Neurobiology and Biomedical Engineering, Duke University, Durham, North Carolina

The Somatosensory System: Deciphering the Brain's Own Body Image

Randall J. Nelson, PhD, Professor of Anatomy and Neurobiology, University of Tennessee Health Sciences Center, Memphis, Tennessee

The Superior Colliculus: New Approaches for Studying Sensorimotor Integration

William C. Hall, PhD, Department of Neuroscience, Duke University, Durham, North Carolina

Adonis Moschovakis, PhD, Department of Basic Sciences, University of Crete, Heraklion, Greece

New Concepts in Cerebral Ischemia
Rick C.S. Lin, PhD, Professor of Anatomy, University of Mississippi Medical Center, Jackson, Mississippi

DNA Arrays: Technologies and Experimental Strategies
Elena Grigorenko, PhD, Technology Development Group, Millennium Pharmaceuticals, Cambridge, Massachusetts

Methods for Alcohol-Related Neuroscience Research
Yuan Liu, PhD, National Institute of Neurological Disorders and Stroke, National Institutes of Health, Bethesda, Maryland
David M. Lovinger, PhD, Laboratory of Integrative Neuroscience, NIAAA, Nashville, Tennessee

Primate Audition: Behavior and Neurobiology
Asif A. Ghazanfar, PhD, Princeton University, Princeton, New Jersey

Methods in Drug Abuse Research: Cellular and Circuit Level Analyses
Barry D. Waterhouse, PhD, MCP-Hahnemann University, Philadelphia, Pennsylvania

Functional and Neural Mechanisms of Interval Timing
Warren H. Meck, PhD, Professor of Psychology, Duke University, Durham, North Carolina

Biomedical Imaging in Experimental Neuroscience
Nick Van Bruggen, PhD, Department of Neuroscience Genentech, Inc.
Timothy P.L. Roberts, PhD, Associate Professor, University of Toronto, Canada

The Primate Visual System
John H. Kaas, Department of Psychology, Vanderbilt University, Nashville, Tennessee
Christine Collins, Department of Psychology, Vanderbilt University, Nashville, Tennessee

Neurosteroid Effects in the Central Nervous System
Sheryl S. Smith, PhD, Department of Physiology, SUNY Health Science Center, Brooklyn, New York

Modern Neurosurgery: Clinical Translation of Neuroscience Advances
Dennis A. Turner, Department of Surgery, Division of Neurosurgery, Duke University Medical Center, Durham, North Carolina

Sleep: Circuits and Functions
Pierre-Hervé Luppi, Université Claude Bernard, Lyon, France

Methods in Insect Sensory Neuroscience
Thomas A. Christensen, Arizona Research Laboratories, Division of Neurobiology, University of Arizona, Tuscon, Arizona

Motor Cortex in Voluntary Movements
Alexa Riehle, INCM-CNRS, Marseille, France
Eilon Vaadia, The Hebrew University, Jerusalem, Israel

Neural Plasticity in Adult Somatic Sensory-Motor Systems
Ford F. Ebner, Vanderbilt University, Nashville, Tennessee

Advances in Vagal Afferent Neurobiology
Bradley J. Undem, Johns Hopkins Asthma Center, Baltimore, Maryland
Daniel Weinreich, University of Maryland, Baltimore, Maryland

The Dynamic Synapse: Molecular Methods in Ionotropic Receptor Biology
Josef T. Kittler, University College, London, England
Stephen J. Moss, University College, London, England

Animal Models of Cognitive Impairment
Edward D. Levin, Duke University Medical Center, Durham, North Carolina
Jerry J. Buccafusco, Medical College of Georgia, Augusta, Georgia

The Role of the Nucleus of the Solitary Tract in Gustatory Processing
Robert M. Bradley, University of Michigan, Ann Arbor, Michigan

Brain Aging: Models, Methods, and Mechanisms
David R. Riddle, Wake Forest University, Winston-Salem, North Carolina

Neural Plasticity and Memory: From Genes to Brain Imaging
Frederico Bermudez-Rattoni, National University of Mexico, Mexico City, Mexico

Serotonin Receptors in Neurobiology
Amitabha Chattopadhyay, Center for Cellular and Molecular Biology, Hyderabad, India

TRP Ion Channel Function in Sensory Transduction and Cellular Signaling Cascades
Wolfgang B. Liedtke, MD, PhD, Duke University Medical Center, Durham, North Carolina
Stefan Heller, PhD, Stanford University School of Medicine, Stanford, California

Methods for Neural Ensemble Recordings, Second Edition
Miguel A.L. Nicolelis, MD, PhD, Professor of Neurobiology and Biomedical Engineering, Duke University Medical Center, Durham, North Carolina

Biology of the NMDA Receptor
Antonius M. VanDongen, Duke University Medical Center, Durham, North Carolina

Methods of Behavioral Analysis in Neuroscience
Jerry J. Buccafusco, PhD, Alzheimer's Research Center, Professor of Pharmacology and Toxicology, Professor of Psychiatry and Health Behavior, Medical College of Georgia, Augusta, Georgia

In Vivo Optical Imaging of Brain Function, Second Edition
Ron Frostig, PhD, Professor, Department of Neurobiology, University of California, Irvine, California

Fat Detection: Taste, Texture, and Post Ingestive Effects
Jean-Pierre Montmayeur, PhD, Centre National de la Recherche Scientifique, Dijon, France
Johannes le Coutre, PhD, Nestlé Research Center, Lausanne, Switzerland

The Neurobiology of Olfaction
Anna Menini, PhD, Neurobiology Sector International School for Advanced Studies, (S.I.S.S.A.), Trieste, Italy

Neuroproteomics
Oscar Alzate, PhD, Department of Cell and Developmental Biology, University of North Carolina, Chapel Hill, North Carolina

Translational Pain Research: From Mouse to Man
Lawrence Kruger, PhD, Department of Neurobiology, UCLA School of Medicine, Los Angeles, California
Alan R. Light, PhD, Department of Anesthesiology, University of Utah, Salt Lake City, Utah

Advances in the Neuroscience of Addiction
Cynthia M. Kuhn, Duke University Medical Center, Durham, North Carolina
George F. Koob, The Scripps Research Institute, La Jolla, California

Neurobiology of Huntington's Disease: Applications to Drug Discovery
Donald C. Lo, Duke University Medical Center, Durham, North Carolina
Robert E. Hughes, Buck Institute for Age Research, Novato, California

Neurobiology of Sensation and Reward
Jay A. Gottfried, Northwestern University, Chicago, Illinois

The Neural Bases of Multisensory Processes
Micah M. Murray, CIBM, Lausanne, Switzerland
Mark T. Wallace, Vanderbilt Brain Institute, Nashville, Tennessee

Neurobiology of Depression
Francisco López-Muñoz, University of Alcalá, Madrid, Spain
Cecilio Álamo, University of Alcalá, Madrid, Spain

Astrocytes: Wiring the Brain
Eliana Scemes, Albert Einstein College of Medicine, Bronx, New York
David C. Spray, Albert Einstein College of Medicine, Bronx, New York

Dopamine–Glutamate Interactions in the Basal Ganglia
Susan Jones, University of Cambridge, United Kingdom

Alzheimer's Disease: Targets for New Clinical Diagnostic and Therapeutic Strategies
Renee D. Wegrzyn, Booz Allen Hamilton, Arlington, Virginia
Alan S. Rudolph, Duke Center for Neuroengineering, Potomac, Maryland

The Neurobiological Basis of Suicide
Yogesh Dwivedi, University of Illinois at Chicago

Transcranial Brain Stimulation
Carlo Miniussi, University of Brescia, Italy
Walter Paulus, Georg-August University Medical Center, Göttingen, Germany
Paolo M. Rossini, Institute of Neurology, Catholic University of Rome, Italy

Spike Timing: Mechanisms and Function
Patricia M. Di Lorenzo, Binghamton University, Binghamton, New York
Jonathan D. Victor, Weill Cornell Medical College, New York City, New York

Neurobiology of Body Fluid Homeostasis: Transduction and Integration
Laurival Antonio De Luca Jr., São Paulo State University–UNESP, Araraquara, Brazil
Jose Vanderlei Menani, São Paulo State University–UNESP, Araraquara, Brazil
Alan Kim Johnson, The University of Iowa, Iowa City, Iowa

Neurobiology of Chemical Communication
Carla Mucignat-Caretta, University of Padova, Padova, Italy

Itch: Mechanisms and Treatment
E. Carstens, University of California, Davis, California
Tasuku Akiyama, University of California, Davis, California

Translational Research in Traumatic Brain Injury
Daniel Laskowitz, Duke University, Durham, North Carolina
Gerald Grant, Duke University, Durham, North Carolina

TRANSLATIONAL RESEARCH IN TRAUMATIC BRAIN INJURY

Edited by
Daniel Laskowitz
Duke University
Durham, North Carolina, USA

Gerald Grant
Stanford University
Stanford, California, USA

CRC Press
Taylor & Francis Group
Boca Raton London New York

CRC Press is an imprint of the
Taylor & Francis Group, an **informa** business

CRC Press
Taylor & Francis Group
6000 Broken Sound Parkway NW, Suite 300
Boca Raton, FL 33487-2742

© 2016 by Taylor & Francis Group, LLC
CRC Press is an imprint of Taylor & Francis Group, an Informa business

No claim to original U.S. Government works

Printed on acid-free paper
Version Date: 20150730

International Standard Book Number-13: 978-1-4665-8491-4 (Hardback)

This book contains information obtained from authentic and highly regarded sources. While all reasonable efforts have been made to publish reliable data and information, neither the author[s] nor the publisher can accept any legal responsibility or liability for any errors or omissions that may be made. The publishers wish to make clear that any views or opinions expressed in this book by individual editors, authors or contributors are personal to them and do not necessarily reflect the views/opinions of the publishers. The information or guidance contained in this book is intended for use by medical, scientific or health-care professionals and is provided strictly as a supplement to the medical or other professional's own judgement, their knowledge of the patient's medical history, relevant manufacturer's instructions and the appropriate best practice guidelines. Because of the rapid advances in medical science, any information or advice on dosages, procedures or diagnoses should be independently verified. The reader is strongly urged to consult the relevant national drug formulary and the drug companies' and device or material manufacturers' printed instructions, and their websites, before administering or utilizing any of the drugs, devices or materials mentioned in this book. This book does not indicate whether a particular treatment is appropriate or suitable for a particular individual. Ultimately it is the sole responsibility of the medical professional to make his or her own professional judgements, so as to advise and treat patients appropriately. The authors and publishers have also attempted to trace the copyright holders of all material reproduced in this publication and apologize to copyright holders if permission to publish in this form has not been obtained. If any copyright material has not been acknowledged please write and let us know so we may rectify in any future reprint.

Except as permitted under U.S. Copyright Law, no part of this book may be reprinted, reproduced, transmitted, or utilized in any form by any electronic, mechanical, or other means, now known or hereafter invented, including photocopying, microfilming, and recording, or in any information storage or retrieval system, without written permission from the publishers.

For permission to photocopy or use material electronically from this work, please access www.copyright.com (http://www.copyright.com/) or contact the Copyright Clearance Center, Inc. (CCC), 222 Rosewood Drive, Danvers, MA 01923, 978-750-8400. CCC is a not-for-profit organization that provides licenses and registration for a variety of users. For organizations that have been granted a photocopy license by the CCC, a separate system of payment has been arranged.

Trademark Notice: Product or corporate names may be trademarks or registered trademarks, and are used only for identification and explanation without intent to infringe.

Visit the Taylor & Francis Web site at
http://www.taylorandfrancis.com

and the CRC Press Web site at
http://www.crcpress.com

Contents

Series Preface

The Frontiers in Neuroscience Series presents the insights of experts on emerging experimental technologies and theoretical concepts that are, or will be, at the vanguard of neuroscience.

The books cover new and exciting multidisciplinary areas of brain research and describe breakthroughs in fields like visual, gustatory, auditory, olfactory neuroscience as well as aging biomedical imaging. Recent books cover the rapidly evolving fields of multisensory processing, depression, and different aspects of reward.

Each book is edited by experts and consists of chapters written by leaders in a particular field. The books have been richly illustrated and contain comprehensive bibliographies. The chapters provide substantial background material relevant to the particular subject.

The goal is for these books to be the references every neuroscientist uses in order to acquaint themselves with new information and methodologies in brain research. My task as series editor is to produce outstanding products that contribute to the broad field of neuroscience. Now that the chapters are available online, the effort put in by us—the publisher, the book editors, and individual authors—will contribute to the further development of brain research. To the extent that you learn from these books, we will have succeeded.

Sidney A. Simon, PhD
Series Editor

Preface

In recent years, increased recognition of the long-term impact of traumatic brain injury (TBI) in both the civilian and military populations has been brought to the forefront of our national consciousness. At present, TBI remains a significant source of death and permanent disability, contributing to about one-third of all injury-related deaths in the United States. Unfortunately, in the setting of moderate and severe TBI, the development of therapies to improve functional outcomes has been limited. Aside from surgical intervention in select cases, treatment remains largely supportive to prevent secondary injury to the brain. In recent years, there has been increased attention paid to the long-term sequelae of mild TBI and repeated subconcussive injuries, which may lead to progressive deterioration of cognitive and motor function. This cascade is of particular concern in the military, where blast-induced mild TBI has become the signature injury of recent conflicts, and in the setting of the repetitive sports injuries that may occur in young athletes. For both soldiers and athletes, the true incidence and progression of delayed neurological deterioration are difficult to predict.

Although the increased resources for research have recently been brought to bear to improve our understanding of the basic mechanisms of brain injury, the development of new diagnostic and therapeutic approaches has been disappointingly slow. In part, this is due to the complex nature of brain injury and its neurocognitive aftermath. To effectively improve the clinical care of the head-injured patient, multidisciplinary research must be performed to more completely define the epidemiology and causes of TBI, as well as molecular mechanisms of neuronal injury, genetic determinants of recovery, and novel approaches in diagnosis and treatment. We also now recognize the marked overlap of symptoms in patients with TBI and posttraumatic stress disorder. Specialists from neurosurgery, neurology, psychology, psychiatry, and rehabilitation medicine are finally at the same table to better understand the intersection of these diagnoses and formulate a rational treatment plan to address both disorders together.

The current volume attempts to integrate expertise from across specialties to address current perspectives and knowledge gaps in the field of TBI. Topics range from epidemiology of TBI, preclinical modeling, recent insights into the pathophysiology of secondary neuronal injury and chronic traumatic encephalopathy, to clinical research and trial design, diagnostic imaging, and current treatment strategies. Our hope is that the process of bringing together the perspectives of experts across disciplines may aid in the translation of new ideas to clinical practice and ultimately improve the care of the brain-injured patient.

We thank all of the authors for their dedication and commitment to this book. It has been a pleasure to work with all of you.

Editors

Daniel T. Laskowitz, MD, is vice chair of research and professor of neurology, neurobiology, and anesthesiology at Duke University Medical Center, where he serves as director of the Neurovascular Laboratories and the director for the third-year medical school curriculum. He is a graduate of Duke University School of Medicine (1991) and Brown University (1987), where he majored in neuroscience. After completing his neurology residency at the University of Pennsylvania in 1995, he returned to Duke to complete fellowship training in neurocritical care and stroke. He has remained active in both laboratory-based and clinical research, and completed his master's of health science in clinical research in 2003.

Dr. Laskowitz attends the Stroke Service and the Neurosciences Intensive Care Unit, a multidisciplinary unit in which patients with life-threatening neurological diseases, such as stroke, trauma, and intracranial hemorrhage, are cared for. His research focus is on the role of genetic influences on neuroinflammatory responses and recovery from brain injury. These results are translated to clinically relevant small animal models with the ultimate goal of exploring new therapeutic interventions in the clinical setting of stroke, intracranial hemorrhage, and closed head injury. Dr. Laskowitz's translational research also focuses on rapid point of care diagnostic approaches for the treatment of acute stroke.

Gerald Grant, MD, is an associate professor of neurosurgery and neurology at Stanford University. Dr. Grant earned his BS degree in neurosciences and German from Duke University and his MD degree at Stanford University. He completed a residency in neurosurgery and a fellowship in pediatric neurosurgery at the University of Washington in Seattle and won numerous awards for teaching patient care and research. Dr. Grant then entered active duty with the U.S. Air Force and was deployed to Iraq in support of Operation Iraqi Freedom. Dr. Grant attained the rank of lieutenant colonel and received a Meritorious Service Medal for his service and pioneering work on blast concussive injury. He then returned to Duke and became chief of neurotrauma. At Duke he was a site PI for the Boost II trial funded by the NIH and InTRUST Consortium funded by the Department of Defense focusing on the interface of posttraumatic stress disorder and traumatic brain injury.

Dr. Grant is a fellow of the American College of Surgeons and was elected to the National Committee on Trauma for a 6-year term. In 2003, Dr. Grant was recruited back to Stanford University where he is the division chief of pediatric neurosurgery and a vice chair in the Department of Neurosurgery. He is a diplomate of the American Board of Neurological Surgery and is also board-certified in pediatric neurosurgery by the American Board of Pediatric Neurological Surgery. Dr. Grant runs a translational laboratory funded by the NIH focusing on the blood–brain barrier and is investigating novel ways to improve drug delivery into the brain. He also holds leadership positions on several executive committees of national organizations.

Contributors

Shawn K. Acheson
Duke University Medical Center
and
Research and Development/Mental Health Services
Durham Veterans Affairs Medical Center
and
Veterans Integrated Service Networks
Mid-Atlantic Mental Illness Research Education and Clinical Center
Durham, North Carolina

Jason M. Bailie
General Dynamics Information Technology
Frederick, Maryland

Michael Bell
University of Pittsburgh School of Medicine
Pittsburgh, Pennsylvania

Ellen R. Bennett
Duke University Medical Center
Durham, North Carolina

Matthew Campbell
Trinity College Dublin
Dublin, Ireland

Bruce P. Capehart
Duke University Medical Center
and
Research and Development/Mental Health Services
Durham Veterans Affairs Medical Center
and
Veterans Integrated Service Networks
Mid-Atlantic Mental Illness Research Education and Clinical Center
Durham, North Carolina

Michael Clark
University of North Carolina at Chapel Hill
Chapel Hill, North Carolina

Wesley R. Cole
General Dynamics Information Technology
Frederick, Maryland

Giulia Curia
University of Modena and Reggio Emilia
Modena, Italy

Raimondo D'Ambrosio
University of Washington
Seattle, Washington

Ramon Diaz-Arrastia
Uniformed Services
University of the Health Sciences
Rockville, Maryland

Kan Ding
Uniformed Services
University of the Health Sciences
Rockville, Maryland

Charlotte E. Dunn
Research and Development/Mental Health Services
Durham Veterans Affairs Medical Center
and
Veterans Integrated Service Networks
Mid-Atlantic Mental Illness Research Education and Clinical Center
Durham, North Carolina

Clifford L. Eastman
University of Washington
Seattle, Washington

Gerald Grant
Stanford University School of Medicine
Stanford, California

Puneet K. Gupta
Uniformed Services
University of the Health Sciences
Rockville, Maryland

Kevin Guskiewicz
University of North Carolina at Chapel Hill
Chapel Hill, North Carolina

Edward D. Hall
University of Kentucky Medical Center
Lexington, Kentucky

Finnian Hanrahan
Trinity College Dublin
Dublin, Ireland

Jonathon R. Howlett
University of California, San Diego
La Jolla, California

Jakob Hufschmidt
Brigham and Women's Hospital
Harvard Medical School
Boston, Massachusetts

Damir Janigro
Cleveland Clinic Lerner College of Medicine
Cleveland, Ohio

Jason D. Kilts
Duke University Medical Center
and
Research and Development/Mental Health Services
Durham Veterans Affairs Medical Center
and
Veterans Integrated Service Networks
Mid-Atlantic Mental Illness Research Education and Clinical Center
Durham, North Carolina

Inga K. Koerte
Brigham and Women's Hospital
Harvard Medical School
Boston, Massachusetts

Kamesh Krishnamurthy
Duke University School of Medicine
Durham, North Carolina

Shivanand P. Lad
Duke University Medical Center
Durham, North Carolina

Daniel T. Laskowitz
Duke University School of Medicine
Durham, North Carolina

Peter Le Roux
Lankenau Medical Center
Wynnewood, Pennsylvania

Peter Leo
Columbia St. Mary's Hospital
Milwaukee, Wisconsin

Alexander P. Lin
Brigham and Women's Hospital
Harvard Medical School
Boston, Massachusetts

Christine E. Marx
Duke University Medical Center
and
Research and Development/Mental Health Services
Durham Veterans Affairs Medical Center
and
Veterans Integrated Service Networks
Mid-Atlantic Mental Illness Research Education and Clinical Center
Durham, North Carolina

Michael McCrea
Medical College of Wisconsin
Milwaukee, Wisconsin

John W. Miller
University of Washington
Seattle, Washington

Rajendra A. Morey
Duke University Medical Center
and
Research and Development/Mental Health Services
Durham Veterans Affairs Medical Center
and
Veterans Integrated Service Networks
Mid-Atlantic Mental Illness Research Education and Clinical Center
Durham, North Carolina

Marc Muehlmann
Ludwig-Maximilian-University
Munich, Germany

Jennifer C. Naylor
Duke University Medical Center
and
Research and Development/Mental Health Services
Durham Veterans Affairs Medical Center
and
Veterans Integrated Service Networks
Mid-Atlantic Mental Illness Research Education and Clinical Center
Durham, North Carolina

Lulit Price
University of Iowa
Iowa City, Iowa

Vikram Puvenna
Cleveland Clinic Lerner College of Medicine
Cleveland, Ohio

Shervin Rahimpour
Duke University Medical Center
Durham, North Carolina

Karin Reuter-Rice
Duke University
Durham, North Carolina

Lawrence J. Shampine
Duke University Medical Center
and
Research and Development/Mental Health Services
Durham Veterans Affairs Medical Center
and
Veterans Integrated Service Networks
Mid-Atlantic Mental Illness Research Education and Clinical Center
Durham, North Carolina

Martha E. Shenton
Brigham and Women's Hospital
and
Harvard Medical School
and
VA Healthcare System
Boston, Massachusetts

Murray B. Stein
University of California, San Diego
La Jolla, California

Erik Su
Johns Hopkins University School of Medicine
Baltimore, Maryland

YouRong Sophie Su
Stanford University School of Medicine
Stanford, California

Steven T. Szabo
Duke University Medical Center
and
Research and Development/Mental Health Services
Durham Veterans Affairs Medical Center
and
Veterans Integrated Service Networks
Mid-Atlantic Mental Illness Research Education and Clinical Center
Durham, North Carolina

Larry A. Tupler
Duke University Medical Center
and
Research and Development/Mental Health Services
Durham Veterans Affairs Medical Center
and
Veterans Integrated Service Networks
Mid-Atlantic Mental Illness Research Education and Clinical Center
Durham, North Carolina

Dennis A. Turner
Duke University Medical Center
Durham, North Carolina

VA Mid-Atlantic MIRECC Workgroup
Veterans Integrated Service Networks
Mid-Atlantic Mental Illness Research Education and Clinical Center
Durham, North Carolina

Anand Veeravagu
Stanford University School of Medicine
Stanford, California

Christy Wilson
Stanford University
Stanford, California

Jin Zhang
Cleveland Clinic Lerner College of Medicine
Cleveland, Ohio

1 Epidemiology

Peter Leo and Michael McCrea

CONTENTS

INTRODUCTION

Traumatic brain injury (TBI) is an important public health concern that is one of the leading causes of death and disability annually around the world.[1,2] Several factors have fueled increasing interest in TBI over the last several years, including rising awareness around the acute and chronic effects of sport-related concussion (SRC) and the reported incidence of head injuries sustained by U.S. military service members deployed to Iraq and Afghanistan. Sporting and military cohorts are now known to be at elevated risk of mild TBI (mTBI), characterized by more subtle neurocognitive and neurobehavioral symptoms that are often challenging to assess and characterized in a clinical setting.

A strong epidemiological framework for TBI is vital to improving our understanding of the injury's occurrence, populations at risk, and effective strategies for injury prevention. The purpose of the current chapter is to review the most recent epidemiological literature on TBI in both civilian and military populations.

The epidemiological literature on TBI is limited by a number of factors, not the least of which is that the definition of a TBI varies across studies. Though there are a number of operational definitions, TBI is commonly and most basically defined as an alteration in brain functioning or the emergence of evidence of brain pathology caused by an external force.[3] One of the methodological limitations of many epidemiological studies over the last 20 to 30 years has been suboptimal or inconsistent classification and inclusion criteria. Many studies have included individuals who sustained an injury to the head that may not have involved any alteration in brain function or physiological damage to the brain. For example, Bazarian and colleagues used emergency department records for a study of mTBI that included individuals with diagnostic codes such as "skull fracture," "intracranial injury of unspecified nature," and "head injury, unspecified."[4] Even though many individuals with these diagnoses likely met the criteria for mTBI, there are others who did not sustain a TBI at all or may have sustained an injury that was more severe than an mTBI.

Population-based epidemiological studies are typically based on hospital admission or discharge records, emergency department records, or death certificates, which creates a selection bias.[5] It has long been held that many patients with less severe injuries do not present for medical treatment and therefore are not included in many of these studies. In addition, changes in hospitalization patterns over the past several years, with fewer less severe patients being hospitalized, has also likely lead to an underestimation of the true incidence of milder forms of TBI (for a review, see the study by McCrea[6]).

TBI severity classification can also vary among different studies. Commonly, clinical indicators related to acute injury characteristics are used to indicate injury severity such as mild, moderate, or severe. The indicators most commonly used are alteration in consciousness, loss of consciousness (LOC), and post-traumatic amnesia (PTA) following the injury. The Glasgow Coma Scale (GCS) is the most commonly used clinical tool for assessing consciousness following an injury.[7] A GCS score of 13 to 15 is considered a mild injury, 9 to 12 is considered a moderate injury, and 8 or less is considered a severe injury. Methods used to classify TBI also have varied utility. The GSC as a highly effective tool to differentiate the severity of TBI has been well founded and the GCS has been shown to be useful in predicting morbidity and mortality in more severe injuries,[7–10] but it has been less useful as a prognostic indicator for mTBI.[11] Historically, not all epidemiological studies have employed a common or standardized classification system, which complicates interpretation of epidemiological data and comparisons across studies.

In terms of injury reporting, there is no single system in the United States or most other countries that tracks the occurrence of head injury over large populations. As a result, a true appreciation for TBI's national or worldwide incidence is lacking. The Centers for Disease Control and Prevention (CDC) is the most comprehensive data source of epidemiological studies of TBI in the United States, as it monitors death records (National Vital Statistics System), hospitalizations (National Hospital Discharge Survey), and emergency department visits (National Hosptial Ambulatory Medical Care).[12] The data for military TBI injuries comes from the Department of Defense (DoD) in collaboration with the Armed Forces Health Surveillance Center.[13] Where available, other sources from well-designed epidemiological studies in the United States and around the world were also reviewed.

The first section of the this chapter will review the civilian literature, including sport-related concussions, and the second section will focus on the military epidemiological literature.

TBI IN CIVILIAN POPULATIONS

Incidence

According to the most recent figures from the CDC, the average estimated incidence of TBI in the United States between the years 2002 and 2006 was 1,691,481 (576.8 per 100,000). Of these 1,364,797 (465.4 per 100,000) were treated in the emergency department (ED) and released; 275,146 (93.8 per 100,000) were hospitalized and discharged alive; and 51,538 (17.6 per 100,000) died.[12] Earlier examination of CDC data between 1995 and 2001 estimated a total of 1,396,000 (506.4 per 100,000) TBIs each year with 1,111,000 (403.1 per 100,000) ED visits, 235,000 (85.2 per 100,000) hospitalizations, and 49,900 (18.1 per 100,000) deaths.[14] These data from the CDC do not include individuals that were treated at outpatient facilities or who did not seek treatment.

Previous reports from the United States that used the 1991 National Health Interview survey of 46,761 households estimated there were approximately 1.5 million (618 per 100,000) individuals who sustained nonfatal brain injuries in the United States and of those approximately 25% did not seek any type of treatment.[15] Comparison between the Sosin and colleagues report[15] and the reports from the CDC data are difficult to make based on methodological differences. However, comparison between the CDC reports from Langlois and colleagues[14] to that of Faul and colleagues[12] clearly shows that the average annual estimate of ED visits and hospitalizations has increased but the rate of TBI related deaths has decreased. This was investigated more closely in a report of death rates from 1997 to 2007 that showed an 8.2% decrease, from 19.3 to 17.8 per 100,000, during this time and the authors noted that this follows a long-standing trend of decreases in TBI-related deaths over time.[16]

A review of European epidemiological studies by Tagliaferri and colleagues found a wide range in the incidence in TBI.[17] The authors reviewed 23 epidemiological reports from various European countries between 1980 and 2003. Methodology varied considerably between the studies, as did the estimated rates of TBI. One study from a province of Spain estimated an annual average to be 91 per 100,000, whereas a study from Sweden estimated 546 per 100,000. The authors removed the extreme reports from the study and found an aggregate of 235 per 100,000 TBIs annually.

The aforementioned studies constitute some of the best epidemiological data on incidence of TBI, but the data likely grossly underestimates the incidence of mTBI.[5] A World Health Organization (WHO) systematic review of the mTBI literature found that 70% to 90% of TBI was mild in nature and that hospital-treated mTBI was approximately 100 to 300 per 100,000 in the studies it reviewed. However, given the undertreatment and reporting of mTBI, the WHO estimated that the true yearly incidence was likely 600 per 100,000.[18]

A population-based study in New Zealand attempted to correct for underreporting by using both prospective and retrospective surveillance systems in an attempt to register all instances of TBI over a 1-year period between March 2010 and February 2011. The study was based in the city of Hamilton, which has a population of 129,249 and the surrounding rural area with a population of 43,956. The investigators reportedly enrolled all healthcare providers (e.g., family physicians, health centers, hospitals, ambulance services), the local prison, community services (e.g., school, sports clubs), and used national healthcare databases and the death registry to assist in identifying cases. Regular contact was kept among these sources throughout the year and each identified case was crosschecked to ensure there were no duplicates. During this time, 1369 TBIs were identified, which the authors concluded to be an annual incidence rate of 790 per 100,000.[19] This rate is considerably higher than most epidemiological studies and the authors asserted that this was due to a larger number of mild traumatic brain injuries captured that typically go unaccounted for in most other epidemiological studies.

Severity

Data regarding the severity of TBI is difficult to ascertain, given that classification systems are not consistent across studies or sometimes even within studies. It is commonly accepted that the majority of TBIs are mild in severity and make up between 70% and 90% of all TBIs.[20] This number is likely an underestimation, given that many mTBIs are thought to go untreated and therefore unreported. In Feigin and colleagues,[19] mTBI made up nearly 95% of the total sample. Another confounding factor when considering injury severity is that individuals who sustained more significant brain injury or other trauma are commonly sedated, which complicates injury classification. In one study of adults receiving inpatient rehabilitation services, 19.2% of the sample was classified as severe, 10.3% was moderate, and 16.4% of the sample was chemically sedated. The injury severity and reason for sedation in those individuals is not entirely clear, although the authors indicated that most were thought to be in the severe range.[21]

In addition to underestimating the number of mild injuries, the current epidemiological literature likely overestimates the number of severe injuries as many studies utilize hospital admission records to classify injury. Similar to Cuthbert and colleagues,[21] studies that utilize hospitalization records report that 20% of injuries are classified as severe[22] while other authors reported that 80% of hospitalized individuals sustained mild injuries, 10% sustained moderate injuries, and 10% were severe. These figures are considerably higher than the report by Feigin and colleagues,[21] and also well above a review of European epidemiological studies that calculated a TBI severity ratio of 22:1.5:1 for mild, moderate, and severe injuries, respectively.

Risk Factors and Characteristics

Gender

A consistent finding across the epidemiological literature is that TBI is much more frequent in males than females. According to the most recent CDC data, males were 1.4 times more likely to sustain a TBI, as they had an estimated average annual TBI rate of 998,176 compared to 693,329 for females. Of the TBIs sustained by males, approximately 17% lead to hospitalization and 4% were fatal injuries. In comparison, approximately 15% of the females were hospitalized and less than 2% died following TBI. The rate of TBI in males was greatest across all age groups.[12] The ratio reported by the CDC is largely consistent with previous findings from United States data, which showed a male to female rate of 1.6:1[23] and also European studies that ranged from 1.46:1[24] to 1.8:1.[25] It should be noted that most European epidemiological studies had a bias toward more severe injuries, which may have inflated the estimate relative to the United States samples. Feigin and colleagues found a male to female rate ratio of 1.67:1 in a population based sample that captured a significant number of more mild injuries, so it may be that the CDC data is a slight underestimate.[19]

Age

According to the most recent data from the CDC, TBI rates are highest among young children age 0 to 4 (1337.3 per 100,000) and older adolescence aged 15 to 19 (896.2 per 100,000). Older adults aged 75 and above also have a high rate of TBI (932 per 100,000) and they account for the highest rate of TBI-associated hospitalizations (339.3 per 100,000) and death (56.6 per 100,000). To give a point of comparison, the rate of hospitalization in the most elderly was over 3.5 times the rate of hospitalization of the total sample (93.8 per 100,000) and the death rate was over 3 times greater than that of the total sample (17.6 per 100,000). This pattern of high rates of TBI in early childhood, late adolescence, and in the elderly has been shown in many population-based studies.[1]

Race and Ethnicity

Some studies show higher incidence of TBI in non-whites compared to whites, but this is somewhat controversial given methodological inconsistencies and the quality of the data used to generate these findings. According to the most recent data from the CDC that looked at emergency department visits, African Americans had the highest rate of TBI (618.6 per 100,000), followed by whites (448.3 per 100,000), and then American Indians/Alaska Natives/Asians or Pacific Islanders (334.7 per 100,000).[12]

Alcohol

The positive association between blood alcohol concentration (BAC) and many types of injuries is well established. In terms of brain injury, it has been shown that a high percentage of individuals seek medical attention following a TBI have a positive BAC, with figures ranging from 56% to 72% in various United States samples.[26] In addition, a large number of those individuals (36% to 55%) were legally intoxicated at time of injury.[27–30] European epidemiological studies also show a strong relationship between alcohol consumption and TBI, though the figures are less consistently high as the data from U.S. populations, with reports ranging from 24% to 51% who were positive for alcohol.[17] Langlois and colleagues reported that alcohol use was reported in 21% of motor vehicle occupants hospitalized with a TBI, and over 12% had a BAC above the legal limit.[31]

Recurrent TBI

Recurrent TBI, especially mTBI, has become a topic of considerable interest in recent years due to concern that TBI increases the risk of cognitive impairment later in life[32] or possible neurodegenerative conditions.[33] Though skepticism exists regarding the link between mTBI and the development of neurodegenerative conditions[34] it has been shown that recurrent mTBI is associated with prolonged recovery and that athletes who have a history of mTBI may be at greater risk of sustaining future injuries.[35] Annegers and colleagues conducted the first population based study of recurrent TBI and found that the relative risk of a second TBI among those with an earlier TBI was 2.8 to 3 times greater than the noninjured sample.[36] Additionally, in those that sustained a second head injury the risk of sustaining a third head injury was 7.8 to 9.3 times that of an initial head injury in the population. Recurrent TBI has been found to be related to alcohol abuse in a number of studies[22] with one study indicating that if an individual's first injury after age 12 was alcohol related they were at a fourfold greater risk of repeat head injury by age 34. Recurrent TBI has long been a concern in sports with a recent epidemiological study of sport concussion in high school athletes in the United States showing that 13.2% of all concussions were recurrent, and consistent with previous findings the symptoms of recurrent concussion took longer to resolve.[37]

External Causes

According to the most recent data from the CDC, falls are the most common cause of TBI, with an estimated annual average of 523,043. The rate of fall-related TBI was greatest among children aged 0 to 4 years (839 per 100,000) and adults age 75 and older (599 per 100,000). Falls accounted for approximately half of the TBIs in children age 0 to 14 and approximately 60% of adults aged 65 and older. The second most common cause is motor vehicle traffic accidents (estimated annual average 292,202), followed by struck by/against events (estimated annual average 169,625) and assault (estimated annual average 169,625).[12] Motor vehicle accidents resulted in the largest percentage of TBI-related deaths at 31.8%.

Few European epidemiological studies have investigated external causes, but most reports indicate that motor vehicle accidents are the most common events leading to TBI, followed by falls.[17] It should be noted that the external cause of TBI varies considerably by country due to a myriad of factors, including economic status. Lower and middle-income countries that had not had access to motor vehicles are showing sharp increases in motor-vehicle-related injuries and deaths, while wealthier countries are developing better safety standards and road laws that have led to a decrease in motor-vehicle-related mortality over the years.[38] These factors need to be better understood to understand the epidemiological impact of TBI around the world.

TBIs related to engagement in sports and recreational activity are a major cause of TBI, but are likely underreported in the literature given that most of the injuries are mild in severity and untreated. Previous epidemiological studies estimated approximately 300,000 sports- and recreation-related TBIs per year.[39] However, it has been noted that Thurman and colleagues[39] only included TBIs for which the person reported a loss of consciousness, which is considered a very small minority (10%) of TBIs related to sports and recreation.[40,41] Langlois and colleagues indicated that if all injuries are taken into account, including those who do not seek medical attention, a more accurate estimate of 1.6 to 3.8 million sports- and recreation-related concussions occur annually.[2] The data on sports-related concussion is likely to improve in the coming years as identification of concussion by athletic trainers, coaches, athletes, and medical personnel improves and more athletes seek treatment. A recent study of emergency departments found that sports-related concussion visits doubled for children aged 8 to 13 and increased by more than 200% in individuals aged 14 to 19 from 1997 to 2007.[42]

Abusive head trauma is the leading cause of serious head injury and death in children age 2 years and younger. Keenan and colleagues prospectively collected data from nine hospital pediatric intensive care units from January 1, 2000, to December 31, 2001.[43] Results showed that the incidence of inflicted TBI in the first 2 years of life was 17 per 100,000 person-years, with infants having a higher incidence (29.7 per 100,000) relative to children in the second year of life (3.8 per 100,000). To determine whether an injury was inflicted the authors relied on confession from the parents or medical/social service determination of abuse so it is likely that these are underestimates of the true incidence. A more recent study using a national sample of inflicted injury in children found a slightly higher rate of 18.7 per 100,000 for children less than 2 years of age.[44]

TBI associated with violence against adults is also underreported and underidentified due to many factors, including low reporting rates in domestic violence and other factors associated with the types of trauma sustained during assaults.[22] Previous data from the CDC estimated that at least 156,000 TBI-related emergency department visits, hospitalizations, and deaths each year are associated with all types of assaults[2] with more recent estimates showing approximately 169,625 TBIs per year. According to these data, assaults account for approximately 10% of total TBIs.[12] A study based on the New Zealand population found that assaults comprised almost 17% of all TBIs with an estimated annual incidence of 132 per 100,000,[19] which is considerably higher than estimates from the United States. According to the CDC data, the highest risk of sustaining a TBI secondary to assault is in individuals age 15 to 44, as this age group comprises 75% of assault-related head injuries.[12]

Lifetime Prevalence

The lifetime prevalence of TBI is less well studied in the epidemiological literature due primarily to the methodology employed by most studies. Lifetime prevalence of TBI refers to the number of individuals who have ever experienced a TBI. In a large-scale study, Winqvist and colleagues used a birth cohort design of individuals born in two provinces of North Finland in 1966.[45] The entire cohort included 12,058 residents who were followed through age 34. Results showed that the average annual incidence of TBI was 118 per 100,000 and by age 35, 3.8% of the cohort had experienced at least one hospitalization for TBI.[45] The study only included hospital and health center discharge records with outpatient services and emergency department visits of less than 24 hours being excluded from analysis, so these data are only capturing more severe injuries. A birth cohort study in New Zealand that used any type of medical attendance where TBI was diagnosed and followed through age 25 found a much higher average annual incidence rate of 1750 per 100,000. By age 25, 31.6% of the cohort had received a TBI for which they received medical care.[46] In a study by McKinlay and colleagues, only 32.9% of the individuals who sustained a TBI were admitted to the hospital for further observation or treatment.[46] Retrospective self-report studies by Anstey and colleagues[47] and Silver and colleagues[48] that looked at TBI with loss of consciousness indicated a lifetime prevalence rate of 5.7% to 8.5%, respectively.

Consequences of TBI

Due in large part to advances in medicine and safety, more people are surviving TBI than ever before. Estimates of individuals in the United States who are living with long-term or lifelong disability associated with a TBI range from 3.2 million to 5.3 million.[2,49] These estimates only include individuals who were hospitalized as a result of their injuries and therefore are biased toward more severe injuries. The Glasgow Outcome Scale (GOS) is an unsophisticated but widely used indicator of disability or residual effects used by hospitals at the time of discharge. The major classifications of the GOS are death, persistent vegetative state (i.e., patient exhibits no obvious cortical function), severe disability (i.e., patient is conscious but requires daily support due to disability), moderate disability (i.e., disabled but independent in regard to daily living needs), and good recovery (i.e., resumption of normal activities) though minor neurologic and psychological deficits may be present.[50] Data from a 14-state CDC-funded TBI surveillance study found that 17% of hospital discharges were reported as having moderate to severe disability based on the GOS.[31] Krause and colleagues estimated that the total number of disabilities from new brain injuries for the year 2000 was 98,560 or a rate of approximately 35 per 100,000 of the population.[26] A population-based sample of persons with TBI from the South Carolina Traumatic Brain Injury Follow-up Registry was used to develop a predictive model of TBI that estimated 43.3% of hospitalized TBI survivors in 2003 experienced TBI-related long-term disability, which in turn estimated that 124,626 Americans per year who likely need rehabilitative or supportive services following a TBI.[51]

Disability associated with mTBI is difficult to calculate given that only about 10% are hospitalized and therefore not considered in many epidemiological studies. However it has been well established that mTBI results in physical, cognitive, psychological, and social dysfunction that leads to lost productivity and decreased quality of life (for a review, see McCrea[6]). The impact on occupation has been reported to be comparable to general trauma patients relative to the duration of time off work and reported problems upon returning to work.[52] Boake and colleagues showed that most mTBI patients treated and released from emergency departments did not return to work until 1 to 3 months after the injury and there was virtually no difference between those who were treated and released from the emergency department (46% working at 1 month, 66% at 3 months, and 68% at 6 months) versus those who were hospitalized (39% working at 1 month, 62% at 3 months, and 71% at 6 months).[52]

Economic Cost

The real economic cost of TBI is difficult to calculate given the number of untreated TBIs that may lead to missed workdays or decreased productivity at work. Additionally, it is difficult to truly appreciate and quantify the impact that TBI has on caregivers and family members. Langlois and colleagues[22] extrapolated data from Finkelstein and colleagues[53] to calculate the estimated economic burden of TBI in 2009 dollars. The authors estimated that the total costs of fatal, hospitalized, and nonhospitalized TBI in the United States totaled more than $221 billion, including $14.6 billion for medical costs, $69.2 billion for work lost costs, and $137 billion for the value of lost quality of life.[22] A report by the CDC indicated that the economic burden of TBI in the United States was estimated to be $56 billion, with mTBI accounting for $16.7 billion of that figure.[54]

TBI IN THE MILITARY POPULATION

TBI has been referred to as the "signature injury" among U.S. service members serving in the conflicts in Iraq (Operation Iraqi Freedom, OIF) and Afghanistan (Operation Enduring Freedom, OEF).[55] The identification of moderate and severe TBI is not complicated even in the war theater given the clear clinical signs and symptoms. Accurate identification of mTBI can be a challenge given the sometimes relatively subtle nature of the symptoms, the lack of diagnostic tools with good sensitivity and specificity that can be used during combat, and the significant symptom overlap with other conditions such as post-traumatic stress disorder (PTSD).[56] One study investigating the co-occurrence of TBI with psychiatric disturbance and pain found that of the service members diagnosed with TBI, 89% also had a psychiatric diagnosis with the most frequent being PTSD at 73%. In the TBI sample, 70% also had a diagnosis of head, back, or neck pain.[57]

Surveillance data of military TBI are primarily available through the Department of Defense (DoD) in cooperation with the Armed Forces Health Surveillance Center. It should be noted that data from the DoD reports comprise TBI data that occurred anywhere in the world U.S. forces are located, not just deployed servicemen. The DoD reported that 80% of the injuries that are reported occur in nondeployed settings with common causes being crashes in privately owned and military vehicles, falls, sport and recreation activities, and military training.[13]

The DoD data subtypes TBI into four classifications according to specific criteria.

1. *Concussion/mTBI*—A confused or disoriented state that lasts less than 24 hours; loss of consciousness (LOC) for up to 30 minutes; post-traumatic amnesia (PTA) lasting less than 24 hours; and structural brain imaging (MRI or CT) that is read to be normal.
2. *Moderate TBI*—A confused or disoriented state lasting over 24 hours; LOC for more than 30 minutes, but less than 24 hours; PTA lasting greater than 24 hours but less than 7 days; and structural brain imaging that shows normal or abnormal results.

3. *Severe TBI*—A confused or disoriented state that lasts more than 24 hours; LOC for more than 24 hours; PTA for more than 7 days; and structural brain imaging yielding normal or abnormal results.
4. *Penetrating TBI*—Any head injury in which the dura mater is penetrated.

Incidence

According to the most recent report by the DoD, there have been a total of 294,172 TBIs from 2000 through the fourth quarter of 2013.[13] The most common injury is concussion/mTBI, which accounted for 242,676 (82.5%) of the injuries, followed by moderate TBIs (23,754; 8.1%), not classifiable (20,433; 6.9%), penetrating (4389; 1.5%), and severe (2920; 1%). When the individual years are considered, the total prewar incidence of TBI was 10,958 for 2000. The years following the commencement of OEF were not considerably different, with subsequent increases of about 3% to 9% each year until 2006, when 17,037 injuries were reported. The number of injuries increased drastically in 2007, when 23,217 injuries were reported, which was an increase of approximately 36% from 2006. Another considerable increase of 22% was recorded the following year (28,877 injuries in 2008). TBI rates continued to increase until a peak in 2011 of 32,625 injuries, and then there was a slight decline in 2012 (the last full year of available data), when 30,406 TBIs were reported.

The distribution of injury type has not drastically changed since the start of OEF, with the vast majority (71% to 86%) of injuries being concussions/mTBI. The increase in TBI since the start of OEF has been driven almost exclusively by large increases in the concussion/mTBI group. According to prewar data in 2000, there were 275 penetrating injuries, 178 severe, 1616 moderate, and 7179 mild TBIs. In comparison, there were 324 penetrating, 261 severe, 1857 moderate, and 27,535 mild TBIs recorded in the year 2011, which had the highest rate of total TBIs. When the year 2011 and 2000 are compared, substantial increases are noted across injury types, including penetrating, severe, and moderate TBI, with increases of approximately 17%, 47%, and 15%, respectively. However, the most substantial increase, and what accounts for the large increase of total TBIs, was concussion/mild TBI injury that showed a 283% increase 2000 to 2011. This increase is probably at least partially accounted for by better identification and diagnosis of these less severe injuries. In any event, it is clear that concussion/mTBI has the largest scale impact on our service members.

External Causes

The mechanisms of injury for military personnel are generally similar to those found in civilians and include motor vehicle accidents, falls, sports/recreation related, assaults, and collisions with objects. According to a review by Langlois and colleagues,[22] mechanism of injury data is typically only available for hospitalizations and is provided by the Armed Forces Health Surveillance Centers (AFHSC). Upon review of the data from the AFHSC, the authors showed that during the prewar period, land transport (~25%) and falls/miscellaneous (~24%) were the most common injury type followed by sports/recreation (9%) and nonbattle assaults (7%) among active-duty servicemen.[22]

Injury data for deployed servicemen is difficult to attain, as this data is not included in the DoD reports. A study of an Army unit (3973 servicemen) following a 1-year deployment in Iraq identified 907 soldiers diagnosed with a TBI. According to survey data, TBI was most commonly caused by blast (88%) followed by motor vehicle accidents (39%), falls (20%), fragments (15.8%), and bullets (3.1%). The servicemen in the study commonly identified more than one cause of their injury, which led to the percentages being greater than 100.[58] Blast injury was clearly the most commonly reported mechanism of injury, and the high rate of blast injury is reported elsewhere.[59,60] It is now commonly held that a dual mechanism (blast and blunt) is common in the military environment.

There are several mechanisms of blast injury described in the literature, including primary, secondary, tertiary, and quaternary. Primary injuries are due to overpressurization or underpressurization shock waves caused by explosions. The shock waves move through the body from solid- and liquid-filled sections to gas-filled organs (e.g., lungs), which causes damage to the organs. Secondary injuries are due to fragments and projectiles from explosions. Tertiary injuries are caused by a displacement of air (blast wind), throwing the individual and causing them to collide with an object. Quaternary injuries are those not classified by the other three and include toxic inhalation, burns, and crush injuries.[61] The exact nature of the blast injury is not typically identified, but it is clear that most are likely caused by secondary and tertiary injuries as these involve a mechanical component similar to other TBI-causing mechanisms.[62] TBI related to primary blast injuries has been reported in the literature,[63] but this remains controversial given the exact mechanism of injury remains unclear.[64] Given the specific context in which blast injuries occur and the lack of reliable data regarding acute injury characteristics and mechanism of injury, these injuries are difficult to study and the epidemiological data is lacking. Future research is required, ideally employing a standard definition of injury and robust injury surveillance systems.

CONCLUSIONS

TBI is one of the most significant public health problems in the United States and worldwide based on incidence, prevalence, healthcare resource utilization, resulting death and disability, and total economic cost. The highest rates of TBI are observed in the very young and very old, males, minorities of low socioeconomic status, and substance abusers. The overwhelming majority of TBIs are categorized as mTBI. Determining the true incidence of mTBI has been hampered by a multitude of methodological factors (e.g., variable injury definitions and criteria, surveillance systems, research setting). Athletes competing in contact/collision sports and military service members are at particularly higher risk of mTBI, and, in some instances, the potential for recurrent injury. Traditional means for classifying TBI severity have limited utility in detecting and categorizing mTBI, which has significant implications to epidemiologic research. Multidimensional definitions that incorporate information on biomechanics, acute injury characteristics, clinical signs, and symptoms result in the most accurate diagnosis of mTBI. These methods provide promise for future studies to further clarify the true epidemiology of TBI, which in turn will guide development of clinical endpoints for diagnostic and outcome studies.

REFERENCES

1. Hyder, A.A. et al. 2007. The impact of traumatic brain injuries: A global perspective. *NeuroRehabilitation* 22(5):341–353.
2. Langlois, J.A., Rutland-Brown, W., Wald, M.M. 2006. The epidemiology and impact of traumatic brain injury: A brief overview. *J Head Trauma Rehabil* 21(5): 375–378.
3. Menon, D.K. et al. 2010. Position statement: Definition of traumatic brain injury. *Arch Phys Med Rehabil* 91(11):1637–1640.
4. Bazarian, J.J. et al. 2005. Mild traumatic brain injury in the United States, 1998–2000. *Brain Inj* 19(2):85–91.
5. Andelic, N. 2013. The epidemiology of traumatic brain injury. *Lancet Neurol* 12(1): 28–29.
6. McCrea, M. 2008. *Mild Traumatic Brain Injury and Postconcussion Syndrome: The New Evidence Base for Diagnosis and Treatment*. Oxford: Oxford University Press.
7. Teasdale, G., Jennett, B. 1974. Assessment of coma and impaired consciousness. A practical scale. *Lancet* 2(7872):81–84.
8. Dikmen, S.S. et al. 1995. Neuropsychological outcome at 1-year post head injury. *Neuropsychology* 9(1):80–90.
9. Jane, J.A., Rimel, R.W. 1982. Prognosis in head injury. *Clin Neurosurg* 29:346–352.
10. Klauber, M.R. et al. 1981. The epidemiology of head injury: A prospective study of an entire community—San Diego County, California, 1978. *Am J Epidemiol* 113(5): 500–509.
11. Cappa, K.A., Conger, J.C., Conger, A.J. 2011. Injury severity and outcome: A meta-analysis of prospective studies on TBI outcome. *Health Psychol* 30(5):542–560.
12. Faul, M. et al. 2010. *Traumatic Brain Injury in the United States: Emergency Department Visits, Hospitalizations and Deaths 2002–2006*. Atlanta, GA: Centers for Disease Control and Prevention, National Center for Injury Prevention and Control.
13. Department of Defense. 2013. DoD Worldwide Numbers for TBI. Retrieved April 15, 2014. Available at http://dvbic.dcoe.mil/dod-worldwide-numbers-tbi.
14. Langlois, J.A., Rutland-Brown, W., Thomas, K.E. 2004. *Traumatic Brain Injury in the United States: Emergency Department Visits, Hospitalizations, and Deaths*. Atlanta, GA: US Center for Disease Control and Prevention.
15. Sosin, D.M., Sniezek, J.E., Thurman, D.J. 1996. Incidence of mild and moderate brain injury in the United States, 1991. *Brain Inj* 10(1):47–54.
16. Coronado, V.G. et al. 2011. Surveillance for traumatic brain injury-related deaths—United States, 1997–2007. *MMWR Surveill Summ* 60(5):1–32.
17. Tagliaferri, F. et al. 2006. A systematic review of brain injury epidemiology in Europe. *Acta Neurochir* 148(3):255–268; discussion 268.
18. Carroll, L.J. et al. 2004. Prognosis for mild traumatic brain injury: Results of the WHO Collaborating Centre Task Force on Mild Traumatic Brain Injury. *J Rehabil Med* 43(Suppl):84–105.
19. Feigin, V.L. et al. 2013. Incidence of traumatic brain injury in New Zealand: A population-based study. *Lancet Neurology* 12(1):53–64.

20. Cassidy, J.D. et al. 2004. Incidence, risk factors and prevention of mild traumatic brain injury: Results of the WHO Collaborating Centre Task Force on Mild Traumatic Brain Injury. *J Rehabil Med* 43(Suppl):28–60.
21. Cuthbert, J.P. et al. 2014. Epidemiology of adults receiving acute inpatient rehabilitation for a primary diagnosis of traumatic brain injury in the United States. *J Head Trauma Rehabil* 30(2):122–135.
22. Langlois Orman, J.A. et al. 2011. Epidemiology. In *Textbook of Traumatic Brain Injury*, 2nd ed., eds. J.M. Silver, T.W. McAllister, S.C. Yudofsky, xxii. Washington, DC: American Psychiatric Publishing, Inc.
23. Jager, T.E. et al. 2000. Traumatic brain injuries evaluated in U.S. emergency departments, 1992–1994. *Acad Emerg Med* 7(2):134–140.
24. Andersson, E.H. et al. 2003. Epidemiology of traumatic brain injury: A population based study in western Sweden. *Acta Neurol Scand* 107(4):256–259.
25. Santos, M.E., De Sousa, L., Castro-Caldas, A. 2003. *Acta Med Port* 16(2):71–76.
26. Kraus, J.F., Chu, L.C. 2005. Epidemiology. In *Textbook of Traumatic Brain Injury*, vol. 1, J.M. Silver, T.W. McCallister, S.C. Yudofsky, 3–26. Washington, DC: American Psychiatric Publishing, Inc.
27. Kraus, J.F. et al. 1989. Blood alcohol tests, prevalence of involvement, and outcomes following brain injury. *Am J Public Health* 79(3):294–299.
28. Parry-Jones, B.L., Vaughan, F.L., Miles Cox, W. 2006. Traumatic brain injury and substance misuse: A systematic review of prevalence and outcomes research (1994–2004). *Neuropsychol Rehabil* 16(5):537–560.
29. Rimel, R.W. 1981. A prospective study of patients with central nervous system trauma. *J Neurosurg Nurs* 13(3):132–141.
30. Savola, O., Niemela, O., Hillbom, M. 2005. Alcohol intake and the pattern of trauma in young adults and working aged people admitted after trauma. *Alcohol Alcohol* 40(4):269–273.
31. Langlois, J.A. et al. 2003. Traumatic brain injury-related hospital discharges. Results from a 14-state surveillance system, 1997. Morbidity and mortality weekly report. *MMWR Surveill Summ* 52(4):1–20.
32. Guskiewicz, K.M. et al. 2005. Association between recurrent concussion and late-life cognitive impairment in retired professional football players. *Neurosurgery* 57(4): 719–726; discussion 719–726.
33. Stein, T.D., Alvarez, V.E., McKee, A.C. 2014. Chronic traumatic encephalopathy: A spectrum of neuropathological changes following repetitive brain trauma in athletes and military personnel. *Alzheimers Res Ther* 6(1):4.
34. Karantzoulis, S., Randolph, C. 2013. Modern chronic traumatic encephalopathy in retired athletes: What is the evidence? *Neuropsychol Rev* 23(4):350–360.
35. Guskiewicz, K.M. et al. 2003. Cumulative effects associated with recurrent concussion in collegiate football players: The NCAA Concussion Study. *JAMA* 290(19):2549–2555.
36. Annegers, J.F. et al. 1980. The incidence, causes, and secular trends of head trauma in Olmsted County, Minnesota, 1935–1974. *Neurology* 30(9):912–919.
37. Castile, L. et al. 2012. The epidemiology of new versus recurrent sports concussions among high school athletes, 2005–2010. *Br J Sports Med* 46(8):603–610.
38. Maas, A.I., Stocchetti, N., Bullock, R. 2008. Moderate and severe traumatic brain injury in adults. *Lancet Neurol* 7(8):728–741.
39. Thurman, D.J., Branche, C.M., Sniezek, J.E. 1998. The epidemiology of sports-related traumatic brain injuries in the United States: Recent developments. *J Head Trauma Rehabil* 13(2):1–8.
40. McCrea, M. et al. 2003. Acute effects and recovery time following concussion in collegiate football players: The NCAA Concussion Study. *JAMA* 290(19):2556–2563.

41. Schulz, M.R. et al. 2004. Incidence and risk factors for concussion in high school athletes, North Carolina, 1996–1999. *Am J Epidemiol* 160(10):937–944.
42. Bakhos, L.L. et al. 2010. Emergency department visits for concussion in young child athletes. *Pediatrics* 126(3):e550–e556.
43. Keenan, H.T. et al. 2003. A population-based study of inflicted traumatic brain injury in young children. *JAMA* 290(5):621–626.
44. Parks, S. et al. 2012. Characteristics of non-fatal abusive head trauma among children in the USA, 2003–2008: Application of the CDC operational case definition to national hospital inpatient data. *Inj Prev* 18(6):392–398.
45. Winqvist, S. et al. 2007. Traumatic brain injuries in children and young adults: A birth cohort study from northern Finland. *Neuroepidemiology* 29(1–2):136–142.
46. McKinlay, A. et al. 2008. Prevalence of traumatic brain injury among children, adolescents and young adults: Prospective evidence from a birth cohort. *Brain Inj* 22(2):175–181.
47. Anstey, K.J. et al. 2004. A population survey found an association between self-reports of traumatic brain injury and increased psychiatric symptoms. *J Clin Epidemiol* 57(11):1202–1209.
48. Silver, J.M. et al. 2001. The association between head injuries and psychiatric disorders: Findings from the New Haven NIMH Epidemiologic Catchment Area Study. *Brain Inj* 15(11):935–945.
49. Zaloshnja, E. et al. 2008. Prevalence of long-term disability from traumatic brain injury in the civilian population of the United States, 2005. *J Head Trauma Rehabil* 23(6):394–400.
50. Jennett, B., Bond, M. 1975. Assessment of outcome after severe brain damage. *Lancet* 1(7905):480–484.
51. Selassie, A.W. et al. 2008. Incidence of long-term disability following traumatic brain injury hospitalization, United States, 2003. *J Head Trauma Rehabil* 23(2):123–131.
52. Boake, C. et al. 2005. Lost productive work time after mild to moderate traumatic brain injury with and without hospitalization. *Neurosurgery* 56(5):994–1003; discussion 994–1003.
53. Finkelstein, E., Corso, P.S., Miller, T.R. 2006. *The Incidence and Economic Burden of Injuries in the United States*. Oxford: Oxford University Press.
54. National Center for Injury Prevention and Control. 2003. *The Report to Congress on Mild Traumatic Brain Injury in the United States: Steps to Prevent a Serious Public Health problem*. Atlanta, GA. Retrieved from http://www.cdc.gov/ncipc/pub-res/mTBI/mTBIreport.pdf.
55. Hayward, P. 2008. Traumatic brain injury: The signature of modern conflicts. *Lancet Neurol* 7(3):200–201.
56. Iverson, G.L. et al. 2009. Challenges associated with post-deployment screening for mild traumatic brain injury in military personnel. *Clin Neuropsychol* 23(8):1299–1314.
57. Taylor, B.C. et al. 2012. Prevalence and costs of co-occurring traumatic brain injury with and without psychiatric disturbance and pain among Afghanistan and Iraq War Veteran V.A. users. *Med Care* 50(4):342–346.
58. Terrio, H. et al. 2009. Traumatic brain injury screening: Preliminary findings in a US Army Brigade Combat Team. *J Head Trauma Rehabil* 24(1):14–23.
59. Hoge, C.W. et al. 2008. Mild traumatic brain injury in U.S. Soldiers returning from Iraq. *N Engl J Med* 358(5):453463.
60. Mac Donald, C.L. et al. 2011. Detection of blast-related traumatic brain injury in U.S. military personnel. *N Engl J Med* 364(22):2091–2100.
61. DePalma, R.G. et al. 2005. Blast injuries. *N Engl J Med* 352(13):1335–1342.

62. Sayer, N.A. 2012. Traumatic brain injury and its neuropsychiatric sequelae in war veterans. *Annu Rev Med* 63:405–419.
63. Nakagawa, A. et al. 2011. Mechanisms of primary blast-induced traumatic brain injury: Insights from shock-wave research. *J Neurotrauma* 28(6):1101–1119.
64. Hicks, R.R. et al. 2010. Neurological effects of blast injury. *J Trauma* 68(5):1257–1263.

2 Sport-Related Traumatic Brain Injury

Michael Clark and Kevin Guskiewicz

CONTENTS

INTRODUCTION

Sport-related traumatic brain injuries (TBIs) have received significant media coverage in recent years, in part due to an increased body of scientific literature and growing concern surrounding their long-term effects. The major focus is centered on concussions, which are believed to account for 80% of TBI-related visits to emergency departments.[1] With the development and application of advanced concussion assessment tools, including neuropsychological testing, neuroimaging, and balance and gait assessments, there is a rising tide of data that is driving changes in clinical practices and management of concussed patients. In this chapter, we explore the major features of sport-related TBI, with a special emphasis on concussions. This overview will cover definitions of the injury, pathophysiology, biomechanics, epidemiology, clinical management, and future directions in research. The details of each of these sections are taken from the most current and impactful literature in the field; this includes both primary research as well as expert opinion and consensus and position statements of major societies and organizations imparting meaningful guidance to clinicians and healthcare professionals. These include the American Academy of Neurology,[2] American Academy of Pediatrics,[3] American Medical Society for Sports Medicine,[4] National Athletic Trainers' Association,[5] and the 4th International Conference on Concussion in Sport.[6] Furthermore, we present commentary on this literature that is meant to highlight the difficulties of translational research in the field and project the trajectory of the sport-related concussion field in the coming years.

DEFINITIONS AND NOMENCLATURE

To answer the question "what is a concussion?" we must first discuss the recent history of the term and how it has changed over time.* Although medical descriptions of concussion, or *commotio cerebri*, date back to the Corpus Hippocraticum, there has been considerable difficulty in establishing a universal definition of concussion for decades. Several classification and grading schemes[8–11] were used with no consensus as to which classification scheme was most appropriate or useful. This created a problem of inconsistency of definition in the field of concussion research; however, many of these definitions have common elements. Nearly all grading/classification systems use the presence and duration of loss of consciousness and post-traumatic amnesia as major criteria for diagnosis and stratification into varying levels of severity. The ability to predict prognosis following concussion based on post-traumatic amnesia and loss of consciousness has proven difficult and the use of such classification systems has significantly dropped. In sports injury research, a widely used definition has been put forth by the International Conference on Concussion in Sport (ICCS). In the most recent 2012 consensus statement, the authors defined concussion as "a complex pathophysiological process affecting the brain, induced by biomechanical forces."[6] While the description goes on to describe concussion as involving neuropathological changes, the authors emphasize that the acute clinical symptoms reflect a functional disturbance rather than a structural injury. A striking feature of the ICCS definition is the lack of a specific description of a clinical presentation; this was intentional as the nature of concussive injury varies widely, as will be discussed in the succeeding section.

Of note, concussion has often been referred to as a "mild TBI" in a large volume of research literature. Although some authors draw distinctions between the terms,[4] they are often considered synonymous. The term "concussion" has gained favor over mild TBI because of the concern that "mild" TBI may give the impression that the injury does not have serious sequelae.[6] Contrariwise, the authors of one study examining the connotations of the term "concussion" as applied to pediatric patients suggests that clinicians often use "concussion" rather than "mild TBI" to convey to parents that their child's injury is transient and without long-term sequelae.[12] Whichever term is used, it is generally accepted that concussion is a serious injury that requires early recognition and proper management.

SIGNS AND SYMPTOMS

The common signs and symptoms of concussion are listed in Table 2.1 and generally fall into four domains: physical, cognitive, emotional, and sleep related.[13–20] The specific set of symptoms experienced by a given patient can vary, but generally headache, dizziness, feeling "slowed down," and fatigue are the most commonly reported symptoms.[16,19,21,22] The presence of any of these symptoms following a blow to the head (or body in the case of a whiplash mechanism of injury) is sufficient to diagnose

---

* For a thorough discussion of the history and evolution of our understanding of head injury and concussion, we refer readers to the excellent review provided by McCrory and Berkovic, 2001.[7]

TABLE 2.1
Common Signs and Symptoms of Concussion

Physical	Cognitive	Emotional	Sleep-Related
Headache	Loss of consciousness	Irritability	Difficulty falling sleeping
Dizziness	Post-traumatic amnesia	Emotional lability	Sleeping more or less than usual
Nausea	Difficulty concentrating	Depressed mood	Drowsiness
Vomiting	Memory problems		
Balance problems			
Fatigue			

concussion. A meta-analysis of postinjury follow-up studies of concussed athletes showed that neurocognitive performance and balance decrease, and self-reported symptoms increase on initial evaluation.[23] Caution is advised, however, against using any one measure to diagnose concussion. A multimodal approach assessing balance, neurocognitive function, and self-reported symptoms, with comparisons to baseline measures, is the preferred method for evaluating suspected concussion cases. We will examine several assessment instruments later in this chapter.

Concussion is a subset of TBI distinct from moderate and severe forms in that it typically does not result in an *extended period* of loss of consciousness. Clinicians often use the Glasgow Coma Scale (GCS) to distinguish between mild, moderate, and severe TBI, with lower scores indicating a deeper loss of consciousness.[24] Concussions are generally defined by GCS scores between 13 and 15.* Loss of consciousness and post-traumatic amnesia were once erroneously considered "conditio sine qua non" with regard to diagnosing a concussion.[26] Recent evidence, however, suggests loss of consciousness occurs in less than 10% of sport-related injuries and post-traumatic amnesia between 25% and 30% of cases.[16,17,27] This is a major point of misunderstanding among athletes, coaches, and parents, as well as clinicians.†

The natural history of concussion is variable, but most individuals become asymptomatic in 7 to 10 days.[27,29] Neuropsychological performance may be decreased beyond when the athlete no longer reports symptoms,[27,30–35] though the clinical significance of this finding is unclear. Interestingly, for self-reported symptoms, rate of symptom resolution is correlated to number of assessments given,[23] possibly indicating a role for repeated assessments; this phenomenon is not observed with neurocognitive performance, though balance assessments show a learning effect in healthy individuals.[36,37] Symptom exacerbation can occur during physical and/or cognitive exertion, and thus rest for both domains is recommended in the acute setting. The

* It should be noted that the GCS was not intended to supplant a full neurological examination. However, serial evaluations of the GCS can be useful as the progression of scores can be used to predict prognoses (i.e., a high GCS score staying high or a low score becoming high are scenarios indicative of a good prognosis).[25]

† As exemplified in a *New England Journal of Medicine* article in 2007 where the authors erroneously stated: "Concussion refers to an immediate and transient loss of consciousness accompanied by a brief period of amnesia after a blow to the head."[28]

role of rest is unclear as symptoms persist outside of the typical recovery timeframe of 1 to 2 weeks.

PATHOPHYSIOLOGY

Determining the pathophysiological underpinnings of the symptoms of concussion is a major focus in concussion research. Animal models of mild TBI have elucidated a metabolic cascade of events in the acute phase following injury. A large potassium efflux occurs, coinciding with a diffuse release of glutamate.[38] This is believed to be caused by a large number of indiscriminant depolarizations at the time of injury, likely due to sheer and strain forces on neurons. The ionic imbalance is restored through the action of K^+/Na^+ pumps, which require adenosine triphosphate (ATP), the primary energy substrate in the brain. Evidence suggests that action potentials and postsynaptic effects of glutamate account for as much as 80% of base rate energy metabolism in the cerebral cortex.[39,40] Thus, indiscriminate neuronal firing and release of glutamate after concussion creates a massive rise in brain energy demand. This large energy demand coincides with a decrease in cerebral blood flow, leading to an imbalance between cerebral glucose metabolism and perfusion.[41–43] It is important to note that these mechanisms have not been extensively studied in human subjects. However, studies using proton magnetic resonance spectroscopy have found neurometabolite alterations indicative of neural damage and distress following sport-related concussion.[44,45]

Advanced neuroimaging of concussive injury is a growing field within sport-related concussion research. Routine neuroimaging studies, including magnetic resonance imaging (MRI) and cranial tomography (CT), are typically normal in the majority of concussed patients,[46,47] leading many to call concussion a *functional* rather than structural injury. However, using more advanced MRI imaging protocols, particularly diffusion tensor imaging (DTI), microscopic disruptions in white matter tracts can be visualized.[48–52] Differences in image acquisition (including scanner type, time to scan following injury) and study design (voxel- versus tract-based versus region of interest analyses) have complicated synthesizing the results from DTI studies. Additionally, some studies have demonstrated no changes in white matter integrity following sport-related concussion.[53] A meta-analysis by Aoki et al., showed that the most consistent changes occurred in the corpus callosum and were unrelated to time following injury.[54] Specifically, two major metrics of white matter integrity were noted to change: fractional anisotropy (FA) decreased while mean diffusivity (MD) increased. Functional MRI (fMRI) and electroencephalography (EEG) have been useful in determining functional network changes following concussion. Resting state fMRI has shown changes in the default mode network, a neuronal network involved in non-task-related cognition.[55–58] Task-based fMRI studies have shown changes in both episodic[59] and working memory[57] networks. EEG studies have demonstrated differences following concussion,[60–65] with one well-designed, prospective study showing changes at 12 months following injury;[66] however, there are no signature patterns on EEG that can be used to definitively diagnose concussion.

BIOMECHANICS

Linear and rotational head accelerations are hypothesized to be the primary risk factors for concussion during an impact. Both direct and inertial (i.e., whiplash) loading of the head may result in linear and rotational head acceleration. Head acceleration induces strain patterns in brain tissue, which may cause injury. Current science has not identified an exact threshold for concussive injury, and direct measurement of brain dynamics during impact is extremely difficult in humans. Head acceleration, on the other hand, can be more readily measured; its relationship to severe brain injury has been postulated and tested for more than 50 years. Both linear and rotational acceleration of the head play important roles in producing diffuse injuries to the brain. However, the relative contributions of these accelerations to specific injury mechanisms have not been conclusively established. The numerous mechanisms theorized to result in brain injury have been evaluated in cadaveric and animal, surrogate,[67] and computer[68,69] models. Prospective clinical studies combining head impact biomechanics and clinical outcomes have been relatively void in the literature.

Our ongoing studies at the University of North Carolina involving collegiate and high school football players employs a real-time helmet accelerometer data collection methodology to better understand the biomechanics of concussion. Our findings suggest a higher propensity of top-of-the-head impacts and a higher relative risk of concussion for those impacts. In one of our published papers, six of thirteen concussions occurred from impacts to the top of the head, which is in contrast to four, two, and one concussions occurring to the front, right, and back, respectively (Figure 2.1).[70] These findings suggested that football players are concussed by head impacts occurring at a wide range of magnitudes (60.51 to 168.71 g linear acceleration), and that clinical measures of acute symptom severity, balance, and neuropsychological function, all appear to be largely independent of impact magnitude and location. There was no relationship between impact magnitude or location, and clinical outcomes of symptomatology, balance, or neuropsychological performance. The concussions sustained as a result of lower end magnitudes presented with just as many clinical deficits as those with higher end magnitudes. Thus, despite the literature suggesting that high magnitudes of head impact, particularly with high angular acceleration, result in more serious clinical outcomes in cases of moderate or severe TBI,[71,72] the magnitude and location do not predict clinical recovery in cases of sport-related concussion.

Concussions are often referred to structurally as "diffuse axonal injuries" and result in some degree of functional impairment, but differ from more moderate to severe TBI in that the impairment is transient in nature. Diffuse axonal injury, in addition to linear coup-contrecoup mechanisms of injury, can result in disruption to centers of the brain responsible for breathing, heart rate, and consciousness, but more typically result in memory loss, cognitive deficits, balance disturbances, and a host of other somatic symptoms. In the context of sport concussion, the term *impact* typically denotes an injurious blow that makes direct contact with the head. An indirect impact typically refers to an impact that sets the head in motion without directly striking it. Direct head impacts in sport range from helmet-to-helmet collisions, striking an opponent's head with a stick, or being struck in the head by a projectile

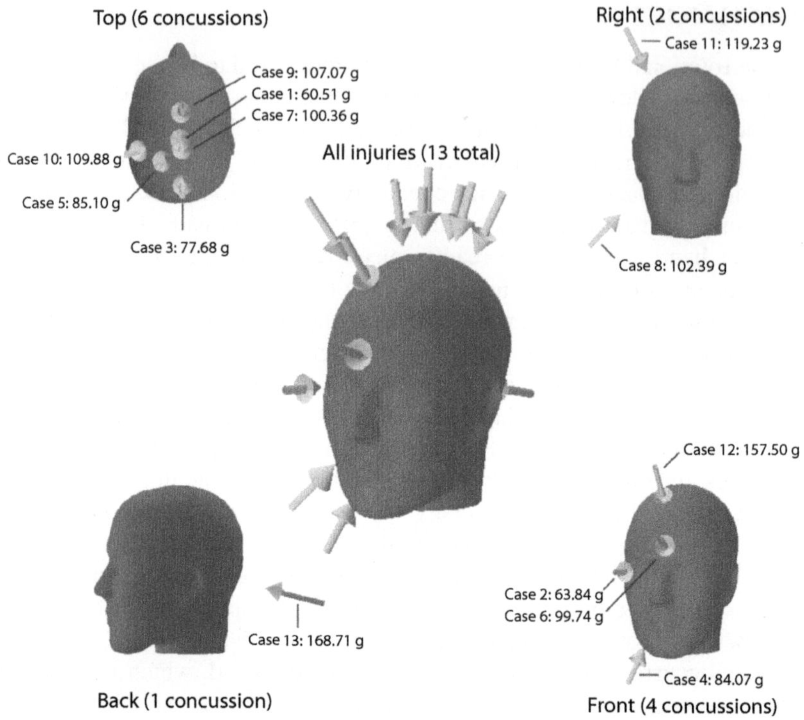

FIGURE 2.1 Helmet accelerometer data showing location and magnitude of impact leading to concussion in cohort of collegiate American football players. (From Guskiewicz, K.M. et al., *Neurosurgery* 61(6): 1244–1252, 2007.)

(e.g., soccer ball, hockey puck). Indirect impacts are most commonly caused by tackling or body checking, and are the result of abruptly stopping an opponent's body from traveling in the direction in which it was originally moving. Direct and indirect impacts are traditionally linear (translational) or angular (rotational) in nature. In real-world activities, there is usually some combination of both linear and angular accelerations associated with direct and indirect impacts.

Many factors are thought to play a role in the body's ability to dissipate head impact forces including individual differences in cerebrospinal fluid levels and function, vulnerability to brain tissue injury, relative musculoskeletal strengths and weaknesses, and the anticipation of an oncoming direct or indirect impact. Few studies have investigated the influence of cervical musculature on head impact biomechanics, and in general, increased cervical strength and muscle size has not been shown to reduce head acceleration and therefore prevent concussion.[73,74] However, a recent study by Schmidt et al.[75] observed that greater cervical muscle stiffness and angular displacement following perturbation reduces an athlete's odds of sustaining higher magnitude impacts. Therefore, suggesting that improving the cervical muscles' ability to quickly contract may be more important than overall muscle strength.

EPIDEMIOLOGY

An estimated 1.6 million to 3.8 million sport-related concussions occur in the United States annually,[76] with an estimated 10 million all-cause TBIs occurring globally each year.[77] As many as 50% of these sport-related concussions go unreported.[78] The factors contributing to the underreporting of concussive injury are multifactorial and include gaps in knowledge concerning concussion symptoms, beliefs that the injury is not serious, and unwillingness to be removed from competition.[78,79] Despite this underreporting, concussion incidence rates have been increasing in the past two decades.[80,81] This may be due, in part, to increased awareness of the signs and symptoms of concussion and the growing concern over their long-term effects leading to increased reporting. A national study of TBIs *with loss of consciousness* indicated that 20% were attributed to sports and recreational activities.[33] Within this sport-related sample (~300,000 total TBIs), 34% did not see a physician and 55% received only outpatient care (including emergency department visits). It is a major concern that concussions without major symptoms, such as loss of consciousness, may go unrecognized and unmanaged. It should be noted that there are barriers to obtaining high-quality epidemiological data about concussion incidence. In addition to underreporting, there is a lack of a widely used surveillance system in youth sports. Furthermore, very little information exists concerning sports below the high school level.

Nevertheless, concussion is a concern at every competitive level in contact sports.[29,82–84] Organized team sports account for approximately half of emergency department (ED) visits for concussion in the 14- to 18-year-old age range.[81] The youth sports accounting for the greatest number of concussions are football, wrestling, girls' soccer, boys' soccer, and girls' basketball.[85–87] Girls are reported to have a greater incidence of concussions when compared to boys playing similar sports.[85,87–89] The reasons for this gender disparity are unknown, but may include differences in cervical neck muscle strength, greater tendency to report symptoms in females, or due to the nature of female versions of the sports themselves.*

Of particular concern is football, which not only has one of the highest rates of concussion but also the largest participation.[91] Highlighting the commonality of concussion in football, in the National Football League (NFL) a concussion occurs almost every other game (incidence of 0.41 per game[92]). In a 2007 study of 2552 retired NFL players, over 60% reported one or more concussions in their playing career with 24% reporting three or more concussions.[93] Recurrent injury is not unique to football; evidence indicates that athletes with a history of concussion are at increased risk (as high as three- to fivefold greater risk) of additional concussions in the future.[83,84,94–96]

EVALUATION

The evaluation of concussion should rely on a detailed history of injury and subjective symptomatology, physical exam, and objective assessment tools of cognitive

* For instance, in ice hockey, females tend to get concussed more frequently than males. One possible explanation is that checking is disallowed in women's hockey, and thus, players may not anticipate contact as often as male skaters.[90]

functioning and balance. The immediate focus of the initial postinjury evaluation is on ruling out more serious injury to neurological structures; these include intracerebral, epidural, subdural, and subarachnoid hemorrhages; skull fractures; and cervical spine injury. Such injuries are emergencies that require immediate medical attention and intervention. The major concerning symptoms are alterations in consciousness, neck pain, loss of sensation or motor control, and abnormal posturing. Two major items of importance in the history of present illness include loss of consciousness and presence of post-traumatic amnesia as these predict presence of neurocranial abnormalities on cranial tomography.[26] Beyond general impressions of consciousness, the physical exam should include a full neurological examination of the 12 cranial nerves, extremity strength, and fine touch sensation as these can be useful to further rule out focal deficits following injury. Additionally, previous history of concussion, history of neurological or psychological disorders, and family history of migraines and headaches may predict a worse prognosis.[97]

Once more serious injuries have been ruled out, definitive diagnosis of concussion relies on self-reported symptoms and objective measures of cognitive functioning and balance. As noted earlier, symptoms may vary but are commonly described as feeling dazed or “in a fog” and with a headache of variable intensity. Concussed patients may be slow to answer questions, repeat questions on answering, or have difficulty orienting to their surroundings. The standard assessment of concussion (SAC) is a well-developed tool for assessing acute cognitive functioning immediately postinjury with 94% sensitivity in detecting concussion.[31,32,98,99] The SAC contains general questions concerning alertness and tasks to assess memory formation, delayed recall, and attention. The Balance Error Scoring System (BESS) is a useful instrument to assess balance immediately postinjury.[100,101] The BESS involves scoring athletes’ ability to hold three different static stances (double leg, single leg, and tandem stance), first on a hard surface and then on a foam pad, each for 20 seconds (Figure 2.2). The SAC and BESS can be easily deployed on the sideline for immediate objective information regarding the mental status of athletes suspected of having a concussion. When used in combination with a graded symptom checklist, this multimodal approach to concussion management has been found to be very sensitive to diagnosing and tracking recovery following the injury.[27,29,30] Figure 2.3 shows typical recovery curves of a graded symptom checklist, SAC, and BESS.

More detailed investigation into cognitive functioning is useful to track recovery and influence return-to-play decisions. Computerized neuropsychological testing systems, such as ImPACT (ImPACT Applications, Inc.; Pittsburgh, Pennsylvania), CNS-Vital Signs (CNS Vital Signs, LLC; Morrisville, North Carolina), Axon Sports (Axon Sports, LLC; Wausau, Wisconsin), combine a variety of paper-and-pencil tests into a single battery with rapid interpretation of results. These batteries give information across a broad range of cognitive domains including memory, attention, reaction time, processing speed, psychomotor speed, and fine motor coordination. The domains that are most sensitive to concussion are delayed memory, memory acquisition, and global cognitive functioning.[33] Using baseline results and reliable change indices, trained clinicians can determine when cognitive functioning has returned to baseline levels. Caution is advised in the use of neuropsychological testing in making decisions regarding return-to-play.[102] Although they are highly

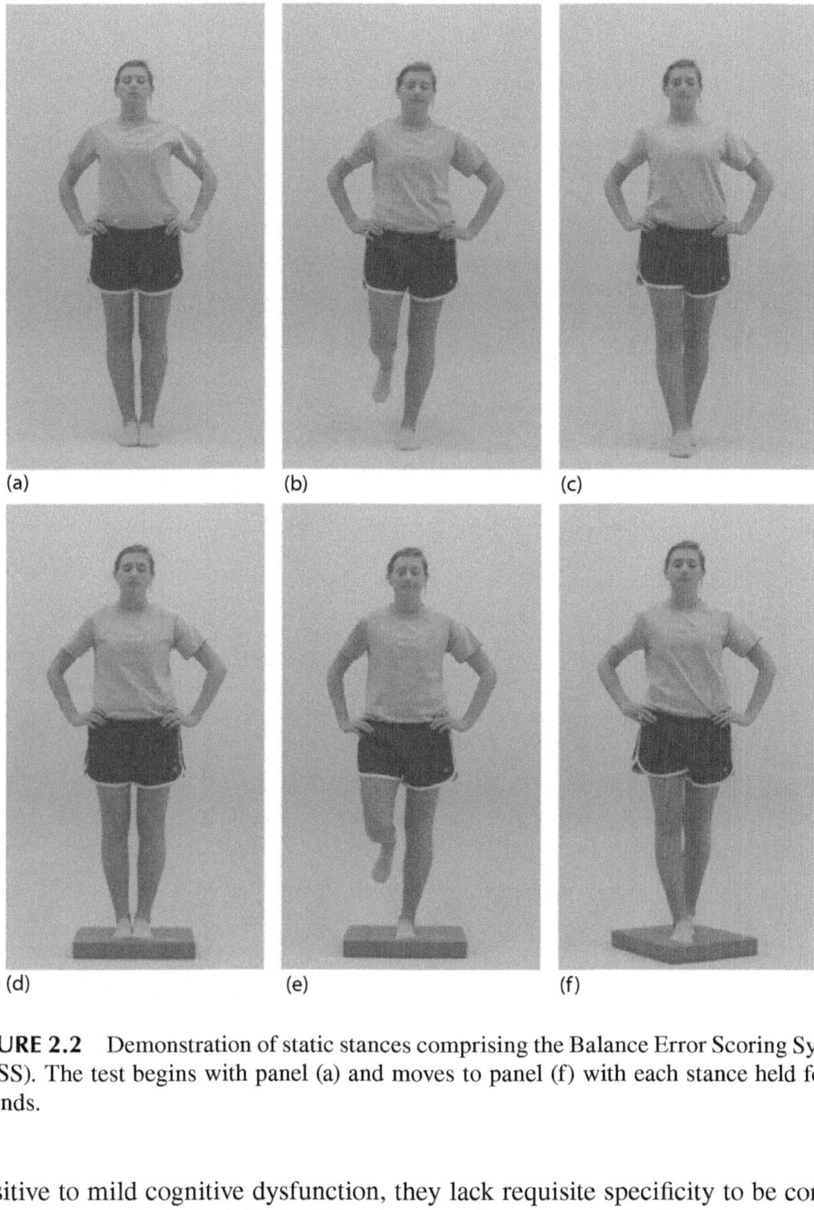

FIGURE 2.2 Demonstration of static stances comprising the Balance Error Scoring System (BESS). The test begins with panel (a) and moves to panel (f) with each stance held for 20 seconds.

sensitive to mild cognitive dysfunction, they lack requisite specificity to be considered a diagnostic tool. Additionally, the domains computerized neuropsychological batteries seek to measure (as listed earlier) are inherently unstable, they can be affected by external factors (drugs, disease states, time of day) and internal factors (motivation, fatigue, mood, level of alertness).[103] However, several of these tests have demonstrated reliability and can be considered a useful tool in the *trained* clinician's armamentarium.[103,104]

Additionally, postural control and balance can be assessed in finer detail using NeuroCom (NeuroCom Inc.; Clackamas, Oregon) force plate-based technology.

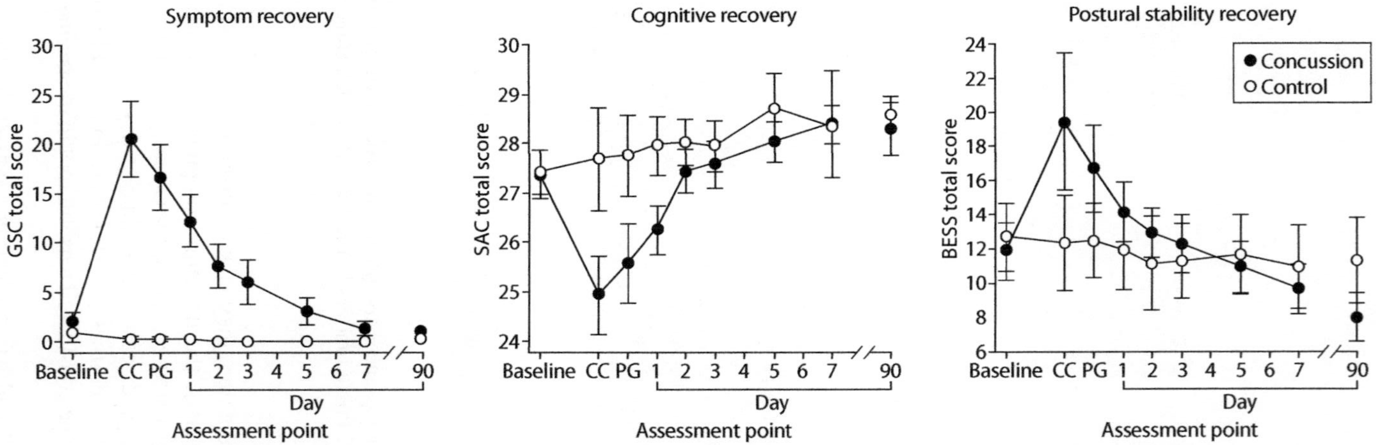

Higher scores on the graded symptom checklist (GSC) indicate more severe symptoms; lower scores on the standardized assessment of concussion (SAC) indicate poorer cognitive performance; and higher scores on the Balance Error Scoring System (BESS) indicate poorer postural stability. Error bars indicate 95% confidence intervals. CC indicates time of concussion; PG, postgame/postpractice. On the BESS, multiple imputation was used to estimate means and 95% confidence intervals for control participants for the CC and PG assessments.

FIGURE 2.3 Typical recovery curves of serial assessments of symptoms, cognitive performance, and balance following a sport-related concussion. (From McCrea, M. et al., *JAMA* 290(19): 2556–2563, 2003).

Specifically the Sensory Organization Test (SOT) is useful to isolate and measure the function of different domains including vestibular, visual, somatosensory, and preferential balance. SOT scores are decreased acutely following concussion and improve over the course of several days.[100,101] Additionally, the SOT can be useful in determining what balance systems, if any, are affected in patients experiencing prolonged recovery. The affected systems can be targeted using either visual or vestibular rehabilitation. As with any injury, full consideration of all patient information should be the basis of management decisions, particularly in regard to the concussed athlete. Neither cognitive performance nor balance is sufficient in determining treatment or return-to-play decisions.

A relatively new domain of interest in initial concussion evaluations and tracking recovery is the oculomotor system. Disruptions in visual performance have been noted following blast-related mild TBI in military service members.[105] One instrument designed to probe these deficits is the King–Devick test. This test requires the subject to read aloud a string of numbers across several test cards as quickly as possible, which requires both saccadic eye movements and rapid processing of information. Some studies have shown promise in detecting head trauma and concussion,[106,107] whereas an emergency department study did not support its use.[108] Further validation and development of the instrument is needed, but there is promise that the King–Devick test will become another useful tool for clinicians facing a concussed athlete.

Last, gait tasks have become increasingly popular due to the known deficits in balance following concussion. By assessing gait, the clinician can see an athlete perform a dynamic and more functionally relevant task than quiet standing or stance-holding as required by the SOT and BESS, respectively. Studies assessing gait postconcussion have found observable deficits.[109–112] Cantena et al. recommend pairing a gait task with an attentional task to obviate deficits in gait.[113–115] Such dual-task paradigms are being studied both for concussion evaluation as well as rehabilitation.[116–119] To date, there are no definitive studies supporting the use of dual-task rehabilitation programs in sport-related concussion.

TREATMENT

Cognitive and physical rest is considered the cornerstone of concussion management.[2–4,6] The extent and duration of rest to maximize benefit is unclear, as only a scarcity of studies address the relationship between activity level after concussion and recovery of concussed athletes; however, they corroborate the principle of rest as an acute treatment.[120–122] The recommendation for rest is primarily based on an absence of literature surrounding interventions that may reduce risk of prolonged symptoms or improve recovery.[3,123] Based on anecdotal evidence, the general recommendation is for complete rest in the immediate postinjury phase for 2 to 3 days followed by a gradual increase in activity as tolerated. For an athlete, this means temporary withdrawal from practice and reductions in cognitive activities (e.g., scholastic work, reading, cellular phone use). As symptoms abate, a progressive, stepwise return-to-play protocol is followed that increases physical activity until full participation can be resumed.

There are concerns regarding the immediate management and return-to-play decisions surrounding youth athletes. Many states have passed legislation that requires immediate removal from play, evaluation by a medical professional, and no same day return-to-play under any circumstance.[124] For many years, sports concussion literature highlighted second-impact syndrome (SIS) as a potentially lethal consequence of concussion mismanagement.[125,126] This catastrophic consequence is described as a loss of autoregulation of cerebral blood flow and thought to be the result of an impact occurring to an athlete while they are still symptomatic following an initial head injury. This second impact does not need to be severe enough to cause a second concussion, but can trigger a sequence of events that leads to nearly 50% mortality and 100% morbidity. However, the fact remains that SIS is a poorly understood phenomenon, due in large part to the rarity of its occurrence with only 35 probable cases identified in a 23-year period.[127,128] If the risk of SIS is directly tied to concussive injury, one might expect a greater number of cases given the large number of estimated unreported concussions and the potential for athletes to receive impacts to the head following an unreported concussion.

Regardless of SIS, there are reasons to keep athletes from returning to play while they are symptomatic. Pediatric patients are slower to recover from injury than adults.[129,130] It is generally accepted that there is unnecessary risk to returning an athlete to play before they are evaluated and cleared by a medical professional. There is a lack of understanding concerning the long-term effects of concussion, particularly where the developing brain is concerned. As previously mentioned, with symptom resolution typically occurring within 1 to 2 weeks, the time course for recovery from concussion is relatively rapid when compared to musculoskeletal injuries that can take weeks or months to heal. In the end, there should be no rush to return an athlete who may suffer from symptom exacerbation and an increased risk of suffering an additional injury. The focus of care should of course be on the general health and well-being of the athlete, not the rapidity of their return-to-play. Particular concern is warranted in the case of recurrent concussion suffered in the context of sports. When an athlete presents with repeat sport-related concussions, clinicians should engage in a frank discussion of risks versus reward with respect to continued participation in their sport.

FUTURE DIRECTIONS IN RESEARCH

Because of the dynamic and interdisciplinary nature of sport-related concussion, there are a large number of research areas and topics that are being aggressively pursued. While each of these avenues of research contributes to our understanding of the nature of the injury, there is a shortlist of priority areas that require meticulous, well-designed studies to move the field forward and make impactful change. These include the long-term effects of concussion, developing the clinical utility of neuroimaging, determining the effect of subconcussive impacts on the developing brain, improving clinical management practices, and making youth sports safer. We will now discuss some of the ongoing studies addressing these issues and the gaps in knowledge that are yet to be addressed.

While relatively few studies have examined the long-term effects of recurrent, sport-related concussion, several concerning trends have emerged, including increased risk of depression,[93] cognitive impairment,[131,132] earlier onset Alzheimer's disease,[42] dementia,[133] and neurodegenerative cause of death.[134] Despite these clinical observations, little is known about the neurophysiological changes associated with recurrent concussions and their impact on neurodegenerative disease. Recently, an association between repetitive head trauma and the development of chronic traumatic encephalopathy (CTE) has been proposed.[135–138] CTE is described as a neurodegenerative disease, distinguishable from Alzheimer's disease and other forms of dementia on the basis of postmortem neuropathology. Histological studies on the brains of deceased contact athletes and military veterans have shown several characteristic findings such as a distinct pattern of hyperphosphorylated tau, reduction in total brain weight, and enlargement of ventricles.[5–10]

The clinical presentation of CTE consists of memory problems, executive dysfunction, and behavioral and personality changes, among others.[137] However, it should be noted that clinical descriptions of CTE are based on informant interviews of deceased patients with CTE confirmed on postmortem histopathology. In vivo evidence of CTE is limited and further study is needed to understand the role of recurrent concussions on neurodegenerative disease. As such, CTE is a controversial diagnosis that is not universally accepted as a distinct pathology due to the largely nonspecific clinical descriptions and lack of clear diagnostic criteria.[132] Nevertheless, preliminary results of a recent study demonstrate the presence of tau deposits in symptomatic retired football players distributed in a pattern consistent with autopsy studies.[139] Additionally, in a mouse model that expresses human tau isoforms, repetitive mild TBI resulted in greater hyperphosphorylated tau deposition; however the authors noted that it did not follow the pattern described in CTE.[140] Whether CTE is a distinct pathology, neurodegenerative disease is a major concern for retired athletes and military veterans. The controversy surrounding CTE highlights our need for structured, prospective investigations of the linkages between neurodegenerative disease and repetitive head trauma.

A major issue in studying the long-term effects of concussion is that the neuroimaging modalities by which we can study the structure and function of the brain in vivo are not yet perfected. Most imaging studies have used a concussed group with comparisons made to a matched control group, often with a small number of subjects (less than 20). Using a single scan, cross-sectional, group approach makes voxelwise analyses troublesome. Within the concussed group, the location of neuronal injury must occur in an approximately identical location in the brain for all the subjects in order to detect a significant difference between patient and the control groups. Given that the nature of concussive injury is heterogeneous based on the mechanics of the injury itself, more sophisticated analytic techniques are needed to adequately assess neuronal injury using imaging. Studies that compare concussed individuals to their preinjury scans are ideal, however, exceedingly difficult to perform, not in small part due to the large number of subjects that would need to be enrolled and the high cost of the scans. Regardless, such prospective investigation is needed to advance the utility of neuroimaging after concussion. Thus far, the clinical application of neuroimaging in concussion is underdeveloped. While CT is useful in ruling out more serious injury, there is little more to recommend imaging following concussion. To translate advanced neuroimaging to the clinic, e.g., DTI, susceptibility weighted imaging, or quantitative EEG, a large atlas of normative data would be needed across a wide range of ages. Injuries would need to be well characterized and studied across several time-points. Once these studies have been performed, the field can address the questions surrounding diagnostic sensitivity and specificity and prognostic ability of neuroimaging.

An interest in subconcussive head impacts has emerged in recent years following reports of CTE pathology in concussion naïve patients. Subconcussive impacts refer to head impacts sustained during collision sports that do not result in clinically apparent symptoms. Prospective study of repetitive subconcussive impacts is more easily implemented because investigators need not wait for concussion cases to present. Bazarian et al. has used DTI to study college football players across a single season and found greater changes in diffusion metrics in those athletes exposed to repetitive head impacts as compared to noncollision sport controls.[141,142] Many of these changes persisted to six months following cessation of collision-sport activity. The findings of the Bazarian et al. studies are worrying, though preliminary. It is unclear whether the observed changes are deleterious or perhaps adaptive, given that in some regions there were increases in white matter integrity (increased FA) and others in which decreases were noticed (reduced FA). These changes were, for the most part, not strongly correlated to changes in neuropsychological testing, indicating that the neuroanatomical changes are clinically undetectable. Other studies employing a similar design found differences in diffusion metrics in club soccer players[143] and college football and ice hockey players.[144]

Aside from imaging, another helpful tool to investigators seeking to further study the effects of subconcussive impacts is accelerometer technology. Helmeted sports, such as football, ice hockey, and lacrosse, have been studied using accelerometers embedded in the helmet; these systems provide information on the linear and rotational accelerations of the helmet and head during impact. Studies utilizing helmet accelerometers have improved our understanding of the in-game risk factors involved in concussive injury including position, sport, and event type (game versus practice).[70,145,146] Furthermore, these systems have been used to determine the biomechanics of concussive injury, though no definitive threshold of head acceleration to predict concussion has been found.[147] Studying the biomechanics of head impacts in youth sports has implications on practice structure, rule revisions, and coaching proper form and technique.[148,149]

In addition to research, accelerometer systems have clinical utility. Our group has used accelerometer systems as a way to monitor athletes during games and practices, and check in with those athletes who experience impacts that are above a cutoff magnitude. This practice has a caveat in that smaller impacts can still cause concussion, and thus, the clinical staff must be vigilant. Another use for the accelerometers entails behavioral correction of athletes who display "poor" head impact profiles (i.e., numerous high magnitude impacts, particularly to the crown of the head) that may be predictive of improper technique. Newer acceleration systems have been developed, such as intraoral devices and adhesive patches that can be applied directly to the head. These systems, though new and in need of validation, allow researchers to study nonhelmeted sports that convey a high concussion risk (e.g., soccer), and may be a more accurate method for directly measuring head acceleration. This will enable a finer, more resolved method to studying youth sports and informing decisions to improve safety and reduce concussion risk.

Although prevention is certainly a major goal in reducing the burden of sport-related concussions, improving management and clinical decision-making is equally important. In approximately 10% of sport-related concussion cases, symptoms persist beyond the normal window of recovery, and can last for weeks or months.[29,95] Although definitions of postconcussion syndrome (PCS) vary depending on the source and there is no universally accepted definition of the syndrome, the common feature of the definitions of PCS is the continued experience of symptoms elicited from the initial injury. The World Health Organization's International Statistical Classification of Diseases (10th revision) definition has a sensitivity and specificity of 73% and 61%, respectively, at one month postinjury.[150]

PCS is poorly understood, as are the risk factors associated with the condition. Initial self-reported symptom severity, prolonged headache, and concentration deficits have been associated with prolonged recovery.[14] Corroborating some of these findings, a 10-year, controlled study of adolescent athletes found that presence of unconsciousness, post-traumatic amnesia, and symptom severity were most strongly associated with prolonged recovery.[29,151] However, neither loss of consciousness nor amnesia are associated with greater deficits or slower recovery of objective measures of postural stability and neurocognitive function.[101] Whether previous concussion history is predictive of developing PCS on repeat injury is in question as contradicting results exist across studies.[152,153] Some evidence suggests that involvement in litigation or workers' compensation may increase the severity and duration of concussion symptoms.[22,154] Recent evidence suggests a role for a prescribed regimen of exercise[155–158] and/or cervicovestibular rehabilitation[159] for treatment of patients with PCS. However, there is concern that reactive intervention may not be optimal and that earlier interventions, before PCS is diagnosed, are warranted. An obstacle to improving acute care of sport-related concussions lies in the fact that the vast majority of injuries self-resolve without observable, lasting impairments. Thus, the objective for interventions acutely following concussion should be to reduce the risk of developing PCS, which as described, is understudied.

CONCLUSION

There has been quite an evolution with respect to concussion management over the last two decades. Much of this can be attributed to research that has advanced our understanding of the biomechanics, pathophysiology, and recovery patterns associated with concussion. This evolution has brought technology and objective testing methods to the forefront of concussion management. Additionally, greater emphasis on education and awareness has certainly played a major part in helping to recognize and effectively manage these injuries. One of the greatest influences that clinicians can have in preventing sport-related concussion, and catastrophic outcomes, is to educate athletes, coaches, and parents about the dangers of playing while symptomatic following a concussion. Reports of the cumulative effects of multiple concussions, as well as the potential for multiple head impacts to affect long-term cognitive health, should lead clinicians to rethink their approach to managing concussion.

Contemporary methods of concussion assessment, involving the use of symptom checklists, neuropsychological testing, and postural stability/balance testing, are indicated for any athlete suspected of having sustained a concussion. Clinicians working with high-risk sports should conduct baseline assessments, including neuropsychological and postural stability tests, prior to the start of the season. Testing should also be conducted following exertional activities that are typically performed prior to a full return to activity. Most important, clinicians must recognize that the recovery and return-to-play considerations involve many factors. More recent research aimed at identifying biomarkers for determining the potential risk factors that predict symptom onset and progression of neurodegenerative disease in athletes who have played contact sports will eventually bring advanced technologies such as neuroimaging and blood biomarkers to our toolbox. With these new tools comes the promise of improving patient outcomes and maximizing safety for athletes with regard to their mental and cognitive health and longevity.

REFERENCES

1. Faul, M. et al. 2010. *Traumatic Brain Injury in the United States: Emergency Department Visits, Hospitalizations and Deaths 2002–2006*. Centers for Disease Control and Prevention, National Center for Injury Prevention and Control, Atlanta, GA.
2. Giza, C.C. et al. 2013. Summary of evidence-based guideline update: Evaluation and management of concussion in sports: Report of the Guideline Development Subcommittee of the American Academy of Neurology. *Neurology* 80(24):2250–2257.
3. Halstead, M.E. et al. 2010. Sport-related concussion in children and adolescents. *Pediatrics* 126(3):597–615.
4. Harmon, K.G. et al. 2013. American Medical Society for Sports Medicine position statement: Concussion in sport. *Br J Sports Med* 47(1):15–26.
5. Broglio, S. et al. 2014. National Athletic Trainers' Association position statement: Management of sport concussion. *J Athl Train* 49(2):245–265.
6. McCrory, P. et al. 2013. Consensus statement on concussion in sport: The 4th International Conference on Concussion in Sport held in Zurich, November 2012. *Br J Sports Med* 47(5):250–258.
7. McCrory, P.R., Berkovic, S.F. 2001. Concussion: The history of clinical and pathophysiological concepts and misconceptions. *Neurology* 57(12):2283–2299.
8. The American Congress of Rehabilitation Medicine. 1993. Definition of mild traumatic brain injury. *J Head Trauma Rehab* 8(3):86.
9. Vos, E. et al. 2002. EFNS guideline on mild traumatic brain injury: Report of an EFNS task force. *Eur J Neurol* 9(3):207–219.
10. Cantu, R.C. 2001. Posttraumatic retrograde and anterograde amnesia: Pathophysiology and implications in grading and safe return to play. *J Athl Train* 36(3):244–248.
11. Colorado Medical Society. 1991. *Report of the Sports Medicine Committee: Guidelines for the Management of Concussions in Sport (Revised)*. Colorado Medical Society, Denver, CO.
12. Dematteo, C.A. et al. 2010. My child doesn't have a brain injury, he only has a concussion. *Pediatrics* 125(2):327–334.
13. Eisenberg, M.A., Meehan 3rd, W., Mannix, R. 2014. Duration and course of post-concussive symptoms. *Pediatrics* 133(6):999–1006.
14. Makdissi, M. et al. 2010. Natural history of concussion in sport: Markers of severity and implications for management. *Am J Sports Med* 38(3):464–471.

15. Stovner, L.J. et al. 2009. Headache after concussion. *Eur J Neurol* 16(1):112–120.
16. Dikmen, S. et al. 2010. Rates of symptom reporting following traumatic brain injury. *J Int Neuropsychol Soc* 16(3):401–411.
17. Meehan, W., d'Hemecourt, P., Comstock, R.D. 2010. High school concussions in the 2008–2009 academic year: Mechanism, symptoms, and management. *Am J Sports Med* 38(12):2405–2409.
18. Englander, J. et al. 1992. Mild traumatic brain injury in an insured population: Subjective complaints and return to employment. *Brain Inj* 6(2):161–166.
19. Lucas, S. et al. 2013. A prospective study of prevalence and characterization of headache following mild traumatic brain injury. *Cephalalgia* 34(2):93–102.
20. Lovell, M.R. 2006. Measurement of symptoms following sports-related concussion: Reliability and normative data for the post-concussion scale. *Appl Neuropsychol* 13(3):166–174.
21. Kraus, J. et al. 2005. Physical complaints, medical service use, and social and employment changes following mild traumatic brain injury: A 6-month longitudinal study. *J Head Trauma Rehabil* 20(3):239–256.
22. Paniak, C. et al. 2002. A longitudinal study of the relationship between financial compensation and symptoms after treated mild traumatic brain injury. *J Clin Exp Neuropsychol* 24(2):187–193.
23. Broglio, S., Puetz, T.W. 2008. The effect of sport concussion on neurocognitive function, self-report symptoms and postural control: A meta-analysis. *Sports Med* 38(1): 53–67.
24. Teasdale, G., Jennett, B. 1974. Assessment of coma and impaired consciousness. A practical scale. *Lancet* 2(7872):81–84.
25. Jennett, B. et al. 1977. Severe head injuries in three countries. *J Neurol Neurosurg Psychiatry* 40(3):291–298.
26. Smits, M. et al. 2007. A history of loss of consciousness or post-traumatic amnesia in minor head injury: "Conditio sine qua non" or one of the risk factors? *J Neurol Neurosurg Psychiatry* 78(12):1359–1364.
27. McCrea, M. et al. 2003. Acute effects and recovery time following concussion in collegiate football players: The NCAA Concussion Study. *JAMA* 290(19):2556–2563.
28. Ropper, A.H., Gorson, K.C. 2007. Concussion. *N Eng J Med* 356(2):166–172.
29. McCrea, M. et al. 2013. Incidence, clinical course, and predictors of prolonged recovery time following sport-related concussion in high school and college athletes. *J Int Neuropsychol Soc* 19(1):22–33.
30. McCrea, M. et al. 2005. Standard regression-based methods for measuring recovery after sport-related concussion. *J Int Neuropsychol Soc* 11(1):58–69.
31. Echemendia, R.J. et al. 2001. Neuropsychological test performance prior to and following sports-related mild traumatic brain injury. *Clin J Sport Med* 11(1):23–31.
32. Barr, W.B., McCrea, M. 2001. Sensitivity and specificity of standardized neurocognitive testing immediately following sports concussion. *J Int Neuropsychol Soc* 7(6): 693–702.
33. Belanger, H.G., Vanderploeg, R.D. 2005. The neuropsychological impact of sports-related concussion: A meta-analysis. *J Int Neuropsychol Soc* 11(4):345–357.
34. Collins, M.W. et al. 1999. Relationship between concussion and neuropsychological performance in college football players. *JAMA* 282(10):964–970.
35. Mathias, J.L., Beall, J.A., Bigler, E.D. 2004. Neuropsychological and information processing deficits following mild traumatic brain injury. *J Int Neuropsychol Soc* 10(2):286–297.
36. Wrisley, D.M. et al. 2007. Learning effects of repetitive administrations of the sensory organization test in healthy young adults. *Arch Phys Med Rehabil* 88(8):1049–1054.

37. Valovich, T.C., Perrin, D.H., Gansneder, B.M. 2003. Repeat administration elicits a practice effect with the balance error scoring system but not with the standardized assessment of concussion in high school athletes. *J Athl Train* 38(1):51–56.
38. Katayama, Y. et al. 1990. Massive increases in extracellular potassium and the indiscriminate release of glutamate following concussive brain injury. *J Neurosurg* 73(6):889–900.
39. Attwell, D., Iadecola, C. 2002. The neural basis of functional brain imaging signals. *Trends Neurosci* 25(12):621–625.
40. Attwell, D., Laughlin, S.B. 2001. An energy budget for signaling in the grey matter of the brain. *J Cereb Blood Flow Metab* 21(10):1133–1145.
41. Giza, C.C., Hovda, D.A. 2001. The neurometabolic cascade of concussion. *J Athl Train* 36(3):228–235.
42. Prins, M.L. et al. 2010. Repeat traumatic brain injury in the juvenile rat is associated with increased axonal injury and cognitive impairments. *Dev Neurosci* 32(5–6):510–518.
43. Barkhoudarian, G., Hovda, D.A., Giza, C.C. 2011. The molecular pathophysiology of concussive brain injury. *Clin Sports Med* 30(1):33–48.
44. Vagnozzi, R. et al. 2010. Assessment of metabolic brain damage and recovery following mild traumatic brain injury: A multicentre, proton magnetic resonance spectroscopic study in concussed patients. *Brain* 133(11):3232–3242.
45. Gasparovic, C. et al. 2009. Neurometabolite concentrations in gray and white matter in mild traumatic brain injury: An 1H-magnetic resonance spectroscopy study. *J Neurotrauma* 26(10):1635–1643.
46. Jagoda, A.S. et al. 2008. Clinical policy: Neuroimaging and decisionmaking in adult mild traumatic brain injury in the acute setting. *Ann Emerg Med* 52(6):714–748.
47. Davis, G.A. et al. 2009. Contributions of neuroimaging, balance testing, electrophysiology and blood markers to the assessment of sport-related concussion. *Br J Sports Med* 43(1):058123.
48. Cubon, V.A. et al. 2011. A diffusion tensor imaging study on the white matter skeleton in individuals with sports-related concussion. *J Neurotrauma* 28(2):189–201.
49. Grossman, E.J. et al. 2012. Thalamus and cognitive impairment in mild traumatic brain injury: A diffusional kurtosis imaging study. *J Neurotrauma* 29(13):2318–2327.
50. Grossman, E.J. et al. 2013. Cognitive impairment in mild traumatic brain injury: A longitudinal diffusional kurtosis and perfusion imaging study. *Am J Neuroradiol* 34(5):951–957.
51. Kumar, R. et al. 2009. Serial changes in the white matter diffusion tensor imaging metrics in moderate traumatic brain injury and correlation with neuro-cognitive function. *J Neurotrauma* 26(4):481–495.
52. Messe, A. et al. 2011. Diffusion tensor imaging and white matter lesions at the subacute stage in mild traumatic brain injury with persistent neurobehavioral impairment. *Hum Brain Mapp* 32(6):999–1011.
53. Ilvesmaki, T. et al. 2014. Acute mild traumatic brain injury is not associated with white matter change on diffusion tensor imaging. *Brain* 137(Pt 7):1876–1882.
54. Aoki, Y. et al. 2012. Diffusion tensor imaging studies of mild traumatic brain injury: A meta-analysis. *J Neurol Neurosurg Psychiatry* 83(9):870–876.
55. Palacios, E.M. et al. 2013. Resting-state functional magnetic resonance imaging activity and connectivity and cognitive outcome in traumatic brain injury. *JAMA Neurol* 70(7): 845–851.
56. Sharp, D.J. et al. 2011. Default mode network functional and structural connectivity after traumatic brain injury. *Brain* 134(Pt 8):2233–2247.
57. Palacios, E.M. et al. 2012. White matter integrity related to functional working memory networks in traumatic brain injury. *Neurology* 78(12):852–860.
58. Zhang, K. et al. 2012. Default mode network in concussed individuals in response to the YMCA physical stress test. *J Neurotrauma* 29(5):756–765.

59. Ford, J.H., Giovanello, K.S., Guskiewicz, K.M. 2013. Episodic memory in former professional football players with a history of concussion: An event-related functional neuroimaging study. *J Neurotrauma* 30(20):1683–1701.
60. Scrascia, F. et al. 2014. Relationship among diffusion tensor imaging, EEG activity, and cognitive status in mild cognitive impairment and Alzheimer's disease patients. *J Alzheimers Dis* 38(4):939–950.
61. Virji-Babul, N. et al. 2014. Changes in functional brain networks following sports related concussion in adolescents. *J Neurotrauma* 1;31(23):1914–1919.
62. Thatcher, R. 2006. Electroencephalography and mild traumatic brain injury. In *Foundations of Sport-Related Brain Injuries*, eds. S. Slobounov, W. Sebastianelli, 241–265. Springer, New York.
63. Thompson, J.G. 2006. EEG changes and balance deficits following concussion: One piece of the puzzle. In *Foundations of Sport-Related Brain Injuries*, eds. S. Slobounov, W. Sebastianelli, 341–374. Springer, New York.
64. Nuwer, M.R. et al. 2005. Routine and quantitative EEG in mild traumatic brain injury. *Clin Neurophysiol* 116(9):2001–2025.
65. Duff, J. 2004. The usefulness of quantitative EEG (QEEG) and neurotherapy in the assessment and treatment of post-concussion syndrome. *Clin EEG Neurosci* 35(4):198–209.
66. Slobounov, S., Sebastianelli, W., Hallett, M. 2012. Residual brain dysfunction observed one year post-mild traumatic brain injury: Combined EEG and balance study. *Clin Neurophysiol* 123(9):1755–1761.
67. Ono, K., Kanno, M. 1996. Influences of the physical parameters on the risk to neck injuries in low impact speed rear-end collisions. *Accid Anal Prev* 28(4):493–499.
68. Viano, D.C., Lovsund, E.R. 1999. Biomechanics of brain and spinal-cord injury: Analysis of neuropathologic and neurophysiology experiments. *J Crash Prev Injury Control* 1(1):35–43.
69. Viano, D.C. et al. 1989. Injury biomechanics research: An essential element in the prevention of trauma. *J Biomech* 22(5):403–417.
70. Guskiewicz, K.M. et al. 2007. Measurement of head impacts in collegiate football players: Relationship between head impact biomechanics and acute clinical outcome after concussion. *Neurosurgery* 61(6):1244–1252.
71. Holbourn, A.H.S. 1945. The mechanics of brain injuries. *Brit Med Bull* 3(6):147–149.
72. Ommaya, A.K., Gennarelli, T.A. 1974. Cerebral concussion and traumatic unconsciousness. Correlation of experimental and clinical observations of blunt head injuries. *Brain* 97(4):633–654.
73. Mihalik, J. et al. 2011. Does cervical muscle strength in youth ice hockey players affect head impact biomechanics? *Clin J Sport Med* 21(5):416–421.
74. Mansell, J. et al. 2005. Resistance training and head-neck segment dynamic stabilization in male and female collegiate soccer players. *J Athl Train* 40(4):310–319.
75. Schmidt, J.D. et al. 2014. The influence of cervical muscle characteristics on head impact biomechanics in football. *Am J Sports Med* 42(9):2056–2066.
76. Langlois, J.A., Rutland-Brown, W., Wald, M.M. 2006. The epidemiology and impact of traumatic brain injury: A brief overview. *J Head Trauma Rehabil* 21(5):375–378.
77. Hyder, A.A. et al. 2007. The impact of traumatic brain injuries: A global perspective. *NeuroRehabilitation* 22(5):341–353.
78. McCrea, M. et al. 2004. Unreported concussion in high school football players: Implications for prevention. *Clin J Sport Med* 14(1):13–17.
79. Register-Mihalik, J.K. et al. 2013. Knowledge, attitude, and concussion-reporting behaviors among high school athletes: A preliminary study. *J Athl Train* 48(5):645–653.
80. Marin, J.R. et al. 2014. Trends in visits for traumatic brain injury to emergency departments in the united states. *JAMA* 311(18):1917–1919.

81. Bakhos, L.L. et al. 2010. Emergency department visits for concussion in young child athletes. *Pediatrics* 126(3):e550–e556.
82. Thurman, D.J., Branche, C.M., Sniezek, J.E. 1998. The epidemiology of sports-related traumatic brain injuries in the United States: Recent developments. *J Head Trauma Rehabil* 13(2):1–8.
83. Guskiewicz, K.M. et al. 2000. Epidemiology of concussion in collegiate and high school football players. *Am J Sports Med* 28(5):643–650.
84. Schulz, M.R. et al. 2004. Incidence and risk factors for concussion in high school athletes, North Carolina, 1996–1999. *Am J Epidemiol* 160(10):937–944.
85. Marar, M. et al. 2012. Epidemiology of concussions among United States high school athletes in 20 sports. *Am J Sports Med* 40(4):747–755.
86. Powell, J.W., Barber-Foss, K.D. 1999. Traumatic brain injury in high school athletes. *JAMA* 282(10):958–963.
87. Lincoln, A.E. et al. 2011. Trends in concussion incidence in high school sports: A prospective 11-year study. *Am J Sports Med* 39(5):958–963.
88. Dick, R.W. 2009. Is there a gender difference in concussion incidence and outcomes? *Br J Sports Med* 43(Suppl 1):i46–i50.
89. Gessel, L.M. et al. 2007. Concussions among United States high school and collegiate athletes. *J Athl Train* 42(4):495–503.
90. Hootman, J.M., Dick, R., Agel, J. 2007. Epidemiology of collegiate injuries for 15 sports: Summary and recommendations for injury prevention initiatives. *J Athl Train* 42(2):311–319.
91. Daneshvar, D.H. et al. 2011. The epidemiology of sport-related concussion. *Clin Sports Med* 30(1):1–17.
92. Pellman, E.J. et al. 2004. Concussion in professional football: Epidemiological features of game injuries and review of the literature—Part 3. *Neurosurgery* 54(1):81–94.
93. Guskiewicz, K.M. et al. 2007. Recurrent concussion and risk of depression in retired professional football players. *Med Sci Sports Exerc* 39(6):903–909.
94. Noble, J.M., Hesdorffer, D.C. 2013. Sport-related concussions: A review of epidemiology, challenges in diagnosis, and potential risk factors. *Neuropsychol Rev* 23(4):273–284.
95. Guskiewicz, K.M. et al. 2003. Cumulative effects associated with recurrent concussion in collegiate football players: The NCAA concussion study. *JAMA* 290(19):2549–2555.
96. Benson, B.W. et al. 2011. A prospective study of concussions among National Hockey League players during regular season games: The NHL-NHLPA Concussion Program. *CMAJ* 183(8):905–911.
97. Morgan, C.D. et al. 2014. 114 Risk factors for post-concussion syndrome in an exclusively sport-related concussion group: Case control study. *Neurosurgery* 1(196):88.
98. McCrea, M. 2001. Standardized mental status testing on the sideline after sport-related concussion. *J Athl Train* 36(3):274–279.
99. McCrea, M. et al. 1997. Standardized assessment of concussion in football players. *Neurology* 48(3):586–588.
100. Riemann, B.L., Guskiewicz, K.M. 2000. Effects of mild head injury on postural stability as measured through clinical balance testing. *J Athl Train* 35(1):19–25.
101. Guskiewicz, K.M., Ross, S.E., Marshall, S.W. 2001. Postural stability and neuropsychological deficits after concussion in collegiate athletes. *J Athl Train* 36(3):263–273.
102. Randolph, C., McCrea, M., Barr, W.B. 2005. Is neuropsychological testing useful in the management of sport-related concussion? *J Athl Train* 40(3):139–152.
103. Gualtieri, C.T., Johnson, L.G. 2006. Reliability and validity of a computerized neurocognitive test battery, CNS Vital Signs. *Arch Clin Neuropsychol* 21(7):623–643.
104. Resch, J. et al. 2013. ImPact test-retest reliability: Reliably unreliable? *J Athl Train* 48(4):506–511.

105. Capo-Aponte, J.E. et al. 2012. Visual dysfunctions and symptoms during the subacute stage of blast-induced mild traumatic brain injury. *Mil Med* 177(7):804–813.
106. Galetta, K.M. et al. 2011. The King–Devick test as a determinant of head trauma and concussion in boxers and MMA fighters. *Neurology* 76(17):1456–1462.
107. Galetta, K.M. et al. 2011. The King–Devick test and sports-related concussion: Study of a rapid visual screening tool in a collegiate cohort. *J Neurol Sci* 309(1–2):34–39.
108. Silverberg, N.D. et al. 2014. Assessment of mild traumatic brain injury with the King–Devick Test(R) in an emergency department sample. *Brain Inj* 5:1–4.
109. Catena, R.D., van Donkelaar, P., Chou, L.S. 2009. Different gait tasks distinguish immediate vs. long-term effects of concussion on balance control. *J Neuroeng Rehabil* 6:25.
110. Martini, D.N. et al. 2011. The chronic effects of concussion on gait. *Arch Phys Med Rehabil* 92(4):585–589.
111. Buckley, T.A. et al. 2013. Altered gait termination strategies following a concussion. *Gait Posture* 38(3):549–551.
112. Fait, P. et al. 2009. Alterations to locomotor navigation in a complex environment at 7 and 30 days following a concussion in an elite athlete. *Brain Inj* 23(4):362–369.
113. Catena, R.D., van Donkelaar, P., Chou, L.S. 2007. Cognitive task effects on gait stability following concussion. *Exp Brain Res* 176(1):23–31.
114. Catena, R.D., van Donkelaar, P., Chou, L.S. 2011. The effects of attention capacity on dynamic balance control following concussion. *J Neuroeng Rehabil* 8:8.
115. Catena, R.D., van Donkelaar, P., Chou, L.S. 2007. Altered balance control following concussion is better detected with an attention test during gait. *Gait Posture* 25(3):406–411.
116. Ingriselli, J.M. et al. 2014. Outcomes, utility, and feasibility of single task and dual task intervention programs: Preliminary implications for post-concussion rehabilitation. *J Sci Med Sport* 17(6):580–585.
117. Dorman, J.C. et al. 2015. Tracking postural stability of young concussion patients using dual-task interference. *J Sci Med Sport* 18(1):2–7.
118. Cossette, I., Ouellet, M.C., McFadyen, B.J. 2014. A preliminary study to identify locomotor-cognitive dual tasks that reveal persistent executive dysfunction after mild traumatic brain injury. *Arch Phys Med Rehabil* 95(8):1594–1597.
119. Howell, D.R., Osternig, L.R., Chou, L.S. 2014. Return to activity after concussion affects dual-task gait balance control recovery. *Med Sci Sports Exerc* 5:5.
120. Majerske, C.W. et al. 2008. Concussion in sports: Postconcussive activity levels, symptoms, and neurocognitive performance. *J Athl Train* 43(3):265–274.
121. Brown, N.J. et al. 2014. Effect of cognitive activity level on duration of post-concussion symptoms. *Pediatrics* 133(2):e299–e304.
122. Moser, R.S., Glatts, C., Schatz, P. 2012. Efficacy of immediate and delayed cognitive and physical rest for treatment of sports-related concussion. *J Pediatr* 161(5):922–926.
123. Silverberg, N.D., Iverson, G.L. 2013. Is rest after concussion "the best medicine?": Recommendations for activity resumption following concussion in athletes, civilians, and military service members. *J Head Trauma Rehabil* 28(4):250–259.
124. National Conference of State Legislatures. 2013. Traumatic Brain Injury Legislation, July. Cited August 2014. Available at http://www.ncsl.org/research/military-and-veterans-affairs/traumatic-brain-injury-legislation.aspx.
125. Wetjen, N.M., Pichelmann, M.A., Atkinson, J.L. 2010. Second impact syndrome: Concussion and second injury brain complications. *J Am Coll Surg* 211(4):553–557.
126. Cantu, R.C., Gean, A.D. 2010. Second-impact syndrome and a small subdural hematoma: An uncommon catastrophic result of repetitive head injury with a characteristic imaging appearance. *J Neurotrauma* 27(9):1557–1564.
127. Bey, T., Ostick, B. 2009. Second impact syndrome. *West J Emerg Med* 10(1):6–10.
128. Cantu, R.C. 1998. Second-impact syndrome. *Clin Sports Med* 17(1):37–44.

129. Field, M. et al. 2003. Does age play a role in recovery from sports-related concussion? A comparison of high school and collegiate athletes. *J Pediatr* 142(5):546–553.
130. Lovell, M.R. et al. 2003. Recovery from mild concussion in high school athletes. *J Neurosurg* 98(2):296–301.
131. Guskiewicz, K.M. et al. 2005. Association between recurrent concussion and late-lifecognitive impairment in retired professional football players. *Neurosurgery* 57(4): 719–726.
132. Randolph, C. 2014. Is chronic traumatic encephalopathy a real disease? *Curr Sports Med Rep* 13(1):33–37.
133. Weir, D.R., Jackson, J.S., Sonnega, A. 2009. *National Football League Player Care Foundation Study of Retired Players*. Ann Arbor, MI: University of Michigan.
134. Lehman, E.J. et al. 2012. Neurodegenerative causes of death among retired National Football League players. *Neurology* 79(19):1970–1974.
135. McKee, A.C. et al. 2009. Chronic traumatic encephalopathy in athletes: Progressive tauopathy after repetitive head injury. *J Neuropathol Exp Neurol* 68(7):709–735.
136. Stern, R.A. et al. 2011. Long-term consequences of repetitive brain trauma: Chronic traumatic encephalopathy. *PMR* 3(10 Suppl 2):S460–S467.
137. Stern, R.A. et al. 2013. Clinical presentation of chronic traumatic encephalopathy. *Neurology* 81(13):1122–1129.
138. McKee, A.C. et al. 2012. The spectrum of disease in chronic traumatic encephalopathy. *Brain* 136(Pt 1):43–64.
139. Small, G.W. et al. 2013. PET scanning of brain tau in retired National Football League players: Preliminary findings. *Am J Geriatr Psychiatry* 21(2):138–144.
140. Ojo, J.O. et al. 2013. Repetitive mild traumatic brain injury augments tau pathology and glial activation in aged hTau mice. *J Neuropathol Exp Neurol* 72(2):137–151.
141. Bazarian, J.J. et al. 2014. Persistent, long-term cerebral white matter changes after sports-related repetitive head impacts. *PLoS One* 9(4):e94734.
142. Bazarian, J.J. et al. 2012. Subject-specific changes in brain white matter on diffusion tensor imaging after sports-related concussion. *Magn Reson Imaging* 30(2):171–180.
143. Koerte, I.K. et al. 2012. White matter integrity in the brains of professional soccer players without a symptomatic concussion. *JAMA* 308(18):1859–1861.
144. McAllister, T.W. et al. 2014. Effect of head impacts on diffusivity measures in a cohort of collegiate contact sport athletes. *Neurology* 82(1):63–69.
145. Mihalik, J. et al. 2007. Measurement of head impacts in collegiate football players: An investigation of positional and event-type differences. *Neurosurgery* 61(6):1229–1235.
146. Reed, N. et al. 2010. Measurement of head impacts in youth ice hockey players. *Int J Sports Med* 31(11):826–833.
147. Guskiewicz, K.M., Mihalik, J. 2011. Biomechanics of sport concussion: Quest for the elusive injury threshold. *Exerc Sport Sci Rev* 39(1):4–11.
148. Cobb, B.R. et al. 2013. Head impact exposure in youth football: Elementary school ages 9–12 years and the effect of practice structure. *Ann Biomed Eng* 41(12):2463–2473.
149. Daniel, R.W., Rowson, S., Duma, S.M. 2012. Head impact exposure in youth football. *Ann Biomed Eng* 40(4):976–981.
150. Kashluba, S., Casey, J.E., Paniak, C. 2006. Evaluating the utility of ICD-10 diagnostic criteria for postconcussion syndrome following mild traumatic brain injury. *J Int Neuropsychol Soc* 12(1):111–118.
151. Grubenhoff, J.A. et al. 2014. Acute concussion symptom severity and delayed symptom resolution. *Pediatrics* 134(1):54–62.
152. Slobounov, S. et al. 2007. Differential rate of recovery in athletes after first and second concussion episodes. *Neurosurgery* 61(2):338–344.
153. Lau, B. et al. 2009. Neurocognitive and symptom predictors of recovery in high school athletes. *Clin J Sport Med* 19(3):216–221.

154. McKinlay, W.W., Brooks, D.N., Bond, M.R. 1983. Post-concussional symptoms, financial compensation and outcome of severe blunt head injury. *J Neurol Neurosurg Psychiatry* 46(12):1084–1091.
155. Leddy, J.J. et al. 2010. A preliminary study of subsymptom threshold exercise training for refractory post-concussion syndrome. *Clin J Sport Med* 20(1):21–27.
156. Leddy, J.J., Willer, B. 2013. Use of graded exercise testing in concussion and return-to-activity management. *Curr Sports Med Rep* 12(6):370–376.
157. Baker, J.G. et al. 2012. Return to full functioning after graded exercise assessment and progressive exercise treatment of postconcussion syndrome. *Rehabil Res Pract* 2012:705309.
158. Gagnon, I. et al. 2009. Active rehabilitation for children who are slow to recover following sport-related concussion. *Brain Inj* 23(12):956–964.
159. Schneider, K.J. et al. 2014. Cervicovestibular rehabilitation in sport-related concussion: A randomised controlled trial. *Br J Sports Med* 48(17):1294–1298.

3 Diffuse Axonal Injury

Erik Su and Michael Bell

CONTENTS

Traumatic brain injury (TBI) is the leading cause of death in the United States in people between the ages of 1 and 44 years and occurs in hundreds of thousands of subjects yearly. Recently, the importance of apparently mild injuries has been recognized as a public health crisis for soldiers in the combat theater, children and young adults in sport activities, and others throughout their normal life. Our understanding of the pathology of TBI is rudimentary despite years of study. This chapter will summarize an important aspect of TBI pathology—diffuse axonal injury (DAI)—that is increasingly recognized as an important cause of long-term disability and mortality.

DAI describes a process of widespread axonal damage in the aftermath of acute or repetitive TBI, leading to deficits in cerebral connectivity that may or may not recover over time. It is a component of injury in 40% to 50% of hospital admissions for traumatic brain injury (TBI)[1] and one of the most common pathologies in all closed-head trauma.[2–5] DAI is typically characterized by coma without focal lesion[6,7] on presentation and pathologically defined by axonal damage in multiple regions of the brain parenchyma, often causing impairments in cognitive, autonomic motor, and sensory function by virtue of disrupted neuronal connectivity. Areas commonly affected include axons in the brainstem, parasagittal white matter near the cerebral cortex, and corpus callosum.[1] Though DAI is often characterized as a structural disease, where the fundamental brain architecture is irreparably disrupted on a microscopic scale, many aspects of DAI are influenced not only by physical factors but by perturbations in any number of pathways including metabolism, electrochemistry, and inflammation among others. DAI preceded by a period of relative lucidity diagnosed with imaging findings has also been described.[8,9] This may correlate with animal models of DAI demonstrating direction dependence of trauma influencing coma, however, lower extremity long bone orthopedic intervention factored heavily in these clinical cases, and imaging findings in these patients' brains may have been confounded by emboli mimicking DAI findings.

DAI was first described in comatose trauma subjects who demonstrated scattered axonal injury in the cerebrum, cerebellum, and brainstem on postmortem examination.[10] Advances in histopathology during the 1950s further revealed the extent of axonal injury associated with DAI at the cellular level. Through the 1960s additional studies would describe an early histopathological pattern of scattered axonal injury characterized by worsening swelling and distortion of normal architecture into retraction bulbs and helical structures. Extensive characterization of these histologic findings in trauma victims subject to large acceleration/deceleration forces led to the initial definitions of "diffuse axonal injury" described in 1982 by Adams and colleagues.[2] The Adams classification is used to grade DAI from a pathology standpoint and is described (Table 3.1).[11]

TABLE 3.1
Adams Classification of DAI

	Pathological Findings
Grade I (Mild DAI)	Microscopic changes in white matter of cerebral cortex, corpus callosum, brainstem, and occasionally the cerebellum
Grade II (Moderate DAI)	Grossly evident focal lesions in the corpus callosum
Grade III (Severe DAI)	Additional focal lesions in the dorsolateral quadrants of the rostral brainstem (involving superior cerebellar peduncle)

With increasingly sophisticated methods for neurological evaluation, the understanding of DAI as a diffuse process has evolved. It is now instead considered a regional process influencing multiple disparate brain regions, initiated by a global and significant high-energy traumatic insult.[1]

CLINICAL FEATURES

Some degree of DAI is likely in patients suffering moderate to severe TBI with loss of consciousness,[1] with initial Glasgow Coma Scale (GCS) assessment perhaps reflecting functional impairment of the brainstem and the reticular activating system within the midbrain. Seminal work by Genarelli et al. indicates that DAI can be the sole factor in causing coma after TBI.[12] In such circumstances, improvements in responsiveness and alertness may develop slowly a protracted course over weeks to months of intensive rehabilitation. DAI has been historically defined using histopathological methods,[1] though recent advances in neuroimaging have considerably improved in vivo diagnosis. In terms of functional outcome DAI is likely the most common cause of severe impairment after TBI. Disruptions in consciousness were initially attributed specifically to brainstem injury, however, coma after DAI is also frequently associated with axonal damage in cerebral white matter as well.[13] Persistent cognitive and memory deficits, seen in TBI in general, are prominent in these patients with deficits in information processing.[14,15] Magnetic resonance imaging (MRI) data concurrently demonstrate a dose-dependent effect of DAI lesions in the brain on cognitive impairment.[16–19]

Hypothalamic injury and panhypopituitarism have been associated with DAI, possibly due to shear injury across the pituitary stalk from the same high kinetic energy forces that cause DAI.[20–23] Additionally dopaminergic pathways in the anteroventral third ventricular region mediate arginine vasopressin release and may be disrupted in DAI. This may contribute to sodium and free water derangements seen complicating post-TBI management.[23]

INITIAL INJURY AND PRIMARY AXOTOMY

The initial traumatic insult causes dynamic deformation of the brain parenchyma, thereby putting long-tract structures such as axons and blood vessels at risk for stretch and shear injuries. Distortion of the axonal cytoskeleton subsequently disrupts normal axonal transport mechanisms, leading to accumulation of transport products in injured regions[24] and alterations in neuronal homeostasis. Until recently, the axonopathy that occurs subsequent to trauma has been characterized as the progressive formation of axonal varicosities within 2 to 3 hours after injury with disconnection by 6 to 12 hours.[25] More recently, another form of axonal injury has been described characterized by increased axolemmal permeability, mitochondrial swelling, and cytoskeletal compaction damaging microtubule and neurofilament structures.[26,27] Notably this form of injury is not associated with the characteristic axonal swelling associated histologically with injury, and reversed axonal transport has been considered as a possible mechanism for the observed disturbances in axon function without swelling.[28] Shearing of axonal fibers leading to complete disconnection after trauma, or primary axotomy, can cause a more pronounced accumulation of transport products in the injury referred to as an "axonal bulb" (also referred to as a "retraction ball") (Figure 3.1a). The appearance of these bulbs can appear in

(a)

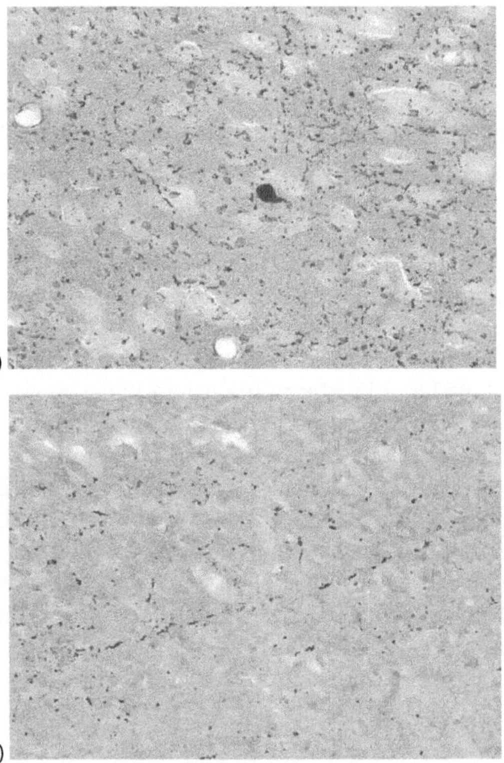

(b)

FIGURE 3.1 (See color insert.) (a) Silver-stained tissue of juvenile rat brain tissue at 40× following impact-acceleration TBI revealing axonal retraction bulb at center. (b) Silver-stained tissue of juvenile rat brain tissue at 40× following impact-acceleration TBI revealing linear degenerating axonal tracts. (Image courtesy of Vassilis Koliatsos, MD/Leyan Xu, MD.)

transected tissue for years following injury reflecting complete and persistent axonal separation,[29] even when varicosities of presumably intact axons resolve, though a full understanding of the temporal course of this process remains to be done. Despite the apparent correlation of traumatic force magnitude with degree of axonal damage, primary axotomy is considered a minor contributor compared to secondary mechanisms in the overall axonopathy seen after TBI.[6,30]

Stretch injury without complete axotomy is considered a greater contributor to the pathology seen in DAI than injury with axotomy. In vitro studies of neuronal preparations reveal that immediately subsequent to stretch injury axonal arrangement is distorted and some axons become undulated and twisted due to cytoskeletal damage (Figure 3.1b). Internodal regions of the axon appear particularly vulnerable, whether due to specific mechanical features of this region, a lack of association with supportive oligodendrocytes,[30] or perhaps the density of transmembrane ion channels in this area.[31] Changes to the cytoskeleton are prominent in the wake of trauma, as neurofilament morphology in some structures changes within 15 minutes after injury[25] with subsequent disruption of axonal transport and accumulation of transport products. Though axonal transport is commonly mediated by the microtubule cytoskeleton as opposed to neurofilaments, neurofilament morphology is considered a mirror of concurrent microtubular integrity.[25] Supporting this, a loss of microtubules following trauma has also been noted in models of TBI and axonal injury.[32] Ultimately, the use of the microtubule stabilizing medication taxol ameliorates microtubule damage after experimental TBI,[33] but cytotoxic effects of the drug preclude its use as a therapy for DAI at this time.[34] TBI damage decreases axonal elasticity and prevents recovery of the axon to its original structural conformation.[33] The individual axon response to injury appears idiosyncratic to some degree and governed by multiple factors. Though large-scale traumatic forces would likely render roughly uniform mechanical stresses across microscopic axon bundles, initial histological examination reveals scattered axonopathy in affected tissue. One possible explanation is that myelination varies between neighboring axons, and finely myelinated small axons appear more vulnerable to injury.[25] Histologically visible findings likely underestimate injury in neighboring fibers with varying degrees of injury, as the proportion of histologically abnormal axons seen in trauma models and patients is incongruent with the magnitude of DAI symptoms seen after TBI. Also suggests some axons may appear morphologically normal but are functionally incapacitated after trauma.[35] Due to the progressive nature of DAI early after trauma, imaging and histological findings are also significantly affected by time of assessment. Staining techniques such as hematoxylin and eosin and silver staining delineating structural features are of limited benefit as they may underestimate the burden of gross axonal pathology.[10] Rather, immunohistochemical stains identifying axonal transport products such as amyloid precursor protein (APP), which can be identified in damaged axons within 2 hours after injury, may be more useful. Commonly seen in neurodegenerative diseases in older patients,[36] APP positive damaged axons appear in high frequency in TBI patients who die soon after injury, even in children.[37,38] Therefore importantly, other clinical pathologies affecting axonal transport should be considered in addition to injury mechanisms in the histological evaluation of DAI.[39]

SECONDARY AXOTOMY AND DISRUPTED NEURONAL HOMEOSTASIS

Traumatic brain injury is intrinsically heterogeneous and therefore axonal injury does not occur in a vacuum, rather multiple injury cascades from oxidative, excitotoxic, and inflammatory pathways occur as well and affect the evolution of axonal pathology. Following trauma, axons that have not been ruptured by primary axotomy may have been injured from physical stretch due to deformation of the brain. Axonal stretch injures cells through several mechanisms. In addition to direct cytoskeletal damage, stretch disrupts membrane permeability and precipitates depolarization.[40] This in turn alters the electrochemistry of the damaged axon and triggers the release of excitatory neurotransmitters such as glutamate to concentrations as much as 50 times normal in TBI.[41,42] This surge is worsened by dysfunction of normal glutamate reuptake processes by neighboring astrocytes in injury. Glutamate acting on NMDA and AMPA receptors increases cytosolic calcium influx and concentration. This in turn engages secondary messenger systems involving calcium, alters transmembrane ion gradients, and the osmotic load of the neuronal cytosol among other mechanisms that precipitate additional neuronal injury. Calcium ion increases seem dependent on two sources: (1) entry of sodium secondary to depolarization and reversal of the Na+/Ca2+ membrane antiporter[31,43,44] and (2) release of calcium from the endoplasmic reticulum.[45] Other possible sources of altered axonal electrophysiology include increased sodium ion channel permeability,[31,44,46] decreased Na-K ATPase activity resulting in ATP starvation of the neuron,[47–49] and ectopic distribution of channels.[50] Calcium induces changes in mitochondria leading to the opening of the mitochondrial permeability transition pore. This ultimately leads to neutralization of the electrochemical gradient essential for mitochondrial function, permits water influx, and instigates mitochondrial demise.[51] Calcium-mediated proteolysis by calpain and caspases in turn leads to further damage to the axonal cytoskeletal as well as ion channel structures.[43,51–53] The effects of these calcium-dependent enzymes are demonstrated in the lysis of the cytoskeletal protein spectrin in disparate products of proteolysis characteristic for each enzyme. Calcineurin inhibition with cyclosporine A and tacrolimus and calpain inactivation has been demonstrated to mitigate axonal injury in vivo[53–58] and may offer potential therapies in the future. Oxidative stress and disrupted neuronal energy metabolism secondary to mitochondrial injury and eventual destruction also plays an important role in neuronal cell death. This is due to cytoskeletal injury, protein alteration, and metabolic failure, a process partially ameliorated by free-radical scavengers.[59] Neuroinflammatory processes and microglial activation also contribute to local injury processes and endure long after the initial insult.[29,60,61]

Axons not severed in primary axotomy may proceed to breakage via secondary axotomy. Secondary axotomy is a process of rapidly progressive axonal deterioration, breakage, and retraction occurring following but not at the time of injury. Axonal bulbs form and the axon assumes a helical appearance as the proximal damaged axon retracts. Concurrently the severed distal axonal fragment begins Wallerian degeneration by 24 hours, a process of progressive lysis and disintegration.[25] By 72 hours the damaged, separated segment is consumed by activated glia. Recent evidence suggests that unmyelinated axons are also at higher risk for secondary disconnection compared to myelinated axons as demonstrated in in vitro models where these smaller fibers are more prone to secondary disconnection than larger myelinated axons.[62,63] Myelin sheath integrity changes during the evolution of DAI after injury, with progressive demyelination occurring in affected areas and white matter atrophy up to 1 year after injury.[64] Of note, oligodendrocytes in apoptosis have been observed in affected tissue following TBI and may reflect myelin degeneration in progress. Appearance of persistent myelin basic protein (MBP) in the cerebrospinal fluid (CSF) of patients following trauma[65] suggests an ongoing sheath degeneration consistent with the appearance of sheath globoids seen in conjunction with axonal bulbs after DAI.[66]

NEUROINFLAMMATION

The primary inflammatory response seen in the brain after DAI is mediated by microglia, as it is in other TBI. In a study by Oehmichen et al.[67] immunohistochemical labeling with β-APP for axonal damage and CD68 for microglia identified colocalization of the labels in half of the patients who survived 5 to 15 days in areas of the brain—thereby demonstrating moderate microglial infiltration in areas of axonal injury.[68] Microglia migrate rapidly into injured areas, and activated microglia extend cytoplasmic processes toward injured axons so as to isolate damaged structures. Thalamic infiltration with astrocytes has been noted at 4 to 8 hours after injury in animal DAI models, with maximal injury markers evident at 48 hours to 2 weeks after injury. Concomitant microglia activation has also been observed in the cortex and hippocampus at 4 hours, with upregulation of MHC Class II epitopes in white matter 24 hours following injury. Macrophages localize to the meninges and perivascular spaces within 24 to 48 hours and persist for up to 2 weeks, however, infiltration into the parenchyma is more limited.[69]

Cytokines involved in neuroinflammation include mediators commonly seen in TBI such as the IL-1 family, IL-6, IL-10, and TNF-α. The IL-1 family includes IL-1α, IL-1β, and IL-18, and is a well-studied constellation of cytokines that stimulate lymphocytes and macrophages, as well as trigger further inflammatory mediators. A study by Lu et al.[70] demonstrates an increase in cortical IL-1α and β following an impact-acceleration TBI in rats. Hans et al.[71] also using an impact acceleration model noted IL-6 activity increases 1 hour following histologically confirmed DAI peaking at 2 to 4 hours then receding to normal in 24 hours. It was noted that IL-6 mRNA and expression was highest in regions of axonal damage. IL-6 in particular is important in regulation of inflammation and activity of granulocytes, lymphocytes, and NK cells as well as inducing release of soluble TNFR and IL-1 receptor antagonists. In a fluid-percussion model of DAI, Kita and colleagues[72] identified increasing TNF-α concentrations in the brainstem and corpus callosum over the first 3 hours after trauma. TNF-α is well known as a pro-inflammatory, proapoptotic cytokine with effects on macrophage/monocyte/NK cell stimulation, as well as secretion of platelet activation factor, ICAM, thromboxane A2, prostaglandin E2 as well as endogenous nitric oxide.[69] TNF-α has been detected in lysosomes of microglia, astrocytes, and oligodendroglia in other models of DAI as well.[73,74] A factor that has been demonstrated to exacerbate the inflammatory response is hypoxia.[74] Given that respiratory compromise can occur through impact apnea, mechanical asphyxiation, inhalation injury or other mechanisms associated with TBI, hypoxia compounding DAI likely influences injury evolution following moderate-severe TBI.

The leukocyte adhesion molecule ICAM-1 is another potential mediator of post-DAI secondary injury in that it is upregulated following DAI[75,76] 16 hours after injury, peaking 4 days after injury. Lymphocyte adhesion associated chemokines MIP-2 and MCP-1 are also elevated after focal TBI, however, only MCP-1 was increased after DAI in this experimental DAI model. MCP-1 is associated with monocyte recruitment from the bloodstream and interestingly MIP-2 is also a neutrophil attractant.

EVOLUTION OF DAI AND PLASTICITY

Historically, DAI has been described as a process that evolved in days to weeks following trauma, with peak neuropathology occurring at 1 to 2 days after injury.[25] Histopathological and imaging evidence,[77] however, suggests that trauma incites a protracted process of axonal degeneration and impaired transport over months in a progressive, Alzheimer's-like neurodegenerative process.[29,78] Likewise, the processes of axonal recovery, regrowth, and neuronal plasticity are poorly understood in the healing patient. After experimental DAI in mice NG2 positive oligodendrocyte precursors initially decrease, but then proliferate by three days after injury in the damaged corpus callosum and subventricular zone.[79] Mature oligodendrocytes demonstrate a similar tendency in total numbers, which decrease during the first 3 days following injury but appear higher at 7 days after injury. Despite diffuse axonal damage, frank demyelination is not seen in the first week after injury, and redundant or disordered myelin may persist.[80] It can be difficult to assess myelin-sheath-associated proteins in the tissue as a marker of injury due to retention and ongoing breakdown of damaged sheath.[80] Regrowth of myelin with oligodendrocyte recovery likely influences other recovery and plasticity processes, however, the extent of this interaction is not well characterized.[79] Basic recovery processes such as restoration of cytoskeletal structures with microtubule and neurofilament turnover and recovery of normal ionic homeostasis in the injured axon have been postulated,[10] and it is possible that axonal varicosities seen with disrupted axonal transport following TBI can resolve with restoration of normal axonal function. The utility of biomarkers to predict recovery is unclear, as the thresholds for marker detection may involve injury beyond the possibility for axon recovery.[40] Axonal regrowth is normally inhibited by resident myelin sheath proteins such as Nogo-A, oligodendrocyte-myelin glycoprotein (OMgp), and myelin-associated protein (MAP).[68,81,82] Binding of these mediators to Nogo receptors (NgR) results in Rho A mediated inhibition of axonal growth and prevents axonal elongation. Though this is likely protective in normal neuronal architecture, it has been postulated that this process impedes axonal regeneration after trauma. Interference with the Nogo signaling pathway using a NgR inhibitor and other medications in in vitro and in vivo models releases axonal growth inhibition and has been shown to restore some recovery of spinal cord function after hemisection in a rat model.[68] Further exploration in cerebral axonal disconnection is necessary to determine the manipulability of this pathway for treating DAI.[77] Altogether parameters governing the extent and degree of these recovery processes, determinants of injured axonal survival or death, as well as their direct effect on outcome are important subjects for further investigation.[83]

SPECIAL CONSIDERATIONS: BLAST INJURY, CHRONIC TRAUMATIC ENCEPHALOPATHY, AND ABUSIVE HEAD TRAUMA

Interest in blast injury has increased markedly in recent times due to use of explosives as weapons and for breaching structures in theater. Blast injury causes a unique constellation of symptoms, including DAI, due to the high amount of kinetic energy transferred to patients after injury. Multisystem injury can be expected in addition to direct blast injury, including penetrating injury due to shrapnel, and respiratory injury from thermal and toxic exposure. Explosive blasts release energy in the form of acoustic, light, thermal, and electromagnetic energy, which can all potentially interact with a patient.[84] A mathematically idealized blast wave initially develops with a near instantaneous increase in local air pressure to peak levels, followed by an exponential decrease in pressure that approaches nadir below baseline atmospheric pressure, and then recovers to baseline over roughly twice the time required to reach peak pressure.[84] Blast forces can reflect off surfaces and redirect as compound waves toward the patient in a closed space, and lead to successive buffeting of the patient from multiple directions.[84] The primary blast passes on the order of milliseconds through tissue and loses less energy perpetuating through noncompressible fluids and body tissue in comparison to air, which dampens blast forces. Additional injury results from physical displacement of the victim and subsequent contact with other structures as the force of the explosive accelerates the body. This type of additional trauma has been described as a "blast plus" component of injury, possibly incurring additional acceleration-deceleration injuries to the brain or other organs. Though helmets reduce direct deceleration injury in military and civilian trauma, they do not usually stop rotational acceleration/deceleration that is particularly injurious to axons.[85,86] Chronic low level repetitive blast events may also cause some level of cognitive disturbance, as has been demonstrated in military door breachers.[87] Progressive advances in military blast injury management highlight inefficiencies in civilian blast TBI care[88] and have contributed to improvements in postblast DAI management.[89,90]

Chronic traumatic encephalopathy (CTE) was first described in boxers in 1928[91] and has received increasing attention due to increased recognition and association with contact sports prone to repeated head injury. Though predominantly described in American football and boxing, it has also been described in soccer, hockey,[92,93] and mixed martial arts[94] though the true burden of CTE is unknown in multiple sports. Because its pathophysiology is a long and usually slow progressive course evolving from frequent repetitive head injuries, it is frequently diagnosed well into its chronic phase when persistent long-term neurological and behavioral effects predominate. Brain histopathology performed on CTE-symptomatic ex-boxers postmortem reveals scattered intracellular accumulation of microtubule associated tau protein and neuropil threads[95–97] and similar findings have been found in postmortem examinations of professional American football players.[98,99] Clinical findings include expected cognitive and neuropsychiatric disturbances with disinhibition,[100] and focal neurological signs may include speech problems, ataxia, spasticity, and extrapyramidal symptoms.[98] Recent advancements in imaging present an ominous picture of boxing, with 76% of boxers having imaging abnormalities consistent with DAI occurring in a dose-dependent fashion that correlates with career length and number of bouts.[101] CTE is also unique as a TBI syndrome in that it rarely involves penetrating injury or skull deformation, and the pathology arises primarily from global acceleration-deceleration forces causing neuropathology primarily through DAI. Consequently, CTE is not well modeled by most current preclinical TBI models. CTE and Alzheimer's disease (AD) bear similarities histopathologically in terms of progressive axonal degeneration. A number of studies appear to link the history of a single reported TBI with eventual AD, as well as acceleration of dementia onset.[102–112] Given that the apolipoprotein E4 (ApoE4) phenotype is associated with increased severity of AD pathology, TBI patients manifesting ApoE4 are also predisposed to future axonopathic neurodegeneration demonstrated in boxers with severe CTE.[113]

Abusive head trauma (AHT) in infants and young children has also been a topic of increasing interest due to the injury pattern and advances in fathoming its unique characteristics. AHT victims frequently suffer significant repetitive injury potentially over time, and a multitude of injury mechanisms in addition to axonal shear and focal lesions. Over 90% of patients who die of AHT also develop subdural hematoma.[114] Mechanistically effects of a proportionally larger head with less neck control than an adult would seem to predispose children to greater acceleration-deceleration forces[115] reinforced by adult data that brain injury is worse in adult patients when the neck is limp.[116] This implicates axonal injury as an important component in the pathology observed. Stresses on the axon likely entail direct shear as well as internal stresses from centripetally displaced axonal cytoplasm within the axon course. This differs from direct contact injury such as that seen in contusion.[1] Frequently a period of medical neglect follows injury, during which the AHT victim may become hypopneic from multiple mechanisms including depressed mental status and direct injury to medullary respiratory centers. Moreover, the physical exam may be complicated by a varied, underreported history of partially healed repeated injuries of the brain and other organs. A series by Geddes et al.[96] examined 53 cases of inflicted head injury and found that whereas children over a year of age demonstrated DAI patterns similar to adults, infants under the age of 1 year manifested infrequent axonal pathology, however, vascular injury and cerebral edema was more prominent concomitant with anoxic injury. Thirty-one percent of patients do demonstrate axonal injury in the cervical cord or the craniocervical junction consistent with shaking, however, providing a causal relationship for medullary damage in AHT and associated respiratory failure.[40] The overall low amount of cerebral axonal damage is puzzling given the magnitude of functional deficits seen, and some authors[115] propose that myelination protects the mature axon from injury since most axonal damage when seen is in intermodal areas. Several have also postulated that the immature axon may be less susceptible to axonal damage or that shaking injuries mechanistically were insufficient to cause axonal injury.[117,118]

MECHANISTICS

Animal, fabricated plastic, and *in silico* models have all been instrumental in characterizing the mechanical forces that cause DAI pathology. As can be expected there are a multitude of external forces imposed on the brain in TBI leading to rotational, tensile, and compressive strains on the tissues. Moreover the size and complexity of the human brain are important factors as well, due to its inertia, relative freedom of movement on the cervical spine, and variable density of the grey and white matter resulting in separation at significant shear stress. Consequently, only an exact model of the human brain's unique geometry and environment adequately models shear forces important for human TBI. The falx separating the cerebral hemisphere and tentorium in particular alters shear wave propagation in the brain and can amplify local forces on the brain near the falx's insertion points.[119–121] Even at the cellular level cytoplasm is displaced by traumatic forces within the axon and can directly damage the cytoskeleton.[25,122] In particular the threadlike structure of axons and small blood vessels in the brain expectedly places them at exquisite risk for shear injury in the brain under deformational stress. Though DAI is attributed to white matter injury, myelination increases the dynamic modulus of white matter axons and makes them stiffer. Per unit length, they are therefore, less susceptible to shear stress than corresponding grey matter structures, a factor balanced by white matter axons being longer and therefore more prone to DAI.[119,123] Evidence suggests normal compliance of the cellular membrane and cytoskeleton permits some pliability in low velocity brain deformation. Under higher, rapidly applied stresses, however, axons become stiffer, brittle, and susceptible to shear.[44,124–127]

The substance of the brain is analogous to a semisolid of heterogeneous densities and composition. Therefore, any model must account for this in simulating the shear forces that characterize DAI.[1] When force is applied to a semisolid it is deformed in a manner consistent with its elasticity, viscosity, and plasticity. The tissue response to deformation subsequently depends on the duration and amplitude of the force application. When this force is distributed through the tissue by cells in contact with each other, shear forces develop within that cause permanent damage when the magnitude of shear exceeds the tissues' elastic threshold. Longer durations of force application increase the amount of energy transferred to the brain and maintain a degree of physical displacement that rents the tissue, therefore causing more damage. Acceleration-deceleration at high speed or in repetitive shaking injuries can result in significant directional forces lasting over 1 second in duration and place the tissue in peril. Moreover high kinetic energy head injury frequently involves multiple directions of acceleration-deceleration that compounds shear forces on the brain. Given that the dynamic modulus, or tolerance of physical stress, of threadlike axons and vascular structures is direction dependent, compounded shear directions in the brain can potentially injure multiple vulnerable long axon pathways. In particular, angular movement causes greater injury than translational movement, and angular head motion in the coronal plane causes greater injury than sagittal movement.[12] In repetitive injury, initial damage caused by DAI may also alter the compliance of affected tissue. In a study of rodent impact-acceleration DAI the modulus of injured axons decreased after initial injury, implying that on successive impacts greater deformation of axons occurs, thereby causing axonopathy.[128]

ANIMAL MODELS

The structural complexity of the human brain contributes to difficulty in modeling of the multitude of dynamic forces acting on it in DAI. Gyrencephalic models best represent the complex dynamics of TBI from an anatomical standpoint; apparatuses that consistently cause reproducible injury with high kinetic stresses are resource intensive and limited by the types of injury caused. Presently, lissencephalic models in smaller animals are more accessible to investigators due to their economy, reproducibility, and capacity for titration of injury magnitude.[129]

GYRENCEPHALIC MODELS

A nonhuman primate model of DAI was first described by Genarelli in 1982 utilizing high velocity acceleration-deceleration kinetics.[12,130] Notably, the investigation also indicated DAI's strong association with post-traumatic coma independent of mass effect from intracranial hemorrhage. In this inertial acceleration study apparatus, the animal body was restrained in a frame, whereas the head was secured to a rotating armature that caused rapid acceleration and sudden deceleration about the cervical axis through a designated plane. Though primates are reasonably analogous to human injury by virtue of similar brain structure and size (approximately 95 g), significantly higher rotational acceleration was required to cause the same expected mechanical shear seen in human TBI[121,131] as the lighter brains of smaller animals cannot be accelerated by this apparatus to velocities necessary to yield axonal injury.[132] Subsequent investigations in this laboratory have been performed on miniature swine, which have similarly sized gyrencephalic brains and are more accessible than primate animals.[129] This model brings the head through a biphasic centroidal rotation of 100 degrees over 20 ms and yields diffuse lesions consistent with DAI. Even with extensive white matter injury in these models, ultimately consciousness was largely dependent on brainstem pathology suggesting location of insult is more important than degree of total injury.[133]

BLAST INJURY

Blast injury research in animal models has advanced considerably in the past decade pursuant to the extensive use of explosive devices in warfare. A number of models have been described, including open air blast where animals are arrayed around a central explosive,[134] and models of enclosed vehicle spaces to simulate explosive attacks.[135] Open field models, especially those that simulate vehicle or confined space situations where the blast environment is simulated, provide realistic situation-specific injury models.[136] The most straightforward and commonly used apparatuses for blast injury, however, particularly for small animals, are blast tube models.[137] Blast tubes produce calibrated blast peak pressure levels of 20 kPa–350 kPa and replicate the kinetics of blast overpressure to the brain.[138] A compressed gas shock tube used to produce this type of injury (Figure 3.2d) consists of a compression chamber separated from an expansion chamber by Mylar sheets, which rupture at a predetermined pressure.[139] Within the expansion apparatus an animal can be positioned

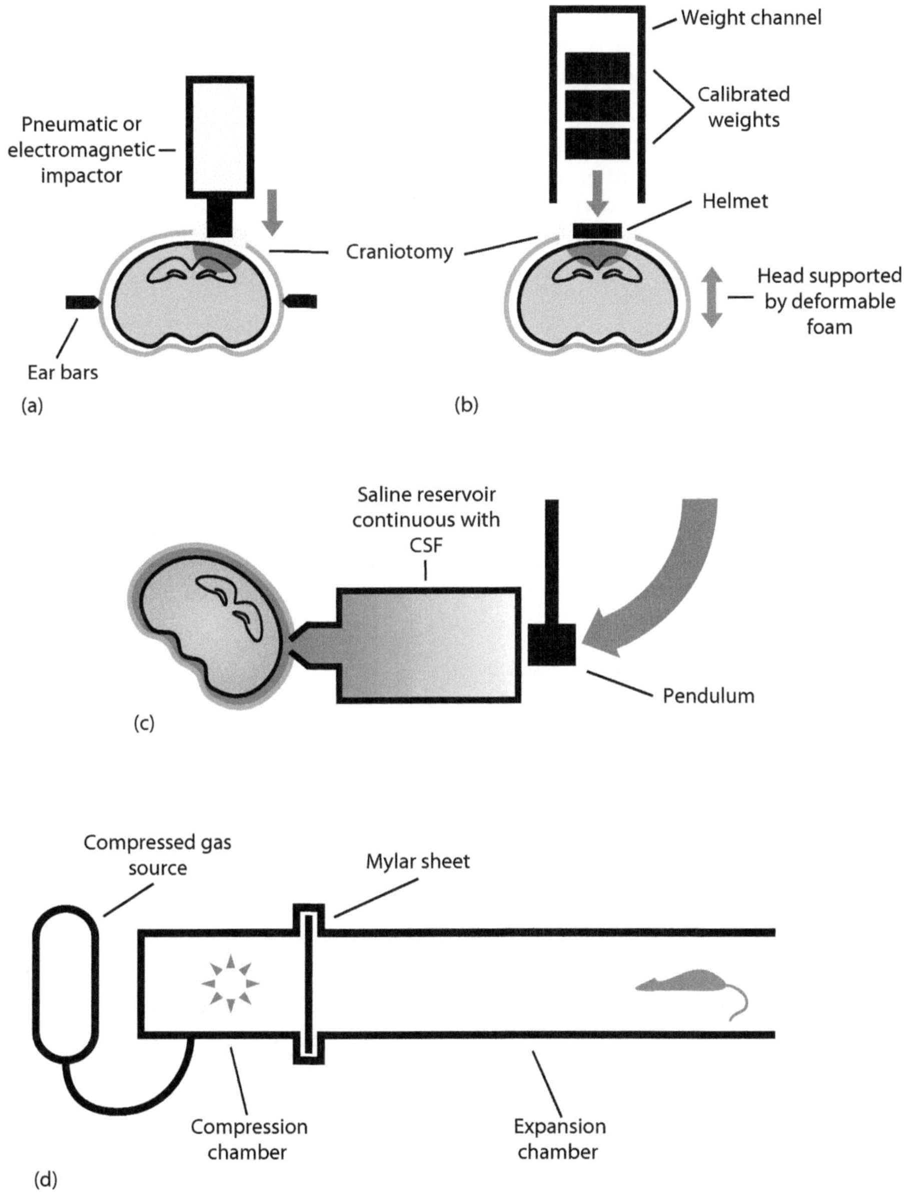

FIGURE 3.2 (See color insert.) Major experimental models of traumatic brain injury used in simulating DAI in lissencephalic animals: (a) controlled cortical impact, (b) impact acceleration, (c) fluid percussion, (d) compressed gas blast tube.

at different angles to the shock wave direction to determine directional effects. Compressed air or helium is pressurized behind the Mylar membrane, and upon rupture the shock wave progresses down the expansion chamber toward the animal. Altogether this model is reproducible and relatively safe to investigators compared to explosive-driven models, though it does not replicate effects of heat and toxic organic compound containing exhaust on the subject. Explosive-driven blast tubes are more commonly used with larger animals models partly due to the amount of energy needed to replicate a larger blast.[140] The explosive charge is placed in the tube prior to the expansion area naturally without a separation membrane.

LISSENCEPHALIC MODELS

Lissencephalic models, primarily rodent models, benefit from simple mechanical designs, economy of scale, relative safety profile for investigators, and the freedom to work with genetically manipulated murine colonies. A disadvantage is that they do not adequately model complex human neuropathology after injury due to the lower cortical complexity seen in rodents.[141] Certainly human craniospinal angle, brain and skull geometry, brain functional topography, gray/white matter mass ratio, and junctional surfaces, as well as cortical complexity are not adequately replicated in these models.

Controlled Cortical Impact Model

The Controlled Cortical Impact (CCI) model (Figure 3.2a) uses a pneumatic or electromagnetically driven impactor to deliver a direct percussive injury traditionally to an animal's exposed dura or brain.[132,142] The animal's skull position in three axes is fixed by mouth and ear bars, and the subject can remain intubated through the injury to ameliorate trauma-induced apnea. However, this means the brain is not subject to global acceleration-deceleration forces seen in DAI permitted by free head movement. Injury magnitude can be titrated with adjustments in impactor velocity, displacement, and duration of deformation. Pitfalls of this injury include that it causes a focal injury primarily, with marked contusion and hemorrhage. It also violates the skull integrity, alters skull compliance, and likely changes intracranial pressure dynamics. Settings needed for large white matter injury also causes significant contusion and tissue destruction in overlying cortex. Modifications of this model to simulate diffuse injury have been employed using a rubber tipped impactor against a unrestrained, intact skull,[143,144] as well as a flat impactor against the bregma of a partially restrained skull with blunt, rubber covered ear bars, though this results in skull fractures.[79]

Impact Acceleration Model

The impact acceleration or weight drop model (Figure 3.2b) involves letting a weight fall freely within a low friction guide tube to strike an animal's head, typically a rodent, either midline between the lambda and bregma or laterally.[141,145,146] In the midline approach there is a risk of skull fracture and concurrent contusion injury, and it is generally recommended that in this approach the skull is exposed and a metal helmet disc be affixed to the skull to distribute the impact. There are a variety of surfaces on which the rodent rests. Classically the device places the animal restrained on a foam surface of standard elastance. This permits free movement of the head with the weight drop in the direction of the weight's fall. The animal should be restrained, as the weight impact can eject the animal from the device. Mass of the weight and distance of weight fall can be adjusted to vary degree of injury. Additionally the surface on which the rodent rests has been changed in some models to tissue paper or aluminum foil so as to alter recoil characteristics after impact to simulate other types of injury.[73] Axonal injury is induced in the brainstem with this injury model using the midline approach, however, in distributions different from human DAI in part due to rodent brain and skull geometry. Pitfalls of this model include the potential for multiple strikes by the weight after initial contact, which can be avoided by pulling a brake line attached to the upper surface of the weight immediately after impact. Additionally, because the animal's head is not secured to the surface there is the potential for inconsistency between trials if the head is not placed in the same location or if the weight trajectory is not consistent either due to poor fit within the guide tube or excessive distance between the guide tube and the skull.

Fluid Percussion Model

In the fluid percussion model (Figure 3.2c) the animal's skull is exposed and trephinated for access for the injury device. Through a surgically inserted saline-filled reservoir continuous with the rodent's CSF space, a percussive shock wave is directed through the closed system that distributes force over the cortical surface.[147] This access point can enter the otherwise closed cranium either at midline or laterally.[148,149] The magnitude of injury is titratable by altering the volume of saline displacement, though causes of error may include air in the fluid system or animal, or compliant structural weaknesses in the fluid system that diffuse impact energy.[150] It is mechanistically more similar to contusion or extraaxial hemorrhage as opposed to DAI, though the model transfers kinetic energy to a broad area of brain and some axonal injury does occur.[150] The fact that generalized acceleration-deceleration of the brain does not occur in this model is a limitation in replicating DAI, and axonal injury in this model occurs in peripheral areas, as opposed to the brainstem and other central areas.[141]

Linear and Angular Acceleration-Deceleration

Li and colleagues have described an adult rat model of combined linear (transverse) and angular acceleration-deceleration closed-head DAI.[151] This novel device is composed of a rotating head helmet affixed to the animal using ear bars on the moving portion of the device. The helmet rotates through 75 degrees coronally as it is pneumatically driven over a geared track over 4.68 ms, also producing 1.57 cm of translational lateral motion. Following injury with this model, rats demonstrated APP positive staining in the corpus callosum, brainstem, subcortical and periventricular white matter, hippocampus, and thalamus, and notable subarachnoid hemorrhages were present.[151] Electron microscopy and diffusion tensor imaging[152,153] have also confirmed changes consistent with DAI in this model. At the time of this publication this device had only been described in use by a single group of investigators.

In Vitro DAI Models

Consistent replication of trauma forces on isolated neurons is also important in DAI research, and the majority of mechanistic findings in axonal injury have been elucidated in isolated cell preparations.[150] In vitro preparations of neurons and associated glia may be prepared on elastic membranes that simulate primary injury via stretching of the elastic in one or more directions. A single axis stretch can be used to injure cell preparations and organotypic tissues such as prepared hippocampal slices[154,155] in a pattern consistent with pure shear. A biaxial stretch pattern is more commonly described with equal stretch in two dimensions without isolating direction of stretch or shear. In these models, the elastic membrane can be placed over a percussion chamber that uses fluid or gas to rapidly displace the membrane in direction perpendicular to the plane of the membrane. A nonequal biaxial stretch model that stretches in one direction and shears in another can also be used.[90] In these models combined cell culture preparations with neurons and glia, as well as organotypic slices, have been used.

Optic Nerve Stretch Model

Basic molecular mechanisms are readily simplified using the optic nerve stretch model, which mechanistically allows reproducible stretch injury in vivo along the major axis of parallel axon bundles in the second cranial nerve as a central nervous system structure. Histological analysis is facilitated by the straightforward structure of the large nerve, and maintenance of the animal postinjury permits long duration studies (as long as 3 months has been described).[156] Much of the current understanding in the basic subcellular mechanisms of traumatic axonal stretch has been elucidated using this model in rodents.[31,125] Pitfalls include anatomical differences between nerve and parenchymal white matter, differences in neighboring cell populations from the parenchyma, and the mechanistically simple single-direction stretch injury that does not incorporate shear across axons or oscillations.

BIOMARKER ASSAYS

The various pathological processes described earlier characterizing DAI evolution also produce an array of potential biomarkers for monitoring, some of which are temporally useful for monitoring specific processes. An advantage of using biological samples for assessment of brain injury is sample portability and typically a low risk to the patient. Characteristics of an optimal biomarker include temporal relevance to injury state, and a short half-life and processing time are important such that current status is adequately reflected. Another benefit of timely injury state reporting by a biomarker is that it can assess therapy effect. Other optimal characteristics importantly include specificity to the injury of interest.

Calcium-Dependent Proteolysis and αII Spectrin Breakdown Products

Calpain and caspase are calcium-dependent enzymes involved in cytoskeletal breakdown following TBI. With significant increases in intracellular calcium immediately following injury, calpain activity rises early after TBI. A second peak occurs with progression of secondary axonal injury processes to necrosis and apoptosis.[157–159] Some evidence suggests necrotic cell death may be related to glutamate exposure as well as environmental calcium concentrations with elevated glutamate and calcium concentrations favoring necrosis.[160–162] Caspases are also active following TBI reflecting apoptotic mechanisms initiated by intrinsic and extrinsic pathway mechanisms, as well as by oxidative stress. Caspase-3 activity rises between 8 and 12 hours after injury peaking at 24 hours before decreasing to lower levels by 3 days.

Axonal damage mediated by calpain and caspases is reflected in patterned breakdown of αII spectrin breakdown characterized by known protease cleavage points, and calpain and caspase activity can be attributed based on predictable lengths of αII spectrin breakdown product (SBDP) produced after cleavage. αII spectrin is a component of the cytoskeleton and connects axolemmal components to the presynaptic terminal, stabilizing the nodal structure of myelinated axons. Calpain lyses the spectrin cytoskeleton into SBDP 145 and SBDP 150, and caspase-3 renders SBDP120. Consequently, the relative contribution of each of these calcium-dependent proteases makes cytoskeletal breakdown potentially distinguishable. SBDP increases in mice as quickly as within 15 minutes of insult[163] and in rats peaks at 24 hours after injury with a progressive decline through 3 days after TBI.[164,165] Following injury most calpain-generated SBDP is found associated with axonal bulbs, and can geographically be found in the neocortex and subcortical white matter at the gray–white matter junction. Concomitant decreases in intact αII spectrin can be seen in the corpus callosum at this time, recovering to normal titers by 7 days after experimental injury.[166] In humans, SBDP is seen to rise within 6 hours postinjury and peak concentrations can be seen by 2 to 3 days after injury.[167] A comparative study of SBDP concentrations after adult TBI in 38 patients demonstrated predominant elevations in SBDP 145 and 150 3 hours after injury, whereas SBDP 120 did not manifest as high an elevation after injury and suggested that calpain-mediated cytoskeletal breakdown predominates over caspase-3 processes after TBI.[168] Further studies assessing other markers of comparative caspase and calpain activation are necessary to determine whether relative SBDP concentrations reveal differences in the proteases' relative activity.

Neurofilament Markers

As neurofilament protein changes feature prominently in histological changes seen after TBI, the protein subunits that comprise these fibers are useful cytoskeletal injury markers. The NF heteropolymer is comprised of light chain (NF-L, 68 kDa), medium chain (NF-M, 160 kDa), and heavy chain (NF-H, 200 kDa) subunits.[163,169,170] The subunits are transported within the axon, and in axonal injury accumulate in areas of disrupted transport.[171–173] Phosphorylation of the subunits influences the speed of transport and controls axonal structure and organization.[174] Following TBI, NF sidearms compact progressively and weaken structurally as they undergo proteolysis by caspase and calpain.[26] NF-L is one of the first subunits to be degraded within a few hours after injury. In pig and rat experimental models of TBI, phosphorylated NF-L decreases between 0.5 and 6 hours, as well as in the corpus callosum of TBI patients.[175] Immunohistochemistry labeling NF-L demonstrates accumulation of NF-L in axonal bulbs and varicosities, becoming progressively more conspicuous between 1 and 3 days postinjury.[176] Phosphorylated NF-H also decreases 1 day after injury and up to 80% in the corpus callosum of TBI patients.[175] NF-H demonstrates a bimodal peak pattern at both 12 and 48 hours, likely reflecting primary and secondary injury effects on the cytoskeleton. Comparatively NF-M has not been demonstrated to be a TBI biomarker, demonstrating modest immunohistochemical detectability at 24 hours after injury, though this increased by 3 days.[177]

Amyloid Precursor Protein, Amyloid β, and Microtubule-Associated Protein Tau

Beta amyloid precursor protein (APP) is a transmembrane glycoprotein synthesized in neurons and plays a role in cell adhesion, growth, and response to injury.[178] It has neurotrophic functions including promotion of axonal sprouting, neurite outgrowth, and synaptogenesis essential for repair after injury.[16] Because it is transported in the axon via fast axonal transport mechanisms, it accumulates in areas of disrupted transport rapidly after injury.[179] In animal models it shows up as early as 15 minutes after the initial insult. Consequently, it has been well described in surgical pathology specimens as a marker of axonal damage in TBI patients.[36] Factors that may affect APP accumulation include rate of APP production[24,180] and whether axonal transport is functional between the injury and the neuronal body such that accumulation is possible. Additionally, axonal transport velocity decreases with age and relative tissue concentrations of previously accumulated APP may vary.[178] Even in trauma, APP only appears in affected tissue 30% to 50% of the time compared to αII spectrin breakdown products.[181] Furthermore, it is somewhat nonspecific for DAI as it appears in patients with diffuse axonopathy from other causes such as metabolic[182] and ischemic causes.[179,183] The burden of APP positive lesions seen in the brain does not necessarily predict outcome as well, as a large prospective series of trauma patients did not find an association between lesions and 6-month outcome after TBI except in children and young adults.[184] Therefore, an adequate clinical history is necessary to interpret APP if it is to be used as a TBI biomarker in infrequently obtained ex vivo tissue.

Amyloid β is a peptide that normally exists extracellularly in monomeric form. Following injury it polymerizes into plaques neurotoxic to neighboring tissue. Rapid escalation of the marker is seen in the brain within the first day following injury in experimental animals, and may continue to be seen through day 14.[176] These plaques are readily identifiable on immunohistology and can last for years following insult, though regression is possible secondary to enzyme mechanisms for catabolism. A factor affecting appearance of the lesions is their perseverance, as amyloid β positive lesions that appear in areas of axonal injury after severe TBI, may regress over the first few years after injury.[29] It is primarily used in TBI as an ex vivo histological marker of axonal injury, as serum concentrations are roughly 100-fold lower than in the CSF[185] and do not change with injury.[186]

Microtubule-associated protein tau (MAP-tau) is a microtubule bound protein that plays a role in cytoskeletal organization and axonal transport. MAP-tau exists in different states of phosphorylation, and hyperphosphorylated tau is commonly associated with Alzheimer's disease.[16] After axonal injury and the initiation of microtubular cytoskeletal breakdown, MAP-tau is cleaved by activated calpain and caspase-3 into cleaved tau (c-tau) fragments that subsequently mediate amyloid β-induced tau hyperphosphorylation. These hyperphosphorylated tau complexes begin to accumulate in the tissue as neurofibrillary tangles and subsequently trigger apoptotic cell death.[30] Rapidly after injury, soluble c-tau fragments can be found in the CSF in 1 hour postinjury and rise precipitously at 24 hours, with subsequent decline over 3 days.[187–189] CSF concentrations of c-tau may also be useful as an outcome marker, as higher quantities are negatively associated with functional outcome.[187,188] In a comparison study of 28 severe TBI patients to medical neurology and nonneurology patients,[187] TBI patients manifested c-tau increases of ~40,000-fold over concentrations in controls and correlated with Glasgow Outcome Score at discharge. Moreover, CSF concentrations of c-tau demonstrate axonal damage in patients with DAI with more sensitivity than a CT scan,[187,188] however the biomarker in serum has not been shown to be similarly selective[173,174] likely due to the blood–brain barrier (BBB).

Glial Fibrillary Acidic Protein

Glial fibrillary acidic protein (GFAP) is an acidic filament protein located within astrocytes and otherwise not found outside the central nervous system (CNS). Following experimental DAI in rats, increased GFAP mRNA expression has been noted for at least 11 days following injury.[190] Following initial injury, it increases markedly to peak within the first 24 hours after injury and then progressively declines to baseline by approximately a week, though patients with unfavorable outcome still demonstrated higher GFAP concentrations than favorable outcome patients at 11 to 14 days.[191] In several series, serum concentrations have been demonstrated to be predictive of ultimate survival,[191–194] as well as correlate with imaging findings. A comparison study of GFAP and S-100β in 92 severe TBI patients demonstrated that though both markers were comparable in their correlation mortality, GFAP was better able to discriminate between severe disability and persistent vegetative state than S-100β.[194,195] However, a recent study assessing day 1 GFAP among other markers after all TBI requiring computed tomography (CT) imaging of the head did not find similar outcome discrimination.[193] This may suggest that GFAP elevation may only be most informative in patients with sufficient brain injury as to have astrocyte injury from secondary hypoxia-ischemia or other mechanisms.

Myelin Basic Protein

Myelin basic protein (MBP) is a major constituent of the white matter and comprises approximately 30% of it. Closely associated in axonal sheath layers, it is readily released into the CSF following injury and can appear as early as 2 hours after TBI in rats.[196] Following injury it is lysed by calpain, matrix metalloproteinases, and lysosomal proteases and by-products of its dissolution can be detected as well. Though MBP is readily found in the CSF, the BBB appears to impede its translocation into the serum and therefore it may be difficult to access as an acute indicator of axonal injury.[65]

S-100β

S-100β is a low molecular weight (10.5 kDa) acidic calcium binding protein found in glial and Schwann cells of white matter.[197] It is unclear whether S-100β may be protective or detrimental, as it may stimulate glial repair processes but may also be linked to calcium influx.[198] As a biomarker, clinical studies have found elevations in S-100β early in injury are tied to eventual functional outcome.[199–201] S-100β also cannot diffuse across intact BBB, however, and serum concentrations are heavily dependent on BBB permeability.[202,203] Importantly, there is a potential for confounding in multisystem trauma with musculoskeletal injury as S-100β is also found in adipocytes and chondrocytes. Some trauma literature has yielded conflictual results regarding the utility of the marker in predicting outcome[204–207] particularly in severe extracranial injury.[208] Therefore, careful interpretation is likely indicated in trauma patients with multisystem injury.

Neuron-Specific Enolase

Neuron specific enolase is housed in the cytoplasm of neurons and can also be found in circulating erythrocytes and platelets. Within the axon it travels via the slow axonal transport system. It is not normally seen extracellularly and can be found in injured tissue as soon as 1.5 hours after injury. Though hemolysis will contaminate serum and possibly hemorrhagic CSF samples with hematogenous NSE, corrections exist for correcting hemolysis effect on NSE in serum samples.[209] As a biomarker NSE has correlates with Glasgow Outcome Score within the first year following pediatric TBI.[210] NSE outcome correlations appear to best match early peak values, and therefore the value of NSE might be an assessment of magnitude of early injury in the patient. However, the marker's half-life is 20 hours, and over the course of early injury it is likely not useful for revealing acute changes in status.

Ubiquitin Carboxy-Terminal Hydroxylase L1

Ubiquitin carboxy-terminal hydroxylase L1 (UCH-L1) is expressed in neurons as a small cysteine protease that hydrolyzes the C-terminal bond between ubiquitin and polypeptides. Several studies have demonstrated significant UCH-L1 increases in the CSF of severely injured patients within an hour after injury.[193,211,212] Moreover, the magnitude of UCH-L1 increase was seen to be larger in patients with worse 6-month outcome and postinjury complications.[211] The BBB also restricts translocation of UCH-L1 and therefore serum concentrations are highly dependent on BBB integrity, though serum concentrations are proportional to increases in CSF concentrations.[197]

NEUROIMAGING IN DIFFUSE AXONAL INJURY

With improvements in imaging technologies, radiographic analysis of the brain following TBI is gaining importance in outcome prediction, as has been demonstrated in predicative models based on head computed tomography.[213,214] Neuroradiology data in the clinical setting is typically evaluated in qualitative terms,[215] though this is fallible as critical pathology indicative of outcome is missed to more than an insignificant degree even among blinded expert neuroradiologists interpreting CT and magnetic resonance (MR) data in TBI patients.[216,217]

Computed Tomography of the Brain

Head CT frequently does not identify pathology associated with DAI.[218] Only 10% of patients with DAI demonstrate hemorrhagic punctate lesions of the corpus callosum and gray–white matter junctions of the cerebrum and pontine–mesencephalic junction near the cerebellar peduncles. Weeks after injury, atrophic changes may occur in the white matter, dependent on the degree and topography of injury. Areas where atrophy may appear more prominent include areas of high white matter concentration and periventricular parenchyma recedes and ventriculomegaly may develop[219] without sulcal effacement or other evidence of hydrocephalus. Clear advantages of neuroimaging with CT remain, however, particularly in terms of image acquisition speed and scanner accessibility in most acute trauma environments.

Conventional Magnetic Resonance Imaging (MRI)

Like CT, MR neuroimaging is based on tomographic reconstruction of the brain in imaging slices acquired sequentially, but in the case of MR, data is reconstructed from the magnetic resonance properties of hydrogen atoms in a powerful magnetic field. MRI is more sensitive for the diffuse and physically small pathology found in DAI as opposed to head CT.[216,220,221] Shearing forces associated with DAI transect small blood vessels running in parallel with axons, resulting in detectable microscopic hemorrhages forming in these areas. Gradient echo and T_2^* (T_2 star) weighted MR imaging detects signal dropout caused by iron-containing heme groups in slow-moving blood. Distributions of these microhemorrhages in areas associated with axonal injury such as the corpus callosum, brainstem, and other white matter tracts strongly suggest an imaging diagnosis of DAI. DAI microhemorrhages typically appear as punctate signal-free lesions in the white matter that "bloom" or appear slightly larger than their true anatomic size due to iron-induced magnetic field distortion. Consequently, signal loss caused by punctate hemorrhages from DAI can be visualized for years after injury though lesions fade over time.[222] Confounders that may appear to be microhemorrhages include air and calcium deposits; however, these are uncommon in areas affected by DAI. Density of T_2^* lesions have been associated with severity of injury in terms of admission GCS[223–225] as well as maximum ICP during admission and 3 month Glasgow Outcome Score.[226] An additional factor is magnet field strength, as 3 Tesla MRI demonstrates practically twice the sensitivity of 1.5 Tesla devices for microhemorrhages,[227] and using different MRI

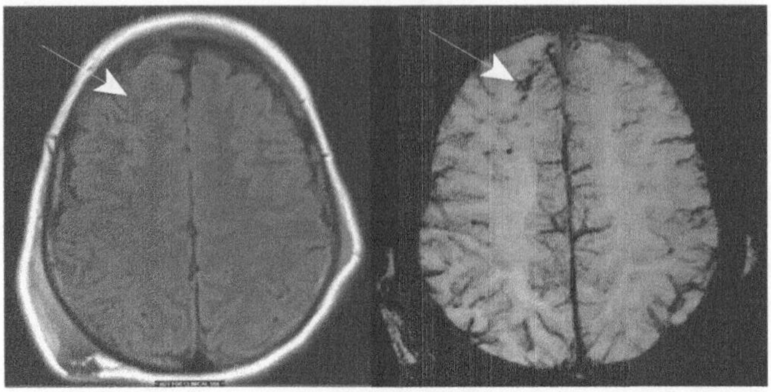

FIGURE 3.3 MRI imaging of a 2-year-old patient with DAI after being ejected from a motor vehicle. T_2-weighted FLAIR imaging appears on the left, with corresponding susceptibility weighted imaging (SWI) slice appearing on right. Punctate areas of increased T_2 signal appear in the white matter of the right frontal and parietal lobes on the left, with some corresponding dotlike areas of signal loss in the right frontal white matter on the SWI image (arrows).

devices utilizing different pulse sequence methods, even of the same field strength between manufacturers, can complicate interpretation between serial exams.

Punctate areas of T_2 signal hyperintensity in the white matter and at gray–white matter junctions in the frontal or occipital lobes are also suggestive of DAI (Figure 3.3).[217,228] These nonhemorrhagic lesions are optimally visualized in high-resolution T_2-weighted sequences with or without fluid attenuated inversion recovery (FLAIR) suppressing CSF signals. Even in the absence of microhemorrhages conventional MRI has been demonstrated superior to CT in evaluating DAI.[229] With regard to outcome, nonhemorrhagic lesions may not necessarily be associated with worsening outcome suggesting different mechanisms for their appearance.[229] In terms of long-term follow-up imaging, atrophy and gliosis as a consequence of DAI may be also be evaluated with conventional MRI.

Susceptibility Weighted Imaging

Similarly to T_2^* weighted imaging, susceptibility weighted imaging (SWI) is a high-resolution three-dimensional imaging sequence that produces images based on local magnetic field aberrations.[230] However, it is much more sensitive than T_2^*, detecting traumatic microhemorrhages at roughly 6 times the sensitivity of T_2^*, and renders hemorrhage volume approximately twice the size seen on conventional MRI gradient echo imaging.[230] Though the physics of the sequence are different, it similarly accentuates hemorrhage, calcification, air, or other features causing aberrations in magnetic susceptibility. In patients with severe DAI and resultant cerebral edema, venous stasis in SWI is an ominous sign and prominently appears as dark areas of susceptibility artifacts outlining engorged veins coursing through the brain parenchyma. In severe cerebral edema and associated venous stasis, delayed transit of venous blood causes pronounced SWI signal decrease in the distribution of medullary cerebral

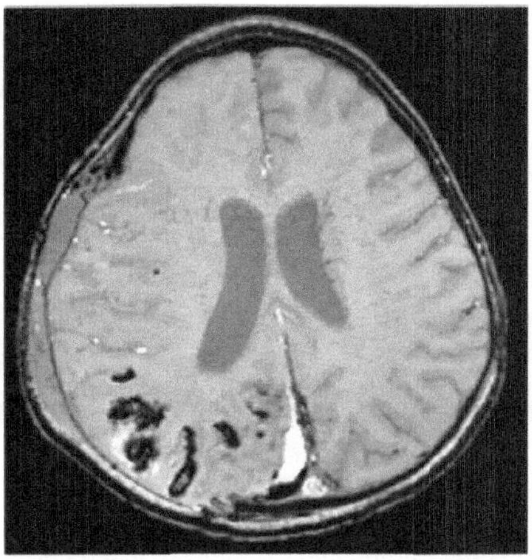

FIGURE 3.4 SWI revealing diffuse vascular injury in an infant after abusive head trauma, appearing as serpentine areas of dark absent signal in the right occipital lobe.

veins, appearing as irregular radially distributed rays of signal loss through the white matter, and is described as diffuse vascular injury (Figure 3.4).[231]

Diffusion Weighted Imaging (DWI) and Diffusion Tensor Imaging (DTI)

Diffusion imaging is performed with serial acquisition of signal data after an initial coding from radiofrequency pulse, thereby identifying fluid movement within a magnetic gradient. DWI has demonstrated improved detection of nonhemorrhagic lesions over conventional MRI in areas of cytotoxic or vasogenic edema. In a series of 74 adult DAI patients, DWI-derived apparent diffusion coefficient (ADC) imaging revealed a higher burden of high intensity ADC signals in the corpus callosum and brainstem, and the magnitude of abnormal signal correlated with duration of coma.[232] In children, identification of deep frontal and temporal white matter and basal ganglia diffusion abnormality was associated with poorer outcome in a subset of severe pediatric TBI patients.[233] DTI is a further modification of this imaging technique where multiple diffusion vectors are measured in three dimensions, and diffusion magnitude can be determined in multiple directions for a given cube-shaped voxel of space in the brain. Data is aggregated from individual voxels arranged in a three-dimensional grid representing the brain in imaged space, with each voxel typically measuring approximately 2 to 3 mm on a side and between 6 and 64 or more diffusion directions measured. Each voxel is assigned vectors measuring a magnitude of diffusion in three perpendicular cardinal directions, with the principal vector being the major axis of diffusion within the voxel space (Figure 3.5). Regarding utility of DTI, data continues to emerge[234,235] on its utility in TBI and it is more sensitive to change following TBI than high-resolution conventional MRI.[236] Multiple metrics have been derived for interpreting DTI data. Axial diffusivity is the magnitude of the major axis of diffusion inside the voxel and may indicate relative integrity of fluid filled tubular structures such as axons.[218,237] Radial diffusivity is the mean magnitude

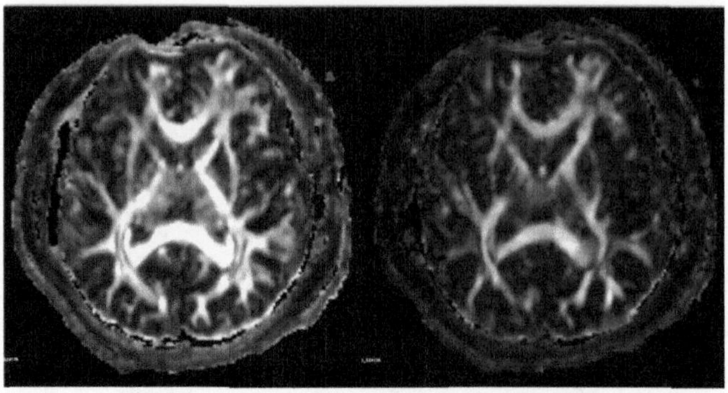

FIGURE 3.5 **(See color insert.)** Diffusion tensor imaging images of relative fractional anisotropy in an 8-year-old patient after TBI from being struck by a motor vehicle. The left grayscale image reflects FA magnitude and the right image incorporates direction data, with green labeling identifying anterior-posterior tensor alignment, red identifying the left-right direction, and blue corresponding to the cephalad-caudad direction. Note decreased FA magnitude in the R frontal white matter.

of the two minor axes perpendicular to the major axis, and some authors suggest it may reflect integrity of tubular structures, such as the myelin sheath of axons.[236] Fractional anisotropy (FA) is an approximation of the overall directionality of flow within a voxel, and a value between 0 and 1 is derived from the three eigenvectors of a voxel, with higher values reflecting greater anisotropy or directionality. Given that a standard clinical voxel averages a volume of approximately 8 mm^3 and may contain several hundreds of thousands of neurons, these presumptions are potentially prone to inaccuracy particularly in regions with large numbers of crossing structures or varying axon density.[238,239] Despite data suggesting the utility of DTI in describing anatomical changes associated with outcome in significant DAI, DTI have not consistently shown similar significance in mild TBI.[240,241] Tractography is an additional postprocessing technique that utilizes DTI data and "seeds" placed in white matter regions of interest in the brain to track approximated white matter tracts. From these seeds and user-determined tracing thresholds, diffusion tensors in each voxel are used to trace out estimated diffusion tracts. When seeds are placed within white matter, these tracts are presumed to represent bundles of continuous axons. DTI and tractography data must be interpreted considering operator seed placement as well as the effects of multiple cellular entities in addition to axons on diffusion, such as astrocytes, crossing fibers, and vascular structures among others.[242] It is worth considering that DTI metrics generalize the myriad diffusivities of hundreds of thousands of cells in a single voxel into just three eigenvectors.

Strategies have emerged for improving DTI data resolution, including adjustments for complex fluid dynamics in biological systems, such as in diffusion kurtosis imaging (DKI).[243,244] Other methodologies involve increased sampling of diffusion directions such that multiple principal vectors can be derived, as seen in high angular resolution diffusion imaging (HARDI)[245] and diffusion spectrum imaging (DSI),[246] which permit high-resolution tractography as seen in high-definition fiber tracking

(HDFT).[247] The utility of these methods in assessing DAI in heterogeneous TBI patients will rely on future imaging investigations. Variability between patients is important in clinical practice given accurate postinjury DTI interpretation requires some estimate of preinjury DTI conditions. A number of studies have indicated reduced anisotropy (evaluated with FA) in DAI-vulnerable white matter voxels (genu and splenium of the corpus callosum, internal capsule) is associated with postinjury cognitive deficits,[248–254] however, significant sample and methodology variance between these studies preclude their generalizability.[255,256] Deciphering the predictive accuracy of DTI on TBI patient outcome therefore is still an important topic of ongoing research. This may be more challenging in DAI than in focal injury patients, since an unaffected hemisphere can often serve as a control in focal injury patients.[247] In contrast, DAI patients will likely require aggregate DTI atlas data from large-scale population studies to generalize DTI findings.[257] Presently, this information is in development,[257,258] particularly in children where myelination steadily progresses throughout neurodevelopment[259] and DTI metrics vary over brain maturation. Moreover the effects of age, subacute or chronic medical conditions on DTI findings over the human lifespan are still being elucidated. Therefore, interpreting abnormal findings in trauma patients across the spectrum of age will require a much more complete picture of normal DTI than is currently available.[260]

Magnetization Transfer Imaging

Magnetization transfer imaging (MTI) identifies chemical alterations in tissue caused by injury, based on associated changes in tissue magnetic moment. The protons of large organic macromolecules are selectively saturated by a radiofrequency pulse that is off-resonance from free water protons,[261] and the magnetic moments of these macromolecule protons are "transferred" to water surrounding the tissue. These free water protons are interrogated thereby producing the MTI data.[261] The methodology appears to be useful in TBI, as areas of decreased magnetization transfer ratio (MTR) have been identified in the white matter of trauma patients in the absence of findings on conventional MRI.[262–267] MTI in TBI patients also demonstrates abnormal findings in trauma patients with ongoing cognitive deficits despite having "normal" conventional MRI sequences.[262,264] Histological confirmation of MTI-related lesions in animal models has been performed,[263] though the exact physiology underlying these changes is not clear and likely multifactorial.[261] A study describing MTI and MR spectroscopy of the corpus callosum for long-term follow-up of TBI patients has also been performed, revealing a lack of association between MTI and outcome in this investigation, though proton MRS demonstrated that decreased N-acetylaspartate to creatinine ratio (as an indicator of decreased mitochondria from neuronal loss) might suggest poor outcome. Pitfalls of MTI include lengthy study time and susceptibility to motion artifact, as well as qualitative interpretation susceptible to variability between interpreters.[261]

Morphometric and Volumetric Analysis

Morphometric analysis entails isolation and rendering of specific brain structure volumes by virtue of their contrast from surrounding tissue, and can be based on automatic parcellation or manual three-dimensional tracing based on obtained MR images. Morphology and volumes of these structures in turn reveal disease effects, and is ideally suited to analyzing both the direct physical injuries incurred by TBI as well as evolution of brain appearance over time. Morphometric analysis has associated anatomic injury with specific deficits[268–271] and readily captures volume changes in atrophic structures as they evolve after injury. A study examining volumetric analysis based on T_2-weighted FLAIR images in 24 adults with severe DAI found volume of affected white matter correlated significantly with eventual GOSE (Glasgow Outcome Scale-Extended) at 6 months.[272] Strangman and colleagues[273] also found that longitudinal changes in the volume of high connectivity areas such as the hippocampus, thalamus, lateral prefrontal cortex, and cingulate cortex are predictive of neurorehabilitation outcome. In a more topology oriented approach, Kinnunen and colleagues[274] utilized tract-based spatial statistics (TBSS) to characterize risk of specific deficits after TBI based on white matter changes. Given that morphometry is highly dependent on accurate tissue definition, it is easily confounded by measurement error. Edema in particular affects the appearance of structures in both CT and MRI, and the heterogeneous nature of TBI and DAI makes parcellation of structures difficult for both manual and automated systems especially when the brain is severely deformed from pathology.[275]

Functional MRI

Functional MRI (fMRI) findings are based on the detection of elevated blood oxygen level dependent (BOLD) signals presumably in regions of the brain parenchyma with increased metabolic demand.[276,277] Determination of where the brain is most active is based on relative oxygen content differences detected in blood delivery and return. In trauma, frontoparietal activity in the brain is attenuated in DAI survivors compared to controls in working memory tasks, particularly in the right middle and superior frontal lobe cortex,[278] and fMRI may identify functional compensation and forecast response to learning strategies in patients rehabilitating from DAI.[144,279] Over time, fMRI has been demonstrated to reveal compensatory regional activation well into recovery after TBI.[280,281] There is enthusiasm among imaging specialists[275] that identification of active brain combined with DTI structural data will permit connectivity mapping of the brain as it recovers from injury. In the acute phase, issues with fMRI include that BOLD signals can be confounded by blood pooling from other sources in trauma and also perfusion effects caused by increased ICP. With careful consideration of these effects fMRI may prove a useful methodology for profiling outcomes of brain recovery after DAI in the future.

ELECTROPHYSIOLOGY AND MAGNETOENCEPHALOGRAPHY

DAI's effects on long axons and cerebral connectivity affects neural electrophysiology after injury, manifest in evoked potentials, electroencephalography, as well as magnetoencephalography (MEG). As a functional test of brain activity, MEG records magnetic flux at the head surface caused by neuronal depolarization in shallow cortical structures below the scalp.[261] The methodology has high temporal and spatial resolution and is a promising means of identifying abnormal cortical areas. In TBI patients, MEG demonstrates changes such as reduced background activity persistently after injury.[282,283] Given that abnormal EEG activity (and therefore MEG as well) in gray matter may be a consequence of white matter de-afferentation by axonal damage,[284] regionally abnormal MEG findings may identify affected cortical areas after DAI. In a case series of 10 mild TBI patients with no CT or MR radiographic evidence of injury in nine, combined analysis using DTI and MEG revealed gray matter MEG lesions topographically near areas of reduced anisotropy identified with DTI.[283] MEG machines are not common even in large medical centers, however, and clinical expertise in interpretation, as well as clinical access for patients who cannot easily get into the device, may be challenging. Appropriate MEG helmet sizing for pediatric patients may also be challenging as there will be increased distance between a child's skull and biomagnetometer coils compared to an adult's head.[285]

Electrophysiologic methods identifying altered evoked potentials in DAI patients may also provide insight into patterns of disrupted cerebral connectivity following injury. DAI survivors may demonstrate increased resting corticospinal motor thresholds necessary to initiate MEPs, and reduced area under the MEP waveform indicating damaged cerebral connectivity.[286] In a study of 52 adult survivors of TBI associated with motor vehicle accidents and chronic functional impairment from DAI, DTI of the corticospinal tract, and motor evoked potential (MEP) evaluation with transcranial magnetic stimulation,[287] revealed that patients who did not generate MEP readings in response to stimulation had lower corticospinal tract FA than DAI patients who did generate MEPs as well as uninjured controls. Though electrophysiology is informative, imaging may provide better lesion localization not captured using these electrophysiological methods. Further research may suggest whether serial evoked potential data is informative regarding functional recovery after TBI.

THERAPEUTICS

Calcineurin Modulators

As mentioned earlier, calcineurin (CN) modulators are a potential therapy for axonal injury by virtue of CN's responsiveness to calcium as a calcium/calmodulin phosphatase, and possible role in calcium-induced axonal damage following TBI and excitotoxic injury. Cyclosporin A (CsA) initially received attention as a axonal injury modulator in part because its capacity suppresses calcineurin-mediated formation of the mitochondrial transition pore, averting mitochondrial failure and subsequent apoptotic cell death. In a rodent model of impact acceleration DAI CsA administration 30 minutes after injury resulted in decreased loss of mitochondrial *N*-acetylaspartate, and treated animals had higher ATP production than untreated controls signifying mitochondrial preservation.[288] Histologically CsA pretreatment decreases APP positive axonal swellings in rats following impact acceleration DAI.[56,57] Time dependency of CsA dosing in relationship to TBI seems important as well, given that administration of CsA 1 hour after midline fluid percussion injury in rodents worsens axon function determined by compound action potential (CAP), and in comparison when CsA was given 15 minutes postinjury, CAP is preserved.[289] Toxicity of CsA has also been a concern as a DAI therapy since it can cause life-threatening seizures, and drug delivery to the brain is restricted by intact BBB thereby making brain concentrations in TBI unpredictable.

Tacrolimus (FK506) has also been investigated in multiple preclinical series as a potential TBI and DAI therapy, and histopathological evidence postinjury suggests a beneficial role for the agent in preserving axonal transport and reducing APP accumulation.[55,290,291] Functionally, however, the role of tacrolimus does not affect neurofilament sidearm compaction in the progression to axonal disconnection[54] and therefore does not protect against all histopathological effects of axonal injury. Additionally, despite beneficial effects of tacrolimus in juvenile animal DAI models, reducing histological axonal damage, it functionally did not facilitate recovery of CAP deficits in a modified CCI model for juvenile rat DAI.[292]

Stem Cell Therapy

The limited regenerative capacity of the brain makes the potential for stem cell therapy very attractive and the concept of introducing pluripotential cells into the CNS to regenerate brain has been explored sporadically since the end of the nineteenth century.[293] In addition to possibly reconstituting damaged neurons, astrocytes, and oligodendrocytes among other brain cells, stem cells could also potentially be neuroprotective by maintaining the penumbral cell milieu through trophic, anti-inflammatory, direct chemotactic, or other mechanisms, thereby limiting secondary injury.[293] Stem cell delivery routes vary, as intracerebral administration optimizes transplant localization but poses an invasive risk, intravenous delivery suffers from pulmonary first-pass sequestration and poor target specificity, and intraarterial delivery may result in pathologic emboli.[294] Multiple studies have been performed demonstrating disparate results in animal models utilizing varied methodologies, but stem cell differentiation into neurons appears rare compared to other cell types.[295] Recently, Riess and colleagues intracerebrally transplanted murine embryonic stem cells in rats following a fluid-percussion model of DAI, and found that functional outcome appeared improved, but only 1 of 10 animals at 7 weeks retained transplanted cells, while 2 animals developed tumors.[296] Stem cells from more restricted lineages may also be practical and have less tumorigenic potential.[7] Examples include neurogenic stem cells derived from remnants of primitive embryonic rests in the walls of the ventricular system (subventricular zone[295]) and hippocampus among other structures,[293] or bone-marrow-derived stem cells.[294] Another potential strategy is the use of oligodendrocyte precursor cells to target remyelination of injured axons.[297] In humans, a clinical trial administered human embryonic stem cells intracranially to a cohort of comatose patients and reported some improved functional outcomes compared to a matched group of comatose controls.[298] In this study, coma etiologies included 15 DAI patients of the total 25 treated patients, some with signs of frontal lobe atrophy on MR imaging. Further clinical investigations into human patient safety and functional outcomes after neurological stem cell therapies are necessary to determine the scope of utility of stem cell therapy in the clinical setting.

Recombinant Human Erythropoetin

Erythropoetin (EPO) as an erythrogenesis stimulator is found in the CSF throughout the life cycle and likely has a role in the basic homeostasis of the CNS.[299] The role of EPO in the CNS is unclear, but it appears to reduce neuronal apoptosis possibly by EPO-receptor initiated autophosphorylation of Jak-2 and subsequent activation of antiapoptotic pathways through PI3K-Akt/protein kinase B/bcl-2, NF-KB, and RAS mechanisms among others.[300] Animal models have demonstrated improved preservation of neuronal architecture in experimental TBI treated with EPO,[301,302] and improved in-hospital mortality has been demonstrated in a retrospective clinical TBI study[303] and trauma subset of a Phase II trial on critically ill adults.[304] Following this Phase II trial, a Phase III trial of EPO administration in ICU patients also demonstrated a significant reduction in mortality (adjusted hazard ratio 0.40 at day 140, 95% CI 0.23–0.69), however, a significant increase in incidence of thrombotic complications accompanied this.[305] An additional issue potentiating side effects is that the intact BBB impedes EPO diffusion into the brain and it is difficult to predict optimal drug dosing in TBI.[299] Despite animal TBI model evidence of axonal preservation after injury with EPO, EPO has been implicated in inducing neuronal calcium ion influx, and therefore could theoretically potentiate calcium-mediated neuronal injury.[306,307]

Docosahexanoic Acid

Docosahexanoic acid (DHA) is well known as an omega-3 fatty acid, major constituent of the brain and neuronal phospholipid membranes, and potential anti-inflammatory specifically decreasing cyclooxygenase activity.[308,309] In an impact acceleration rodent model of DAI, rats supplemented with 10 or 40 mg/kg/d DHA containing feed on postinjury day 1 manifested significantly lower APP positive stained corticospinal tract axons and decreased caspase-3 activation than untreated TBI animals, with observed concentrations in treated animals similar to those seen in uninjured animals.[308,309] Functional and outcome studies have not been performed in DAI models, but in a clinical trial could be readily implemented in compatible TBI management protocols permitting early feeds.

SUMMARY

In total DAI is a complex process that merely begins with stretching and shearing of axons, at times proceeding to a persistent syndrome of cerebral disconnection and longstanding functional impairment. Research has improved our understanding of the mechanisms of axonopathy in DAI, its biomarkers, and potential therapeutic targets to treat it. Rapid technological innovations in biomarkers and neuroimaging, especially in MRI, are increasing our understanding of DAI's evolution following TBI, however, there remain significant hurdles with regard to balancing data granularity and generalizable interpretation methods given the heterogeneous nature of TBI. With increasing scientific attention on brain connectivity, a working knowledge of DAI and its potential future therapies will also likely see significant advances in the near future.

REFERENCES

1. Meythaler, J.M. et al. 2001. Current concepts: Diffuse axonal injury-associated traumatic brain injury. *Arch Phys Med Rehabil* 82(10):1461–1471.
2. Adams, J.H. et al. 1982. Diffuse axonal injury due to nonmissile head injury in humans: An analysis of 45 cases. *Ann Neurol* 12(6):557–563.
3. Povlishock, J.T. et al. 1983. Axonal change in minor head injury. *J Neuropathol Exp Neurol* 42(3):225–242.
4. Povlishock, J.T., Katz, D.I. 2005. Update of neuropathology and neurological recovery after traumatic brain injury. *J Head Trauma Rehabil* 20(1):76–94.
5. Smith, D.H., Meaney, D.F. 2000. Axonal damage in traumatic brain injury. *Neuroscientist* 6(6):483–495.
6. Christman, C.W. et al. 1994. Ultrastructural studies of diffuse axonal injury in humans. *J Neurotrauma* 11(2):173–186.
7. Li, X.-Y., Feng, D.-F. 2009. Diffuse axonal injury: Novel insights into detection and treatment. *J Clin Neurosci* 16(5):614–619.
8. Corbo, J., Tripathi, P. 2004. Delayed presentation of diffuse axonal injury: A case report. *Ann Emerg Med* 44(1):57–60.
9. Gieron, M.A., Korthals, J.K., Riggs, C.D. 1998. Diffuse axonal injury without direct head trauma and with delayed onset of coma. *Pediatr Neurol* 19(5):382–384.
10. Johnson, V.E., Stewart, W., Smith, D.H. 2013. Axonal pathology in traumatic brain injury. *Exp Neurol* 246(C):35–43.
11. Adams, H. et al. 1977. Diffuse brain damage of immediate impact type. Its relationship to "primary brain-stem damage" in head injury. *Brain* 100(3):489–502.
12. Gennarelli, T.A. et al. 1982. Diffuse axonal injury and traumatic coma in the primate. *Ann Neurol* 12(6):564–574.
13. Gaetz, M. 2004. The neurophysiology of brain injury. *Clin Neurophysiol* 115(1):4–18.
14. Meyers, C.A. et al. 1983. Early versus late lateral ventricular enlargement following closed head injury. *J Neurol Neurosurg Psychiatry* 46(12):1092–1097.
15. Wilson, J.T. et al. 1988. Early and late magnetic resonance imaging and neuropsychological outcome after head injury. *J Neurol Neurosurg Psychiatry* 51(3):391–396.
16. Blennow, K., Hardy, J., Zetterberg, H. 2012. The neuropathology and neurobiology of traumatic brain injury. *Neuron* 76(5):886–899.
17. Lipton, M.L. et al. 2008. Multifocal white matter ultrastructural abnormalities in mild traumatic brain injury with cognitive disability: A voxel-wise analysis of diffusion tensor imaging. *J Neurotrauma* 25(11):1335–1342.
18. Niogi, S.N. et al. 2008. Extent of microstructural white matter injury in postconcussive syndrome correlates with impaired cognitive reaction time: A 3T diffusion tensor imaging study of mild traumatic brain injury. *AJNR Am J Neuroradiol* 29(5): 967–973.
19. Wilde, E.A. et al. 2008. Diffusion tensor imaging of acute mild traumatic brain injury in adolescents. *Neurology* 70(12):948–955.
20. Baxter, D. et al. 2013. Pituitary dysfunction after blast traumatic brain injury. *Ann Neurol* 74(4):527–536.
21. Jeong, J.H. et al. 2010. Negative effect of hypopituitarism following brain trauma in patients with diffuse axonal injury. *J Neurosurg* 113(3):532–538.
22. Sundaram, N.K., Geer, E.B., Greenwald, B.D. 2013. The impact of traumatic brain injury on pituitary function. *Endocrinol Metab Clin North Am* 42(3):565–583.
23. Richmond, E., Rogol, A.D. 2013. Traumatic brain injury: Endocrine consequences in children and adults. *Endocrine*.
24. Singleton, R.H. et al. 2002. Traumatically induced axotomy adjacent to the soma does not result in acute neuronal death. *J Neurosci* 22(3):791–802.

25. Povlishock, J.T., Christman, C.W. 1995. The pathobiology of traumatically induced axonal injury in animals and humans: A review of current thoughts. *J Neurotrauma* 12(4):555–564.
26. Pettus, E.H. et al. 1994. Traumatically induced altered membrane permeability: Its relationship to traumatically induced reactive axonal change. *J Neurotrauma* 11(5):507–522.
27. Farkas, O., Povlishock, J.T. 2007. Cellular and subcellular change evoked by diffuse traumatic brain injury: A complex web of change extending far beyond focal damage. *Prog Brain Res* 161:43–59.
28. Stone, J.R. et al. 2004. Impaired axonal transport and altered axolemmal permeability occur in distinct populations of damaged axons following traumatic brain injury. *Exp Neurol* 190(1):59–69.
29. Chen, X.-H. et al. 2009. A lack of amyloid beta plaques despite persistent accumulation of amyloid beta in axons of long-term survivors of traumatic brain injury. *Brain Pathol* 19(2):214–223.
30. Li, J. et al. 2010. Biomarkers associated with diffuse traumatic axonal injury: Exploring pathogenesis, early diagnosis, and prognosis. *J Trauma* 69(6):1610–1618.
31. Stys, P.K. 1998. Anoxic and ischemic injury of myelinated axons in CNS white matter: From mechanistic concepts to therapeutics. *J Cereb Blood Flow Metab* 18(1):2–25.
32. Maxwell, W.L., Povlishock, J.T., Graham, D.L. 1997. A mechanistic analysis of nondisruptive axonal injury: A review. *J Neurotrauma* 14(7):419–440.
33. Tang-Schomer, M.D. et al. 2010. Mechanical breaking of microtubules in axons during dynamic stretch injury underlies delayed elasticity, microtubule disassembly, and axon degeneration. *FASEB J* 24(5):1401–1410.
34. Adlard, P.A., King, C.E., Vickers, J.C. 2000. The effects of taxol on the central nervous system response to physical injury. *Acta Neuropathol* 100(2):183–188.
35. Tomei, G. et al. 1990. Morphology and neurophysiology of focal axonal injury experimentally induced in the guinea pig optic nerve. *Acta Neuropathol* 80(5):506–513.
36. Ikonomovic, M.D. et al. 2004. Alzheimer's pathology in human temporal cortex surgically excised after severe brain injury. *Exp Neurol* 190(1):192–203.
37. Roberts, G.W. et al. 1991. Beta A4 amyloid protein deposition in brain after head trauma. *Lancet* 338(8780):1422–1423.
38. Roberts, G.W. et al. 1994. Beta amyloid protein deposition in the brain after severe head injury: Implications for the pathogenesis of Alzheimer's disease. *J Neurol Neurosurg Psychiatry* 57(4):419–425.
39. Ommaya, A.K. 1995. Head injury mechanisms and the concept of preventive management: A review and critical synthesis. *J Neurotrauma* 12(4):527–546.
40. Medana, I.M. 2003. Axonal damage: A key predictor of outcome in human CNS diseases. *Brain* 126(3):515–530.
41. Bullock, R. et al. 1995. Evidence for prolonged release of excitatory amino acids in severe human head trauma. Relationship to clinical events. *Ann N Y Acad Sci* 765:290–297; discussion 298.
42. Bullock, R. et al. 1998. Factors affecting excitatory amino acid release following severe human head injury. *J Neurosurg* 89(4):507–518.
43. Iwata, A. et al. 2004. Traumatic axonal injury induces proteolytic cleavage of the voltage-gated sodium channels modulated by tetrodotoxin and protease inhibitors. *J Neurosci* 24(19):4605–4613.
44. Wolf, J.A. et al. 2001. Traumatic axonal injury induces calcium influx modulated by tetrodotoxin-sensitive sodium channels. *J Neurosci* 21(6):1923–1930.
45. Staal, J.A. et al. 2010. Initial calcium release from intracellular stores followed by calcium dysregulation is linked to secondary axotomy following transient axonal stretch injury. *J Neurochem* 112(5):1147–1155.

46. Rosenberg, L.J., Teng, Y.D., Wrathall, J.R. 1999. Effects of the sodium channel blocker tetrodotoxin on acute white matter pathology after experimental contusive spinal cord injury. *J Neurosci* 19(14):6122–6133.
47. Fink, D.J., Datta, S., Mata, M. 1994. Isoform specific reductions in Na+,K(+)-ATPase catalytic (alpha) subunits in the nerve of rats with streptozotocin-induced diabetes. *J Neurochem* 63(5):1782–1786.
48. Tavalin, S.J., Ellis, E.F., Satin, L.S. 1997. Inhibition of the electrogenic Na pump underlies delayed depolarization of cortical neurons after mechanical injury or glutamate. *J Neurophysiol* 77(2):632–638.
49. Ahmed, S.M. et al. 2000. Stretch-induced injury alters mitochondrial membrane potential and cellular ATP in cultured astrocytes and neurons. *J Neurochem* 74(5): 1951–1960.
50. Kornek, B. et al. 2001. Distribution of a calcium channel subunit in dystrophic axons in multiple sclerosis and experimental autoimmune encephalomyelitis. *Brain* 124 (Pt 6):1114–1124.
51. Büki, A. et al. 2000. Cytochrome c release and caspase activation in traumatic axonal injury. *J Neurosci* 20(8):2825–2834.
52. Büki, A., Koizumi, H., Povlishock, J.T. 1999. Moderate posttraumatic hypothermia decreases early calpain-mediated proteolysis and concomitant cytoskeletal compromise in traumatic axonal injury. *Exp Neurol* 159(1):319–328.
53. Saatman, K.E. et al. 2000. Behavioral efficacy of posttraumatic calpain inhibition is not accompanied by reduced spectrin proteolysis, cortical lesion, or apoptosis. *J Cereb Blood Flow Metab* 20(1):66–73.
54. Marmarou, C.R., Povlishock, J.T. 2006. Administration of the immunophilin ligand FK506 differentially attenuates neurofilament compaction and impaired axonal transport in injured axons following diffuse traumatic brain injury. *Exp Neurol* 197(2): 353–362.
55. Singleton, R.H. et al. 2001. The immunophilin ligand FK506 attenuates axonal injury in an impact-acceleration model of traumatic brain injury. *J Neurotrauma* 18(6): 607–614.
56. Okonkwo, D.O. et al. 1999. Cyclosporin A limits calcium-induced axonal damage following traumatic brain injury. *Neuroreport* 10(2):353–358.
57. Okonkwo, D.O., Povlishock, J.T. 1999. An intrathecal bolus of cyclosporin A before injury preserves mitochondrial integrity and attenuates axonal disruption in traumatic brain injury. *J Cereb Blood Flow Metab* 19(4):443–451.
58. Staal, J.A. et al. 2007. Cyclosporin-A treatment attenuates delayed cytoskeletal alterations and secondary axotomy following mild axonal stretch injury. *Devel Neurobio* 67(14):1831–1842.
59. Deng-Bryant, Y. et al. 2008. Neuroprotective effects of tempol, a catalytic scavenger of peroxynitrite-derived free radicals, in a mouse traumatic brain injury model. *J Cereb Blood Flow Metab* 28(6):1114–1126.
60. Loane, D.J., Byrnes, K.R. 2010. Role of microglia in neurotrauma. *Neurotherapeutics* 7(4):366–377.
61. Gentleman, S.M. et al. 2004. Long-term intracerebral inflammatory response after traumatic brain injury. *Forensic Sci Int* 146(2–3):97–104.
62. Reeves, T.M., Phillips, L.L., Povlishock, J.T. 2005. Myelinated and unmyelinated axons of the corpus callosum differ in vulnerability and functional recovery following traumatic brain injury. *Exp Neurol* 196(1):126–137.
63. Staal, J.A., Vickers, J.C. 2011. Selective vulnerability of non-myelinated axons to stretch injury in an in vitro co-culture system. *J Neurotrauma* 28(5):841–847.
64. Bramlett, H.M., Dietrich, W.D. 2003. Synuclein aggregation: Possible role in traumatic brain injury. *Exp Neurol* 184(1):27–30.

65. Su, E. et al. 2012. Increased CSF concentrations of myelin basic protein after TBI in infants and children: Absence of significant effect of therapeutic hypothermia. *Neurocrit Care* 17(3):401–407.
66. Ng, H.K., Mahaliyana, R.D., Poon, W.S. 1994. The pathological spectrum of diffuse axonal injury in blunt head trauma: Assessment with axon and myelin strains. *Clin Neurol Neurosurg* 96(1):24–31.
67. Oehmichen, M., Theuerkauf, I., Meissner, C. 1999. Is traumatic axonal injury (AI) associated with an early microglial activation? Application of a double-labeling technique for simultaneous detection of microglia and AI. *Acta Neuropathol* 97(5):491–494.
68. GrandPré, T., Li, S., Strittmatter, S.M. 2002. Nogo-66 receptor antagonist peptide promotes axonal regeneration. *Nature* 417(6888):547–551.
69. Lin, Y., Wen, L. 2013. Inflammatory response following diffuse axonal injury. *Int J Med Sci* 10(5):515–521.
70. Lu, K.-T. et al. 2005. Extracellular signal-regulated kinase-mediated IL-1-induced cortical neuron damage during traumatic brain injury. *Neurosci Lett* 386(1):40–45.
71. Hans, V.H. et al. 1999. Experimental axonal injury triggers interleukin-6 mRNA, protein synthesis and release into cerebrospinal fluid. *J Cereb Blood Flow Metab* 19(2):184–194.
72. Kita, T. et al. 2000. The role of tumor necrosis factor-alpha in diffuse axonal injury following fluid-percussive brain injury in rats. *Int J Legal Med* 113(4):221–228.
73. Khuman, J. et al. 2011. Tumor necrosis factor alpha and Fas receptor contribute to cognitive deficits independent of cell death after concussive traumatic brain injury in mice. *J Cereb Blood Flow Metab* 31(2):778–789.
74. Yan, E.B. et al. 2011. Post-traumatic hypoxia exacerbates neurological deficit, neuroinflammation and cerebral metabolism in rats with diffuse traumatic brain injury. *J Neuroinflammation* 8(1):147.
75. Rancan, M. et al. 2001. Upregulation of ICAM-1 and MCP-1 but not of MIP-2 and sensorimotor deficit in response to traumatic axonal injury in rats. *J Neurosci Res* 63(5):438–446.
76. Rhodes, J.K.J., Sharkey, J., Andrews, P.J.D. 2009. The temporal expression, cellular localization, and inhibition of the chemokines MIP-2 and MCP-1 after traumatic brain injury in the rat. *J Neurotrauma* 26(4):507–525.
77. Büki, A., Povlishock, J.T. 2005. All roads lead to disconnection?—Traumatic axonal injury revisited. *Acta Neurochir (Wien)* 148(2):181–194.
78. Blumbergs, P.C., Jones, N.R., North, J.B. 1989. Diffuse axonal injury in head trauma. *J Neurol Neurosurg Psychiatry* 52(7):838–841.
79. Sullivan, G.M. et al. 2013. Oligodendrocyte lineage and subventricular zone response to traumatic axonal injury in the corpus callosum. *J Neuropathol Exp Neurol* 72(12):1106–1125.
80. Ludwin, S.K. 1990. Oligodendrocyte survival in Wallerian degeneration. *Acta Neuropathol* 80(2):184–191.
81. Grandpré, T., Strittmatter, S.M. 2001. Nogo: A molecular determinant of axonal growth and regeneration. *Neuroscientist* 7(5):377–386.
82. Wang, T. et al. 2012. The role of Nogo-A in neuroregeneration: A review. *Brain Res Bull* 87(6):499–503.
83. Keyvani, K., Schallert, T. 2002. Plasticity-associated molecular and structural events in the injured brain. *J Neuropathol Exp Neurol* 61(10):831–840.
84. Magnuson, J., Leonessa, F., Ling, G.S.F. 2012. Neuropathology of explosive blast traumatic brain injury. *Curr Neurol Neurosci Rep* 12(5):570–579.
85. Richter, M. et al. 2001. Head injury mechanisms in helmet-protected motorcyclists: Prospective multicenter study. *J Trauma* 51(5):949–958.
86. Pellman, E.J. et al. 2003. Concussion in professional football: Reconstruction of game impacts and injuries. *Neurosurgery* 53(4):799–812; discussion 812–814.

87. Tate, C.M. et al. 2013. Serum brain biomarker level, neurocognitive performance, and self-reported symptom changes in soldiers repeatedly exposed to low-level blast: A breacher pilot study. *J Neurotrauma* 30(19):1620–1630.
88. Bochicchio, G.V. et al. 2008. Blast injury in a civilian trauma setting is associated with a delay in diagnosis of traumatic brain injury. *Am Surg* 74(3):267–270.
89. DuBose, J.J. et al. 2011. Isolated severe traumatic brain injuries sustained during combat operations: Demographics, mortality outcomes, and lessons to be learned from contrasts to civilian counterparts. *J Trauma* 70(1):11–16; discussion 16–18.
90. Chen, Y.C., Smith, D.H., Meaney, D.F. 2009. In-vitro approaches for studying blast-induced traumatic brain injury. *J Neurotrauma* 26(6):861–876.
91. Martland, H.S. 1928. Punch drunk. *JAMA* 91(15):1103–1107.
92. McKee, A.C. et al. 2014. The neuropathology of sport. *Acta Neuropathol* 127(1):29–51.
93. Powell, J.W., Barber-Foss, K.D. 1999. Traumatic brain injury in high school athletes. *JAMA* 282(10):958–963.
94. Shin, W. et al. 2013. Diffusion measures indicate fight exposure-related damage to cerebral white matter in boxers and mixed martial arts fighters. *Am J Neuroradiol.*
95. Dale, G.E. et al. 1991. Neurofibrillary tangles in dementia pugilistica are ubiquitinated. *J Neurol Neurosurg Psychiatry* 54(2):116–118.
96. Geddes, J.F. et al. 1999. Neuronal cytoskeletal changes are an early consequence of repetitive head injury. *Acta Neuropathol* 98(2):171–178.
97. Tokuda, T. et al. 1991. Re-examination of ex-boxers' brains using immunohistochemistry with antibodies to amyloid beta-protein and tau protein. *Acta Neuropathol* 82(4):280–285.
98. McKee, A.C. et al. 2009. Chronic traumatic encephalopathy in athletes: Progressive tauopathy after repetitive head injury. *J Neuropathol Exp Neurol* 68(7):709–735.
99. Omalu, B. et al. 2011. Emerging histomorphologic phenotypes of chronic traumatic encephalopathy in American athletes. *Neurosurgery* 69(1):173–183.
100. Jordan, B.D. 2000. Chronic traumatic brain injury associated with boxing. *Semin Neurol* 20(2):179–185.
101. Orrison, W.W. et al. 2009. Traumatic brain injury: A review and high-field MRI findings in 100 unarmed combatants using a literature-based checklist approach. *J Neurotrauma* 26(5):689–701.
102. Fleminger, S. et al. 2003. Head injury as a risk factor for Alzheimer's disease: The evidence 10 years on; A partial replication. *J Neurol Neurosurg Psychiatry* 74(7):857–862.
103. Graves, A.B. et al. 1990. The association between head trauma and Alzheimer's disease. *Am J Epidemiol* 131(3):491–501.
104. Guo, Z. et al. 2000. Head injury and the risk of AD in the MIRAGE study. *Neurology* 54(6):1316–1323.
105. Mortimer, J.A. et al. 1985. Head injury as a risk factor for Alzheimer's disease. *Neurology* 35(2):264–267.
106. Mortimer, J.A. et al. 1991. Head trauma as a risk factor for Alzheimer's disease: A collaborative re-analysis of case-control studies. EURODEM Risk Factors Research Group. *Int J Epidemiol* 20(Suppl 2):S28–S35.
107. O'Meara, E.S. et al. 1997. Head injury and risk of Alzheimer's disease by apolipoprotein E genotype. *Am J Epidemiol* 146(5):373–384.
108. Plassman, B.L. et al. 2000. Documented head injury in early adulthood and risk of Alzheimer's disease and other dementias. *Neurology* 55(8):1158–1166.
109. Salib, E., Hillier, V. 1997. Head injury and the risk of Alzheimer's disease: A case control study. *Int J Geriatr Psychiatry* 12(3):363–368.
110. Nemetz, P.N. et al. 1999. Traumatic brain injury and time to onset of Alzheimer's disease: A population-based study. *Am J Epidemiol* 149(1):32–40.

111. Sullivan, P., Petitti, D., Barbaccia, J. 1987. Head trauma and age of onset of dementia of the Alzheimer type. *JAMA* 257(17):2289–2290.
112. Schofield, P.W. et al. 1997. Alzheimer's disease after remote head injury: An incidence study. *J Neurol Neurosurg Psychiatry* 62(2):119–124.
113. Jordan, B.D. et al. 1997. Apolipoprotein E epsilon4 associated with chronic traumatic brain injury in boxing. *JAMA* 278(2):136–140.
114. Case, M.E. 2008. Inflicted traumatic brain injury in infants and young children. *Brain Pathol* 18(4):571–582.
115. Case, M.E. 2008. Accidental traumatic head injury in infants and young children. *Brain Pathol* 18(4):583–589.
116. Cantu, R.C., Gean, A.D. 2010. Second-impact syndrome and a small subdural hematoma: An uncommon catastrophic result of repetitive head injury with a characteristic imaging appearance. *J Neurotrauma* 27(9):1557–1564.
117. Geddes, J.F. et al. 2001. Neuropathology of inflicted head injury in children. II. Microscopic brain injury in infants. *Brain* 124(Pt 7):1299–1306.
118. Duhaime, A.C. et al. 1998. Nonaccidental head injury in infants—The "shaken-baby syndrome." *N Engl J Med* 338(25):1822–1829.
119. Clayton, E.H., Genin, G.M., Bayly, P.V. 2012. Transmission, attenuation and reflection of shear waves in the human brain. *J R Soc Interface* 9(76):2899–2910.
120. Meaney, D.F. et al. 1995. Biomechanical analysis of experimental diffuse axonal injury. *J Neurotrauma* 12(4):689–694.
121. Margulies, S.S., Thibault, L.E., Gennarelli, T.A. 1990. Physical model simulations of brain injury in the primate. *J Biomech* 23(8):823–836.
122. Povlishock, J.T. 2013. Pathobiology of traumatically induced axonal injury in animals and man. *Ann Emerg Med* 22(6):980–986.
123. Wright, R.M. et al. 2013. A multiscale computational approach to estimating axonal damage under inertial loading of the head. *J Neurotrauma* 30(2):102–118.
124. Smith, D.H. et al. 1999. High tolerance and delayed elastic response of cultured axons to dynamic stretch injury. *J Neurosci* 19(11):4263–4269.
125. Gennarelli, T.A. et al. 1989. Axonal injury in the optic nerve: A model simulating diffuse axonal injury in the brain. *J Neurosurg* 71(2):244–253.
126. Metz, H., McElhaney, J., Ommaya, A.K. 1970. A comparison of the elasticity of live, dead, and fixed brain tissue. *J Biomech* 3(4):453–458.
127. Laksari, K., Shafieian, M., Darvish, K. 2012. Constitutive model for brain tissue under finite compression. *J Biomech* 45(4):642–646.
128. Shafieian, M., Darvish, K.K., Stone, J.R. 2009. Changes to the viscoelastic properties of brain tissue after traumatic axonal injury. *J Biomech* 42(13):2136–2142.
129. O'Connor, W.T., Smyth, A., Gilchrist, M.D. 2011. Animal models of traumatic brain injury: A critical evaluation. *Pharmacol Ther* 130(2):106–113.
130. Graham, D.I., Adams, J.H., Gennarelli, T.A. 1988. Mechanisms of non-penetrating head injury. *Prog Clin Biol Res* 264:159–168.
131. Holbourn, A. 1943. Mechanics of head injuries. *Lancet* 242(6267):438–441.
132. Lighthall, J.W. 1988. Controlled cortical impact: A new experimental brain injury model. *J Neurotrauma* 5(1):1–15.
133. Browne, K.D. et al. 2011. Mild traumatic brain injury and diffuse axonal injury in swine. *J Neurotrauma* 28(9):1747–1755.
134. Rubovitch, V. et al. 2011. A mouse model of blast-induced mild traumatic brain injury. *Exp Neurol* 232(2):280–289.
135. Bauman, R.A. et al. 2009. An introductory characterization of a combat-casualty-care relevant swine model of closed head injury resulting from exposure to explosive blast. *J Neurotrauma* 26(6):841–860.

136. Säljö, A. et al. 2008. Neuropathology and pressure in the pig brain resulting from low-impulse noise exposure. *J Neurotrauma* 25(12):1397–1406.
137. Ling, G. et al. 2009. Explosive blast neurotrauma. *J Neurotrauma* 26(6):815–825.
138. Hicks, R.R. et al. 2010. Neurological effects of blast injury. *J Trauma* 68(5): 1257–1263.
139. Long, J.B. et al. 2009. Blast overpressure in rats: Recreating a battlefield injury in the laboratory. *J Neurotrauma* 26(6):827–840.
140. Gyorgy, A. et al. 2011. Time-dependent changes in serum biomarker levels after blast traumatic brain injury. *J Neurotrauma* 28(6):1121–1126.
141. Xiong, Y., Mahmood, A., Chopp, M. 2013. Animal models of traumatic brain injury. *Nat Rev Neurosci* 14(2):128–142.
142. Lighthall, J.W., Goshgarian, H.G., Pinderski, C.R. 1990. Characterization of axonal injury produced by controlled cortical impact. *J Neurotrauma* 7(2):65–76.
143. Shitaka, Y. et al. 2011. Repetitive closed-skull traumatic brain injury in mice causes persistent multifocal axonal injury and microglial reactivity. *J Neuropathol Exp Neurol* 70(7):551–567.
144. Maruichi, K. et al. 2009. Graded model of diffuse axonal injury for studying head injury-induced cognitive dysfunction in rats. *Neuropathology* 29(2):132–139.
145. Marmarou, A. et al. 1994. A new model of diffuse brain injury in rats. Part I: Pathophysiology and biomechanics. *J Neurosurg* 80(2):291–300.
146. Foda, M.A., Marmarou, A. 1994. A new model of diffuse brain injury in rats. Part II: Morphological characterization. *J Neurosurg* 80(2):301–313.
147. Thibault, L.E. et al. 2013. Biomechanical aspects of a fluid percussion model of brain injury. *J Neurotrauma* 9(4):311–322.
148. McIntosh, T.K. et al. 1987. Traumatic brain injury in the rat: Characterization of a midline fluid-percussion model. *Cent Nerv Syst Trauma* 4(2):119–134.
149. Thompson, H.J. et al. 2005. Lateral fluid percussion brain injury: A 15-year review and evaluation. *J Neurotrauma* 22(1):42–75.
150. Wang, H.-C., Ma, Y.-B. 2010. Experimental models of traumatic axonal injury. *J Clin Neurosci* 17(2):157–162.
151. Li, X.Y. et al. 2010. Diffuse axonal injury induced by simultaneous moderate linear and angular head accelerations in rats. *Neuroscience* 169(1):357–369.
152. Li, J. et al. 2012. Exploring temporospatial changes in glucose metabolic disorder, learning, and memory dysfunction in a rat model of diffuse axonal injury. *J Neurotrauma* 29(17):2635–2646.
153. Li, J. et al. 2011. Quantitative evaluation of microscopic injury with diffusion tensor imaging in a rat model of diffuse axonal injury. *Eur J Neurosci* 33(5):933–945.
154. Lusardi, T.A. et al. 2004. Effect of acute calcium influx after mechanical stretch injury in vitro on the viability of hippocampal neurons. *J Neurotrauma* 21(1):61–72.
155. Pfister, B.J. et al. 2003. An in vitro uniaxial stretch model for axonal injury. *Ann Biomed Eng* 31(5):589–598.
156. Mohammed Sulaiman, A. et al. 2011. Stereology and ultrastructure of chronic phase axonal and cell soma pathology in stretch-injured central nerve fibers. *J Neurotrauma* 28(3):383–400.
157. Saatman, K.E., Creed, J., Raghupathi, R. 2010. Calpain as a therapeutic target in traumatic brain injury. *Neurotherapeutics* 7(1):31–42.
158. Saatman, K.E. et al. 2003. Traumatic axonal injury results in biphasic calpain activation and retrograde transport impairment in mice. *J Cereb Blood Flow Metab* 23(1):34–42.
159. Agoston, D.V. et al. 2009. Proteomic biomarkers for blast neurotrauma: Targeting cerebral edema, inflammation, and neuronal death cascades. *J Neurotrauma* 26(6): 901–911.

160. Ankarcrona, M. et al. 1995. Glutamate-induced neuronal death: A succession of necrosis or apoptosis depending on mitochondrial function. *Neuron* 15(4):961–973.
161. Zipfel, G.J. et al. 2000. Neuronal apoptosis after CNS injury: The roles of glutamate and calcium. *J Neurotrauma* 17(10):857–869.
162. Choi, D.W. 1995. Calcium: Still center-stage in hypoxic-ischemic neuronal death. *Trends Neurosci* 18(2):58–60.
163. Serbest, G. et al. 2007. Temporal profiles of cytoskeletal protein loss following traumatic axonal injury in mice. *Neurochem Res* 32(12):2006–2014.
164. Huh, J.W. et al. 2006. Regionally distinct patterns of calpain activation and traumatic axonal injury following contusive brain injury in immature rats. *Dev Neurosci* 28(4–5):466–476.
165. Thompson, S.N. et al. 2006. Relationship of calpain-mediated proteolysis to the expression of axonal and synaptic plasticity markers following traumatic brain injury in mice. *Exp Neurol* 201(1):253–265.
166. Reeves, T.M. et al. 2010. Proteolysis of submembrane cytoskeletal proteins ankyrin-G and αII-spectrin following diffuse brain injury: A role in white matter vulnerability at nodes of Ranvier. *Brain Pathol* 20(6):1055–1068.
167. Farkas, O. et al. 2005. Spectrin breakdown products in the cerebrospinal fluid in severe head injury—Preliminary observations. *Acta Neurochir (Wien)* 147(8):855–861.
168. Brophy, G.M. et al. 2009. alphaII-Spectrin breakdown product cerebrospinal fluid exposure metrics suggest differences in cellular injury mechanisms after severe traumatic brain injury. *J Neurotrauma* 26(4):471–479.
169. Anderson, K.J. et al. 2008. The phosphorylated axonal form of the neurofilament subunit NF-H (pNF-H) as a blood biomarker of traumatic brain injury. *J Neurotrauma* 25(9):1079–1085.
170. Van Geel, W.J.A., Rosengren, L.E., Verbeek, M.M. 2005. An enzyme immunoassay to quantify neurofilament light chain in cerebrospinal fluid. *J Immunol Methods* 296(1–2):179–185.
171. Park, E. et al. 2007. Heavy neurofilament accumulation and α-spectrin degradation accompany cerebellar white matter functional deficits following forebrain fluid percussion injury. *Exp Neurol* 204(1):49–57.
172. Letournel, F. et al. 2006. Neurofilament high molecular weight–green fluorescent protein fusion is normally expressed in neurons and transported in axons: A neuronal marker to investigate the biology of neurofilaments. *Neuroscience* 137(1):103–111.
173. Teunissen, C.E., Dijkstra, C., Polman, C. 2005. Biological markers in CSF and blood for axonal degeneration in multiple sclerosis. *Lancet Neurol* 4(1):32–41.
174. Sihag, R.K. et al. 2007. Role of phosphorylation on the structural dynamics and function of types III and IV intermediate filaments. *Exp Cell Res* 313(10):2098–2109.
175. McCracken, E. et al. 1999. Calpain activation and cytoskeletal protein breakdown in the corpus callosum of head-injured patients. *J Neurotrauma* 16(9):749–761.
176. Hamberger, A. et al. 2003. Redistribution of neurofilaments and accumulation of beta-amyloid protein after brain injury by rotational acceleration of the head. *J Neurotrauma* 20(2):169–178.
177. Chen, X.H. et al. 1999. Evolution of neurofilament subtype accumulation in axons following diffuse brain injury in the pig. *J Neuropathol Exp Neurol* 58(6):588–596.
178. Smith, D.H., Meaney, D.F., Shull, W.H. 2003. Diffuse axonal injury in head trauma. *J Head Trauma Rehabil* 18(4):307–316.
179. Kaur, B., Rutty, G.N., Timperley, W.R. 1999. The possible role of hypoxia in the formation of axonal bulbs. *J Clin Pathol* 52(3):203–209.
180. Chen, Y., Tang, B.L. 2006. The amyloid precursor protein and postnatal neurogenesis/neuroregeneration. *Biochem Biophys Res Commun* 341(1):1–5.

181. Stone, J.R., Singleton, R.H., Povlishock, J.T. 2001. Intra-axonal neurofilament compaction does not evoke local axonal swelling in all traumatically injured axons. *Exp Neurol* 172(2):320–331.
182. Reichard, R.R., Smith, C., Graham, D.I. 2005. The significance of beta-APP immunoreactivity in forensic practice. *Neuropathol Appl Neurobiol* 31(3):304–313.
183. Reichard, R.R. et al. 2003. Beta-amyloid precursor protein staining in nonhomicidal pediatric medicolegal autopsies. *J Neuropathol Exp Neurol* 62(3):237–247.
184. Teasdale, G.M. et al. 1997. Association of apolipoprotein E polymorphism with outcome after head injury. *Lancet* 350(9084):1069–1071.
185. Raby, C.A. et al. 1998. Traumatic brain injury increases beta-amyloid peptide 1-42 in cerebrospinal fluid. *J Neurochem* 71(6):2505–2509.
186. Olsson, A. et al. 2004. Marked increase of beta-amyloid(1-42) and amyloid precursor protein in ventricular cerebrospinal fluid after severe traumatic brain injury. *J Neurol* 251(7):870–876.
187. Zemlan, F.P. et al. 2002. C-tau biomarker of neuronal damage in severe brain injured patients: Association with elevated intracranial pressure and clinical outcome. *Brain Res* 947(1):131–139.
188. Zemlan, F.P. et al. 1999. Quantification of axonal damage in traumatic brain injury: Affinity purification and characterization of cerebrospinal fluid tau proteins. *J Neurochem* 72(2):741–750.
189. Cengiz, P. et al. 2008. Cerebrospinal fluid cleaved-tau protein and 9-hydroxyoctadecadienoic acid concentrations in pediatric patients with hydrocephalus. *Pediatr Crit Care Med* 9(5):524–529.
190. Bodjarian, N. et al. 1997. Strong expression of GFAP mRNA in rat hippocampus after a closed-head injury. *Neuroreport* 8(18):3951–3956.
191. Nylén, K. et al. 2006. Increased serum-GFAP in patients with severe traumatic brain injury is related to outcome. *J Neurol Sci* 240(1–2):85–91.
192. Wiesmann, M. et al. 2010. Outcome prediction in traumatic brain injury: Comparison of neurological status, CT findings, and blood levels of S100B and GFAP. *Acta Neurol Scand* 121(3):178–185.
193. Diaz-Arrastia, R. et al. 2014. Acute biomarkers of traumatic brain injury: Relationship between plasma levels of ubiquitin C-terminal hydrolase-L1 and glial fibrillary acidic protein. *J Neurotrauma* 31(1):19–25.
194. Pelinka, L.E. et al. 2004. GFAP versus S100B in serum after traumatic brain injury: Relationship to brain damage and outcome. *J Neurotrauma* 21(11):1553–1561.
195. Svetlov, S.I. et al. 2013. Biomarkers of blast-induced neurotrauma: Profiling molecular and cellular mechanisms of blast brain injury. *J Neurotrauma* 26(6):913–921.
196. Ottens, A.K. et al. 2008. Proteolysis of multiple myelin basic protein isoforms after neurotrauma: Characterization by mass spectrometry. *J Neurochem* 104(5):1404–1414.
197. Jeter, C.B. et al. 2013. Biomarkers for the diagnosis and prognosis of mild traumatic brain injury/concussion. *J Neurotrauma* 30(8):657–670.
198. Hurley, R.A. et al. 2004. Traumatic axonal injury: Novel insights into evolution and identification. *J Neuropsychiatry Clin Neurosci* 16(1):1–7.
199. de Kruijk, J.R. et al. 2001. S-100B and neuron-specific enolase in serum of mild traumatic brain injury patients. A comparison with health controls. *Acta Neurol Scand* 103(3):175–179.
200. Savola, O., Hillbom, M. 2003. Early predictors of post-concussion symptoms in patients with mild head injury. *Eur J Neurol* 10(2):175–181.
201. Undén, J., Romner, B. 2009. A new objective method for CT triage after minor head injury—Serum S100B. *Scand J Clin Lab Invest* 69(1):13–17.
202. Bazarian, J.J. et al. 2006. Serum S-100B and cleaved-tau are poor predictors of long-term outcome after mild traumatic brain injury. *Brain Inj* 20(7):759–765.

203. Ingebrigtsen, T., Romner, B. 2003. Biochemical serum markers for brain damage: A short review with emphasis on clinical utility in mild head injury. *Restor Neurol Neurosci* 21(3–4):171–176.
204. Kleindienst, A. et al. 2007. The neurotrophic protein S100B: Value as a marker of brain damage and possible therapeutic implications. *Prog Brain Res* 161:317–325.
205. Kleindienst, A., Ross Bullock, M. 2006. A critical analysis of the role of the neurotrophic protein S100B in acute brain injury. *J Neurotrauma* 23(8):1185–1200.
206. Piazza, O. et al. 2007. S100B is not a reliable prognostic index in paediatric TBI. *Pediatr Neurosurg* 43(4):258–264.
207. Nygren de boussard, C. et al. 2004. S100 in mild traumatic brain injury. *Brain Inj* 18(7):671–683.
208. Ohrt-Nissen, S. et al. 2011. How does extracerebral trauma affect the clinical value of S100B measurements? *Emerg Med J* 28(11):941–944.
209. Berger, R., Richichi, R. 2009. Derivation and validation of an equation for adjustment of neuron-specific enolase concentrations in hemolyzed serum. *Pediatr Crit Care Med* 10(2):260–263.
210. Berger, R.P. et al. 2007. Serum biomarker concentrations and outcome after pediatric traumatic brain injury. *J Neurotrauma* 24(12):1793–1801.
211. Papa, L. et al. 2010. Ubiquitin C-terminal hydrolase is a novel biomarker in humans for severe traumatic brain injury. *Crit Care Med* 38(1):138–144.
212. Berger, R.P. et al. 2012. Serum concentrations of ubiquitin C-terminal hydrolase-L1 and αII-spectrin breakdown product 145 kDa correlate with outcome after pediatric TBI. *J Neurotrauma* 29(1):162–167.
213. Murray, G.D. et al. 2007. Multivariable prognostic analysis in traumatic brain injury: Results from the IMPACT study. *J Neurotrauma* 24(2):329–337.
214. MRC CRASH Trial Collaborators et al. 2008. Predicting outcome after traumatic brain injury: Practical prognostic models based on large cohort of international patients. *BMJ* 336(7641):425–429.
215. Tollard, E. et al. 2009. Experience of diffusion tensor imaging and 1H spectroscopy for outcome prediction in severe traumatic brain injury: Preliminary results. *Crit Care Med* 37(4):1448–1455.
216. Lee, H. et al. 2008. Focal lesions in acute mild traumatic brain injury and neurocognitive outcome: CT versus 3T MRI. *J Neurotrauma* 25(9):1049–1056.
217. Laalo, J.P., Kurki, T.J., Tenovuo, O.S. 2014. Interpretation of magnetic resonance imaging in the chronic phase of traumatic brain injury: What is missed in the original reports? *Brain Inj* 28(1):66–70.
218. Provenzale, J.M. 2010. Imaging of traumatic brain injury: A review of the recent medical literature. *AJR Am J Roentgenol* 194(1):16–19.
219. Reider-Groswasser, I. et al. 1997. CT findings in persistent vegetative state following blunt traumatic brain injury. *Brain Inj* 11(12):865–870.
220. Jones, N.R. et al. 1998. Correlation of postmortem MRI and CT appearances with neuropathology in brain trauma: A comparison of two methods. *J Clin Neurosci* 5(1): 73–79.
221. Kim, J.J., Gean, A.D. 2011. Imaging for the diagnosis and management of traumatic brain injury. *Neurotherapeutics* 8(1):39–53.
222. Imaizumi, T. et al. 2003. Dynamics of dotlike hemosiderin spots associated with intracerebral hemorrhage. *J Neuroimaging* 13(2):155–157.
223. Tong, K.A. et al. 2003. Hemorrhagic shearing lesions in children and adolescents with posttraumatic diffuse axonal injury: Improved detection and initial results. *Radiology* 227(2):332–339.
224. Yanagawa, Y. et al. 2000. A quantitative analysis of head injury using T2*-weighted gradient-echo imaging. *J Trauma* 49(2):272–277.

225. Scheid, R. et al. 2003. Diffuse axonal injury associated with chronic traumatic brain injury: Evidence from T2*-weighted gradient-echo imaging at 3 T. *AJNR Am J Neuroradiol* 24(6):1049–1056.
226. Yanagawa, Y. et al. 2009. Relationship between maximum intracranial pressure and traumatic lesions detected by T2*-weighted imaging in diffuse axonal injury. *J Trauma* 66(1):162–165.
227. Luccichenti, G. et al. 2010. 3 Tesla is twice as sensitive as 1.5 Tesla magnetic resonance imaging in the assessment of diffuse axonal injury in traumatic brain injury patients. *Funct Neurol* 25(2):109–114.
228. Viswanathan, A. 2006. Cerebral microhemorrhage. *Stroke* 37(2):550–555.
229. Paterakis, K. et al. 2000. Outcome of patients with diffuse axonal injury: The significance and prognostic value of MRI in the acute phase. *J Trauma* 49(6):1071–1075.
230. Tong, K.A. et al. 2008. Susceptibility-weighted MR imaging: A review of clinical applications in children. *AJNR Am J Neuroradiol* 29(1):9–17.
231. Iwamura, A. et al. 2012. Diffuse vascular injury: Convergent-type hemorrhage in the supratentorial white matter on susceptibility-weighted image in cases of severe traumatic brain damage. *Neuroradiology* 54(4):335–343.
232. Zheng, W.B. et al. 2006. Coma duration prediction in diffuse axonal injury: Analyses of apparent diffusion coefficient and clinical prognostic factors. *Conf Proc IEEE Eng Med Biol Soc* 1:1052–1055.
233. Galloway, N.R. et al. 2008. Diffusion-weighted imaging improves outcome prediction in pediatric traumatic brain injury. *J Neurotrauma* 25(10):1153–1162.
234. Mac Donald, C.L. et al. 2011. Detection of blast-related traumatic brain injury in U.S. military personnel. *N Engl J Med* 364(22):2091–2100.
235. Wang, J.Y. et al. 2008. Diffusion tensor tractography of traumatic diffuse axonal injury. *Arch Neurol* 65(5):619–626.
236. Bendlin, B.B. et al. 2008. Longitudinal changes in patients with traumatic brain injury assessed with diffusion-tensor and volumetric imaging. *Neuroimage* 42(2):503–514.
237. Kou, Z. et al. 2010. The role of advanced MR imaging findings as biomarkers of traumatic brain injury. *J Head Trauma Rehabil* 25(4):267–282.
238. Oouchi, H. et al. 2007. Diffusion anisotropy measurement of brain white matter is affected by voxel size: Underestimation occurs in areas with crossing fibers. *AJNR Am J Neuroradiol* 28(6):1102–1106.
239. Sarlls, J.E., Pierpaoli, C. 2009. In vivo diffusion tensor imaging of the human optic chiasm at sub-millimeter resolution. *Neuroimage* 47(4):1244–1251.
240. Tang, C.Y. et al. 2012. Diffuse disconnectivity in TBI: A resting state fMRI and DTI study. *Transl Neurosci* 3(1):9–14.
241. Zhang, K. et al. 2010. Are functional deficits in concussed individuals consistent with white matter structural alterations: Combined FMRI & DTI study. *Exp Brain Res* 204(1):57–70.
242. Zhang, Y. et al. 2010. Atlas-guided tract reconstruction for automated and comprehensive examination of the white matter anatomy. *Neuroimage* 52(4):1289–1301.
243. Jensen, J.H., Helpern, J.A. 2010. MRI quantification of non-Gaussian water diffusion by kurtosis analysis. *NMR Biomed* 23(7):698–710.
244. Lu, H. et al. 2006. Three-dimensional characterization of non-Gaussian water diffusion in humans using diffusion kurtosis imaging. *NMR Biomed* 19(2):236–247.
245. Tuch, D.S. et al. 2002. High angular resolution diffusion imaging reveals intravoxel white matter fiber heterogeneity. *Magn Reson Med* 48(4):577–582.
246. Wedeen, V.J. et al. 2005. Mapping complex tissue architecture with diffusion spectrum magnetic resonance imaging. *Magn Reson Med* 54(6):1377–1386.
247. Shin, S.S. et al. 2012. High-definition fiber tracking for assessment of neurological deficit in a case of traumatic brain injury: Finding, visualizing, and interpreting small sites of damage. *J Neurosurg* 116(5):1062–1069.

248. Chan, J.H.M. et al. 2003. Diffuse axonal injury: Detection of changes in anisotropy of water diffusion by diffusion-weighted imaging. *Neuroradiology* 45(1):34–38.
249. Lee, Z.I. et al. 2003. Diffusion tensor magnetic resonance imaging of microstructural abnormalities in children with brain injury. *Am J Phys Med Rehabil* 82(7):556–559.
250. Jang, S.H. 2011. Diffusion tensor imaging studies on corticospinal tract injury following traumatic brain injury: A review. *NeuroRehabilitation* 29(4):339–345.
251. Arfanakis, K. et al. 2002. Diffusion tensor MR imaging in diffuse axonal injury. *AJNR Am J Neuroradiol* 23(5):794–802.
252. Bazarian, J.J. et al. 2007. Diffusion tensor imaging detects clinically important axonal damage after mild traumatic brain injury: A pilot study. *J Neurotrauma* 24(9): 1447–1459.
253. Miles, L. et al. 2008. Short-term DTI predictors of cognitive dysfunction in mild traumatic brain injury. *Brain Inj* 22(2):115–122.
254. Mayer, A.R. et al. 2010. A prospective diffusion tensor imaging study in mild traumatic brain injury. *Neurology* 74(8):643–650.
255. Hulkower, M.B. et al. 2013. A decade of DTI in traumatic brain injury: 10 Years and 100 articles later. *AJNR Am J Neuroradiol* 34(11):2064–2074.
256. Marquez de la Plata, C.D. et al. 2010. Diffusion tensor imaging biomarkers for traumatic axonal injury: Analysis of three analytic methods. *J Int Neuropsychol Soc* 17(1):24–35.
257. Mori, S. et al. 2013. Atlas-based neuroinformatics via MRI: Harnessing information from past clinical cases and quantitative image analysis for patient care. *Annu Rev Biomed Eng* 15(1):71–92.
258. Faria, A.V. et al. 2010. Atlas-based analysis of neurodevelopment from infancy to adulthood using diffusion tensor imaging and applications for automated abnormality detection. *Neuroimage* 52(2):415–428.
259. Paus, T. et al. 2001. Maturation of white matter in the human brain: A review of magnetic resonance studies. *Brain Res Bull* 54(3):255–266.
260. Metting, Z. et al. 2007. Structural and functional neuroimaging in mild-to-moderate head injury. *Lancet Neurol* 6(8):699–710.
261. Hunter, J.V. et al. 2012. Emerging imaging tools for use with traumatic brain injury research. *J Neurotrauma* 29(4):654–671.
262. Bagley, L.J. et al. 2000. Magnetization transfer imaging of traumatic brain injury. *J Magn Reson Imaging* 11(1):1–8.
263. McGowan, J.C. et al. 1999. Diffuse axonal pathology detected with magnetization transfer imaging following brain injury in the pig. *Magn Reson Med* 41(4):727–733.
264. McGowan, J.C. et al. 2000. Magnetization transfer imaging in the detection of injury associated with mild head trauma. *AJNR Am J Neuroradiol* 21(5):875–880.
265. Duckworth, J.L., Stevens, R.D. 2010. Imaging brain trauma. *Curr Opin Crit Care* 16(2):92–97.
266. Mamere, A.E. et al. 2009. Evaluation of delayed neuronal and axonal damage secondary to moderate and severe traumatic brain injury using quantitative MR imaging techniques. *AJNR Am J Neuroradiol* 30(5):947–952.
267. Sinson, G. et al. 2001. Magnetization transfer imaging and proton MR spectroscopy in the evaluation of axonal injury: Correlation with clinical outcome after traumatic brain injury. *AJNR Am J Neuroradiol* 22(1):143–151.
268. Asikainen, I., Kaste, M., Sarna, S. 1999. Early and late posttraumatic seizures in traumatic brain injury rehabilitation patients: Brain injury factors causing late seizures and influence of seizures on long-term outcome. *Epilepsia* 40(5):584–589.
269. Englander, J. et al. 2003. The association of early computed tomography scan findings and ambulation, self-care, and supervision needs at rehabilitation discharge and at 1 year after traumatic brain injury. *Arch Phys Med Rehabil* 84(2):214–220.

270. Sidaros, A. et al. 2008. Diffusion tensor imaging during recovery from severe traumatic brain injury and relation to clinical outcome: A longitudinal study. *Brain* 131(2):559–572.
271. Porto, L. et al. 2011. Morphometry and diffusion MR imaging years after childhood traumatic brain injury. *Eur J Paediatr Neurol* 15(6):493–501.
272. Marquez de la Plata, C. et al. 2007. Magnetic resonance imaging of diffuse axonal injury: Quantitative assessment of white matter lesion volume. *J Neurotrauma* 24(4):591–598.
273. Strangman, G.E. et al. 2010. Regional brain morphometry predicts memory rehabilitation outcome after traumatic brain injury. *Front Hum Neurosci* 4:182.
274. Kinnunen, K.M. et al. 2011. White matter damage and cognitive impairment after traumatic brain injury. *Brain* 134(2):449–463.
275. Irimia, A. et al. 2012. Neuroimaging of structural pathology and connectomics in traumatic brain injury: Toward personalized outcome prediction. *Neuroimage Clin* 1(1):1–17.
276. Belanger, H.G. et al. 2007. Recent neuroimaging techniques in mild traumatic brain injury. *J Neuropsychiatry Clin Neurosci* 19(1):5–20.
277. Hillary, F.G. et al. 2002. Functional magnetic resonance imaging technology and traumatic brain injury rehabilitation: Guidelines for methodological and conceptual pitfalls. *J Head Trauma Rehabil* 17(5):411–430.
278. Sanchez-Carrion, R. et al. 2008. Frontal hypoactivation on functional magnetic resonance imaging in working memory after severe diffuse traumatic brain injury. *J Neurotrauma* 25(5):479–494.
279. Ueno, H. et al. 2009. Brain activations in errorless and errorful learning in patients with diffuse axonal injury: A functional MRI study. *Brain Inj* 23(4):291–298.
280. Palacios, E.M. et al. 2013. Resting-state functional magnetic resonance imaging activity and connectivity and cognitive outcome in traumatic brain injury. *JAMA Neurol* 70(7):845.
281. Christodoulou, C. et al. 2001. Functional magnetic resonance imaging of working memory impairment after traumatic brain injury. *J Neurol Neurosurg Psychiatry* 71(2):161–168.
282. Luo, Q. et al. 2013. Complexity analysis of resting state magnetoencephalography activity in traumatic brain injury patients. *J Neurotrauma* 30(20):1702–1709.
283. Huang, M.-X. et al. 2009. Integrated imaging approach with MEG and DTI to detect mild traumatic brain injury in military and civilian patients. *J Neurotrauma* 26(8):1213–1226.
284. Gloor, P., Ball, G., Schaul, N. 1977. Brain lesions that produce delta waves in the EEG. *Neurology* 27(4):326–333.
285. Gaetz, W., Otsubo, H., Pang, E.W. 2008. Magnetoencephalography for clinical pediatrics: The effect of head positioning on measurement of somatosensory-evoked fields. *Clin Neurophysiol* 119(8):1923–1933.
286. Bernabeu, M. et al. 2009. Abnormal corticospinal excitability in traumatic diffuse axonal brain injury. *J Neurotrauma* 26(12):2185–2193.
287. Yasokawa, Y.-T. et al. 2007. Correlation between diffusion-tensor magnetic resonance imaging and motor-evoked potential in chronic severe diffuse axonal injury. *J Neurotrauma* 24(1):163–173.
288. Signoretti, S. et al. 2004. The protective effect of cyclosporin A upon N-acetylaspartate and mitochondrial dysfunction following experimental diffuse traumatic brain injury. *J Neurotrauma* 21(9):1154–1167.
289. Colley, B.S., Phillips, L.L., Reeves, T.M. 2010. The effects of cyclosporin-A on axonal conduction deficits following traumatic brain injury in adult rats. *Exp Neurol* 224(1):241–251.

290. Suehiro, E. et al. 2001. The immunophilin ligand FK506 attenuates the axonal damage associated with rapid rewarming following posttraumatic hypothermia. *Exp Neurol* 172(1):199–210.
291. Fujita, M. et al. 2011. The combination of either tempol or FK506 with delayed hypothermia: Implications for traumatically induced microvascular and axonal protection. *J Neurotrauma*.
292. DiLeonardi, A.M., Huh, J.W., Raghupathi, R. 2012. Differential effects of FK506 on structural and functional axonal deficits after diffuse brain injury in the immature rat. *J Neuropathol Exp Neurol* 71(11):959–972.
293. Schouten, J.W. et al. 2004. A review and rationale for the use of cellular transplantation as a therapeutic strategy for traumatic brain injury. *J Neurotrauma* 21(11):1501–1538.
294. Harting, M.T. et al. 2008. Cell therapies for traumatic brain injury. *Neurosurg Focus* 24(3–4):E18.
295. Salman, H., Ghosh, P., Kernie, S.G. 2004. Subventricular zone neural stem cells remodel the brain following traumatic injury in adult mice. *J Neurotrauma* 21(3):283–292.
296. Riess, P. et al. 2007. Embryonic stem cell transplantation after experimental traumatic brain injury dramatically improves neurological outcome, but may cause tumors. *J Neurotrauma* 24(1):216–225.
297. Kulbatski, I. et al. 2008. Glial precursor cell transplantation therapy for neurotrauma and multiple sclerosis. *Prog Histochem Cytochem* 43(3):123–176.
298. Seledtsov, V.I. et al. 2006. Cell therapy of comatose states. *Bull Exp Biol Med* 142(1):129–132.
299. Velly, L. et al. 2010. Erythropoietin 2nd cerebral protection after acute injuries: A double-edged sword? *Pharmacol Ther* 128(3):445–459.
300. Mammis, A., McIntosh, T.K., Maniker, A.H. 2009. Erythropoietin as a neuroprotective agent in traumatic brain injury: Review. *Surg Neurol* 71(5):527–531; discussion 531.
301. Hellewell, S.C. et al. 2013. Erythropoietin improves motor and cognitive deficit, axonal pathology, and neuroinflammation in a combined model of diffuse traumatic brain injury and hypoxia, in association with upregulation of the erythropoietin receptor. *J Neuroinflammation* 10:156.
302. Yatsiv, I. et al. 2005. Erythropoietin is neuroprotective, improves functional recovery, and reduces neuronal apoptosis and inflammation in a rodent model of experimental closed head injury. *FASEB J* 19(12):1701–1703.
303. Talving, P. et al. 2010. Erythropoiesis stimulating agent administration improves survival after severe traumatic brain injury: A matched case control study. *Ann Surg* 251(1):1–4.
304. Corwin, H.L. et al. 2002. Efficacy of recombinant human erythropoietin in critically ill patients: A randomized controlled trial. *JAMA* 288(22):2827–2835.
305. Corwin, H.L. et al. 2007. Efficacy and safety of epoetin alfa in critically ill patients. *N Engl J Med* 357(10):965–976.
306. Yamamoto, M. et al. 2004. Effect of erythropoietin on nitric oxide production in the rat hippocampus using in vivo brain microdialysis. *Neuroscience* 128(1):163–168.
307. Tubbs, R.S. et al. 2007. Does the neuroprotective agent erythropoietin amplify diffuse axonal injury in its early stages? *Med Hypotheses* 69(6):1385–1386.
308. Mills, J.D. et al. 2011. Omega-3 fatty acid supplementation and reduction of traumatic axonal injury in a rodent head injury model. *J Neurosurg* 114(1):77–84.
309. Bailes, J.E., Mills, J.D. 2010. Docosahexaenoic acid reduces traumatic axonal injury in a rodent head injury model. *J Neurotrauma* 27(9):1617–1624.

4 Blood–Brain Barrier Pathophysiology following Traumatic Brain Injury

Lulit Price, Christy Wilson, and Gerald Grant

CONTENTS

In the past, the neuron has been the focal point of brain injury research. However, in recent years, the pathophysiology of the blood–brain barrier (BBB) has become increasingly more studied. The BBB's primary role is to create a restrictive barrier between the central nervous system and the rest of the body, barring the entry of unwanted blood-borne factors. The BBB's microvessels are made up of endothelial cells linked by tight junctions, but neighboring glial cells such as astrocytes and microglial are also critical to BBB function. These glial cells physically interact with endothelial cells, communicating via paracrine signaling. Together, this special group of interacting cells is known as the neurovascular unit (NVU).[1]

BLOOD–BRAIN BARRIER: THE NEUROVASCULAR UNIT

The NVU is defined in the literature as "a physiologic entity that is structurally defined by interactions occurring between endothelial cells, pericytes, smooth muscle cells, astrocytes, and neurons."[2] Its foundation is composed of the basal lamina deep to the endothelial cell monolayer; the latter cells are interspersed with tight junction protein complexes that regulate paracellular transport. Transporters and receptors on the luminal and basolateral surfaces of these endothelial cells are also important to the NVU as they mediate solute traffic and aid in the communication between various cells. Other cells included in the neurovascular unit are pericytes, smooth muscle cells, circulating white blood cells, microglia, and astrocytes, whose end-foot processes line most of the vascular walls and some synapses.[3]

Recent reviews in the literature have divided the BBB into three functional barriers based on their major functions: physical, transport, and metabolic.[4] Protein complexes between cells called adherens junctions and tight junctions compose the "physical" barrier that prevents paracellular diffusion. This ensures that substances either remain in the bloodstream to exit the central nervous system (CNS) or diffuse transcellularly into the CNS. Adherens junctions not only join endothelial cells but also contribute to the proper formation of tight junctions. Tight junctions are composed of claudins, occludins, and other transmembrane proteins held together by zona occludens proteins that scaffold these complexes to the cytoskeleton. The "transport" barrier is composed of transport proteins, receptors for endocytosis, and ATP-binding proteins on the endothelial cells that promote the influx of nutrients and ions and the efflux of toxins. Finally, the "metabolic" barrier encompasses enzymes such as P450 and monoamine oxidase in order to metabolize molecules that enter endothelial cells.[4,5] These three major functions of the blood–brain barrier demonstrate that the NVU is both structurally and physiologically dynamic.

TRAUMATIC BRAIN INJURY: PRIMARY AND SECONDARY INJURY

There are two general stages in the development of brain injury: primary damage and secondary damage. These stages are also known as immediate and delayed dysfunction. Traumatic brain injury (TBI) is a direct result of direct or indirect mechanical forces on the brain that results in acute changes such as shearing injuries, contusions, and hematomas. These forces on the brain all encompass the *primary injury*. Vascular and parenchymal damage in the brain contribute to what is known in the literature as BBB breakdown. Once this barrier is disrupted, secondary changes such as edema, inflammation, and hyperexcitability often occur. Several negative effects occur, setting off a cascade of additional complications that begin hours to days after injury. Some of the delayed responses to TBI include neovascularization, impaired cerebral blood flow, glial cell dysfunction, and cell degeneration, and all of these pathophysiological events are vulnerable to secondary injury (i.e., hypoxia, hypotension). A TBI in a patient may result in coma or death, seizures, or cognitive and behavioral disabilities. This secondary damage is the focus of current TBI research.[6] BBB breakdown and its downstream effects, namely edema, will be the focus of this chapter.

TBI AND THE BBB

What do the BBB and the NVU have to do with TBI? There is a consensus in the literature that the BBB is an integral part of the neurovascular unit. It maintains homeostasis in the brain by regulating uptake of molecules and helps regulate cerebral blood flow. Traumatic brain injury causes primary damage to the brain parenchyma, damaging cerebral microvessels and leads to NVU pathophysiology.

The secondary damage from TBI begins with and revolves around the cerebral vasculature. From animal TBI models, we know that trauma disrupts the structural and physiological integrity of vessels. Disruption of the walls of microvessels in the BBB activates the coagulation cascade. Intravascular coagulation leads to ischemia in the areas surrounding the impact site resulting in severely decreased blood flow; this is known as the "no-reflow" phenomenon.[7–10] Since the integrity of the BBB is compromised after injury, blood-borne factors such as thrombin, albumin, and fibrinogen, can now enter the brain. These factors are thought by some to cause microglia activation, proliferation, and its production of pro-inflammatory factors.[11]

It is important to note that although the BBB is often referred to as open or closed, it is a dynamic structure and it can change between these states depending on what proteins are expressed. Permeability of the BBB is largely regulated by the expression of certain tight junction proteins on microvessel endothelial cells. TBI disrupts expression of tight junction proteins as well as interactions between the NVU components. BBB opening can be assessed via IgG or Evans blue extravasation from the vessels. One study showed that TBI promotes BBB opening as early as the day after injury, while others show that the BBB can remain open for as long as 30 days.[12,13] Another study showed high levels of IgG extravasation near the injury site at 1 and 3 days postinjury but much lower levels at 7 days in a controlled cortical impact TBI model with juvenile rats. No extravasation was seen at 60 and 180 days from injury.[5] MRI studies using a diffuse closed head injury model have shown that the initial BBB opening occurs as early 30 minutes postinjury.[14]

BBB permeability can also be assessed by measuring the expression of tight junction proteins such as claudin-5. Claudin-5 plays an important role in regulating paracellular transport in the BBB and is highly expressed on the brain microvascular endothelium. Its level of expression on the BBB is known to fluctuate postinjury.[15] It is also well known from rat TBI models that BBB permeability to albumin and other high molecular weight molecules are biphasic, peaking at 4 to 6 hours and 2 to 3 days after injury.[16–19] It is thought that the first peak in permeability across the BBB is due to an influx of neutrophils and an increase in factors/molecules that lead to BBB dysfunction. The cause of the second peak in BBB permeability is currently unknown but is hypothesized to be a component of the brain's response to injury.[6] Experiments repeatedly demonstrate that claudin-5 expression then increases 1 to 2 weeks after injury and can remain elevated as much as 4 to 8 weeks after initial injury.[15,20]

ROLE OF BLOOD-BORNE FACTORS THAT AFFECT BBB FUNCTION FOLLOWING TBI

In the normal brain, the neurovascular unit interacts with glial cells and vascular endothelial cells via paracrine signaling. After injury, astrocytes and microglial quickly react by producing and releasing a myriad of molecules that affect the communication between the different components of the BBB. Some of the molecules that contribute to the dysfunction of the BBB include TGF-beta, reactive oxygen species (ROS), VEGF, MMPs, and glutamate. Their presence and abnormally high levels in the brain lead to the disruption of BBB integrity primarily by destroying tight junction proteins or decreasing their expression. This then leads to increased BBB permeability, facilitating the development of cerebral edema and other secondary changes. We will discuss each of these factors individually, presenting evidence for their role in the increased BBB permeability we observe after TBI.

There is evidence that matrix metalloproteinases (MMPs) and VEGF lead to increased BBB permeability via the destruction of tight junction proteins. MMPs are produced by many cerebrovascular endothelial cells, astrocytes, microglia, neurons, and leukocytes especially isoforms 2, 3, and 9. They directly target and degrade basal lamina and tight junction proteins on the BBB, promoting vasogenic edema.[21–23] Brain tissue from TBI patients show increased synthesis and elevated levels of MMP 2 and 9 in plasma and cerebrospinal fluid (CSF).[24–26] Furthermore, deletion of the mmp9 gene in TBI mice resulted in a decrease in brain damage and improved function.[27,28] Tissue inhibitors of MMPs, endogenous regulators of MMPs, have been found to be decreased in animal models of TBI. VEGFA is a member of the VEGF family that has been associated with increased vascular permeability to low molecular weight molecules. It is also an important mitogen for vascular endothelial cells. Astrocyte upregulation of VEGFA synthesis only a few hours after injury was observed in brain tissue from TBI patients and animals. In turn, VEGFA downregulates claudin-5 expression, resulting in BBB "opening."[29–32]

Oxidative stress is now known to be a post-TBI event that produces molecules that can lead to BBB dysfunction. Traumatic injury to the brain promotes lipid peroxidation of cell membranes through which certain reactive oxygen species (ROS) are produced. There is evidence that an ROS called 4-HNE significantly increases BBB permeability.[33] Normal cells use endogenous molecules like glutathione to protect themselves from ROS. One study showed that glutathione depletion leads to a significant increase in paracellular BBB permeability to low molecular weight markers.[34] Another study demonstrated that claudin-5 and occludin actually get redistributed and degraded when brain endothelial monolayers were exposed to a mixture of ROS.[35] Adding to the list of molecules that contribute to BBB permeability after TBI is glutamate that gets released from parenchymal brain cells in large amounts. By binding to its mGluR receptor, it increases endothelial monolayer permeability in vitro. Interestingly, these two molecules' actions may be linked as there is some evidence that glutamate increases ROS production to promote apoptosis.[36]

TGF-beta is a molecule with very important cell functions such as proliferation and differentiation and it is produced in many cell types, including platelets, astrocytes, and microglia. Large amounts of TGF-beta get released from platelets in the latent form after vascular wall damage. Cryogenic brain injury studies have shown that there is an increase in both TGF-beta levels after injury and in the expression of TGF-beta receptors on vascular endothelium.[37–39] There are conflicting studies about whether TGF-beta really increases BBB permeability. Some believe that TGF-beta increases tyrosine phosphorylation to reduce claudin-5 and VE-cadherin expression; others believe that it plays a role in maintaining the barrier as it helps stabilize endothelial cell and pericyte interaction via N-cadherin.[40–42]

Paradoxically, many of these molecules initially destroy the BBB after TBI but are later involved in repair processes. Although they seem like good therapeutic targets, blocking their action in the acute phase could lead to problems at a later time. For example, experiments show that glutamate stimulates the activity of heme oxygenase, a protectant against glutamate toxicity, in endothelial cells.[36,43]

IMPORTANCE IN DEVELOPMENT OF CEREBRAL EDEMA FOLLOWING TBI

Cerebral edema is a very important secondary consequence of TBI. It is crucial to understand the development of cerebral edema because it is one of the main factors affecting the morbidity and mortality of TBI patients. Post-traumatic cerebral edema leads to the expansion of brain volume against an enclosed skull and an increase in intracranial pressure (ICP). Not only can increased ICP cause herniation, but it also decreases cerebral perfusion pressure, promoting cerebral ischemia in a brain with already tenuous blood supply.[6,44–47]

As outlined earlier, we know that BBB permeability changes occur after TBI. Cerebral edema results from a combination of endothelial cell damage, tight junction disruption, and abnormal transcellular transport.[48] It is important to note that there are two types of edema that can come about, depending on the type of injury. The majority of TBI and blast injuries result in predominantly *vasogenic edema*—a pathological increase in vascular permeability, in this case, due to vessel damage. This leads to interstitial accumulation of plasma-derived, osmotically active molecules/solutes like plasma proteins followed by water.[1] Ischemic brain injuries tend to result in *cytotoxic edema*, which is caused by changes in cell metabolism and the failure of membrane associated pumps and ion transporters. This results in cellular accumulation of osmotically active molecules followed by water.[1]

There is some evidence that aquaporins may also play a role in the development of post-CNS injury edema, but results from studies have been conflicting. There are many isoforms of this special water transporter but AQP4 is the one most studied. It is located on astrocyte end-foot processes and is believed to play a role in water homeostasis in the NVU.[49] There have not been many studies with TBI animal models so the role of AQP4 remains unclear. However, in summary, AQP4 seems to aid in cytotoxic edema but helps resolve vasogenic edema after the initial opening of the BBB after injury.[44]

STRATEGIES TO REDUCE CEREBRAL EDEMA AFTER TBI

The traditional approach/first-line treatment to treating cerebral edema is to use mannitol, an osmotic diuretic that draws water out of cells and into the vasculature to be excreted by the body. Neurologic and renal side effects limit its repeated administration. Moreover, it is only effective for short periods of time and mortality can continue to increase if brain swelling continues past 24 hours. Hypertonic saline, decompressive craniotomy, and external ventricular drainage are the only other options for decreasing cerebral edema.[48] Steroids, despite their anti-inflammatory effects, have not been shown to be effective in reducing cerebral edema due to TBI. Thus, more effective and safe methods for decreasing edema are needed in order to prevent adverse outcomes from TBI.

According to current TBI management guidelines, decreasing intracranial pressure while optimizing cerebral perfusion pressure caused by TBI can be accomplished using mannitol or hypertonic saline (HS). Mannitol works by expanding blood plasma and drawing water into the vessels to improve cerebral blood flow.[50] HS decreases cerebral water content by creating a gradient across the BBB, shifting water from brain cells to the vasculature.[50] As a result, intravascular volume increases, improving blood flow. Historically, mannitol has been used more frequently than HS, although more recently 3% or 23% hypertonic saline has been used, particularly in combat casualty care. There have been numerous studies attempting to tease out which therapy is more beneficial, but results thus far are conflicting and inconclusive. Both therapies shift fluid from extravascular compartments into the vasculature but hypertonic saline expands intravascular volume while mannitol decreases it by promoting diuresis. Surprisingly, they both have anti-inflammatory properties such as decreasing neutrophil–endothelial cell interactions. A recent study performed by Marks et al. attempted to find a difference between mannitol and hypertonic saline in their effect on microvascular permeability.[51] Using a simulated TBI model involving exogenous IL-1beta, they did not find a significant difference in the ability of either therapy to reduce neutrophil–endothelial interactions.[52] Furthermore, when the BBB is compromised following TBI, these agents may not be as effective since the osmotic gradient may no longer exist in severe areas of injury.

Recent studies have shown that by transiently opening the BBB after TBI, it is possible to decrease post-traumatic cerebral edema. Using a cold-induced cortical impact model to produce focal injury, researchers have demonstrated paradoxically that a transient and size-selective modulation in the BBB promotes enhanced movement of water from the brain parenchyma to the blood vessels, decreasing brain swelling.[48] Campbell et al. have shown that siRNA knockdown of claudin-5 increases the efflux of low molecular weight molecules (less than 1 kDa) across the barrier between 24 and 72 hours after injection of siRNA. It has been postulated that the extracellular water follows these molecules and in turn decreases vasogenic edema.[48] To evaluate larger molecules, the extravasation of Evan's blue dye in the parenchyma, which strongly binds to albumin (70 kDa), was grossly observed. There was a significant decrease in the amount of Evan's blue dye in the peri-injury area in the mice that received claudin-5 targeted siRNA compared to their nontargeted counterparts. The claudin-5 targeted siRNA group also had a significant decrease in their percentage water content as well as the lesion volume 72 hours postinjury.[48]

NOVEL THERAPEUTIC STRATEGIES TO STABILIZE BBB AFTER TBI

Secondary neuronal injury after TBI may be exacerbated by BBB dysfunction. The ideal strategy would be to prevent BBB breakdown and stabilize it, thereby protecting the brain from factors released from damaged blood vessels that cause further damage. Inflammation, excitotoxicity, oxidative stress, edema, and neuronal injury are all components of secondary neuronal injury after the primary damage caused by TBI. Attempts have been made at curtailing the effects of two endogenous mediators in the formation of brain edema: bradykinin and VEGF (vascular endothelial growth factor). Bradykinin is a pro-inflammatory peptide that promotes vasodilation and vascular permeability, and antagonists at its receptor decrease cerebral edema in animals and humans.[53,54] VEGF, a promoter of angiogenesis, increases vascular permeability in hypoxic conditions by modulating tight junction proteins.[55] Studies have shown that blocking VEGF decreases BBB permeability in vivo and that corticosteroids may regulate VEGF expression in the midst of BBB injury.

Numerous hormones including ghrelin, neural growth factor, and progesterone have recently been found to have neuroprotective effects following traumatic brain injury and all three have an effect on the BBB or water flux across the BBB.

Ghrelin is a hormone known for its role in regulating hunger and satiety, but its newly discovered neuroprotective effects are thought to be mediated by maintaining oxidant/anti-oxidant balance, inhibiting pro-inflammatory mediators, and inhibiting neuronal apoptosis.[56] Ghrelin is believed to inhibit apoptosis by upregulating a mitochondrial uncoupling protein called UCP-2, stabilizing mitochondria. Experiments with TBI, stroke, and Parkinson's animal models demonstrate that higher levels of UCP-2 are neuroprotective as this results in decreased activation of caspase-3, an apoptotic protein.[57] Lopez et al. also showed that treating mice with neuropeptide ghrelin preceding and 1 hour after TBI resulted in decreased BBB permeability 6 hours after TBI. Not only did ghrelin treatment significantly decrease BBB permeability, measured by 70kD FITC-dextran extravasation 24 hours after injury, but it also significantly decreased post-TBI apoptosis of injured tissue. This group also showed that this decrease in vascular permeability was linked to a decrease in perivascular aquaporin-4 (AQP-4) expression after ghrelin treatment.[58] This is a significant finding because perivascular levels of AQP-4 receptors increase significantly 6 hours after TBI. AQP-4 receptors are preferentially located on astrocyte end-foot processes adjacent to cerebral microvessels and play a role in the formation of cerebral edema.[59] Their work with ghrelin thus far not only suggests a link between BBB breakdown and apoptosis of injured tissue but also that decreased BBB permeability after TBI limits tissue injury.[57,58]

A hormone called *nerve growth factor* (NGF) is currently being studied for its potential to reduce cellular cerebral edema in mice after TBI via downregulation of AQP-4. NGF was delivered intranasally to bypass the BBB in order to preferentially distribute it in the brain, limiting systemic side effects. Cerebral water content measured at 12, 24, and 72 hours post-TBI coincided with reduced expression of AQP-4.[60]

Preliminary studies involving *progesterone* demonstrates that it too may have neuroprotective effects.[61,62] Not only does cranial edema result from increased BBB permeability after TBI but also from inflammation and secondary brain injury that persists hours later.[1] The latter process is mediated by neutrophil and endothelial cell activation in the microcirculation, leading to microvascular breakdown. Neutrophils are an important part of the inflammatory host response, exerting cytotoxic and phagocytic effects on tissues. However, excessive or inappropriate activation of these cells can lead to unwanted destruction of the microvasculature and surrounding tissue, and in turn exacerbating cerebral edema.[64,65] Progesterone, an important steroid hormone, has been found to inhibit neuronal apoptosis and decrease free radical and cytokine production.[65] Pascaul et al. found that treating TBI rats with progesterone decreased neutrophil rolling and adhesion to the endothelium of the BBB, and decreased cerebral vascular leakage and edema 36 hours after trauma.[65] Data from other studies suggests that neutrophils are recruited to the BBB 12 to 36 hours after TBI, creating a window of opportunity to intervene and attenuate this aspect of secondary brain injury.[66,67] Progesterone's effect on neutrophils is thought to decrease polymorphonuclear (PMN) respiratory burst and oxygen radials as well as decreased PMN adhesion receptor expression. Its effect on the endothelial cells has not yet been studied. A large multicenter RTCs Phase III clinical trial, ProTECT III, just completed investigated progesterone as a potential treatment to improve survival and cognitive recovery post-TBI.[65] Unfortunately the trial was negative.

Tchantchou and Zhang found that inhibiting the degradation of the most abundant *endocannabinoid* in the brain results in decreased BBB dysfunction, brain edema, lesion volume, neuronal death, and better behavioral performance after TBI.[68–72] While these neuroprotective effects of exogenous cannabinoids have been known for some time, psychotropic side effects from global activation of the CB1 receptor make them a poor candidate for treatment in humans. However, new research suggests that it is possible to augment the protective effects of endocannabinoids naturally produced after injury. Endocannabinoids would, instead, activate CB1 and CB2 receptors, expressed on neural cell and inflammatory cells, in specific sites and after a specific event (TBI), minimizing undesirable effects. 2-Arachidonylglycerol (2-AG), the endocannabinoid of interest, increases immediately after TBI.[68] It is known to be neuroprotective due its anti-oxidant, anti-inflammatory, and anti-excitotoxic effects.[68,73,74] Benefits are short-lived due to 2-AG's rapid degradation by the enzymes monoacylglycerol (MAGL) and alpha/beta hydroxyl domain 6 (ABHD6).[75] In this study, mice underwent controlled cortical impact TBI and were given an ABHD6 inhibitor. On a gross level, animals that received the enzyme inhibitor exhibited decreased motor coordination and fine motor deficits. TBI-induced brain lesions were also significantly smaller at 1 and 3 weeks after injury in mice that received the inhibitor via cannabinoid receptior-1 (CB1) blockage. Microscopically, the inhibitor also decreased TBI-induced BBB permeability measured by Evans-blue extravasation and ICAM-1 immunoreactivity. The inhibitor also decreased pro-inflammatory molecules iNOS and COX2 that are usually expressed in glial cells after injury. Additionally, these authors found that chronic administration of the inhibitor reversed the increase in lesional volume that occurs post-TBI. In conclusion, enhancing the action of endocannabinoids in the brain have the potential to decrease BBB breakdown as well as secondary neuronal injury.

STEM CELLS AND BBB RESCUE FOLLOWING TBI

An innovative study by Menge et al. suggests that mesenchymal stem cells (MSCs) may mitigate BBB breakdown following TBI.[76] Human mesenchymal stem cells produce a molecule called tissue inhibitor of matrix metalloproteinase-3 (TIMP3), an inhibitor of soluble MMP. Using a TBI mouse model, they found that recombinant TIMP3 inhibits BBB permeability in TBI mice by blocking VEGF-A mediated breakdown of endothelial cell adherens and tight junctions in vitro and in vivo. The proposed mechanism involves the fact that 90% of intravenously delivered MSCs travel to the lungs, more specifically the pulmonary endothelium. It is believed that interaction between the stem cells and pulmonary endothelial cells promote production of TIMP3 by both cells. Systemic spread of TIMP3 allows for it to stabilize distant vasculature beds such as the BBB. Experiments by Menge et al. demonstrate that TIMP3 inhibits receptor-mediated VEGF-A binding to the VEGFR2 receptor, preventing vascular permeability. Additionally, recombinant TIMP3 increased adherens junctions in the mouse brain after TBI and downregulation of TIMP3 in TBI mice did not result in this protective phenomenon. This study adds to the growing body of work that human stem cells have therapeutic implications for diseases characterized by vascular instability.[76]

CONCLUSIONS

The pathophysiology of the cascade of events that encompass secondary brain injury following TBI is complex. BBB disruption is a pathological hallmark of severe TBI and is associated with neuroinflammatory events contributing to brain edema and cell death. There is a strong temporal and spatial association between the degree of BBB disruption and the ability of circulating inflammatory cells to quickly migrate to the area of CNS injury. The BBB, however, may also be protective by promoting efflux of water across the BBB to mitigate cerebral edema after severe TBI. The therapeutic window of BBB modulation after TBI remains unknown and is further complicated by the biphasic disruption of the BBB following TBI. Further elucidation of the dynamics of BBB dysfunction after TBI would provide important information to guide the selection of therapeutic agents and timing of treatment. However, since neuroinflammation is a common theme across all severities of TBI, strategies aimed at preserving or repairing the BBB in the early period may be beneficial to improve patient outcome, decrease seizure risk, and promote brain health.

ACKNOWLEDGMENT

Supported by K08 NS075144-06 award to GG from the NIH/NINDS.

REFERENCES

1. Chodobski, A. et al. 2011. Blood-brain barrier pathophysiology in traumatic brain injury. *Transl Stroke Res* 2(4):492–516.
2. Iadecola, C., Nedergaard, M. 2007. Glial regulation of the cerebral microvasculature. *Nat Neurosci* 10(11):1369–1376.
3. Neuwelt, E.A. et al. 2011. Engaging neuroscience to advance translational research in brain barrier biology. *Nat Rev Neurosci* 12(3):169–182.
4. Abbott, N.J., Ronnback, L., Hansson, E. 2006. Astrocyte-endothelial interactions at the blood-brain barrier. *Nat Rev Neurosci* 7(1):41–53.
5. Pop, V., Badaut, J. 2011. A neurovascular perspective for long-term changes after brain trauma. *Transl Stroke Res* 2(4):533–545.
6. Shloshberg, D. et al. 2010. Blood-brain barrier breakdown as a therapeutic target in traumatic brain injury. *Nat Rev Neurol* 6:393–403.
7. Schwarzmaier, S.M. et al. 2010. Temporal profile of thrombogenesis in the cerebral microcirculation after traumatic brain injury in mice. *J Neurotrauma* 27(1):121–130.
8. del Zoppo, G.J., Mabuchi, T. 2003. Cerebral microvessel responses to focal ischemia. *J Cereb Blood Flow Metab* 23(8):879–894.

9. Schröder, M.L. et al. 1998. Regional cerebral blood volume after severe head injury in patients with regional cerebral ischemia. *Neurosurgery* 42(6):1276–1280.
10. von Oettingen, G. et al. 2002. Blood flow and ischemia within traumatic cerebral contusions. *Neurosurgery* 50(4):781–788.
11. Nimmerjahn, A., Kirchhoff, F., Helmchen, F. 2005. Resting microglial cells are highly dynamic surveillants of brain parenchyma in vivo. *Science* 308(5726):1314–1318.
12. Beaumont, A. et al. 2006. Bolus tracer delivery measured by MRI confirms edema without blood-brain barrier permeability in diffuse traumatic brain injury. *Acta Neurochir Suppl* 96:171–174.
13. Strbian, D. et al. 2008. The blood-brain barrier is continuously open for several weeks following transient focal cerebral ischemia. *Neuroscience* 153(1):175–181.
14. Barzó, P. et al. 1996. Magnetic resonance imaging-monitored acute blood-brain barrier changes in experimental traumatic brain injury. *J Neurosurg* 85:1113–1121.
15. Nag, S., Venugopalan, R., Stewart, D.J. 2007. Increased caveolin-1 expression precedes decreased expression of occludin and claudin-5 during blood-brain barrier breakdown. *Acta Neuropathol* 114(5):459–469.
16. Shapira, Y. et al. 1993. Blood-brain barrier permeability, cerebral edema, and neurologic function after closed head injury in rats. *Anesth Analg* 77(1):141–148.
17. Baldwin, S.A. et al. 1996. Blood-brain barrier breach following cortical contusion in the rat. *J Neurosurg* 85(3):476–481.
18. Hicks, R.R., Baldwin, S.A., Scheff, S.W. 1997. Serum extravasation and cytoskeletal alterations following traumatic brain injury in rats. Comparison of lateral fluid percussion and cortical impact models. *Mol Chem Neuropathol* 32(1–3):1–16.
19. Başkaya, M.K. et al. 1997. The biphasic opening of the bloodbrain barrier in the cortex and hippocampus after traumatic brain injury in rats. *Neurosci Lett* 226(1):33–36.
20. Lin, J.L. et al. 2010. Ascorbic acid prevents blood-brain barrier disruption and sensory deficit caused by sustained compression of primary somatosensory cortex. *J Cereb Blood Flow Metab* 30(6):1121–1136.
21. Cunningham, L.A., Wetzel, M., Rosenberg, G.A. 2005. Multiple roles for MMPs and TIMPs in cerebralischemia. *Glia* 50(4):329–339.
22. Yang, Y. et al. 2007. Matrix metalloproteinase-mediated disruption of tight junction proteins in cerebral vessels is reversed by synthetic matrix metalloproteinase inhibitor in focal ischemia in rat. *J Cereb Blood Flow Metab* 27(4):697–709.
23. Rosenberg, G.A., Yang, Y. 2007. Vasogenic edema due to tight junction disruption by matrix metalloproteinases in cerebral ischemia. *Neurosurg Focus* 22(5):E4.
24. Rosenberg, G.A. et al. 2001. Immunohistochemistry of matrix metalloproteinases in reperfusion injury to rat brain: Activation of MMP-9 linked to stromelysin-1 and microglia in cell cultures. *Brain Res* 893(1–2):104–112.
25. Gennarelli, T.A. 1994. Animate models of human head injury. *J Neurotrauma* 11(4):357–368.
26. Povlishock, J.T. et al. 1994. Workshop on animal models of traumatic brain injury. *J Neurotrauma* 11(6):723–732.
27. Asahi, M. et al. 2000. Role for matrix metalloproteinase 9 after focal cerebral ischemia: Effects of gene knockout and enzyme inhibition with BB-94. *J Cereb Blood Flow Metab* 20(12):1681–1689.
28. Wang, X. et al. 2000. Effects of matrix metalloproteinase-9 gene knock-out on morphological and motor outcomes after traumatic brain injury. *J Neurosci* 20(18):7037–7042.
29. Ferrara, N., Gerber, H.P., LeCouter, J. 2003. The biology of VEGF and its receptors. *Nat Med* 9(6):669–676.
30. Chodobski, A. et al. 2003. Early neutrophilic expression of vascular endothelial growth factor after traumatic brain injury. *Neuroscience* 122(4):853–867.

31. Suzuki, R. et al. 2003. Very early expression of vascular endothelial growth factor in brain oedema tissue associated with brain contusion. *Acta Neurochir Suppl* 86:277–279.
32. Argaw, A.T. et al. 2009. VEGF-mediated disruption of endothelial CLN-5 promotes blood-brain barrier breakdown. *Proc Natl Acad Sci U S A* 106(6):1977–1982.
33. Mertsch, K., Blasig, I., Grune, T. 2001. 4-Hydroxynonenal impairs the permeability of an in vitro rat blood-brain barrier. *Neurosci Lett* 314(3):135–138.
34. Agarwal, R., Shukla, G.S. 1999. Potential role of cerebral glutathione in the maintenance of blood-brain barrier integrity in rat. *Neurochem Res* 24(12):1507–1514.
35. Schreibelt, G. et al. 2007. Reactive oxygen species alter brain endothelial tight junction dynamics via RhoA, PI3 kinase, and PKB signaling. *FASEB J* 21(13):3666–3676.
36. Parfenova, H. et al. 2006. Glutamate induces oxidative stress and apoptosis in cerebral vascular endothelial cells: Contributions of HO-1 and HO-2 to cytoprotection. *Am J Physiol* 290(5):C1399–C1410.
37. Cook, J.L. et al. 1998. Temporal changes in gene expression following cryogenic rat brain injury. *Mol Brain Res* 55(1):9–19.
38. Fee, D.B. et al. 2004. Traumatic brain injury increases TGFβRII expression on endothelial cells. *Brain Res* 1012(1–2):52–59.
39. Dietrich, W.D., Alonso, O., Halley, M. 1994. Early microvascular and neuronal consequences of traumatic brain injury: A light and electron microscopic study in rats. *J Neurotrauma* 11(3):289–301.
40. Shen, W. et al. 2011. Tyrosine phosphorylation of VE-cadherin and claudin-5 is associated with TGF-β1-induced permeability of centrally derived vascular endothelium. *Eur J Cell Biol* 90(4):323–332.
41. Garcia, C.M. et al. 2004. Endothelial cell-astrocyte interactions and TGFβ are required for induction of blood-neural barrier properties. *Dev Brain Res* 152(1):25–38.
42. Li, F. et al. 2011. Endothelial Smad4 maintains cerebrovascular integrity by activating N-cadherin through cooperation with Notch. *Dev Cell* 20(3):291–302.
43. Parfenova, H., Fedinec, A., Leffler, C.W. 2003. Ionotropic glutamate receptors in cerebral microvascular endothelium are functionally linked to heme oxygenase. *J Cereb Blood Flow Metab* 23(2):190–197.
44. Badaut, J., Ashwal, S., Obenaus, A. 2011. Aquaporins in cerebrovascular disease: A target for treatment of brain edema? *Cerebrovasc Dis* 31(6):521–531.
45. Unterberg, A.W. et al. 2004. Edema and brain trauma. *Neuroscience* 129(4):1021–1029.
46. Klatzo, I. 1985. Brain oedema following brain ischaemia and the influence of therapy. *Br J Anaesth* 57(1):18–22.
47. Helmy, A., Vizcaychipi, M., Gupta, A.K. 2007. Traumatic brain injury: Intensive care management. *Br J Anaesth* 99:32–42.
48. Campbell, M. et al. 2012. Targeted suppression of claudin-5 decreases oedema and improves cognitive outcome following traumatic brain injury. *Nature Commun* 3:849.
49. Neuwelt, E. et al. 2008. Strategies to advance translational research into brain barriers. *Lancet Neurol* 7(1):84–96.
50. The Brain Trauma Foundation. 2007. Guidelines for the management of severe traumatic brain injury, 3rd ed. *J Neurotrauma* 24(Suppl 1):14–20.
51. Marks, J.A. et al. 2012. Similar effects of hypertonic saline and mannitol on the imflammation of the blood-brain barrier microcirculation after brain injury in a mouse. *J Trauma Acute Care Surg.* 73(2):351–357.
52. Cottenceau, V. et al. 2011. Comparison of effects of equimolar doses of mannitol and hypertonic saline on cerebral blood flow and metabolism. *J Neurotrauma* 28:2003–2012.
53. Ivashkova, Y. et al. 2006. Bradykinin B2 receptor antagonism with LF 18–1505T reduces brain edema and improves neurological outcome after closed head trauma in rats. *J Trauma* 61:879–885.

54. Stover, J.F., Dohse, N.K., Unterberg, A.W. 2000. Bradykinin 2 receptor antagonist LF 16–0687Ms reduces posttraumatic brain edema. *Acta Neurochir Suppl* 76:171–175.
55. Servadei, F. et al. 2002. Traumatic subarachnoid hemorrhage: Demographic and clinical study of 750 patients from the European brain injury consortium survey of head injuries. *Neurosurgery* 50:261–267.
56. Ersahin, M. et al. 2010. The anti-inflammatory and neuroprotective effects of ghrelin in subarachnoid hemorrhage-induced oxidative brain damage in rats. *J Neurotrauma* 27:1143–1155.
57. Lopez, N. et al. 2012. Early ghrelin treatment attenuates disruption of the blood brain barrier and apoptosis after traumatic brain injury through a UCP-2 mechanism. *Brain Res* 1489:140–148.
58. Lopez, N. et al. 2012. Ghrelin prevents disruption of the blood-brain barrier after traumatic brain injury. *J Neurotrauma* 29:385–393.
59. Sinard, M. et al. 2003. Signaling at the gliovascular interface. *J Neurosci* 23:9254–9262.
60. Lu, Q. et al. 2013. Intranasal delivery of nerve growth factor attenuates aquaporins-4-induced edema following traumatic brain injury in rats. *Brain Res* 1493:80–89.
61. Wright, D.W. et al. 2007. ProTECT: A randomized clinical trial of progesterone for acute traumatic brain injury. *Ann Emerg Med* 49:391–402.
62. Xiao, G. et al. 2008. Improved outcomes from the administration of progesterone for patients with acute severe traumatic brain injury: A randomized controlled trial. *Crit Care* 12:R61.
63. Rosenfeld, J.V. et al. 2012. Early management of severe traumatic brain injury. *Lancet* 380:1088–1098.
64. Lukaszewicz, A.C., Soyer, B., Payen, D. 2011. Water, water, everywhere: Sodium and water balance and the injured brain. *Curr Opin Anaesthesiol* 24:138–143.
65. Pascual, J. et al. 2013. Neuroprotective effects of progesterone in traumatic brain injury: Blunted in vivo neutrophil activation at the blood brain barrier. *Am J Surg* 206(6):840–845; discussion 845–846.
66. Kim, J.V., Dustin, M.L. 2006. Innate response to focal necrotic injury inside the blood-brain barrier. *J Immunol* 177:5269–5277.
67. Li, S. et al. 2011. In vivo evolution of microvascular inflammation after traumatic brain injury: An intravital microscopy study. *Crit Care Med* 39(Suppl 12):1.
68. Shohami, E. et al. 2011. Endocannabinoids and traumatic brain injury. *Br J Pharmacol* 163:1402–1410.
69. Cohen-Yeshurun, A. et al. 2011. N-arachidonoyl-L-serine is neuroprotective after traumatic brain injury by reducing apoptosis. *J Cereb Blood Flow Metab* 31:1768–1777.
70. Panikashvili, D. et al. 2006. The endocannabinoid 2-AG protects the blood-brain barrier after closed head injury and inhibits mRNA expression of proinflammatory cytokines. *Neurobiol Dis* 22:257–264.
71. Panikashvili, D. et al. 2001. An endogenous cannabinoid (2-AG) is neuroprotective after brain injury. *Nature* 413:527–531.
72. van der Stelt, M. et al. 2001. Exogenous anandamide protects rat brain against acute neuronal injury in vivo. *J Neurosci* 21:8765–8771.
73. Di Marzo, V., Bifulco, M., De Petrocellis, L. 2004. The endocannabinoid system and its therapeutic exploitation. *Nat Rev Drug Discov* 3:771–784.
74. Mackie, K. 2006. Cannabinoid receptors as therapeutic targets. *Annu Rev Pharmacol Toxicol* 46:101–122.
75. Blankman J.L. et al. 2007. A comprehensive profile of brain enzymes that hydrolyze the endocannabinoid 2-arachidonoylglycerol. *Chem Biol* 14(12):1347–1356.
76. Menge, T. et al. 2012. Mesenchymal stem cells regulate blood-brain barrier integrity through timp3 release after traumatic brain injury. *Sci Transl Med* 4(161):1–11.

5 Cellular and Molecular Mechanisms of Secondary Neuronal Injury following Traumatic Brain Injury

Kamesh Krishnamurthy and Daniel T. Laskowitz

CONTENTS

EXCITOTOXICITY: THE MOLECULAR RESPONSE TO HEAD INJURY

Evidence for Increased Extracellular Glutamate after Traumatic Brain Injury (TBI) and Mechanisms of Release

Excitotoxicity—the cascade of intracellular events initiated by excessive stimulation by neurotransmitters leading to intracellular calcium overload—is hypothesized to play an important role in mediating secondary neuronal injury after traumatic brain injury (TBI). While a variety of neurotransmitters could potentially trigger excitotoxic cell injury, glutamate is thought to be the primary contributor because of its potent effect on increasing intracellular calcium through ionotropic receptors (primarily NMDAR, but also kainate and calcium-permeable AMPAR). Microdialysis studies in rodents after both fluid percussion injury and controlled cortical impact have demonstrated a substantial increase in extracellular glutamate proportional to the severity of injury.[1,2] A number of clinical observations also support the hypothesis that TBI increases extracellular glutamate. Multiple reports analyzing glutamate levels in cerebrospinal fluid (CSF) with the elevation often persisting beyond 1 week after injury.[3,4] A prospective microdialysis study in patients suffering severe TBI showed an increase in extracellular glutamate in over 75% of patients; interestingly, values normalized over the course of 120 hours in 60% of these patients, but persistent elevations of extracellular glutamate were associated with increased mortality.[5]

Release of glutamate following trauma is believed to be mediated by the stretching of neuronal membranes, creating micropores that allow for sodium influx. This ionic shift depolarizes the membrane potential, opening voltage-gated calcium channels and triggering neurotransmitter release. Presynaptic glutamate release can then result in depolarization and excessive calcium influx in postsynaptic neurons by binding to calcium-permeable NMDARs and GluR2-lacking AMPARs. This pathological progression is propagated by a positive feedback loop, since the depolarization triggered by the initial glutamate results in additional neurotransmitter release.[6,7] Furthermore, impaired reuptake of extracellular glutamate by astrocytes may exacerbate excitotoxicity after ischemia.[8] This likely plays a role following TBI as well, since postmortem analysis in human TBI patients demonstrated reduced glutamate transporter expression;[9] as discussed later in this section, this has become a promising area of preclinical research.

Interestingly, a recent series of studies using either intracerebral or intravenous administration of membrane-resealing polymers following controlled cortical impact demonstrated reduced blood–brain barrier permeability and brain tissue damage as well as improved functional recovery, but failed to rescue degenerating cells from eventual demise.[10,11] This is likely due to the transient ability of the compounds to reseal cells;[12] studies using more effective membrane-resealing agents are needed to definitively prove the hypothesis that membrane stretch is the trigger of excitotoxic insult following TBI. An alternative hypothesis is that the mechanism by which mechanical stretch triggers excitotoxicity is not solely due to mechanical alterations of the membrane but also mediated by receptor subunits that are particularly sensitive to mechanical injury. Evidence in support of this hypothesis comes from several in vitro studies in which cultured neurons are grown on deformable silastic membranes and subjected to stretch along a single axis.[13] One recent study examined network connectivity in these cultures by analyzing the synchronization of calcium signaling using the calcium indicator GCaMP5. They found that stretch decreases network connectivity, and that higher expression of the NMDAR subunit NR2B (for example, in younger cultures where the ratio of NR2B- vs. NR2A-containing NMDAR is greater) results in a more profound effect.[14]

Summary of Preclinical Studies on Glutamate Receptor Antagonists

Given the robust experimental evidence implicating the role of glutamate excitotoxicity after injury, a number of studies have examined the therapeutic potential of glutamate antagonists. Multiple compounds have been examined, including NMDAR-specific antagonists (MK801, CPP, memantine, riluzole) and non-NMDAR glutamate receptor antagonists (CPP, NBQX). The administration of NMDAR-specific antagonists following injury has been shown to improve neuronal density and reduce cell death in weight-drop[15,16] and controlled cortical impact models,[17] as well as to reduce cortical lesion size in a fluid percussion injury model.[18] In contrast, non-NMDAR antagonists were only effective if administered prior to injury,[15] suggesting that the NMDAR antagonism may represent a more effective therapeutic strategy.

Based on the promise of these findings, the preclinical development of an NMDAR antagonist, traxopradil (CP-101,606), was initiated. In a series of studies using fluid percussion injury in rodents, traxoprodil was found to reduce posttraumatic amnesia, neurological dysfunction, edema, intracranial hypertension, and cortical spreading depression.[19] Unfortunately, despite these promising preclinical findings, a subsequent clinical trial failed to demonstrate benefit in a clinical population assessed with the Neurobehavioral Rating Score.[20]

Novel Concepts in Glutamate Excitotoxicity after TBI: Role of Astrocytic Glutamate Transporters and Gap Junctions

The failure to demonstrate clinical benefit with NMDAR antagonists has led to an increased focus on the identification of new strategies to reduce secondary neuronal injury following trauma. In particular, two mechanisms that have been the subject of increasing interest include astrocytic glutamate transporters and neuronal gap junctions.

During both physiological processes (synaptic transmission) and pathological processes (excitotoxicity), extracellular glutamate is cleared primarily by GLT-1, a transporter protein on astrocytes.[21] A number of studies suggest that this protein is downregulated in preclinical models of TBI. For example, Rao and colleagues demonstrated a significant reduction in protein levels after controlled cortical impact that persisted for up to 7 days.[17] These observations were supported by a similar study using controlled cortical impact that demonstrated an association between the downregulation of GLT-1 levels and the increase in CSF glutamate,[22] suggesting a potential mechanism whereby the reduction of glial glutamate transporters contributes to extra-neuronal glutamate levels and subsequent excitotoxicity. While additional research is currently needed to establish a causal relationship, these findings provide a potential therapeutic target to reduce the toxic levels of glutamate after head injury.

Another evolving area of interest for therapeutic intervention is the role of gap junctions, channels that connect cells and allow the passage of ions and small molecules, including secondary messengers (Ca, IP3, cAMP, etc.). Although gap junctions are expressed in the CNS during development and are known to play critical roles in neuronal differentiation, migration, and circuit formation,[23,24] by postnatal week 4 these channels are widely downregulated. Recent research has suggested that gap junctions may play a role in mediating secondary neuronal injury in the adult brain. For example, data from a fluid percussion model demonstrates upregulation of gap junction expression through the activation of mGluR2 receptors.[25] The increased expression of gap junctions may contribute to the neuronal damage caused by TBI, since it would allow for the spread of destructive second messenger molecules like Ca and pro-apoptotic factors that are increased after insult. Furthermore, this suggests value in exploring the potential uses of mefloquin (a gap junction blocker) or mGluR2 antagonists in targeting this component of excitotoxicity.

Role of Calpains in Excitotoxicity

Increased intracellular calcium causes neuronal damage and death through a variety of mechanisms, including mitochondrial dysfunction and metabolic/oxidative stress, as discussed later in this chapter. Calcium also triggers the activation of calpains, a family of cysteine proteases.[26] Calpain-mediated cell damage is not only thought to be a key contributor of cell death after excitotoxicity but is also a potential target for therapeutic intervention. The interest in calpain activation after TBI was triggered in large part by the discovery of source specificity in cell death initiation. Calcium influx through NMDARs results in neuronal damage, whereas influx through other sources like sodium-calcium exchangers does not. Calpains may mediate this phenomenon.[27]

The calpain family consists of 15 proteases; calpain-1 and calpain-2 are thought to be the specific family members involved in TBI-induced excitotoxicity.[28] The calcium influx triggers hyperactivation of these enzymes, which then act on both structural proteins and glutamate receptors. Calpains cleave key cytoskeletal proteins like MAP2, CRMP2, spectrin, and tau; degradation of these components likely leads to destabilization of the actin cytoskeleton and microtubule network, resulting in structural damage to the neuron.[29] In addition, calpains cleave the mGluR1 receptors. At baseline, these receptors promote both neurotoxicity through the release of calcium from the endoplasmic reticulum (ER) and neuroprotection by activating the

PI3K-AKT pathway; when cleaved, only the neurotoxic cascade can be activated. In this way, calpain activation may exacerbate excitotoxic neuronal injury.[30]

Given these effects, as well as the fact that physiological levels of calpain activation are low relative to other therapeutic targets of TBI (i.e., NMDAR), calpain inhibitors are being evaluated in preclinical studies of TBI. These results are described in a number of reviews (for example, see Ref. 31) and are summarized here. The inhibition of calpains activation can either be indirect—for example, Mg to block NMDAR activation[32] or hypothermia to reduce proteolysis[33]—or by direct inhibition of protease function. Unfortunately, both Mg and hypothermia have failed to demonstrate benefit in clinical trials, and there is conflicting evidence from preclinical studies regarding the palliative effects of calpain inhibitors. For example, intra-arterial infusion of calpain inhibitor II prevents cytoskeletal degradation in a contusion model of TBI[34] but not a weight drop model.[35] Furthermore, administration of AK295, a calpain inhibitor, improved motor and memory function in a fluid percussion model,[36] but did not reduce spectrin breakdown or neuronal death,[37] suggesting dissociation between histological improvement and functional recovery. A more complete understanding of how calpains affect physiology and disease is needed.

GABAergic Excitotoxicity

Secondary neuronal injury mediated by glutamatergic excitotoxicity has been the primary focus of TBI research in large part because glutamate is the most abundant neurotransmitter in the brain and glutamate receptors increase intracellular calcium levels. However, it is important to note that other neurotransmitters are elevated in the extracellular space after brain injury, and increasing evidence suggests that these neurotransmitters may also play a pathological role in specific cell populations. We will focus here on the concept of GABAergic excitotoxicity as it relates to immature adult-born hippocampal neurons.

Hippocampal cell death is found in multiple models of TBI[38,39] and is hypothesized to contribute to neurocognitive deficits that occur in patients following injury. Interestingly, the dentate gyrus within the hippocampus is one of the two mammalian brain regions known to contain neural stem cells.[40] Newborn neurons originate in the subgranular zone and undergo a month-long process of migration and maturation before integrating within the dentate gyrus. Immature adult-born neurons, like their embryonic counterparts, are depolarized by GABA activity because of a reversal in their chloride gradient; furthermore, these neurons also express voltage-gated calcium channels.[41]

These properties are relevant to TBI given the emerging data suggesting a preferential death of immature adult-born neurons after controlled cortical impact.[42,43] The mechanism of cell death in this model was found to be necrotic and not apoptotic,[44] but the underlying properties that contribute to cellular selectivity is not known. Evidence from the field of neonatal anesthesia suggests that GABAergic excitotoxicity contributes to neuronal death after isofluorane exposure in vitro by inducing an increase in intracellular calcium that leads to apoptosis.[45] Although a direct link between isofluorane administration, GABAergic synaptic transmission, and hippocampal cell death has not been demonstrated in vivo, there are reports of isoflurane-mediated apoptosis of hippocampal neurons,[46] and isoflurane is known to enhance GABAa currents[45] so such a connection may be inferred.

Additional work is clearly needed to determine the role of GABAergic excitotoxicity following TBI. For example, microdialysis studies using an open-skull weight drop have reported increased extracellular GABA in cortex,[47] but these observations have not been rigorously quantified. Furthermore, in the studies on isofluorane anesthesia, cell death is mediated by apoptosis, while the controlled cortical impact studies suggests that immature newborn neurons die by necrosis. Additional studies using other methods (closed head injury, fluid percussion injury [FPI]) are needed to fully evaluate whether these effects are model dependent. Understanding the contribution of GABAergic excitotoxicity may prove to be of particular interest because of the current availability of GABA antagonists approved by the U.S. Food and Drug Administration (FDA).

Controversy Regarding the Contribution of Glutamatergic Excitotoxicity to Secondary Neuronal Injury

It is important to note that although it is generally accepted that glutamatergic excitotoxicity contributes to neuronal death following TBI, there remains some disagreement regarding the relative importance of excitotoxicity after brain insult (for review, see Ref. 48). Three pieces of evidence are often cited to cast doubt on the contention that excitotoxic neuronal injury is the major factor in mediating cell death following injury. First, Carbonell and Grady performed fluid percussion injury on two strains of mice: one susceptible and one resistant to kainic acid-induced excitotoxic damage. They found no difference in cell death in CA3, the hippocampal region with the greatest abundance of kainate receptors.[49] Second, the levels of glutamate required for excitotoxicity in vitro greatly exceeds levels found after TBI in vivo.[48] Finally, there is a discrepancy between the time course of increased extracellular glutamate after injury and the therapeutic window of NMDAR antagonists for TBI. For example, Nilsson and colleagues used an open-skull weight drop and reported a peak of extracellular glutamate 5 minutes after injury.[47] For comparison, the preclinical studies using traxoprodil demonstrated benefit when drug administration was initiated 15 minutes after injury and continued for up to 24 hours.[19,50]

These arguments fail to consider several pieces of evidence in support of a role for excitotoxicity following injury. First, it is likely that calcium influx through NMDAR and not excessive activation of kainate receptors is the key mediator of excitotoxicity. Second, it is difficult to quantitatively compare in vitro and in vivo studies of cell death; furthermore, microdialysis studies likely underestimate values of extracellular components, especially in deep brain structures. Finally, while it is true that the peak of extracellular glutamate precedes the typical administration of glutamate receptor antagonists, extracellular glutamate levels remain elevated for at least 24 hours.[47]

However, an examination of the degree and nature of excitotoxic cell death following TBI is critical to therapeutic development of this class of neuroprotectants, and the issue deserves continued attention. For example, it is plausible that dysregulation of NMDAR Mg-blockade may contribute to neuronal death. Given the failure of clinical improvement in a trial of IV magnesium sulfate in acute stroke patients,[51] a more complete understanding of temporal and spatial characteristics of this process may help guide new therapeutic interventions. Furthermore, it is plausible that the beneficial effect of glutamate receptor antagonists or glutamate release inhibitors (such as lamotrigine) is not due to inhibition of glutamate receptors directly but rather a reduction in neuronal metabolic demand. The contribution of metabolic dysfunction to secondary neuronal injury will be discussed in a subsequent section and potentially offers an additional opportunity for treatment.

Therapeutic Targets of Excitotoxicity

The pathophysiology of excitotoxicity after TBI presents a variety of potential therapeutic targets, some of which are summarized in Figure 5.1. Preventing the increase in extracellular glutamate—using either glutamate scavengers, effective membrane resealing agents, or increasing glutamate transporters on astrocytes—can help block this pathological process at its source. However, these interventions would likely be ineffective in the clinical setting since they would need to be administered concurrent with injury. More precise blockade of neurotransmitter receptors is another option. Studies are ongoing using pH-sensitive antagonists (as discussed in subsequent

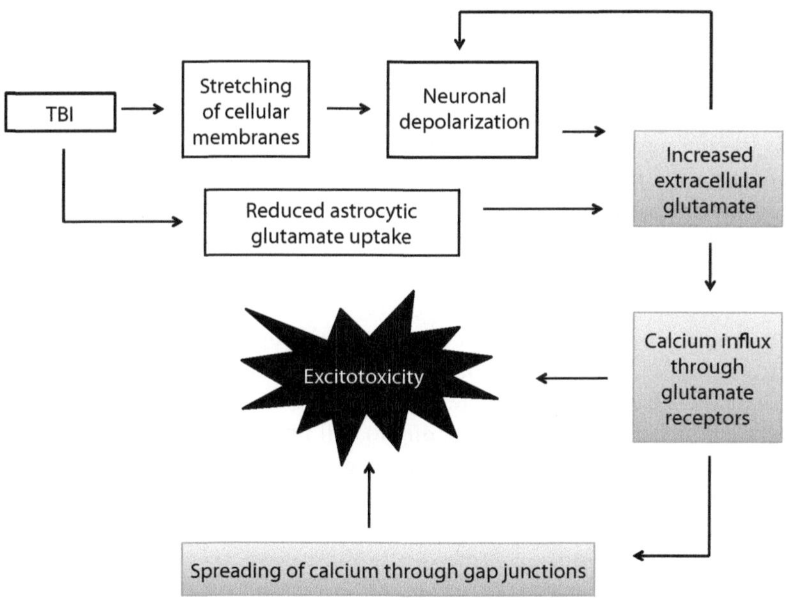

FIGURE 5.1 Schematic of excitotoxic process following TBI. The mechanical forces of TBI lead to stretching of cellular membranes and disrupting ionic gradients, leading to neuronal depolarization and release of glutamate. Combined with reduced astrocytic glutamate uptake, the extracellular glutamate activates glutamate receptors which causes continued neuronal depolarization. Activation of glutamate receptors leads to calcium influx which results in cellular toxicity, a process that is worsened through spreading of calcium through gap junctions. Potential therapeutic targets to prevent this process are highlighted.

sections, the metabolic dysfunction after TBI is associated with a local tissue acidosis); blockade of GABA receptors using flumazenil may effectively prevent death of immature newborn neurons. In vivo, intracellular calcium chelation is unlikely to be effective, but blocking the enzymes activated by the increased intracellular calcium (i.e., calpains) may be feasible. For this to occur, an understanding of the cellular and subcellular locales of calpain activation, as well the neuroanatomic and functional consequences of this activation, is required. Thinking "outside the box" may also open new avenues for treatment. For example, gap junctions may not appear an obvious target for neuroprotection after TBI, but their blockade may help prevent the spread of calcium, reducing neuronal death and improving functional outcomes.

APOPTOSIS: THE CELLULAR RESPONSE TO HEAD INJURY

Introduction

Initial neuronal death following acute brain injury typically occurs by necrosis, which is challenging to target therapeutically because of the narrow therapeutic time window. Necrosis occurs on a timescale of minutes, while hospitalization for moderate-to-severe TBI often takes much longer and the diagnosis of mild TBI is often delayed due to the lack of effective diagnostic biomarkers.[52] However, evidence from clinical studies and preclinical models support the presence of a second wave of delayed cell death that occurs by apoptosis, the evolutionarily conserved mechanism of programmed cell death. Understanding the triggers and pathways of apoptosis can provide targets for neuroprotection. Furthermore, necrosis and apoptosis likely represent a spectrum rather than discrete categories, and developing an appreciation for the overlap of these processes may identify opportunities for neuroprotective strategies.

Overview of the Apoptosis Pathways

The term *apoptosis* was originally coined in 1972 by Kerr and colleagues, and describes a "programmed" pattern of cell death characterized by specific morphological changes. These changes include chromatin condensation, cell shrinkage, and membrane blebbing, ultimately leading to cell phagocytosis without inflammation.[53] A critical component of development as well as normal physiology, apoptosis can be initiated in either a caspase-dependent or caspase-independent mechanism. Furthermore, the caspase-dependent pathways can be activated either intrinsically (via intracellular mechanisms) or extrinsically.[54]

Caspase-Dependent Apoptosis

Caspases are a family of proteases that form the foundation for the apoptotic cascade. They are broadly divided into two categories: executioners, which cleave proteins and cause cell death, and initiators, which cleave and activate executioner caspases. Common initiator caspases include caspase-2, caspase-8 and caspase-9; the critical executioner caspase is caspase-3, though caspase-12, caspase-6, and caspase-7 also fall in this category.[55]

The intrinsic caspase-dependent apoptotic pathway is triggered by mitochondrial damage (Figure 5.2). In the context of TBI, the most common cause is the excessive intracellular calcium levels induced by excitotoxicity; mitochondria sequester calcium but quickly become overwhelmed, resulting in mitochondrial damage and a change in the organelle membrane potential that impairs cellular respiration, producing reactive oxygen species. Mitochondrial damage results in the extrusion of cytochrome-c. Cytochrome-c then binds to a complex with the protein Apaf1 and caspase-9, producing a structure termed the *apoptosome*. The apoptosome then cleaves the executioner caspase, caspase-3, which triggers cellular apoptosis. Cytochrome-c can also cause a feed-forward cascade that pushes the neuron toward apoptosis by interacting with IP3R on the ER to trigger additional calcium release. Finally, caspase-3 can be activated directly after NMDAR-mediated excitotoxicity triggered by TBI, via cleavage by calpains.[56]

A number of preclinical studies have demonstrated that in addition to caspase-3, caspase-12 is activated due to ER stress and results in apoptosis.[57] However, the evidence for caspase-12 activation after injury in humans is lacking. More research is needed before pursuing ER-stress induced caspase-12 activation as a target, but this finding highlights the importance of selecting clinically relevant animal models and validating preclinical findings in the human population.

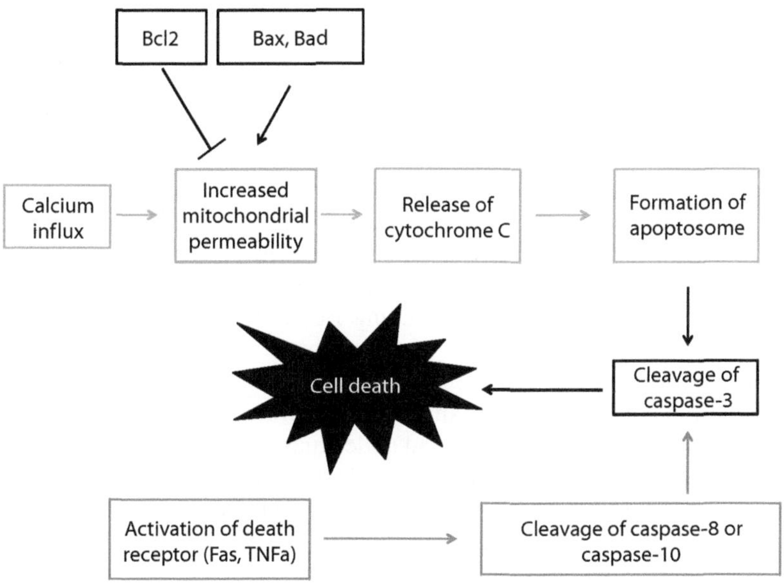

FIGURE 5.2 Schematic of apoptosis following TBI. The intrinsic pathway is diagrammed on the top of the figure, while the extrinsic pathway is diagrammed on the bottom. For the intrinsic pathway, calcium influx into neurons leads to increased mitochondrial permeability, the release of cytochrome C and other pro-apoptotic factors that form the apoptosome that cleaves caspase-3, resulting in cell death. This process is regulated by both pro-apoptotic (i.e., Bax and Bad) and anti-apoptotic (i.e., Bcl2) factors. For the extrinsic pathway, activation of death receptors leads to the cleavage of initiator caspases (caspase-8 and caspase-10), which then cleave caspase-3 and produce cell death.

Extrinsic caspase-dependent apoptosis is triggered by the binding of Fas to FasL or TNF alpha binding to the TNF alpha-receptor. This binding then initiates the cleavage of either the initiator caspase-8 or caspase-10, which then activates the executioner caspase-3.[54]

Caspase-Independent Apoptosis

The surprising failure to prevent apoptosis either genetically (via knockout of the caspases) or pharmacologically (via pan-caspase inhibitors) provides strong support for a caspase-independent mechanism of apoptosis.[58] It is thought that these mechanisms evolved to protect organisms against carcinogenesis when caspase-dependent mechanisms fail.[59] At least two mechanisms are known to trigger caspase-independent apoptosis. The first is triggered by a change in mitochondrial membrane potential, which leads to the extrusion of apoptosis-inducing factor (AIF) that then migrates to the nucleus and causes programmed DNA damage. The second involves the activation of the protein PARP, which in situations of low energy levels results in apoptosis. Both mechanisms have been shown to occur in preclinical models of TBI.[60]

Regulation of Apoptosis

Apoptosis can be regulated by a variety of proteins and signaling cascades, and the therapeutic manipulation of apoptosis remains an active area of translational research (Figure 5.2). The primary protein family that regulates apoptosis is the Bcl-2 family. These function primarily by regulating mitochondrial membrane permeability.[61] Specifically, the proteins Bax and Bad increase mitochondrial permeability and thereby facilitate cyotochrome-c release and apoptosome formation. In the context of TBI, increased Bax and Bad is thought to occur secondary to elevations in the tumor suppressor gene p53, which is elevated in a variety of preclinical models of TBI including FPI,[62] contusion,[63] and weight drop.[64]

In addition to the pro-apoptotic proteins Bax and Bad, there are a number of anti-apoptotic regulatory proteins, including Bcl-2. This protein can prevent both caspase-dependent and independent apoptosis by reducing membrane permeability, thereby preventing the release of cytochrome-c and AIF.[60] Bcl-2 expression increases within hours after injury, and remains elevated in cortex and hippocampus for several days. Interestingly, Clark and colleagues identified a correlation between cellular Bcl-2 expression and neuroprotection after a rat model of contusion by demonstrating that neurons in both cortex and hippocampus that were Bcl-2 positive were predominantly TUNEL-negative. This suggests that Bcl-2 plays a critical role in neuroprotection after head injury.[65]

Evidence for Apoptosis after TBI

The first evidence for apoptosis following TBI comes from human brain specimens of patients with severe injury who required surgical decompression for refractory intracranial hypertension. The samples showed positive TUNEL staining, as well as cleavage of the key apoptosis proteins caspase-1 and caspase-3.[66] More recent studies using improved histological techniques as well as transcriptional analysis have demonstrated increased levels of caspase-3 and caspase-8.[60] It is important that there is a correlation between apoptotic protein levels and clinical outcomes; patients with detectable levels of both Bcl-2 and Bax had better outcomes than patients with high Bax but low Bcl-2.[67,68] Fas and FasL levels in the CSF also increase after injury;[69] thus, apoptotic markers have the potential to serve as biomarkers for the diagnosis of TBI.

Preclinical studies have also provided evidence for apoptosis after injury. Moderate fluid percussion injury results in increased caspase-3 mRNA and decreased Bcl-2 levels for up to 72 hours after injury.[70] In contusion models, a number of studies have demonstrated acute increases in both caspase-3 mRNA and protein levels.[71,72] While some closed head injury studies have shown decreases in Bcl-2 and increased Bax levels, others have utilized the variability in the model to demonstrate an inverse correlation between Bcl-2 levels and TUNEL staining.[60] In aggregate, these studies provide evidence for the occurrence of apoptosis by activation of the intrinsic pathway; however, additional data has suggested that activation of the extrinsic pathway may also play a role. Qiu and colleagues used a contusion model in mice and a combination of immunoprecipitation and immunohistochemistry to demonstrate Fas–FasL binding.[73] Zhang and colleagues have also reported caspase-independent apoptosis via nuclear localization of AIF in a rat contusion model.[74] Taken together, these preclinical studies provide convincing evidence for the occurrence of apoptosis after TBI, and support targeting apoptosis for therapeutic intervention.

Necroptosis

Previously, apoptosis and necrosis have been considered distinct forms of cell death, characterized by their molecular biology, cellular morphology, and presence of inflammation. Apoptosis occurs via molecular "programs" (i.e., signaling cascades), results in nuclear and DNA breakdown and the formation of apoptotic bodies, and produces cell death with minimal inflammation. In contrast, necrosis is the result of cellular injury from external factors, results in cell and organelle swelling and membrane lysis, and is associated with an inflammatory response.[75] However, it is becoming increasingly evident that there is often overlap between apoptotic and necrotic mechanisms, which is consistent with observations that that PARP activation (a target of caspase-3) in a cell with sufficient energy in the form of NAD+ would undergo apoptosis while an energy-depleted cell would undergo necrosis.[60] However, the full extent of the overlap was not fully appreciated until the recent discovery of "necroptosis," or programmed—i.e., triggered by intracellular signaling cascades—necrosis.

The process of necroptosis was first described in initial observations made by Vercammen and colleagues, who noted in vitro the increased sensitivity of L929 murine cells to TNF-triggered cell death despite the presence of a pan-caspase inhibitor.[76] This clarified previous observations that cells exposed to TNF had both apoptotic and necrotic morphologies.[77] It had been previously shown that the protein RIPK1 was necessary for TNF-induced apoptosis.[78] Subsequently, Degterev and colleagues developed necrostatin-1 (Nec-1) a potent inhibitor of RIPK1 with outstanding specificity.[79] This has led to a more "user-friendly" definition of necroptosis as cell death that is inhibited by necrostatin-1.

The regulation of necroptosis and apoptosis involves a complex interplay between caspase-8, RIPK1, and the regulatory protein FLIP. Activation of both mechanisms of cell death can occur via any of the receptors containing a TNF-receptor-associated death domain or Fas-associated death domain (i.e., TNFR, Fas, etc.), but the most extensively studied cascade is initiated by TNF binding TNFR, which will be discussed here. At baseline, TNF–TNFR binding prevents both necroptosis and apoptosis by upregulating NFkB. This protein ubiquinates RIPK1, preventing necroptosis, and increases FLIP, which binds to activated caspase-8, forms a heterodimer, and prevents apoptosis directly. Furthermore, the FLIP-caspase-8 complex prevents necroptosis indirectly by blocking RIPK1.[80] Following cell injury, in the absence of FLIP and presence of caspase-8, TNF–TNFR initiates apoptosis. However, in the scenario where TNF binds TNFR in a cell with a reduced quantity of caspase-8, RIPK1 complexes with the proteins MLKL and RIPK3 (the "executioner" protein of necroptosis), forming a necrosome. The necrosome then induces cell death.[80,81]

Although the presence of necroptosis after TBI has been clearly demonstrated in animal models, a number of questions remain that may help inform new translational strategies. First, necrosome formation has been shown using genetic (caspase 8-null mice and viral knockdown of caspase-8) and pharmacological (pan-caspase inhibitors) scenarios.[82] However, it remains unclear what causes caspase-deficiency in a genetically normal animal or human following injury. Second, the nature of the mechanism by which a necrosome results in necrosis is not well defined and needs to be further understood. Finally, the nature of the switch of TNF-TNFR signaling from one that is cytoprotective (via upregulation of NFkB) to one that is neurotoxic (initiating apoptosis and necroptosis) needs to be defined.

MITOCHONDRIAL INJURY: AN ORGANELLE BRIDGE FROM MOLECULAR TO CELLULAR PATHOLOGY

Given our current understanding that calcium-mediated excitotoxicity is the principle molecular mechanism of secondary neuronal injury triggering apoptosis, mitochondria serve as a "bridge" linking these two pathological processes and provides a potential target for therapy after head injury. Reducing mitochondrial injury after TBI is an appealing strategy for multiple reasons. First, it serves a central role in the intercommunication of various other organelles, including endoplasmic reticulum and lysosomes. Second, mitochondrial injury caused by excitotoxicity can initiate cell death either by apoptosis or necrosis. Third, mitochondrial injury leads to a "cellular injury crisis" that accelerates neuronal death after global insults such as TBI. Finally, mitochondrial injury triggers the formation of reactive oxygen species (ROS) that can cause extensive secondary cellular damage.

Structure and Function of Mitochondria

Mitochondria are organelles that contain two membrane systems, resulting in the envelopment of distinct lumens. The most exterior of these membranes, the outer membrane, contains proteins required for molecular transport into and out of the organelle. These proteins, called porins, allow for free transport of molecules smaller than 5 kDa; any larger proteins require specific transport systems to move between the cytoplasm and mitochondrial lumens. Between the outer and inner membrane is the intermembrane space.[54] The permeability of the mitochondria, and in particular the outer mitochondrial membrane, is a critical variable in cell outcome following injury. Contained within the intermembrane space is a variety of proteins (discussed in the last section on apoptosis) that trigger cell death when released into the cytoplasm; these include cytochrome-c, AIF, and smac/Diablo, among others. The inner mitochondrial membrane contains the necessary protein machinery for cellular energy generation. These include protein transport machinery, components of oxidative phosphorylation, and ATP synthase. Across the inner membrane is an electrochemical gradient that drives the electron transport chain, which allows for the generation of the majority of cellular ATP. It is important to note that the formation of this gradient requires that the inner mitochondrial membrane be largely impermeable; in fact, as discussed later, increased mitochondrial permeability is a key sign of mitochondrial injury and trigger of cell death. Degeneration of the inner mitochondrial membrane leads to a reduced ability to generate ATP and a cellular energy crisis.[83]

Interconnectivity of Mitochondria and Other Organelles

There is a dynamic interplay between mitochondria and other cellular organelles that participate in secondary neuronal injury. In particular, the interaction between mitochondria and the endoplasmic reticulum (ER) plays a critically important role after cellular injury. For many years, ER vesicles found in purified mitochondria fractions were thought to be caused by technical artifact; however, advanced electron and light microscopy techniques have since demonstrated that the presence of a "mitochondria-associated ER membrane" that couples the two organelles. Under physiological conditions, these connections couple lipid transport and calcium exchange between the two compartments, though the precise purpose of this coupling remains unclear. Under pathological conditions, the calcium exchange between the two organelles can serve as a trigger for apoptosis.[84] Mitochondria are also intimately linked to lysosomes, and increases in mitochondrial membrane permeability in cancer cells can trigger the release of lysosomal proteases that can trigger cell death.[85] Understanding the role of these interconnections in normal physiology and after neuronal injury will be critical to targeting these interactions for therapy.

Subcellular Causes of Mitochondrial Injury

The mechanisms surrounding mitochondrial injury remain incompletely defined; however, it is generally thought that the point of no return is the increase in mitochondrial membrane permeability, termed *mitochondrial outer membrane permeabilization*.[86] This leads to the loss of the electrochemical gradient between the mitochondrial membranes as well as the swelling of the organelle with water.[87] The ultimate consequences of these injuries involve the disruption of ATP gradient (discussed in a subsequent section) as well as the release of pro-apoptotic factors such as cytochrome-c, AIF, and SMAC/Diablo that initiate the apoptotic cascade as previously discussed.

What are the processes by which a cell may progress toward the point of no return? There are three prevailing hypotheses, which need not be mutually exclusive and are often described interchangeably in the literature. The first mechanism is mediated by pro-apoptotic proteins that either oligormerize and produce direct membrane permeabilization or increase membrane permeability indirectly through phospholipid oxidation.[88] The second involves the active formation of pores within the mitochondrial membrane. The third involves a sudden increase in the permeability of the inner mitochondrial membrane.[87] The mechanism by which mitochondrial injury occurs may determine cell fate, since the first two trigger apoptosis, whereas the latter initiates necrosis;[89] evidence in the literature exists for all three processes occurring following TBI.[90–92]

Mitochondrial Injury and Reactive Oxygen Species

A critical mechanism by which mitochondrial injury after TBI causes secondary neuronal damage is through the generation of free radical species, including reactive oxygen species (ROS) and reactive nitrogen species (RNS). Free radicals contain unpaired electrons and are therefore highly reactive; they can also initiate chain reactions that produce widespread damage. A variety of ROS and RNS are generated after TBI. Among the ROS, the superoxide radical $\left(O_2^-\right)$, hydroxyl radical (OH), and hydrogen peroxide (H_2O_2) species are known to be increased following TBI and stroke. Superoxide radical may cause the most tissue damage, as it can react in cytoplasm with hydrogen ions to generate a more potent ROS, OH_2. This reaction is favored in acidic environments, and mitochondrial failure after head injury results in a cellular energy crisis that results in the initiation of lactic acidosis and subsequent decreased pH that ultimately favors this reaction. Among the RNS, peroxynite (ONOO–) is highly reactive, known to be produced following TBI,[93] and can undergo reactions with cellular proteins and lipids.[94] This RNS is upregulated by nitric oxide synthase, and inhibition of NOS following TBI has been shown to improve both histological and behavioral outcomes in preclinical studies.[95]

The consequences of free radicals include lipid peroxidation. This results in breakdown of both cellular and organelle membranes, cellular edema and release of calcium, proteases, lysosomal enzymes, and ultimately cell death.[96] Cell death can occur through either frank necrosis or apoptosis through the activation of PARP (previously discussed in the apoptosis section). Other sublethal but damaging cellular consequences include protein degradation (often causing synaptic dysfunction); this dysfunction may lead to cognitive impairment after brain injury and perhaps even neurodegeneration.[97] To combat these reactive species, cells have a number of antioxidant molecules, which either remove the reactive species or inhibit their generation directly or indirectly (for example, by sequestering metal ions needed for their production). A diverse spectrum of antioxidants, including progesterone, vitamin E, and superoxide dismutase mimetic drugs are under study.

Cyclosporine as a Therapeutic for Mitochondrial Injury

Despite the critical role that mitochondrial damage plays in secondary neuronal injury after TBI, the inherent complexity of this organelle makes it a challenging therapeutic target. Interestingly, one of the more promising candidates is the immunosuppressant cyclosporine A (CsA). This extract from a Norwegian fungus was not initially considered for its neuroprotective properties, as it has limited CNS penetration through a healthy blood–brain barrier. Instead, it was used as an immunosuppressant primarily in the transplant setting, where it would bind to the protein cyclophilin A; the ultimate downstream consequence of this interaction was the inhibition of cytokines necessary for lymphocyte activation.[98]

Despite this canonical role, there were early hints that CsA might have direct neuroprotective properties. A compound with a similar mechanism of action, tacrolimis, was found to be neuroprotective in vitro.[99] Another study demonstrated that CsA promotes regeneration of spinal cord neurons after transection.[100] But the seminal discovery suggesting that that CsA might have clinically translatable neuroprotective properties came from a German case report in which a liver transplant patient on chronic CsA had an incidental TBI and displayed remarkable recovery.[101] Since then, several studies in animal models of TBI were pursued to validate potential neuroprotective effects and elucidate mechanisms of action. Early studies used intrathecal delivery of CsA in an impact-acceleration model in rats and found a reduction in cytoskeletal damage, axonal disruption, and APP accumulation in the acute phases following injury.[102,103] Subsequent studies suggested that the likely mechanism of action was through prevention of formation of mitochondrial membrane pores, thereby preventing cellular energy crisis and the generation of ROS.[104–107] Mechanistically, CsA is believed to bind Cyclophilin D, the cyclophilin located in mitochondria.[108] This complex then (1) inhibits NOS, preventing the formation of RNS like peroxynite; (2) sequesters BAD and BAX, preventing the release of cytochrome C from mitochondria; and (3) prevents the formation of the mitochondrial permeability transition pore.[109] CsA is currently in clinical trials. The unexpected finding that this long-used immunosuppressant may prove efficacious in preventing secondary neuronal injury emphasizes how clinical observation can inform basic science research.

EXCITOTOXICITY-INDUCED ENDOPLASMIC RETICULUM STRESS AND TBI: AN EMERGING THERAPEUTIC TARGET

Whereas most preclinical and clinical studies on secondary neuronal injury after TBI have focused on the role of excitotoxicity-mediated apoptosis and mitochondria, an emerging body of evidence suggests that ER stress which is associated with the intracellular accumulation of misfolded proteins may also contribute to the pathophysiology of TBI.[110] A variety of stimuli that occur following head injury are known to trigger ER stress, including increased NO and free radical formation,[111] metabolic dysfunction, and shifts in cellular pH.[112] However, a more recent hypothesis based on electron microscopy findings suggests that TBI leads to excitotoxicity-induced disruption of the ER membrane.[113] This phenomenon is known to be associated with other injuries that lead to excitotoxicity, such as seizures.[114] Singleton and colleagues found that disruption of the ER membrane occurs within hours of a moderate fluid percussion injury and is most evident in neurons with axonal injury.[113]

Misfolded proteins form toxic aggregates. In order to prevent this, cells undergoing ER stress initiate the unfolded protein response (UPR), a conserved set of signaling cascades designed to synthesize chaperones and proteases to clear misfolded proteins while stopping general protein translation to reduce the protein-load on the ER, (Figure 5.3). If the stress is prolonged, the UPR ultimately triggers apoptosis.[115] For a complete examination of the signaling pathways associated with the UPR, a number of review articles are available (e.g., Refs. 110 and 116).

The three primary proteins that initiate the UPR (PERK, IRE1a, and ATF6) remain inactivated by binding to the chaperone protein Grp78 on the ER membrane. As a molecular chaperone, Grp78 rapidly binds to misfolded proteins in order to assist in returning them to their appropriate confirmation; this results in the disinhibition of the initiator proteins. PERK and IRE1a dimerize and undergo phosphorylation, while ATF6 translocates to the Golgi where it is cleaved into its active

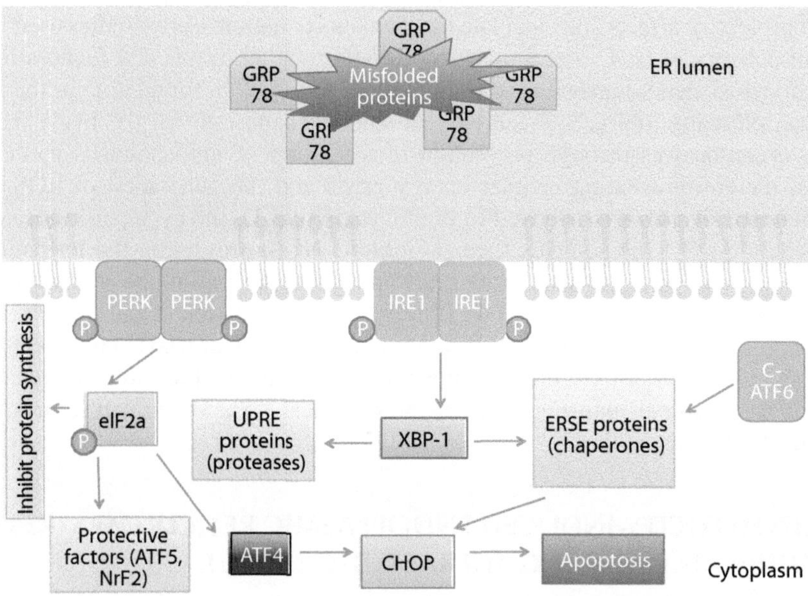

FIGURE 5.3 Diagram of the unfolded protein response. Accumulation of misfolded proteins in the ER lumen leads to the recruitment of the chaperone GRP78, disinhibiting PERK, IRE1, and ATF6. ATF6 is translocated to the Golgi, where it is cleaved into its active form and serves as transcription factors for protein chaperones. PERK and IRE1 dimerize and phosphorylate. P-PERK phosphorylates eIF2a, leading to the inhibition of protein synthesis and the transcription of protective factors. P-IRE1 cleaves the mRNA XBP-1, which then functions as a transcription factor for both proteases and chaperones. This "adaptive" arm of the UPR serves to clear the ER of unfolded proteins. If the ER stress is prolonged, the upregulation of CHOP serves as a "molecular timer" to trigger apoptosis.

form. Cleaved ATF6 acts as a transcription factor for protein chaperones (including Grp78); P-IRE1a cleaves Xbp-1, which then acts as a transcription factor for both protein chaperones and proteases. Together, these pathways reduce the burden of misfolded proteins within the cell (Figure 5.3).[115,116]

There are a complex cascade of molecular signals that regulate the initiation and termination of the unfolded protein response (Figure 5.3). For example, P-PERK acts as a kinase to phosphorylate the initiation factor eIF2a; p-eIF2a inhibits general protein synthesis, reducing the burden on the ER during a period of stress. However, p-eIF2a signaling is more complicated as it also upregulates ATF4, a transcription factor. ATF4 ultimately increases the expression of Gadd34, a protein that participates in the dephosphorylation of p-eIF2a. Gadd34 and Grp78 help "turn off" the UPR once the ER stress is alleviated by dephosphorylating eIF2a and inhibiting the initiator proteins PERK, IRE1a, and ATF6. However, ATF4 also acts as a cellular "timer" by increasing levels of CHOP; CHOP downregulates the anti-apoptotic protein Bcl-2 and upregulates caspase-12, triggering apoptosis if the cell does not respond to the stress in a timely manner.[115,116]

A number of preclinical injury models have demonstrated activation of the UPR in vivo. For example, Singleton and colleagues showed disruption of the ER membrane and elevated levels of p-eIF2a at 24 hours following a fluid percussion injury.[113] A more recent study demonstrated an increase in Grp78 mRNA levels within 4 hours of injury and that ER stress after TBI worsened with hypoxia.[117] Larner and colleagues demonstrated an increase in caspase-12 mRNA proportional to injury severity.[57] Using a severe contusion model, Begum and colleagues demonstrated that p-eIF2a, p-IRE1a, and ATF4 remained elevated 3 weeks after injury; interestingly, CHOP levels declined after 1 week, suggesting that ER-stress-induced apoptosis occurs within the first week following TBI but that adaptive functions of the UPR may remain active for several weeks.[118] Moderate closed head injury using weight drop demonstrated an acute increase in cleaved XBP1 in cortex, hippocampus, and striatum. Finally, although repeat mild head injury results in tau accumulation,[119] a scenario known to result in ER stress,[120] little work has been done to characterize the role of the UPR in this setting. However, Rubovitch and colleagues showed that a single mild injury increased activated ATF6 from 1 hour to 3 days after injury.[121] Taken together, these results provide evidence for ER stress occurring after single injury in multiple models, and lend support to the hypothesis that chronic ER stress may occur after repeat mild TBI.

A recent body of literature suggests that augmenting levels of p-eIF2a is protective during ER stress, providing a potential therapeutic target. A chemical-genetic approach demonstrated that pre-emptive activation of PERK and subsequent induction of p-eIF2a protected cells exposed to ER stress;[122] siRNA knockdown of the constitutively active p-eIF2a phosphatase produced the same effect.[123] Furthermore, a recent screen of protective compounds for cells exposed to ER stress identified potential neuroprotection by the small molecule salubrinal. Salubrinal was found to increase p-eIF2a by inhibiting its specific phosphatase,[124] and studies of this compound in various disease models, including Huntington's, Alzheimer's, Parkinson's, and ALS have all demonstrated neuroprotection.[115,120,125,126] Salubrinal was initially felt to represent a selective phosphatase inhibitor, but is now known to have off-target effects.[127] Protective effects of salubrinal have been demonstrated following excitotoxicity-induced ER stress. Concomitant administration of salubrinal with the chemoconvulsant kainic acid reduced neuronal death both in vitro and in vivo[114] and reduced mortality in a model of continuous seizure activity.[128] This pharmacological increase in p-eIF2a has also been shown to reduce brain damage in stroke models.[129,130] Interestingly, the augmentation of p-eIF2a need not occur after the insult; long-term salubrinal treatment did not cause any baseline cell death in HT22 cells or cultured neurons but still protected cells from subsequent glutamate toxicity.[131] These findings suggest the therapeutic potential of salubrinal in the prophylactic treatment of individuals at high risk for traumatic brain injury.

There are two mechanisms by which p-eIF2a may mediate protective effects following ER stress. First, the inhibition of general protein synthesis allows the cell to conserve energy while chaperones and proteases clear unfolded proteins.[110] Second, it was also recently shown that p-eIF2a upregulates the transcription factor ATF5 in response to ER stress,[132] and that ATF5 KO abolished the protective effects of salubrinal administration after kainic acid both in vitro and in vivo;[133] this suggests that the protective effects of p-eIF2a are mediated by upregulation of specific transcription factors, and that ATF5 may be the downstream link between p-eIF2a and neuroprotection. A more complete understanding of the activation of the UPR following TBI and the precise mechanism(s) by which salubrinal exerts its neuroprotective effects may help facilitate the translation of this small molecule to clinical trials.

NEUROINFLAMMATION AND TBI: BALANCING NEUROPROTECTION VERSUS NEUROTOXICITY

Neuroinflammation after TBI

Although neuroinflammation represents an important component of secondary neuronal injury following TBI, it is also clear that glial activation is associated with adaptive effects that enhance recovery. The CNS is generally considered to be an "immunologically privileged" site, devoid of the circulating immune cells because these cells are not permeable to the blood–brain barrier.[134] However, following traumatic injury there is a breakdown of the blood–brain barrier, due not only to shear forces but also the generation of ROS that damage endothelial cells.[135] The result is infiltration of the CNS parenchyma by neutrophils, macrophages, and lymphocytes. This infiltration has been seen in both preclinical and clinical studies.[136] Once infiltrated, these cells are thought to release neurotoxic cytokines that activate immune cells native to the CNS—microglia and astrocytes.[137] These cells have both protective and maladaptive effects following injury. Microglia initially fuse to form an enclosed area, perhaps containing local bleeding or toxic effects of cellular degradation.[138] However, if microglia are overactivated they can release ROS that potentiate cellular damage, as previously described.[139] Similarly, astrogliosis was previous thought to be primarily detrimental, resulting in secondary tissue injury, impairment in neural circuitry, and consequences such as post-traumatic epilepsy; however, a recent study demonstrated that preventing astrocytosis in a photothrombotic model of injury worsened the extent of cortical damage.[140]

Cytokines and Neuroinflammation

These findings suggest that inflammatory mediators released by glia may have both protective and toxic effects following injury. In particular, studies in the CSF of patients with severe TBI and biochemistry studies in animal models have primarily focused on two proinflammatory cytokines: IL-1B and TNFa.

In both a weight drop and fluid percussion injury model, pharmacological inhibition of the receptor for IL-1B reduced cell death, both at the site of injury and in hippocampus.[141,142] Furthermore, Tehranian and colleagues used a genetic approach to overexpress an IL-1R antagonist and demonstrated improved functional recovery after a weight-drop injury.[143] The mechanism by which IL-1B mediates neuroinflammation is not fully understood, though one possibility is its ability to activate cyclooxygenase, an enzyme central for inflammatory signaling.[144] Several other studies support the hypothesis that IL-1B plays a role in excitotoxicity. For example, direct infusion of NMDA into rats was found to increase IL-1B levels in cortical microglia and astrocytes,[145] and simultaneous infusion of IL-1 receptor bodies (molecules that mimic the IL-1R but have no functional properties and therefore serve to sequester ligand) reduce lesion size.[146] In vitro studies have shown that application of IL-1B to cultured neurons increases phosphorylation of NMDAR subunits[147] and application to human astrocyte cultures reduced expression of GLT-1[148] providing potential mechanisms linking IL-1B to excitotoxicity. Whether this finding occurs in vivo has yet to be investigated.

TNFa is also elevated following TBI in preclinical models and in the CSF of patients following severe brain injury.[149–151] TNFa is produced not just by immune cells but also by neurons[152] and alterations in the signaling pathways triggered by this molecule have been found in various CNS diseases, including Alzheimer's[153] and ALS.[154] TNFa is spliced by matrix metaloproteases upon release, and the spliced peptides can bind to different receptors to produce varying effects, including cellular apoptosis as well as activation of microglia and astrocytes.[154] Furthermore, TNFa has been shown to potentiate glutamatergic excitotoxicity both in vitro[155,156] as well as in vivo following kainic acid administration.[157] The precise mechanism(s) by which this occurs requires further elucidation but appears to be due to glial glutamate release.[158]

Given these observations, a number of preclinical studies have explored the possibility that inhibition of TNFa would be associated with neuroprotection. It was shown in a closed head injury model that inhibition of TNFa signaling improved behavioral outcomes, reduced neuroinflammation, and also reduced blood–brain barrier breakdown and resultant edema. However, the picture is not straightforward, since studies in TNFa nulls demonstrated worsened blood–brain barrier damage and both hippocampal and motor functional outcomes after TBI.[104,105] Similarly, TNF receptor knockouts had worsened histological damage after controlled cortical impact.[104,105] This may be due to differences in the acute versus chronic effect of TNFa, since beam balance scores following controlled cortical impact were worse in the day following injury but demonstrated more rapid recovery in the TNFa null.[159] These studies suggest that timing is extremely important when employing strategies that modulate inflammatory pathways and highlight the importance of rigorous preclinical testing to inform rational clinical trial design. Targeting inflammatory processes is an attractive therapeutic approach because a number of anti-inflammatory therapies are already FDA approved for autoimmune diseases, transplant surgery, and other settings. These drugs can potentially be repurposed for the treatment of head injury, but understanding the temporal progression and effect of neuroinflammation following TBI is critical.

Emerging Concepts in Neuroinflammation: The Role of Apolipoprotein E

It has become increasingly clear that neuroinflammation plays a critical, albeit complex, role in mediating both adaptive cellular responses to injury and exacerbating secondary neuronal injury. An emerging concept that may be of therapeutic value is the role of endogenous CNS proteins that are upregulated after injury, such as apolipoprotein E (apoE). There are three common protein isoforms of apoE, designated apoE2, apoE3, and apoE4, which differ by single amino acid interchanges at positions 112 and 158. Originally described in the context of cholesterol metabolism,[160] the finding that apoE4 was associated with increased risk of developing sporadic and late onset familial Alzheimer's disease[161,162] led to an intense research focus on the neurobiology of this protein. The brain produces a large amount of apoE,[163] which is upregulated after injury, suggesting a role in mediating acute injury responses. Interestingly, clinical observations have suggested that, in addition to its association with Alzheimer's disease, the apoE4 isoform is associated with poor functional outcome in a variety of acute brain injuries, including traumatic brain injury, intracranial hemorrhage, and subarachnoid hemorrhage.[164–168]

The mechanisms by which apoE alters brain injury remain incompletely defined, largely due to the variety of biological roles that this protein plays in the brain, but increasing evidence suggests that, in addition to its well-known roles in mediating amyloid deposition, apoE also modulates glial activation and neuroinflammatory responses. The knockout of murine apoE is associated with increased neuroinflammation and oxidative stress, and worsened histological and functional outcomes.[169,170] Interestingly, these effects appear to be isoform-specific, as the presence of the apoE4 protein isoform is associated with more robust CNS and systemic inflammatory responses in a number of preclinical models (reviewed in Ref. 171);[172] these results are also borne out in clinical studies.[173,174] Although the mechanisms by which ApoE modulates neuroinflammatory responses remain to be fully defined, a series of studies suggest that the anti-inflammatory effects of apoE are mediated by microglial LRP receptor.[175,176] The neuroprotective effects of endogenous apoE also suggest that augmenting the protective effects of the apoE3 isoform may serve as a novel therapeutic strategy. Although the apoE holoprotein does not cross the blood–brain barrier,[177] smaller peptides designed from the receptor binding region of apoE have been designed, and have demonstrated efficacy in a number of murine models, including TBI, stroke, subarachnoid hemorrhage, and intracerebral hemorrhage.[178–183]

SUMMARY

Preclinical research has allowed us to gain insight into the complex biochemical and cellular pathways that mediate primary and secondary neuronal injury after brain

TABLE 5.1
Candidate Therapeutic Targets for Prevention of Secondary Neuronal Injury

Candidate Therapeutic or Therapeutic Target	Additional Preclinical Studies Needed
	Glutamate Excitotoxicity
Membrane resealing agents	(1) Understanding of the respective contribution of membrane damage vs. mechanosensitive ion channels to glutamate influx; (2) development of more potent resealing agents.
Glutamate receptor antagonists	(1) Increased understanding of the process by which glutamate excitotoxocity causes cell death, with particular attention to the role of increased metabolic demand vs. calcium influx; (2) investigation of efficacy NR2B specific antagonists; (3) investigation of efficacy of pH-sensitive antagonists.
Glutamate transporter upregulation	(1) Establishment of causal relationship between downregulation of glutamate receptor transporters following injury and increased extracellular glutamate; (2) development of efficacious glutamate transport upregulators.
Gap junction inhibitors	Preclinical studies examining efficacy and safety of these blockers in vivo following injury.
Calpain inhibitors	(1) Resolution of conflicting preclinical data on these compounds; specifically, an examination of the model dependent (FPI vs. contusion) effect; (2) improved understanding of calpain activation following injury.
GABAR antagonists (Flumazenil)	Further study on the effect of TBI on cell populations vulnerable to GABAergic excitotoxicity; for example, identification of mechanisms of cell death (apoptosis vs. necrosis).
	Apoptosis
Combination therapy (for example, apoptosome inhibitor and pro-apoptotic Bcl-2 inhibitors)	Given the variety of potential therapeutic targets and the complexity of the various signaling cascades, a multifaceted approach to targeting apoptosis may be efficacious but needs to be studied in animal models.
RIPK1 inhibitors and other necroptosis inhibitors	Understanding the molecular mechanisms of necroptosis at greater detail. For example, what regulates the TNF-TNFR signaling switch from neuroprotection to neurotoxicity? Furthermore, how can necrosome formation occur in the absence of caspase-8?
	Mitochondria
Inhibitors of mitochondrial–lysosomal interactions	Improved understanding of this interaction in the context of TBI, including time course and relative contribution to injury.
	ER Stress
Salubrinal	(1) Characterization of ER stress post TBI; (2) determination of relative importance of ER stress in secondary neuronal injury following TBI; (3) determination of off-target effects of this drug.

(*Continued*)

TABLE 5.1 (*Continued*)
Candidate Therapeutic Targets for Prevention of Secondary Neuronal Injury

Candidate Therapeutic or Therapeutic Target	Additional Preclinical Studies Needed
	Inflammation
TNFa inhibitors	Increased understanding of the effect of the acute vs. chronic effect of TNFa signaling following TBI.
IL1-B inhibitors	Understanding of the signaling cascades mediated by IL1-B that contribute to an aberrant inflammatory response.

trauma. A more complete understanding of these fundamental mechanisms will enable the identification of new therapeutic strategies and will help rationally inform the design of early clinical trials in this area of compelling unmet clinical need (Table 5.1).

REFERENCES

1. Faden, A.I. et al. 1989. The role of excitatory amino acids and NMDA receptors in traumatic brain injury. *Science* 244(4906):798–800.
2. Palmer, A.M. et al. 1994. Increased transmitter amino acid concentration in human ventricular CSF after brain trauma. *Neuroreport* 6(1):153–156.
3. Hong, Z. et al. 2001. Excitatory amino acids in cerebrospinal fluid of patients with acute head injuries. *Clin Chem* 47(8):1458–1462.
4. Baker, A.J. et al. 1993. Excitatory amino acids in cerebrospinal fluid following traumatic brain injury in humans. *J Neurosurg* 79(3):369–372.
5. Chamoun, R. et al. 2010. Role of extracellular glutamate measured by cerebral microdialysis in severe traumatic brain injury. *J Neurosurg* 113(3):564.
6. Werner, C., Engelhard, K. 2007. Pathophysiology of traumatic brain injury. *Br J Anaesth* 99(1):4–9.
7. Greve, M.W., Zink, B.J. 2009. Pathophysiology of traumatic brain injury. *Mt Sinai J Med* 76(2):97–104.
8. Beppu, K. et al. 2014. Optogenetic countering of glial acidosis suppresses glial glutamate release and ischemic brain damage. *Neuron* 81(2):314–320.
9. Beschorner, R. et al. 2007. Expression of EAAT1 reflects a possible neuroprotective function of reactive astrocytes and activated microglia following human traumatic brain injury. *Histol Histopathol* 22(5):515–526.
10. Whalen, M.J. et al. 2007. Acute plasmalemma permeability and protracted clearance of injured cells after controlled cortical impact in mice. *J Cereb Blood Flow Metab* 28(3):490–505.
11. Mbye, L.H. et al. 2011. Kollidon VA64, a membrane-resealing agent, reduces histopathology and improves functional outcome after controlled cortical impact in mice. *J Cereb Blood Flow Metab* 32(3):515–524.
12. Miller, B.F. et al. 2014. The pharmacokinetics and pharmacodynamics of Kollidon VA64 dissociate its protective effects from membrane resealing after controlled cortical impact in mice. *J Cereb Blood Flow Metab* 34(8):1347–1353.
13. Tavalin, S.J., Ellis, E.F., Satin, L.S. 1995. Mechanical perturbation of cultured cortical neurons reveals a stretch-induced delayed depolarization. *J Neurophysiol* 74(6):2767–2773.

14. Patel, T.P. et al. 2014. Single-neuron NMDA receptor phenotype influences neuronal rewiring and reintegration following traumatic injury. *J Neurosci* 34(12):4200–4213.
15. Ikonomidou, C., Turski, L. 1996. Prevention of trauma-induced neurodegeneration in infant and adult rat brain: Glutamate antagonists. *Metab Brain Dis* 11(2):125–141.
16. Goda, M. et al. 2002. Both MK801 and NBQX reduce the neuronal damage after impact-acceleration brain injury. *J Neurotrauma* 19(11):1445–1456.
17. Rao, V.L. et al. 2001. Antisense knockdown of the glial glutamate transporter GLT-1 exacerbates hippocampal neuronal damage following traumatic injury to rat brain. *Eur J Neurosci* 13(1):119–128.
18. Zhang, C. et al. 1998. Riluzole attenuates cortical lesion size, but not hippocampal neuronal loss, following traumatic brain injury in the rat. *J Neurosci Res* 52(3):342–349.
19. Okiyama, K. et al. 1997. Effects of the novel NMDA antagonists CP-98,113, CP-101,581 and CP-101,606 on cognitive function and regional cerebral edema following experimental brain injury in the rat. *J Neurotrauma* 14(4):211–222.
20. Yurkewicz, L. et al. 2005. The effect of the selective NMDA receptor antagonist traxoprodil in the treatment of traumatic brain injury. *J Neurotrauma* 22(12):1428–1443.
21. Yi, J.-H., Hazell, A.S. 2006. Excitotoxic mechanisms and the role of astrocytic glutamate transporters in traumatic brain injury. *Neurochem Int* 48(5):394–403.
22. Van Landeghem, F.K.H. et al. 2001. Early expression of glutamate transporter proteins in ramified microglia after controlled cortical impact injury in the rat. *Glia* 35(3):167–179.
23. Hartfield, E.M. et al. 2011. Connexin 36 expression regulates neuronal differentiation from neural progenitor cells. *PLoS One* 6(3):e14746.
24. Todd, K.L., Kristan, W.B., French, K.A. 2010. Gap junction expression is required for normal chemical synapse formation. *J Neurosci* 30(45):15277–15285.
25. Wang, Y. et al. 2012. Neuronal gap junction coupling is regulated by glutamate and plays critical role in cell death during neuronal injury. *J Neurosci* 32(2):713–725.
26. Araújo, I.M. et al. 2010. Calpains and delayed calcium deregulation in excitotoxicity. *Neurochem Res* 35(12):1966–1969.
27. Sattler, R. et al. 1998. Distinct influx pathways, not calcium load, determine neuronal vulnerability to calcium neurotoxicity. *J Neurochem* 71(6):2349–2364.
28. Huang, Y., Wang, K.K.W. 2001. The calpain family and human disease. *Trends Mol Med* 7(8):355–362.
29. Liu, J., Liu, M.C., Wang, K.K.W. 2008. Calpain in the CNS: From synaptic function to neurotoxicity. *Sci Signal* 1(14):re1.
30. Xu, W. et al. 2007. Calpain-mediated mGluR1α truncation: A key step in excitotoxicity. *Neuron* 53(3):399–412.
31. Saatman, K.E., Creed, J., Raghupathi, R. 2010. Calpain as a therapeutic target in traumatic brain injury. *Neurotherapeutics* 7(1):31–42.
32. Saatman, K.E. et al. 2003. Traumatic axonal injury results in biphasic calpain activation and retrograde transport impairment in mice. *J Cereb Blood Flow Metab* 23(1):34–42.
33. Liu, L., Yenari, M.A. 2007. Therapeutic hypothermia: Neuroprotective mechanisms. *Front Biosci* 12:816–825.
34. Posmantur, R. et al. 1997. A calpain inhibitor attenuates cortical cytoskeletal protein loss after experimental traumatic brain injury in the rat. *Neuroscience* 77(3):875–888.
35. Haranishi, Y. et al. 2005. Moderate hypothermia, but not calpain inhibitor 2, attenuates the proteolysis of microtubule-associated protein 2 in the hippocampus following traumatic brain injury in rats. *Eur J Anaesthesiol* 22(2):140–147.
36. Saatman, K.E. et al. 1996. Calpain inhibitor AK295 attenuates motor and cognitive deficits following experimental brain injury in the rat. *Proc Natl Acad Sci U S A* 93(8):3428–3433.

37. Saatman, K.E. et al. 2000. Behavioral efficacy of posttraumatic calpain inhibition is not accompanied by reduced spectrin proteolysis, cortical lesion, or apoptosis. *J Cereb Blood Flow Metab* 20(1):66–73.
38. Anderson, K.J. et al. 2005. Regional distribution of fluoro-jade B staining in the hippocampus following traumatic brain injury. *Exp Neurol* 193(1):125–130.
39. Sato, M. et al. 2001. Neuronal injury and loss after traumatic brain injury: Time course and regional variability. *Brain Res* 917(1):45–54.
40. Mongiat, L.A., Schinder, A.F. 2011. Adult neurogenesis and the plasticity of the dentate gyrus network. *Eur J Neurosci* 33(6):1055–1061.
41. Deng, W., Aimone, J.B., Gage, F.H. 2010. New neurons and new memories: How does adult hippocampal neurogenesis affect learning and memory? *Nat Rev Neurosci* 11(5):339–350.
42. Gao, X. et al. 2008. Selective death of newborn neurons in hippocampal dentate gyrus following moderate experimental traumatic brain injury. *J Neurosci Res* 86(10):2258–2270.
43. Gao, X., Chen, J. 2009. Conditional knockout of brain-derived neurotrophic factor in the hippocampus increases death of adult-born immature neurons following traumatic brain injury. *J Neurotrauma* 26(8):1325–1335.
44. Zhou, H. et al. 2012. Moderate traumatic brain injury triggers rapid necrotic death of immature neurons in the hippocampus. *J Neuropathol Exp Neurol* 71(4):348.
45. Zhao, Y.L. et al. 2011. GABAergic excitotoxicity injury of the immature hippocampal pyramidal neurons' exposure to isoflurane. *Anesth Analg* 113(5):1152–1160.
46. Li, Y. et al. 2013. JNK pathway may be involved in isoflurane-induced apoptosis in the hippocampi of neonatal rats. *Neurosci Lett* 545:17–22.
47. Nilsson, P. et al. 1990. Changes in cortical extracellular levels of energy-related metabolites and amino acids following concussive brain injury in rats. *J Cereb Blood Flow Metab* 10(5):631–637.
48. Obrenovitch, T.P., Urenjak, J. 1997. Is high extracellular glutamate the key to excitotoxicity in traumatic brain injury? *J Neurotrauma* 14(10):677–698.
49. Carbonell, W.S., Grady, M. 1999. Evidence disputing the importance of excitotoxicity in hippocampal neuron death after experimental traumatic brain injury. *Ann N Y Acad Sci* 890(1):287–298.
50. Menniti, F.S. et al. 2000. CP-101,606, an NR2B subunit selective NMDA receptor antagonist, inhibits NMDA and injury induced c-fos expression and cortical spreading depression in rodents. *Neuropharmacology* 39(7):1147–1155.
51. Afshari, D., Moradian, N., Rezaei, M. 2013. Evaluation of the intravenous magnesium sulfate effect in clinical improvement of patients with acute ischemic stroke. *Clin Neurol Neurosurg* 115(4):400–404.
52. Zetterberg, H., Smith, D.H., Blennow, K. 2013. Biomarkers of mild traumatic brain injury in cerebrospinal fluid and blood. *Nat Rev Neurol* 9(4):201–210.
53. Kerr, J.F.R., Wyllie, A.H., Currie, A.R. 1972. Apoptosis: A basic biological phenomenon with wide-ranging implications in tissue kinetics. *Br J Cancer* 26(4):239.
54. Alberts, B. et al. 2013. *Essential Cell Biology*. Garland Science.
55. Tait, S.W.G., Green, D.R. 2010. Mitochondria and cell death: Outer membrane permeabilization and beyond. *Nat Rev Mol Cell Biol* 11(9):621–632.
56. Cao, G. et al. 2007. Critical role of calpain I in mitochondrial release of apoptosis-inducing factor in ischemic neuronal injury. *J Neurosci* 27(35):9278–9293.
57. Larner, S.F. et al. 2004. Increased expression and processing of caspase-12 after traumatic brain injury in rats. *J Neurochem* 88(1):78–90.
58. Elmore, S. 2007. Apoptosis: A review of programmed cell death. *Toxicol Pathol* 35(4):495–516.

59. Bröker, L.E., Kruyt, F.A.E., Giaccone, G. 2005. Cell death independent of caspases: A review. *Clin Cancer Res* 11(9):3155–3162.
60. Zhang, X. et al. 2004. Bench-to-bedside review: Apoptosis/programmed cell death triggered by traumatic brain injury. *Crit Care* 9(1):66.
61. Adams, J.M., Cory, S. 1998. The Bcl-2 protein family: Arbiters of cell survival. *Science* 281(5381):1322–1326.
62. Napieralski, J.A., Raghupathi, R., McIntosh, T.K. 1999. The tumor-suppressor gene, p53, is induced in injured brain regions following experimental traumatic brain injury. *Mol Brain Res* 71(1):78–86.
63. Kaya, S.S. et al. 1999. Apoptosis and expression of p53 response proteins and cyclin D1 after cortical impact in rat brain. *Brain Res* 818(1):23–33.
64. Rachmany, L. et al. 2013. Exendin-4 induced glucagon-like peptide-1 receptor activation reverses behavioral impairments of mild traumatic brain injury in mice. *Age* 35(5):1621–1636.
65. Clark, R.S.B. et al. 1997. Apoptosis-suppressor gene Bcl-2 expression after traumatic brain injury in rats. *J Neurosci* 17(23):9172–9182.
66. Evans, J.P., Scheinker, I.M. 1944. Histologic studies of the brain following head trauma: Late changes, atrophic sclerosis of the white matter. *J Neurosurg* 1:306–320.
67. Ng, I. et al. 2000. Apoptosis occurs after cerebral contusions in humans. *Neurosurgery* 46(4):949–956.
68. Wagner, A.K. et al. 2011. CSF Bcl-2 and cytochrome C temporal profiles in outcome prediction for adults with severe TBI. *J Cereb Blood Flow Metab* 31(9):1886–1896.
69. Lenzlinger, P.M. et al. 2002. Prolonged intrathecal release of soluble Fas following severe traumatic brain injury in humans. *J Neuroimmunol* 122(1):167–174.
70. Yakovlev, A.G. et al. 1997. Activation of CPP32-like caspases contributes to neuronal apoptosis and neurological dysfunction after traumatic brain injury. *J Neurosci* 17(19):7415–7424.
71. Clark, R.S.B. et al. 2000. Caspase-3 mediated neuronal death after traumatic brain injury in rats. *J Neurochem* 74(2):740–753.
72. O'Dell, D.M. et al. 2000. Traumatic brain injury alters the molecular fingerprint of TUNEL-positive cortical neurons in vivo: A single-cell analysis. *J Neurosci* 20(13):4821–4828.
73. Qiu, J. et al. 2002. Upregulation of the Fas receptor death-inducing signaling complex after traumatic brain injury in mice and humans. *J Neurosci* 22(9):3504–3511.
74. Zhang, X. et al. 2002. Intranuclear localization of apoptosis-inducing factor (AIF) and large scale DNA fragmentation after traumatic brain injury in rats and in neuronal cultures exposed to peroxynitrite. *J Neurochem* 82(1):181–191.
75. Kroemer, G. et al. 2009. Classification of cell death: Recommendations of the Nomenclature Committee on Cell Death 2009. *Cell Death Diff* 16(1):3–11.
76. Vercammen, D. et al. 1998. Inhibition of caspases increases the sensitivity of L929 cells to necrosis mediated by tumor necrosis factor. *J Exp Med* 187(9):1477–1485.
77. Laster, S.M., Wood, J.G., and Gooding, L.R. 1988. Tumor necrosis factor can induce both apoptic and necrotic forms of cell lysis. *J Immunol* 141(8):2629–2634.
78. Hsu, H. et al. 1996. TNF-dependent recruitment of the protein kinase RIP to the TNF receptor-1 signaling complex. *Immunity* 4(4):387–396.
79. Degterev, A. et al. 2008. Identification of RIP1 kinase as a specific cellular target of necrostatins. *Nat Chem Biol* 4(5):313–321.
80. Linkermann, A., Green, D.R. 2014. Necroptosis. *N Engl J Med* 370(5):455–465.
81. Dunai, Z., Bauer, P.I., Mihalik, R. 2011. Necroptosis: Biochemical, physiological and pathological aspects. *Pathol Oncol Res* 17(4):791–800.

82. Galluzzi, L. et al. 2011. Programmed necrosis from molecules to health and disease. *Int Rev Cell Mol Biol* 289:1–35.
83. Giza, C.C., Hovda, D.A. 2001. The neurometabolic cascade of concussion. *J Athl Train* 36(3):228.
84. Rizzuto, R. et al. 2009. Ca^{2+} transfer from the ER to mitochondria: When, how and why. *Biochim Biophys Acta (BBA)-Bioenerg* 1787(11):1342–1351.
85. Boya, P. et al. 2003. Mitochondrial membrane permeabilization is a critical step of lysosome-initiated apoptosis induced by hydroxychloroquine. *Oncogene* 22(25):3927–3936.
86. Chipuk, J.E., Green, D.R. 2008. How do BCL-2 proteins induce mitochondrial outer membrane permeabilization? *Trends Cell Biol* 18(4):157–164.
87. Galluzzi, L. et al. 2008. Viral control of mitochondrial apoptosis. *PLoS Pathog* 4(5):e1000018.
88. Chen, R., Yang, L., McIntyre, T.M. 2007. Cytotoxic phospholipid oxidation products cell death from mitochondrial damage and the intrinsic caspase cascade. *J Biol Chem* 282(34):24842–24850.
89. Baines, C.P. et al. 2005. Loss of cyclophilin D reveals a critical role for mitochondrial permeability transition in cell death. *Nature* 434(7033):658–662.
90. Sullivan, P.G. et al. 2005. Mitochondrial permeability transition in CNS trauma: Cause or effect of neuronal cell death? *J Neurosci Res* 79(1–2):231–239.
91. Luo, C.-L. et al. 2013. Poloxamer 188 attenuates in vitro traumatic brain injury-induced mitochondrial and lysosomal membrane permeabilization damage in cultured primary neurons. *J Neurotrauma* 30(7):597–607.
92. Garjan, T.G. et al. 2014. A novel traumatic brain injury model for induction of mild brain injury in rats. *J Neurosci Methods* 233:18–27.
93. Gahm, C., Holmin, S., Mathiesen, T. 2000. Temporal profiles and cellular sources of three nitric oxide synthase isoforms in the brain after experimental contusion. *Neurosurgery* 46(1):169–177.
94. Mésenge, C. et al. 1998. Reduction of tyrosine nitration after Nω-nitro-l-arginine-methylester treatment of mice with traumatic brain injury. *Eur J Pharmacol* 353(1):53–57.
95. Wada, K. et al. 1998. Role of nitric oxide in traumatic brain injury in the rat. *J Neurosurg* 89(5):807–818.
96. Cornelius, C. et al. 2013. Traumatic brain injury: Oxidative stress and neuroprotection. *Antioxid Redox Signal* 19(8):836–853.
97. Povlishock, J.T. 1993. Pathobiology of traumatically induced axonal injury in animals and man. *Ann Emerg Med* 22(6):980–986.
98. Cardenas, M.E., Zhu, D., Heitman, J. 1995. Molecular mechanisms of immunosuppression by cyclosporine, FK506, and rapamycin. *Curr Opin Nephrol Hypertens* 4(6):472–477.
99. Yardin, C. et al. 1998. FK506 antagonizes apoptosis and c-jun protein expression in neuronal cultures. *Neuroreport* 9(9):2077–2080.
100. Palladini, G. et al. 1995. Treatment with cyclosporine A promotes axonal regeneration in rats submitted to transverse section of the spinal cord—II—Recovery of function. *J Hirnforsch* 37(1):145–153.
101. Gogarten, W. et al. 1998. A case of severe cerebral trauma in a patient under chronic treatment with cyclosporine A. *J Neurosurg Anesthesiol* 10(2):101–105.
102. Okonkwo, D.O. et al. 1999. Cyclosporin A limits calcium-induced axonal damage following traumatic brain injury. *Neuroreport* 10(2):353–358.
103. Büki, A., Okonkwo, D.O., Povlishock, J.T. 1999. Postinjury cyclosporin A administration limits axonal damage and disconnection in traumatic brain injury. *J Neurotrauma* 16(6):511–521.

104. Sullivan, P.G. et al. 1999. Exacerbation of damage and altered NF-κB activation in mice lacking tumor necrosis factor receptors after traumatic brain injury. *J Neurosci* 19(15):6248–6256.
105. Sullivan, P.G., Thompson, M.B., Scheff, S.W. 1999. Cyclosporin A attenuates acute mitochondrial dysfunction following traumatic brain injury. *Exp Neurol* 160(1):226–234.
106. Nieminen, A.L. et al. 1996. Cyclosporin A delays mitochondrial depolarization induced by N-methyl-D-aspartate in cortical neurons: Evidence of the mitochondrial permeability transition. *Neuroscience* 75(4):993–997.
107. Scheff, S.W., Sullivan, P.G. 1999. Cyclosporin A significantly ameliorates cortical damage following experimental traumatic brain injury in rodents. *J Neurotrauma* 16(9):783–792.
108. Snyder, S.H. et al. 1998. Neural actions of immunophilin ligands. *Trends Pharmacol Sci* 19(1):21–26.
109. Lulic, D. et al. 2011. A review of laboratory and clinical data supporting the safety and efficacy of cyclosporin A in traumatic brain injury. *Neurosurgery* 68(5):1172–1186.
110. Larner, S.F., Hayes, R.L., Wang, K.K.W. 2006. Unfolded protein response after neurotrauma. *J Neurotrauma* 23(6):807–829.
111. Liu, X., Schnellmann, R.G. 2003. Calpain mediates progressive plasma membrane permeability and proteolysis of cytoskeleton-associated paxillin, talin, and vinculin during renal cell death. *J Pharmacol Exp Ther* 304(1):63–70.
112. Paschen, W., Frandsen, A. 2001. Endoplasmic reticulum dysfunction—A common denominator for cell injury in acute and degenerative diseases of the brain? *J Neurochem* 79(4):719–725.
113. Singleton, R.H. et al. 2002. Traumatically induced axotomy adjacent to the soma does not result in acute neuronal death. *J Neurosci* 22(3):791–802.
114. Sokka, A.-L. et al. 2007. Endoplasmic reticulum stress inhibition protects against excitotoxic neuronal injury in the rat brain. *J Neurosci* 27(4):901–908.
115. Kim, I., Xu, W., Reed, J.C. 2008. Cell death and endoplasmic reticulum stress: Disease relevance and therapeutic opportunities. *Nat Rev Drug Discov* 7(12):1013–1030.
116. Ron, D., Walter, P. 2007. Signal integration in the endoplasmic reticulum unfolded protein response. *Nat Rev Mol Cell Biol* 8(7):519–529.
117. Truettner, J.S. et al. 2007. Subcellular stress response after traumatic brain injury. *J Neurotrauma* 24(4):599–612.
118. Begum, G. et al. 2014. Docosahexaenoic acid reduces ER stress and abnormal protein accumulation and improves neuronal function following traumatic brain injury. *J Neurosci* 34(10):3743–3755.
119. Hawkins, B.E. et al. 2013. Rapid accumulation of endogenous tau oligomers in a rat model of traumatic brain injury possible link between traumatic brain injury and sporadic tauopathies. *J Biol Chem* 288(23):17042–17050.
120. Lee, K.-S. et al. 2010. Activation of PERK signaling attenuates Aβ-mediated ER stress. *PLoS One* 5(5):e10489.
121. Rubovitch, V. et al. 2011. Does IGF-1 administration after a mild traumatic brain injury in mice activate the adaptive arm of ER stress? *Neurochem Int* 58(4):443–446.
122. Lu, P.D. et al. 2004. Cytoprotection by pre-emptive conditional phosphorylation of translation initiation factor 2. *EMBO J* 23(1):169–179.
123. Jousse, C. et al. 2003. Inhibition of a constitutive translation initiation factor 2α phosphatase, CReP, promotes survival of stressed cells. *J Cell Biol* 163(4):767–775.
124. Boyce, M. et al. 2005. A selective inhibitor of eIF2α dephosphorylation protects cells from ER stress. *Science* 307(5711):935–939.

125. Reijonen, S. et al. 2008. Inhibition of endoplasmic reticulum stress counteracts neuronal cell death and protein aggregation caused by N-terminal mutant huntingtin proteins. *Exp Cell Res* 314(5):950–960.
126. Saxena, S., Cabuy, E., Caroni, P. 2009. A role for motoneuron subtype–selective ER stress in disease manifestations of FALS mice. *Nat Neurosci* 12(5):627–636.
127. Huang, X. et al. 2012. Salubrinal attenuates β-amyloid-induced neuronal death and microglial activation by inhibition of the NF-κB pathway. *Neurobiol Aging* 33(5):1007.e9–e17.
128. Kim, J.S. et al. 2014. Salubrinal, ER stress inhibitor, attenuates kainic acid-induced hippocampal cell death. *J Neural Transm* 121(10):1233–1243.
129. Kumar, R. et al. 2003. Dysfunction of the unfolded protein response during global brain ischemia and reperfusion. *J Cereb Blood Flow Metab* 23(4):462–471.
130. Nakka, V.P., Gusain, A., Raghubir, R. 2010. Endoplasmic reticulum stress plays critical role in brain damage after cerebral ischemia/reperfusion in rats. *Neurotox Res* 17(2):189–202.
131. Lewerenz, J., Maher, P. 2009. Basal levels of eIF2α phosphorylation determine cellular antioxidant status by regulating ATF4 and xCT expression. *J Biol Chem* 284(2):1106–1115.
132. Zhou, D. et al. 2008. Phosphorylation of eIF2 directs ATF5 translational control in response to diverse stress conditions. *J Biol Chem* 283(11):7064–7073.
133. Torres-Peraza, J.F. et al. 2013. Protective neuronal induction of ATF5 in endoplasmic reticulum stress induced by status epilepticus. *Brain* 136(4):1161–1176.
134. Gendelman, H.E. 2002. Neural immunity: Friend or foe? *J Neurovirol* 8(6):474–479.
135. Chodobski, A., Zink, B.J., Szmydynger-Chodobska, J. 2011. Blood–brain barrier pathophysiology in traumatic brain injury. *Transl Stroke Res* 2(4):492–516.
136. Morganti-Kossmann, M.C., Yan, E., Bye, N. 2010. Animal models of traumatic brain injury: Is there an optimal model to reproduce human brain injury in the laboratory? *Injury* 41:S10–S13.
137. Kumar, A., Loane, D.J. 2012. Neuroinflammation after traumatic brain injury: Opportunities for therapeutic intervention. *Brain Behav Immun* 26(8):1191–1201.
138. Dardiotis, E. et al. 2008. Alpha-1 antichymotrypsin gene signal peptide a/t polymorphism and primary intracerebral hemorrhage. *Eur Neurol* 59(6):307–314.
139. Block, M.L., Hong, J.-S. 2005. Microglia and inflammation-mediated neurodegeneration: Multiple triggers with a common mechanism. *Prog Neurobiol* 76(2):77–98.
140. Benner, E.J. et al. 2013. Protective astrogenesis from the SVZ niche after injury is controlled by Notch modulator Thbs4. *Nature* 497(7449):369–373.
141. Lu, K.-T. et al. 2005. Effect of interleukin-1 on traumatic brain injury-induced damage to hippocampal neurons. *J Neurotrauma* 22(8):885–895.
142. Toulmond, S., Rothwell, N.J. 1995. Interleukin-1 receptor antagonist inhibits neuronal damage caused by fluid percussion injury in the rat. *Brain Res* 671(2):261–266.
143. Tehranian, R. et al. 2002. Improved recovery and delayed cytokine induction after closed head injury in mice with central overexpression of the secreted isoform of the interleukin-1 receptor antagonist. *J Neurotrauma* 19(8):939–951.
144. Salvolini, E. et al. 2014. Interleukin-1β, cyclooxygenase-2, and hypoxia-inducible factor-1α in asthenozoospermia. *Histochem Cell Biol* 142(5):569–575.
145. Pearson, V.L., Rothwell, N.J., Toulmond, S. 1999. Excitotoxic brain damage in the rat induces interleukin-1β protein in microglia and astrocytes: Correlation with the progression of cell death. *Glia* 25(4):311–323.
146. Relton, J.K., Rothwell, N.J. 1992. Interleukin-1 receptor antagonist inhibits ischaemic and excitotoxic neuronal damage in the rat. *Brain Res Bull* 29(2):243–246.

147. Viviani, B. et al. 2003. Interleukin-1β enhances NMDA receptor-mediated intracellular calcium increase through activation of the Src family of kinases. *J Neurosci* 23(25):8692–8700.
148. Hu, S. et al. 2000. Cytokine effects on glutamate uptake by human astrocytes. *Neuroimmunomodulation* 7(3):153–159.
149. Fan, L. et al. 1996. Experimental brain injury induces differential expression of tumor necrosis factor-α mRNA in the CNS. *Mol Brain Res* 36(2):287–291.
150. Ross, S.A. et al. 1994. The presence of tumour necrosis factor in CSF and plasma after severe head injury. *Br J Neurosurg* 8(4):419–425.
151. Shohami, E. et al. 1996. Inhibition of tumor necrosis factor alpha (TNFα) activity in rat brain is associated with cerebroprotection after closed head injury. *J Cereb Blood Flow Metab* 16(3):378–384.
152. Gahring, L.C. et al. 1996. Neuronal expression of tumor necrosis factor alpha in the OVOUJI fme brain. *Neuroimmunomodulation* 3(5):289–303.
153. Swardfager, W. et al. 2010. A meta-analysis of cytokines in Alzheimer's disease. *Biol Psychiatry* 68(10):930–941.
154. Olmos, G., Lladó, J. 2014. Tumor necrosis factor alpha: A link between neuroinflammation and excitotoxicity. *Mediators Inflamm* 2014:861231.
155. Gelbard, H.A. et al. 1993. Neurotoxic effects of tumor necrosis factor alpha in primary human neuronal cultures are mediated by activation of the glutamate AMPA receptor subtype: Implications for AIDS neuropathogenesis. *Dev Neurosci* 15(6):417–422.
156. Chao, C.C., Hu, S. 1994. Tumor necrosis factor-alpha potentiates glutamate neurotoxicity in human fetal brain cell cultures. *Dev Neurosci* 16(3–4):172–179.
157. Hermann, G.E. et al. 2001. Tumor necrosis factor-α induces cFOS and strongly potentiates glutamate-mediated cell death in the rat spinal cord. *Neurobiol Dis* 8(4):590–599.
158. Takeuchi, H. et al. 2008. Blockade of microglial glutamate release protects against ischemic brain injury. *Exp Neurol* 214(1):144–146.
159. Scherbel, U. et al. 1999. Differential acute and chronic responses of tumor necrosis factor-deficient mice to experimental brain injury. *Proc Natl Acad Sci U S A* 96(15):8721–8726.
160. Rall, S.C. et al. 1982. Structural basis for receptor binding heterogeneity of apolipoprotein E from type III hyperlipoproteinemic subjects. *Proc Natl Acad Sci U S A* 79(15):4696–4700.
161. Corder, E.H. et al. 1995. The apolipoprotein E E4 allele and sex-specific risk of Alzheimer's disease. *JAMA* 273(5):373–374.
162. Strittmatter, W.J., Roses, A.D. 1996. Apolipoprotein E and Alzheimer's disease. *Annu Rev Neurosci* 19(1):53–77.
163. Huang, Y. et al. 2004. Apolipoprotein E. *J Mol Neurosci* 23(3):189–204.
164. Alfieri, A. et al. 2008. Psychosocial and neurocognitive performance after spontaneous nonaneurysmal subarachnoid hemorrhage related to the APOE-ϵ 4 genotype: A prospective 5-year follow-up study: Clinical article. *J Neurosurg* 109(6):1019–1026.
165. Nicoll, J.A.R., Roberts, G.W., Graham, D.I. 1995. Apolipoprotein E epsilon 4 allele is associated with deposition of amyloid β-protein following head injury. *Nat Med* 1(2):135–137.
166. Rannikmäe, K. et al. 2014. APOE associations with severe CAA-associated vasculopathic changes: Collaborative meta-analysis. *J Neurol Neurosurg Psychiatry* 85(3):300–305.
167. Teasdale, G.M. et al. 1997. Association of apolipoprotein E polymorphism with outcome after head injury. *Lancet* 350(9084):1069–1071.
168. Zhou, W. et al. 2008. Meta-analysis of APOE4 allele and outcome after traumatic brain injury. *J Neurotrauma* 25(4):279–290.
169. Chen, Y. et al. 1997. Motor and cognitive deficits in apolipoprotein E-deficient mice after closed head injury. *Neuroscience* 80(4):1255–1262.

170. Lynch, J.R. et al. 2001. Apolipoprotein E modulates glial activation and the endogenous central nervous system inflammatory response. *J Neuroimmunol* 114(1):107–113.
171. Laskowitz, D.T., Vitek, M.P. 2007. Apolipoprotein E and neurological disease: Therapeutic potential and pharmacogenomic interactions. *Pharmacogenomics* 8(8):959–969.
172. Lynch, J.R. et al. 2003. APOE genotype and an ApoE-mimetic peptide modify the systemic and central nervous system inflammatory response. *J Biol Chem* 278(49):48529–48533.
173. Gale, S.C. et al. 2014. APOε4 is associated with enhanced in vivo innate immune responses in human subjects. *J Allergy Clin Immunol* 134(1):127–134.
174. Moretti, E.W. et al. 2005. APOE polymorphism is associated with risk of severe sepsis in surgical patients. *Crit Care Med* 33(11):2521–2526.
175. Pocivavsek, A., Burns, M.P., Rebeck, G.W. 2009. Low-density lipoprotein receptors regulate microglial inflammation through c-Jun N-terminal kinase. *Glia* 57(4):444–453.
176. Choi, H.Y. et al. 2013. APP interacts with LRP4 and agrin to coordinate the development of the neuromuscular junction in mice. *Elife* 2:e00220.
177. Linton, M.F., Atkinson, J.B., Fazio, S. 1995. Prevention of atherosclerosis in apolipoprotein E-deficient mice by bone marrow transplantation. *Science* 267(5200):1034–1037.
178. Hoane, M.R. et al. 2007. The novel apolipoprotein E-based peptide COG1410 improves sensorimotor performance and reduces injury magnitude following cortical contusion injury. *J Neurotrauma* 24(7):1108–1118.
179. Laskowitz, D.T. et al. 2007. COG1410, a novel apolipoprotein E-based peptide, improves functional recovery in a murine model of traumatic brain injury. *J Neurotrauma* 24(7):1093–1107.
180. Lei, B. et al. 2012. Interaction between sex and apolipoprotein E genetic background in a murine model of intracerebral hemorrhage. *Transl Stroke Res* 3(1):94–101.
181. Lynch, J.R. et al. 2005. A novel therapeutic derived from apolipoprotein E reduces brain inflammation and improves outcome after closed head injury. *Exp Neurol* 192(1):109–116.
182. Mesis, R.G. et al. 2006. Dissociation between vasospasm and functional improvement in a murine model of subarachnoid hemorrhage. *Neurosurg Focus* 21(3):1–7.
183. Wang, H. et al. 2013. Apolipoprotein E mimetic peptides improve outcome after focal ischemia. *Exp Neurol* 241:67–74.

6 Neuroinflammation

Finnian Hanrahan and Matthew Campbell

CONTENTS

INTRODUCTION

Traumatic brain injury (TBI) is the leading cause of death among young people and the elderly in the developed world, and results mainly from falls, violence, and road accidents. Across Europe the incidence of TBI is approximately 235/100,000.[1] While many people recover fully from TBI, many others are left with significant mental, behavioral, and motor deficits, and an estimated figure for people living with ongoing effects from a TBI in Europe is over 7 million.[1] In the United States alone, approximately 1.7 million people suffer a TBI each year, with 275,000 being hospitalized and 52,000 cases resulting in death.[2]

Clinically, TBIs are classified as being mild, moderate, or severe. These classifications are commonly arrived at by means of the Glasgow Coma Scale (GCS), which measures eye response (1–4 points), verbal response (1–5 points), and motor response (1–6 points). A combined GCS of $\geq$13 indicates a mild TBI, GCS of 9–12 indicates a moderate TBI, and a GCS of $<$9 indicates a severe TBI. Other scales are occasionally used however, including the Abbreviated Injury Scale, Injury Severity Score, and the length of Post-Traumatic Amnesia.[1]

The types of damage acquired in TBI can be separated into primary and secondary effects. Primary effects are acquired nearly instantly in a TBI as a direct result of damage due to contact or acceleration/deceleration injury, and these can include leaky ion channels, membrane microporation, conformational changes in proteins, and shearing of blood vessels causing hemorrhage.[3] Secondary effects, meanwhile, lie downstream of these primary effects and begin within minutes or hours of the primary insult, and these effects and their underlying molecular pathobiology are all targets for potential treatments for TBI. They include mitochondrial and subsequent energy production impairment, immune cell infiltration and activation, cytokine release, edema formation and resultant increase in intracranial pressure, and apoptosis.[3,4]

CEREBRAL EDEMA

One of the key features of many TBIs is an increase in intracranial pressure (ICP), putting pressure on brain regions surrounding the site of injury and worsening blood flow in the affected brain area. The largest contributing factor toward elevated ICP is intracranial edema—swelling of the brain that, limited by the calvarium, has limited space into which to expand before pathological increases in pressure occur. Previously it had been believed that observed brain swelling was due to engorgement of intracranial blood vessels, but studies measuring the degree of swelling due to water taken on by edema and blood demonstrated that the contribution by water was far in excess of that by blood.[5]

In general, post-TBI edema can be separated into two main types, vasogenic and cytotoxic edema. Vasogenic edema occurs early following TBI primarily at the central core of the injury, where significant numbers of blood vessels are broken and red blood cells and plasma proteins can be observed in edematous fluid leaking into the site of injury. Osmotic brain edema is similar to vasogenic edema in that fluid enters the brain from the blood vessels, but rather than large-scale, region-specific breakdown of the blood–brain barrier (BBB), edema is driven by high osmolality within the contused, necrotic core of the TBI.[6]

THE BLOOD–BRAIN BARRIER

The BBB is a factor of vital importance to consider in TBI. This is due in part because the delicate tissue of the brain requires a highly regulated environment, with concentrations of particular ions, such as Na^+, K^+, and Ca^{2+}, needing to be maintained within narrow ranges.[7] As such, conditions that might be mildly damaging to peripheral organs, such as bacterial or viral infection, inflammation, or complement activation, could be hugely detrimental if they occurred to a large extent in the central nervous system (CNS). In this regard, the necessity for a distinct separation of the CNS from the circulating bloodstream is of fundamental importance, and it is this separation that underlies the key function and presence of the BBB, ensuring strict homeostasis of the brain environment to keep neurons in a homeostatic and functional state.

Transport across the BBB is, to a great extent, confined to transcellular pathways. Some substances can cross the barrier paracellularly but these are usually extremely small or employ specific mechanisms in order to move between endothelial cells, such as T-cell migration across the BBB by means of a process initiated by leucocyte binding to ICAM-1 and ICAM-2 expressed on endothelial cell surfaces.[8] Transcellular transport pathways make use of receptors that recognize and transport the specific molecules that are required for neuronal function. For instance, the brain's energy needs are met by proteins such as GLUT1 which transports glucose across the endothelial cell layer.[9] Peptides that are required, such as hormones, are recognized by specific receptors, for example, PST-1 that recognizes vasopressin.[10] Ion concentrations are kept constant by proteins such as Na^+, K^+, and ATPase, which exchanges sodium and potassium ions.[11] Organic anions, meanwhile, are transported by OAT-family proteins, such as OAT1 which is found in the brush border of the choroid plexus,[12] which functions to make the cerebrospinal fluid that occupies the ventricular space in the brain and surrounds the spinal cord.

As well as enabling the transport of substances required by the brain, the BBB actively effluxes compounds that might be damaging to the delicate neural microenvironment. P-glycoprotein was the first protein observed to transport chemotherapeutic agents out of cancerous cells[13] and is expressed at the BBB[14] where it plays an active role in preventing access of many drugs to the brain. Other transporters participate in this process too, particularly the multidrug-resistance-related protein (MRP) family. MRP-4, for instance, is expressed in endothelial cells of the BBB and prevents accumulation of the anti-cancer drug topotecan[15] or oseltamivir, a drug targeting the influenza virus.[16]

Efflux transporters such as these are major barriers to drug delivery to the brain, as is the severe constriction of the paracellular spaces between endothelial cells of the BBB. This is due to structures called tight junctions (TJ), which are positioned on the apical aspect of the plasma membrane of neighboring endothelial cells to essentially seal the paracellular pore. Each TJ consists of at least three different types of protein—occludin, members of the junction adhesion molecule (JAM) family of proteins, and claudin proteins[17]—but can consist of over 40 individual proteins between support proteins, structural proteins, transport proteins, and other more unique components (such as tricellulin, a homologue of occludin that concentrates at the focal point of three contacting cells).[18] The actin cytoskeleton of the cell appears to be key to TJ formation, as this framework of actin strands bind the PDZ-domain containing support proteins including ZO-1, ZO-2, ZO-3, MAGI-1, PatJ, PALS1, and MUPP1. It is onto these proteins that the transmembrane proteins that form the TJ bind.[18]

Despite the importance of endothelial cells to the BBB, many other cell types and structures in and surrounding the blood vessels of the brain have been found to contribute to its function. In veins and arteries, smooth muscle cells and pericytes surround the vessels and must also be traversed by cells or substances to gain access to the brain, while in capillaries the former cell type is not present. Final to the blood vessel component of the BBB there is the outer vascular basement membrane, another membrane consisting of connective tissue proteins such as collagen and integrins. Surrounding the veins and arteries of the brain is the perivascular compartment, consisting of perivascular fluid and cells occupying the perivascular space, usually leptomeningeal cells and macrophages. Finally, before entering the brain parenchyma proper, the last barrier to cellular or molecular entry to or exit from the brain is the glia limitans. This is formed by a further basement membrane as well as astrocytic and microglial endfeet.[19]

Interestingly, some parts of the brain do not have a well-evolved BBB or have a diminished barrier. Chemokine expression in response to systemic injection of lipopolysaccharide (LPS), interleukin-1β (IL-1β), and tumor necrosis factor-alpha (TNF-α) is confined in the brain to the circumventricular organs, choroid plexus, leptomeninges, and cerebral blood vessels.[20] Together these structures form the interface between systemic circulating molecules and the brain. The circumventricular organs have secretory and sensory functions, such as the area postrema which initiates vomiting in response to toxins in the bloodstream,[21] and in these cases this tissue must be exposed to a variety of substances present in the blood in order to carry out their functions.

A DEGREE OF IMMUNE PRIVILEGE

The relative immune privilege status of the brain, which shares its status with the eye, the testes, and the pregnant uterus, was initially elucidated by studies on tissue transplants to the anterior chamber in the eye. Here the transplanted skin grafts persisted, where in nonimmune privileged tissue they were swiftly rejected.[22] Indeed, it has since been found that transplanted tissue grafts in the brain are in fact rejected, but the process simply takes considerably longer than in the periphery.[23] Response to LPS also takes longer, with monocyte recruitment being delayed to 3 days after intracerebral injection of LPS. As well as this, a 100-fold quantity of LPS is required for monocytes in the brain to reach the same density as those in the skin following injection there.[24]

These studies indicate that there is a significant barrier to immune activity in the brain, and this is in part a result of the many different layers of membranes and cell types that must be traversed in order to access the brain. This is a complex and multistep process, the first of which is the attachment of leukocytes to the endothelial cells of cerebral blood vessels by means of proteins such as ICAM-1 and ICAM-2. A lack of these adhesion molecules results in failure of T cells to transmigrate across endothelial layers in vitro.[25] The processes involved in crossing many of the other layers of the BBB have not been fully elucidated, however, the matrix metalloproteinase (MMP) enzymes appear to be crucial for penetration of the parenchymal basement membrane by means of dystroglycan cleavage.[26] To illustrate the importance of breaching not only the endothelial layer of the BBB—often the only layer considered significant—mice that overexpress the chemotactic factor CCL2 in the brain accumulate immune cells in the perivascular spaces on the brain side of blood vessels. However, these cells do no damage unless they pass into the parenchyma of the brain, and inhibition of MMP enzyme expression here prevents them from progressing further through the BBB.[27]

MORE THAN A BARRIER IN THE WAY?

When first discovered, immune privilege was believed to be a result of physical separation of the CNS from the systemic immune system by means of the BBB and a lack of lymphatics in the brain preventing systemic immune cell monitoring, and the believed docile phenotype of microglia (macrophages resident in the brain). However, despite the challenges present in crossing the BBB, the immune system does appear to have access to the brain in certain circumstances. Circulating, unactivated lymphocytes are observed migrating in the cerebrospinal fluid (CSF) at similar concentrations to lymph in the periphery,[28] and it has also been found that transgenic T cells that are specific for the myelin protein MBP are found in the brain, completely unresponsive to stimulation, even by antigen in vitro following removal from the brain.[29] This means that rather than relying on preventing immune cells from reaching the CNS alone, other methods must be used to evade or suppress the immune system.

A clue in the evasion of the immune system by the CNS is the fact that viruses often persist long term in neurons, without these cells being discovered and lysed by T cells. In an in vivo study, cytotoxic T cells injected directly into the brain did not lyse any neurons infected with a virus or cause any damage to the brain. The likely cause of this is the lack of any significant major histocompatibility complex (MHC) expression observed in neuronal cell lines.[30] As MHCs are the proteins that present antigens found within the cell to the immune system, and mark some cells for death due to infection, this would result in cytotoxic T cells being blind, in effect, to neurons. This is particularly important for neurons, as the loss of even a small number of neurons can result in the loss of a huge number of axonal connections in the brain, or indeed a patch of blindness in the retina. Due to the lack of regenerative ability in these organs losses such as this are devastating. In contrast to neurons, when the virus had infected leptomeningeal and choroid plexus cells even a greatly reduced number of cytotoxic T cells caused major cell death and tissue damage.[30]

NEUROINFLAMMATION

Despite the demanding requirements of immune cell translocation to and activation in the brain, neuroinflammation is still a highly significant feature of many diseases and injuries, with immune cells playing important roles in both short- and long-term TBI pathology. Wallerian axonal degeneration, seen in diffuse axonal injury (DAI) for instance, has been shown to induce monocyte infiltration into the ipsilateral hemisphere 24 h postinjury.[31] A possible reason for this postlesion influx of immune cells might be due to the chemokine CXCL12. This protein is a potent chemotactic agent for T cells and monocytes, and is constitutively expressed at low levels in the brain. As such, any trauma to the brain in the context of cellular necrosis would expose CXCL12 to the circulating immune system, whereas under normal conditions it is sequestered behind the BBB. In fact, not only is it exposed after brain injury, but its expression is actively altered, which attracts immune cells to the injured region specifically, with CXCL12 levels increasing in the penumbra of lesioned brain regions following focal ischemia, and decreasing in uninjured regions, and this change in expression coincided with the influx of immune cells that expressed the CXCL12 receptor.[32] Another mechanism that encourages immune cell infiltration into the brain is the response of endothelial cells to proinflammatory cytokines. When exposed to factors such as TNF and interferon-gamma (IFN-γ), endothelial cells show actin-cytoskeleton restructuring coinciding with fragmentation of ZO-1 staining at the cell periphery.[33] Similarly, application of activated leukocytes onto endothelial cell monolayers led to aberrant adherens junction localization, associated with actin cytoskeleton restructuring and phosphorylation of adherens junction proteins. These changes led to intercellular gap formation and albumin diffusion across the monolayer.[34] It is clear that in order to counter the level of immune privilege conferred by the BBB, the immune system as well as resident cells employ countermechanisms in order to facilitate leukocyte infiltration in cases of neuroinflammation.

Dendritic Cells

As well as monocytes and various T cells, a wide range of immune cells play active roles in the brain. One such cell type is the dendritic cell, an antigen-presenting cell that plays an early role in many injuries and infections. This cell type is normally detected only in the choroid plexus and the meninges in the uninjured brain,[35] but following TBI dendritic cells, which likely differentiate from activated monocytes crossing the barrier, are found in increasing quantities infiltrating injured regions of the brain.[36] Neutrophils have also been observed to enter the CSF in significant numbers following injection into the subarachnoid space of IL-1β, a proinflammatory cytokine that plays an important role in post-TBI inflammation.[37] Once in the brain, or during their transmigration into it, neutrophils may play a role in weakening the BBB, thus enabling further immune cell invasion,[34] and certainly IL-1β-mediated recruitment of neutrophils is often associated with negative consequences, such as increased water content in ischemic brain damage.[38]

Astrocytes

Although not normally considered immune cells, astrocytes play a large role in detecting and modulating responses to damage or infection in the brain. Astrocytes express high levels of Toll-like receptor-3 (TLR-3), as well as lower levels of TLRs -1, -4, -5 and -9, and produce cytokines such as IL-6, CXCL10 and IFN-β in response to their ligation.[39] TLRs were first recognized for their ability to detect pathogen-associated molecular patterns (PAMPs), and these astrocyte-expressed TLRs are capable of detecting a wide range of pathogenic markers such as double-stranded DNA (dsDNA), double-stranded RNA (dsRNA), LPS, and flagellin. However, TLRs, including TLR-3 and TLR-4, are also capable of detecting danger-associated molecular patterns (DAMPS), endogenous molecules that are normally sequestered within the cell membrane of a cell, and whose presentation to TLRs indicates cell death such as that which occurs following TBI. Complicating the role of astrocytes and these TLRs in inflammatory scenarios, TLR-3 activation can result in protective cytokine release such as IL-9, -10 and -11, and together TLR3-induced factors were found to improve neuronal survival in ex vivo human brain slice cultures.[40]

Microglia

The final cell type to be discussed, and probably the most significant post-TBI immune cell, are microglia. This monocytic cell lineage is the brain's resident macrophage, although under normal circumstances it expresses low or undetectable levels of macrophage phenotypic markers such as MHC I and MHC II, CR3, FcR, CD68, LCA and CD4, which are then upregulated upon activation.[41] It is significant to note that as the only immune cell type that is always observed in the brain under normal, noninflammatory conditions, microglia can be activated by pathogens or tissue damage without any alteration in the BBB or other factors compromising the immune privilege of the brain. Microglia survey the brain microenvironment by means of highly motile filopodia, which, following BBB disruption or TBI, rapidly extend and converge upon the site of injury without movement of the cell body. This chemotaxis appears to be directly in response to extracellular ATP, such as that which would be spilled from cells upon lysis, and in fact this chemotaxis can be stimulated by the injection of ATP into the brain parenchyma. Once at the site of injury the microglial protrusions appear to shield healthy tissue from damaged tissue, and activate other microglia in addition to other infiltrating immune cells.[42,43] Similar to astrocytes, microglia express TLRs, and induction of TLR signaling by sensing PAMPs or DAMPs in the brain microenvironment results in microglial activation and subsequent release of proinflammatory cytokines.[39]

As part of this role, microglia also convey inflammatory signals received from the periphery throughout the brain, priming it for response. Systemic LPS (a TLR-4 ligand) induces TNF-α within the exposed circumventricular organs, and this TNF-α expression increase spreads across microglia of the brain parenchyma, even though these areas have not been exposed to LPS directly. Interestingly this LPS-induced microglial activation is not damaging to the CNS in its own right, but only if the inhibitory activity of glucocorticoids within the brain is blocked.[44] This indicates that upon receiving external danger signals microglia actively produce proinflammatory cytokines, but do not enter a fully neuroinflammatory phenotype due to inhibitory signals in the brain. This is significant because in models such as experimental autoimmune encephalomyelitis (EAE), unchecked inflammation in the brain, even in the absence of other nonautonomous mediators of damage, can be hugely damaging to the neural microenvironment.

Systemic immune cells encounter many barriers to entry, activation, and even survival in the brain. However, microglia have already overcome many significant hurdles by being present and viable in the brain, and can activate via their sensory extensions and TLRs. Accordingly, many inhibitory interactions exist between microglia and endogenous brain cells. Astrocytes, for instance, inhibit LPS-stimulated microglial production of IL-12 and other cytokines when in co-culture.[45] Neurons also play an important role in microglial suppression, through mechanisms such as axon-terminus release of CD22, which engage microglial CD45, thereby inhibiting TNF-α production.[46] In an animal model illustrating neuronal inhibition, mice lacking CD200 on neurons demonstrate an increased microglial response to damage and a more severe onset of EAE.[47] These endogenous mechanisms of microglial suppression are vital for brain integrity, but the safeguards can and do break down.

After TBI the danger of neuroinflammatory damage is most significant when inflammation stretches into the chronic phase, and active microglia and macrophages can be detected in brain tissue years after injury.[48] As well as the proinflammatory activity of microglia, which can result in exacerbation of neuronal damage, these cells have important roles in mediating immune suppression and tissue repair themselves. In the case of the former, microglia have been found to express HLA-G, the previously mentioned T-cell-inhibitory factor, particularly in areas affected by neuroinflammation, possibly in an attempt to decrease T-cell-mediated inflammation.[49] Microglia have even been observed fully engulfing infiltrating neutrophils in the brain, and blockage of this process results in higher levels of neuronal damage.[50]

The significant active contribution of neuronal, glial, and CNS-resident microglial cells in immune responses and regulation shakes apart the previously held view that damaging immune activity in the brain was a product of BBB breakdown and exposure of the delicate CNS to the systemic immune system. Instead, a dynamic system is being revealed, where all the different cell types of the brain play a role in either excluding and suppressing systemic immune cells in some circumstance, or attracting and activating them in others. This raises the notion that inappropriate immune activation in the brain could be as much a product of a dysfunction of the brain itself than of the immune system.

ADAPTIVE OR MALADAPTIVE?

There is a double-edged sword when it comes to neuroinflammation. In many situations and models, large amounts of damage are caused by activation of the immune system following CNS insult, whereas at different time points or in different models inflammation can be protective and its inhibition harmful. In further examples, the protective component to neuroinflammation, mice lacking IL-1β that undergo demyelination following loss of oligodendrocytes, fail to remyelinate or repopulate oligodendrocyte populations to the same extent as wild-type mice.[51] Another proinflammatory cytokine that sees prominent expression following TBI is TNF-α, and this too is found to promote the expansion of oligodendrocyte precursor cells as well as remyelination.[52] Clearly, the complexity involved in neuroinflammatory processes remains to be fully elucidated, whether in TBI or other neural conditions with an inflammatory component, but what is certain is that it is a mistake to dismiss neuroinflammation as a solely negative, damaging process. Instead there is a delicate balancing act in the CNS between the need to have immune cells that can fight infections that breach the CNS, as well as phagocytose dead and dying cells before they damage other cells around them, and the need to regulate inflammation in the delicate, nonregenerative neural tissue. Allowing these cells to be inappropriately activated has well-documented disastrous results such as that observed in multiple sclerosis, but preventing immune cells from activating in cases of infection, cell death, or toxin accumulation can be equally deleterious.

PRECLINICAL MODELS OF NEUROINFLAMMATION

The complexities of neuroinflammation and the variation in outcomes as a result of immune activity following TBI are reasons why selection of the appropriate animal model is exceedingly important. In this regard, special attention should be paid to the selection of an appropriate animal model that will re-capitulate some of the key neuroinflammatory events associated with TBI.

Diffuse Axonal Injury

DAI can be modeled on its own in an impact-free model, which involves acceleration of the head in a number of directions: sagittal, oblique, or lateral, or a combination of these. In this way the model simulates the damage sustained in nonejection vehicle accidents where rapid deceleration shakes and rotates the head, or in some assault cases such as shaken baby syndrome. Animals subjected to this model commonly suffer comas, with coma length being proportional to the force of head movement. Lateral head movement was also observed to result in the most severe pathology, indicating that the brain is less pliable under rotational forces along this axis.[53] Factors such as the degree of rotation, rotation speed, and acceleration can all be varied to deliver a broad spectrum of DAI severities. Axonal damage occurs mostly in the absence of vascular disruption,[54] resulting in a model where DAI can be studied in isolation from the wide range of pathological characteristics present in impact models of TBI.

Weight Drop

The weight-drop model of TBI is performed by allowing an object to fall freely through a guide tube onto the skull of the anesthetized animal. The weight of the object and height from which it falls can both be altered to give injuries of different grades, and if desired the incidence of skull fracture can be reduced by gluing a steel disc onto the skull.[55] This latter precaution can be useful as skull fractures can reduce ICP, which is an important aspect of post-TBI pathology. The pathology of this model is broad as one would expect from a nonspecific injury such as this, with features such as DAI, brain stem hemorrhage, edema, limb deficits, respiratory problems, convulsion, and death being reported during the development of the technique.[55,56] This model has enabled the discovery of other pathological features of TBI, such as angiogenesis of permeable blood vessels in the vicinity of the injury, mediated by vascular endothelial growth factor (VEGF) production by astrocytes and immune cells.[57]

Controlled Cortical Impact

Controlled Cortical Impact (CCI) produces a similar type of injury to that acquired in weight-drop TBI, but with the CCI insult being administered by striking the dura of the exposed brain using a pneumatic cylinder.[58] Different regions of the brain can be impacted in this way, producing differing injury pathologies due to the brain regions in the direct vicinity of the impact being the most severely affected. CCI is the most commonly used TBI model, due in large part to the many pathological features that arise including DAI, BBB disruption, neuroinflammation, and subdural and intraperenchymal hematomas. At later time points pathological features of TBI arise, such as edema, alterations in CBF, as well as chronic phase inflammation. One of the main advantages of this model of TBI is the multitude of aspects that arise in human TBIs, which are almost by nature extremely heterogeneous, complex injuries.

Cold-Induced Injury

Cold-induced injury is another model of impact-free brain injury and, similar to the model of DAI, is designed to more closely study a lower number of pathological factors of TBI. Cold-induced injuries are carried out by holding a liquid nitrogen-cooled metal probe against the exposed skull of anesthetized mice, or other animal, for a set period of time. This produces a highly reproducible model of vasogenic edema when blood vessels at the injury site necrose and spill their contents. This region of edema has low heterogeneity and is clearly demarcated allowing clear quantification by MRI or other methods,[59,60] which is why it is chosen in some studies aimed at understanding or attempting to reduce cerebral edema following TBI.[59,61–63] The model does demonstrate other features of TBI however, with widespread cell death in the injured region, significant changes in CBF, marked release of inflammatory cytokines, and motor and memory deficits.[59] One major disadvantage of this model however is that it does not recapitulate a type of TBI that occurs in humans, however, for specific studies it is still advantageous.

Fluid-Percussion Injury

In the fluid-percussion injury model rather than the brain or skull being impacted with a solid object, the force is a rapid increase in pressure delivered by means of water or saline. The fluid is injected onto the dura, exposed by craniectomy, in a fully enclosed system by striking the injector with a hammer on a pendulum, which enables alterations of hammer weight, drop height, and angle for variable injury strengths. Due to the injection of fluid into the enclosed system ICP rises rapidly, displacing and deforming brain tissue. Because the injury is administered by means of an increase in pressure, the brain region that comes in contact with the fluid is not necessarily the most injured region. In fact studies have shown that the brainstem is most vulnerable to the rapid increase in ICP.[64] Damage does occur elsewhere, however, and the fluid injection site does have a significant effect on the level of cell death in various brain regions. For instance, rostral application of fluid pressure results in higher cortical than hippocampal cell death, whereas caudal, medial, or lateral fluid injection results in the opposite damage profile as well as higher levels of reactive astrocytes.[65] Other features of fluid-percussion pathology include hemorrhages, edema, brain atrophy and cavitation, and memory deficits.[65–67]

Blast Injury

Blast injuries have risen to prominence in recent years due to conflicts around the world, as has the field of research into TBI caused by blasts. As well as the risk of shrapnel causing significant damage to the brain, overpressure waves emanating from an explosive device travel through the bodies of individuals nearby applying strong kinetic forces that distort and place high strain onto their brains. Soldiers and civilians who are caught in the vicinity of one or a number of blasts can appear wholly unharmed externally but can suffer from headaches, concussion, brain swelling, post-traumatic stress disorder, chronic pain, and an increased risk of Alzheimer's disease.[68,69] In fact in one study illustrating the prevalence and long-term mental health impact of TBI in war zones, nearly a third of veterans of the recent Iraq and Afghanistan conflicts surveyed, reported suffering more than one type of head injury, and in that group individuals were sixfold as likely to have suffered from post-traumatic stress disorder, fourfold as likely to have depression, and twice as likely to experience alcoholism.[70]

To model these injuries in animals by means of a reproducible, repeatable technique, the most common method is to use a shock tube—a long, cylindrical tube containing two chambers separated by a membrane. Compressed air or helium is pumped into the first chamber, the compression chamber, until the membrane bursts or until the desired pressure is met in which case the membrane is burst by a blade within the shock tube. The animal, restrained at the end of the second chamber, the expansion chamber, receives the blast overpressure waves, with the restraint preventing tertiary blast injuries caused by being thrown against objects by the blast wave.[71] In this model, animals can either be exposed to a single blast or multiple minor blasts, with the intensity of each blast being variable by alteration of the pressure in the compression chamber or distance of the animal from the membrane. Animals exposed to these blasts exhibit neuronal swelling, oxidative stress, immune cell infiltration, caspase-3 cleavage and apoptosis, amyloid precursor protein accumulation, losses of consciousness, hemorrhages on the surface of the brain and within the cortex, and cognitive deficits.[72–75]

The number of different animal models for TBI could be a point of criticism with regard to the difficulty that treatments and elucidation of neuroinflammatory events have encountered at the clinical stage. The majority of published studies in the TBI area demonstrate their in vivo results in only one model, and with each model demonstrating marked differences in pathology in relation to neuroinflammation. This is a cause for concern. As well as this, the most commonly used animal models—mouse and rat—differ markedly in their brain volumes from humans. Large changes in brain size could fundamentally alter the progression of brain injuries, due to the way in which brain tissue compresses, swells, effects of skull thickness, and other factors. Even within a single species significant differences have been observed between background strains in respect to memory and motor functional outcomes following TBI. In some background strains even sham-operated animals were unable to learn certain tasks.[76] In essence, a large problem is the homogeneity of current in vivo testing, where injuries inflicted are as identical as possible, whereas in humans TBI is a highly heterogeneous condition. Significant improvements can be made in preclinical assessment of prospective therapeutics by widening the range of TBI that the treatment is assessed in—different injury types, different background strains, and different sized animal models.

Ongoing Trials

Despite many high-profile setbacks, there are a number of potentially highly efficacious treatments being trialed for TBI. Many of these deal with specific elements of pathology, as a result of the highly complex, heterogeneous nature of TBI. One Phase IV trial that was expected to conclude in December 2013 replaces human growth hormone, which is deficient for many patients following TBI (NCT00957671). Other examples include a feasibility study for levetiracetam to combat post-TBI epilepsy;[77] arginine vasopressin to improve cerebral perfusion pressure in patients (NCT00795366); intravenous delivery of Cerebrolysin, a neurotrophic factor that aids neuronal tissue protection and repair;[78] sodium lactate infusion (NCT01573507) as the brain switches largely from glucose to lactate metabolism following TBI;[79] and using sildenafil (Viagra®) to treat persistent cardiovascular dysfunction following TBI (NCT01762475). A number of studies have already led to promising results in humans, such as amantadine hydrochloride, an organic compound already used to treat Parkinson's disease, that improved Disability Rating Scale results significantly among 184 patients.[80]

Studies such as these are advantageous due to their focus on a single or small number of specific outcome measures and they show clear evidence of beneficial effect, rather than multiple outcome measures or measures that are more difficult to objectively and consistently determine. Despite this, a novel treatment that would treat a large range of post-TBI dysfunctions by targeting an early inflammatory or core damage pathway in TBI pathology would be an exciting development in the field.

Oral or intravenous progesterone is a candidate for a drug that could do this, with promising results being reported in animal experiments.[81] Treatment with this hormone is currently in a number of parallel clinical trials, including a Phase IV for mild TBI in athletes (NCT01809639), and a Phase III in severe TBI patients due to finish at the end of 2013 (NCT01143064). Results of a number of these trials have already been published, with favorable outcomes being reported in patients with DAI at 3 months,[82] and a further study reporting that progesterone together with vitamin D is even better than progesterone alone.[83]

Hypothermia is another treatment for TBI that is showing a lot of promise at treating many outcomes of TBI. Trials on 80 and 136 patients resulted in improvements in ICP, antioxidant enzyme levels, neurological outcome, and mortality.[84,85] However, there has been some variation in clinical results in respect to hypothermia, with another trial, this time in 392 patients, concluding that no change in outcome or mortality had occurred.[86] Despite this, hypothermia still holds much promise for TBI and other neurological conditions, and further trials are underway such as a Phase III trial to assess the efficacy of longer term, milder hypothermia of 34°C–35°C (NCT01886222).

Similar to the application of hypothermia to TBI some other strategies also do not consist of the delivery of a standard drug. Hyperbaric oxygen therapy, for instance, has shown promising preclinical in vivo results[87] and is currently in a number of human trials (NCT01611194, NCT01847755). In addition, the use of autologous bone marrow-derived mononuclear cells is another atypical treatment type, with positive results reducing BBB permeability and improvements in special memory demonstrated in a rat CCI model,[88] and trials for safety and efficacy in humans is ongoing (NCT01575470).

With these in mind it is a very interesting time in the field of TBI research and therapeutic development, but many of the neuroinflammatory factors involved in this heterogeneous, highly complex type injury remain to be elucidated. This is particularly true in the area of neuroinflammation, where evidence points to both beneficial and deleterious effects of immune cell activation following TBI. Until this and other features of TBI are understood in more detail at the molecular level, and with appropriate animal modeling, problems at the clinical level will almost certainly continue to confound and disappoint.

ACKNOWLEDGMENTS

Matthew Campbell's lab is supported by grants from Enterprise Ireland, GlaxoSmithKline (GSK), Science Foundation Ireland (SFI) (12/YI/B2614), the ARVO Foundation for Eye Research, and the Bright Focus Foundation.

REFERENCES

1. Tagliaferri, F. et al. 2006. A systematic review of brain injury epidemiology in Europe. *Acta Neurochir (Wien)* 148(3):255–268; discussion 268.
2. Faul, M. et al. 2010. *Traumatic Brain Injury in the United States: Emergency Department Visits, Hospitalizations and Deaths 2002–2006*. Atlanta, GA: Centers for Disease Control and Prevention, National Center for Injury Prevention and Control.
3. Maas, A.I. et al. 2008. Moderate and severe traumatic brain injury in adults. *Lancet Neurol* 7(8):728–741.
4. Clausen, F. et al. 2011. Neutralization of interleukin-1beta reduces cerebral edema and tissue loss and improves late cognitive outcome following traumatic brain injury in mice. *Eur J Neurosci* 34(1):110–123.
5. Marmarou, A. et al. 2000. Contribution of edema and cerebral blood volume to traumatic brain swelling in head-injured patients. *J Neurosurg* 93(2):183–193.
6. Kawamata, T. et al. 2007. Tissue hyperosmolality and brain edema in cerebral contusion. *Neurosurg Focus* 22(5):E5.
7. Hawkins, B.T., Davis, T.P. 2005. The blood-brain barrier/neurovascular unit in health and disease. *Pharmacol Rev* 57(2):173–185.
8. Steiner, O. et al. 2010. Differential roles for endothelial ICAM-1, ICAM-2, and VCAM-1 in shear-resistant T cell arrest, polarization, and directed crawling on blood-brain barrier endothelium. *J Immunol* 185(8):4846–4855.
9. Pardridge, W.M., Oldendorf, W.H. 1975. Kinetics of blood-brain transport of hexoses. *Biochim Biophys Acta* 382(3):377–392.
10. Banks, W.A. 2006. The CNS as a target for peptides and peptide-based drugs. *Expert Opin Drug Deliv* 3(6):707–712.
11. Betz, A.L. 1983. Sodium transport in capillaries isolated from rat brain. *J Neurochem* 41(4):1150–1157.
12. Sugiyama, Y. et al. 1999. Kinetic and biochemical analysis of carrier-mediated efflux of drugs through the blood-brain and blood-cerebrospinal fluid barriers: Importance in the drug delivery to the brain. *J Control Release* 62(1–2):179–186.
13. Ling, V. 1995. P-glycoprotein: Its role in drug resistance. *Am J Med* 99(6A):31S–34S.
14. Schinkel, A.H. et al. 1994. Disruption of the mouse mdr1a P-glycoprotein gene leads to a deficiency in the blood-brain barrier and to increased sensitivity to drugs. *Cell* 77(4):491–502.
15. Leggas, M. et al. 2004. Mrp4 confers resistance to topotecan and protects the brain from chemotherapy. *Mol Cell Biol* 24(17):7612–7621.
16. Ose, A. et al. 2009. Limited brain distribution of [3R,4R,5S]-4-acetamido-5-amino-3-(1-ethylpropoxy)-1-cyclohexene-1-carboxyl ate phosphate (Ro 64-0802), a pharmacologically active form of oseltamivir, by active efflux across the blood-brain barrier mediated by organic anion transporter 3 (Oat3/Slc22a8) and multidrug resistance-associated protein 4 (Mrp4/Abcc4). *Drug Metab Dispos* 37(2):315–321.

17. Ben-Yosef, T. et al. 2003. Claudin 14 knockout mice, a model for autosomal recessive deafness DFNB29, are deaf due to cochlear hair cell degeneration. *Hum Mol Genet* 12(16):2049–2061.
18. Anderson, J.M., Van Itallie, C.M. 2008. Tight junctions. *Curr Biol* 18(20):R941–R943.
19. Bechmann, I. et al. 2007. What is the blood-brain barrier (not)? *Trends Immunol* 28(1):5–11.
20. Thibeault, I. et al. 2001. Regulation of the gene encoding the monocyte chemoattractant protein 1 (MCP-1) in the mouse and rat brain in response to circulating LPS and proinflammatory cytokines. *J Comp Neurol* 434(4):461–477.
21. Maolood, N., Meister, B. 2009. Protein components of the blood-brain barrier (BBB) in the brainstem area postrema-nucleus tractus solitarius region. *J Chem Neuroanat* 37(3):182–195.
22. Medawar, P.B. 1948. Immunity to homologous grafted skin. III. The fate of skin homografts transplanted to the brain, to subcutaneous tissue, and to the anterior chamber of the eye. *Br J Exp Pathol* 29(1):58–69.
23. Finsen, B.R. et al. 1991. Leukocyte infiltration and glial reactions in xenografts of mouse brain tissue undergoing rejection in the adult rat brain. A light and electron microscopical immunocytochemical study. *J Neuroimmunol* 32(2):159–183.
24. Andersson, P.B. et al. 1992. The acute inflammatory response to lipopolysaccharide in CNS parenchyma differs from that in other body tissues. *Neuroscience* 48(1):169–186.
25. Lyck, R. et al. 2003. T-cell interaction with ICAM-1/ICAM-2 double-deficient brain endothelium in vitro: The cytoplasmic tail of endothelial ICAM-1 is necessary for transendothelial migration of T cells. *Blood* 102(10):3675–3683.
26. Agrawal, S. et al. 2006. Dystroglycan is selectively cleaved at the parenchymal basement membrane at sites of leukocyte extravasation in experimental autoimmune encephalomyelitis. *J Exp Med* 203(4):1007–1019.
27. Toft-Hansen, H. et al. 2006. Metalloproteinases control brain inflammation induced by pertussis toxin in mice overexpressing the chemokine CCL2 in the central nervous system. *J Immunol* 177(10):7242–7249.
28. Seabrook, T.J. et al. 1998. Cerebral spinal fluid lymphocytes are part of the normal recirculating lymphocyte pool. *J Neuroimmunol* 91(1–2):100–107.
29. Brabb, T. et al. 2000. In situ tolerance within the central nervous system as a mechanism for preventing autoimmunity. *J Exp Med* 192(6):871–880.
30. Joly, E. et al. 1991. Viral persistence in neurons explained by lack of major histocompatibility class I expression. *Science* 253(5025):1283–1285.
31. Bechmann, I. et al. 2005. Circulating monocytic cells infiltrate layers of anterograde axonal degeneration where they transform into microglia. *FASEB J* 19(6):647–649.
32. Stumm, R.K. et al. 2002. A dual role for the SDF-1/CXCR4 chemokine receptor system in adult brain: Isoform-selective regulation of SDF-1 expression modulates CXCR4-dependent neuronal plasticity and cerebral leukocyte recruitment after focal ischemia. *J Neurosci* 22(14):5865–5878.
33. Blum, M.S. et al. 1997. Cytoskeletal rearrangement mediates human microvascular endothelial tight junction modulation by cytokines. *Am J Physiol* 273(1 Pt 2):H286–H294.
34. Tinsley, J.H. et al. 1999. Activated neutrophils induce hyperpermeability and phosphorylation of adherens junction proteins in coronary venular endothelial cells. *J Biol Chem* 274(35):24930–24934.
35. Matyszak, M.K., Perry, V.H. 1996. The potential role of dendritic cells in immune-mediated inflammatory diseases in the central nervous system. *Neuroscience* 74(2):599–608.
36. Israelsson, C. et al. 2008. Distinct cellular patterns of upregulated chemokine expression supporting a prominent inflammatory role in traumatic brain injury. *J Neurotrauma* 25(8):959–974.

37. Quagliarello, V.J. et al. 1991. Recombinant human interleukin-1 induces meningitis and blood-brain barrier injury in the rat. Characterization and comparison with tumor necrosis factor. *J Clin Invest* 87(4):1360–1366.
38. Yamasaki, Y. et al. 1995. Interleukin-1 as a pathogenetic mediator of ischemic brain damage in rats. *Stroke* 26(4):676–680; discussion 681.
39. Jack, C.S. et al. 2005. TLR signaling tailors innate immune responses in human microglia and astrocytes. *J Immunol* 175(7):4320–4330.
40. Bsibsi, M. et al. 2006. Toll-like receptor 3 on adult human astrocytes triggers production of neuroprotective mediators. *Glia* 53(7):688–695.
41. Galea, I. et al. 2007. What is immune privilege (not)? *Trends Immunol* 28(1):12–18.
42. Davalos, D. et al. 2005. ATP mediates rapid microglial response to local brain injury in vivo. *Nat Neurosci* 8(6):752–758.
43. Nimmerjahn, A. et al. 2005. Resting microglial cells are highly dynamic surveillants of brain parenchyma in vivo. *Science* 308(5726):1314–1318.
44. Nadeau, S., Rivest, S. 2003. Glucocorticoids play a fundamental role in protecting the brain during innate immune response. *J Neurosci* 23(13):5536–5544.
45. Aloisi, F. et al. 1997. IL-12 production by central nervous system microglia is inhibited by astrocytes. *J Immunol* 159(4):1604–1612.
46. Mott, R.T. et al. 2004. Neuronal expression of CD22: Novel mechanism for inhibiting microglial proinflammatory cytokine production. *Glia* 46(4):369–379.
47. Hoek, R.M. et al. 2000. Down-regulation of the macrophage lineage through interaction with OX2 (CD200). *Science* 290(5497):1768–1771.
48. Chen, X.H. et al. 2009. A lack of amyloid beta plaques despite persistent accumulation of amyloid beta in axons of long-term survivors of traumatic brain injury. *Brain Pathol* 19(2):214–223.
49. Wiendl, H. et al. 2005. Expression of the immune-tolerogenic major histocompatibility molecule HLA-G in multiple sclerosis: Implications for CNS immunity. *Brain* 128(Pt 11):2689–2704.
50. Neumann, J. et al. 2008. Microglia cells protect neurons by direct engulfment of invading neutrophil granulocytes: A new mechanism of CNS immune privilege. *J Neurosci* 28(23):5965–5975.
51. Mason, J.L. et al. 2001. Interleukin-1beta promotes repair of the CNS. *J Neurosci* 21(18):7046–7052.
52. Arnett, H.A. et al. 2001. TNF alpha promotes proliferation of oligodendrocyte progenitors and remyelination. *Nat Neurosci* 4(11):1116–1122.
53. Gennarelli, T.A. et al. 1982. Diffuse axonal injury and traumatic coma in the primate. *Ann Neurol* 12(6):564–574.
54. Smith, D.H. et al. 1997. Characterization of diffuse axonal pathology and selective hippocampal damage following inertial brain trauma in the pig. *J Neuropathol Exp Neurol* 56(7):822–834.
55. Marmarou, A. et al. 1994. A new model of diffuse brain injury in rats. Part I: Pathophysiology and biomechanics. *J Neurosurg* 80(2):291–300.
56. Foda, M.A., Marmarou, A. 1994. A new model of diffuse brain injury in rats. Part II: Morphological characterization. *J Neurosurg* 80(2):301–313.
57. Skold, M.K. et al. 2005. VEGF and VEGF receptor expression after experimental brain contusion in rat. *J Neurotrauma* 22(3):353–367.
58. Dixon, C.E. et al. 1991. A controlled cortical impact model of traumatic brain injury in the rat. *J Neurosci Methods* 39(3):253–262.
59. Campbell, M. et al. 2012. Targeted suppression of claudin-5 decreases cerebral oedema and improves cognitive outcome following traumatic brain injury. *Nat Commun* 3:849.
60. Houkin, K. et al. 1996. Magnetic resonance imaging of cold injury-induced brain edema in rats. *Neurol Med Chir (Tokyo)* 36(2):72–77.

61. Kawai, N. et al. 2003. Treatment of cold injury-induced brain edema with a non-specific matrix metalloproteinase inhibitor MMI270 in rats. *J Neurotrauma* 20(7):649–657.
62. Murakami, K. et al. 1999. Cold injury in mice: A model to study mechanisms of brain edema and neuronal apoptosis. *Prog Neurobiol* 57(3):289–299.
63. Oury, T.D. et al. 1993. Cold-induced brain edema in mice. Involvement of extracellular superoxide dismutase and nitric oxide. *J Biol Chem* 268(21):15394–15398.
64. Thibault, L.E. et al. 1992. Biomechanical aspects of a fluid percussion model of brain injury. *J Neurotrauma* 9(4):311–322.
65. Floyd, C.L. et al. 2002. Craniectomy position affects morris water maze performance and hippocampal cell loss after parasagittal fluid percussion. *J Neurotrauma* 19(3):303–316.
66. Dixon, C.E. et al. 1988. Physiologic, histopathologic, and cineradiographic characterization of a new fluid-percussion model of experimental brain injury in the rat. *J Neurotrauma* 5(2):91–104.
67. Iwamoto, Y. et al. 1997. Investigation of morphological change of lateral and midline fluid percussion injury in rats, using magnetic resonance imaging. *Neurosurgery* 40(1):163–167.
68. Rosenfeld, J.V. et al. 2013. Blast-related traumatic brain injury. *Lancet Neurol* 12(9):882–893.
69. Veitch, D.P. et al. 2013. Military risk factors for cognitive decline, dementia and Alzheimer's disease. *Curr Alzheimer Res* 10(90):907–930.
70. Maguen, S. et al. 2012. The impact of head injury mechanism on mental health symptoms in veterans: Do number and type of exposures matter? *J Trauma Stress* 25(1):3–9.
71. Reneer, D.V. et al. 2011. A multi-mode shock tube for investigation of blast-induced traumatic brain injury. *J Neurotrauma* 28(1):95–104.
72. Ahlers, S.T. et al. 2012. Assessment of the effects of acute and repeated exposure to blast overpressure in rodents: Toward a greater understanding of blast and the potential ramifications for injury in humans exposed to blast. *Front Neurol* 3:32.
73. Cernak, I. et al. 2001. Cognitive deficits following blast injury-induced neurotrauma: Possible involvement of nitric oxide. *Brain Inj* 15(7):593–612.
74. Cernak, I. et al. 2001. Ultrastructural and functional characteristics of blast injury-induced neurotrauma. *J Trauma* 50(4):695–706.
75. Tompkins, P. et al. 2013. Brain injury: Neuro-inflammation, cognitive deficit and MRI in a model of blast induced TBI. *J Neurotrauma* 30(22):1888–1897.
76. Fox, G.B. et al. 1999. Behavioral responses of C57BL/6, FVB/N, and 129/SvEMS mouse strains to traumatic brain injury: Implications for gene targeting approaches to neurotrauma. *J Neurotrauma* 16(5):377–389.
77. Klein, P. et al. 2012. Results of phase 2 safety and feasibility study of treatment with levetiracetam for prevention of posttraumatic epilepsy. *Arch Neurol* 69(10):1290–1295.
78. Masliah, E., Diez-Tejedor, E. 2012. The pharmacology of neurotrophic treatment with Cerebrolysin: Brain protection and repair to counteract pathologies of acute and chronic neurological disorders. *Drugs Today (Barc)* 48(Suppl A):3–24.
79. Timofeev, I. et al. 2013. Extracellular brain pH with or without hypoxia is a marker of profound metabolic derangement and increased mortality after traumatic brain injury. *J Cereb Blood Flow Metab* 33(3):422–427.
80. Giacino, J.T. et al. 2012. Placebo-controlled trial of amantadine for severe traumatic brain injury. *N Engl J Med* 366(9):819–826.
81. Cekic, M. et al. 2012. Progesterone treatment alters neurotrophin/proneurotrophin balance and receptor expression in rats with traumatic brain injury. *Restor Neurol Neurosci* 30(2):115–126.

82. Shakeri, M. et al. 2013. Effect of progesterone administration on prognosis of patients with diffuse axonal injury due to severe head trauma. *Clin Neurol Neurosurg* 115(1):2019–2022.
83. Aminmansour, B. et al. 2012. Comparison of the administration of progesterone versus progesterone and vitamin D in improvement of outcomes in patients with traumatic brain injury: A randomized clinical trial with placebo group. *Adv Biomed Res* 1:58.
84. Polderman, K.H. et al. 2002. Effects of therapeutic hypothermia on intracranial pressure and outcome in patients with severe head injury. *Intensive Care Med* 28(11):1563–1573.
85. Qiu, W. et al. 2007. Effects of therapeutic mild hypothermia on patients with severe traumatic brain injury after craniotomy. *J Crit Care* 22(3):229–235.
86. Clifton, G.L. et al. 2001. Lack of effect of induction of hypothermia after acute brain injury. *N Engl J Med* 344(8):556–563.
87. Lim, S.W. et al. 2013. Microglial activation induced by traumatic brain injury is suppressed by postinjury treatment with hyperbaric oxygen therapy. *J Surg Res* 184(2):1076–1084.
88. Bedi, S.S. et al. 2013. Autologous bone marrow mononuclear cells therapy attenuates activated microglial/macrophage response and improves spatial learning after traumatic brain injury. *J Trauma Acute Care Surg* 75(3):410–416.

7 Neurosteroids and Traumatic Brain Injury

Translating Biomarkers to Therapeutics; Overview and Pilot Investigations in Iraq and Afghanistan Era Veterans

Christine E. Marx, Jennifer C. Naylor, Jason D. Kilts, Charlotte E. Dunn, Larry A. Tupler, Steven T. Szabo, Bruce P. Capehart, Rajendra A. Morey, Lawrence J. Shampine, Shawn K. Acheson, and VA Mid-Atlantic MIRECC Workgroup

CONTENTS

INTRODUCTION

Neurosteroids are endogenous molecules that are enriched in the brain (where they can be synthesized de novo from cholesterol) and also produced in the adrenals, gonads, and other peripheral tissues. Many neurosteroids demonstrate pleiotropic actions that are potentially relevant to traumatic brain injury (TBI) and its therapeutics. This chapter will provide an overview of preclinical data supporting a role for neurosteroids in TBI, followed by a summary of clinical studies to date involving neurosteroids and TBI. Next, the development of neurosteroids as both biomarker candidates and treatment interventions will be discussed. Initial data examining neurosteroid alterations in Iraq and Afghanistan era veterans with a history of blast-related TBI will be presented, and a theoretical framework for the translation of these biomarker efforts to clinical endeavors will be provided. Results from a pilot proof-of-concept neurosteroid intervention in mild TBI will be presented as an example of a potential biomarker-informed translational model, followed by a synopsis of this area and discussion of future directions.

NEUROSTEROIDS IN TRAUMATIC BRAIN INJURY (TBI): OVERVIEW OF PRECLINICAL STUDIES

Multiple preclinical models have examined neurosteroids in TBI. Many of these preclinical investigations have focused on progesterone, which can also be synthesized in the brain in addition to the periphery. As recently reviewed,[1] over 50 preclinical studies involving progesterone in TBI have been reported; among these investigations, progesterone was shown to decrease proinflammatory cytokine expression,[2–6] decrease brain edema,[3,5,7–15] reduce tissue loss and lesion size,[5,16–18] and decrease axonal injury following TBI,[19] among multiple other actions supporting a possible therapeutic role in TBI. In addition, several preclinical investigations utilizing progesterone reported enhanced functional and behavioral recovery following TBI.[20,21]

Relatively fewer investigations have been conducted with neurosteroids other than progesterone, however. It is possible that GABAergic neurosteroids, in particular, may hold promise for TBI treatment, in that they have the potential to mitigate injury-induced excitotoxicity. For example, allopregnanolone (a twice-reduced metabolite of progesterone) is a potent positive allosteric modulator of $GABA_A$ receptors.[22–24] Consistent with this GABAergic action, allopregnanolone demonstrates anticonvulsant effects in a number of seizure models.[25–29] It also decreases apoptosis,[30–32] enhances neurogenesis,[33] and reduces neurodegeneration.[34–36] Preclinical data suggest that allopregnanolone demonstrates pronounced neuroprotective actions in TBI and other central nervous system (CNS) injury models. For example, allopregnanolone reduces ischemic infarcts to a greater degree than progesterone.[37] It also directly inhibits the mitochondrial permeability transition pore, a hypothesized mechanism for its superior neuroprotective effects in animal models of TBI compared to progesterone.[38] Further, the neuroprotective actions of progesterone in other injury models (including kainic acid) appear to be mediated by metabolism to allopregnanolone.[39–41] In addition, pregnanolone (the 5β-stereoisomer of allopregnanolone) also demonstrates positive modulatory actions at $GABA_A$ receptors,[26,42–44] although fewer investigations have been conducted with this neurosteroid (which is present at extremely low/undetectable levels in rodents; in humans, however, serum pregnanolone levels are several-fold higher than serum allopregnanolone levels).[45] Androsterone is another neurosteroid that is a positive $GABA_A$ receptor modulator[44,46,47] and demonstrates anticonvulsant activity in animal models.[48]

In addition to neurosteroids that positively modulate $GABA_A$ receptors (and could theoretically reduce injury-related excitotoxicity), there are other neurosteroid candidates that may be relevant to TBI that do not share this action at $GABA_A$ receptors. Dehydroepiandrosterone sulfate (DHEAS), for example, appears to have beneficial effects on cognitive and behavioral deficits in a rodent model of mild TBI,[49] and it is also neuroprotective in a focal cortical cold lesion paradigm.[50] In addition, DHEAS enhances behavioral and functional recovery following TBI in rats.[51] In contrast to allopregnanolone, pregnanolone, and androsterone, DHEAS is a negative modulator of $GABA_A$ receptors.[52,53] Pregnenolone is another neurosteroid that may merit further attention in TBI, as it enhances learning and memory in rodents,[54] regulates myelination,[55,56] stabilizes microtubules,[57] and enhances neuritic outgrowth.[58] Both pregnenolone and DHEA also appear to decrease astrocytic proliferation and gliosis following penetrating brain injury in male rats.[59] In addition to TBI models that test neurosteroids as possible therapeutic interventions to mitigate injury sequelae, it has been reported that a number of these molecules increase in multiple brain regions following TBI in rats, including pregnenolone, progesterone, and 5α-dihydroprogesterone—a finding hypothesized to potentially represent an endogenous neuroprotective mechanism.[60,61] There are thus multiple additional neurosteroid candidates that merit further exploration as biomarkers and therapeutics in TBI.

NEUROSTEROIDS AND TBI: OVERVIEW OF CLINICAL STUDIES

Clinical Trials

Given extensive preclinical data supporting progesterone as a therapeutic agent in TBI, two proof-of-concept randomized controlled trials (Phase II) examining progesterone in patients with acute moderate and/or severe TBI were conducted by independent research groups and exhibited considerable promise. In one study, mortality at 30 days was more than 50% lower in the group that received intravenous progesterone compared to the placebo group among patients with acute moderate or severe TBI.[62] In the second Phase II randomized controlled trial, patients with acute severe TBI who were randomized to intramuscular progesterone demonstrated significant improvement in the Glasgow Outcome Scale (GOS) and Functional Independence Measure scores compared to the placebo group.[63] Two larger Phase III randomized controlled trials were subsequently conducted, but neither demonstrated a beneficial effect of progesterone compared to placebo in acute TBI. The Progesterone for Traumatic Brain Injury, Experimental Treatment (PROTECT III) study randomized a total of 882 of the planned sample of 1140 participants with moderate, moderate to severe, or severe, acute TBI; this investigation was stopped for futility with regard to the primary endpoint, a prespecified level of improvement in the Extended GOS score 6 months postinjury.[64] In the second Phase III study, titled Study of a Neuroprotective Agent, Progesterone, in Severe Traumatic Brain Injury (SYNAPSE), there was no demonstrated clinical benefit of progesterone for either the primary outcome measure (GOS score at 6 months postinjury) or secondary outcome measures; 1195 participants with severe nonpenetrating TBI were randomized in this investigation.[65] Progesterone was well tolerated in both Phase III investigations (with the exception of phlebitis or thrombophlebitis, which was observed more frequently in the progesterone group in the PROTECT III study; these episodes "were frequently categorized as not serious and were self-limited").[64] Mortality rates were similar in the progesterone and placebo groups in both Phase III investigations.[64,65] Progesterone as an intervention for acute moderate or severe TBI could thus not be confirmed in either of these two Phase III trials. The investigators involved in these Phase III studies cite multiple potential factors that may have impacted results, including possible insensitivity of outcome measures, injury heterogeneity, absence of adequate biomarkers to inform clinical trial design, confounding preexisting conditions, delayed study drug administration, and characteristics of individual patients, among other possible variables.[64,65]

Negative findings from these two Phase III trials examining progesterone in TBI (that were based on a foundation of robust preclinical data from multiple independent laboratories and two positive Phase II trials) raise multiple questions regarding optimal pharmacological development strategies for TBI. There is currently no pharmacological agent for the treatment of acute or chronic TBI approved by the U.S. Food and Drug Administration (FDA). Multiple options are currently under discussion and debate in the scientific community following these recent negative Phase III progesterone studies to attempt to address this urgent clinical need, and clinical trials with other agents are ongoing. It is possible that more extensive dose-finding investigations would be beneficial at earlier development stages. Although very expensive, such upfront investments could theoretically yield more refined and definitive assessments of pharmacological candidates. Given U-shaped dose response curves that are sometimes described in the neurosteroid literature (reviewed by Guennoun et al.[60]), it is possible that optimal therapeutic progesterone concentrations may not have been achieved in either Phase III clinical trial (or at least not achieved by a substantial subset of participants randomized to progesterone), and that progesterone dosing could have been too high or too low. Complicating the dosing question further, if metabolism to other neurosteroids potentially constitutes a mechanism for the neuroprotective actions of progesterone in rodents (as has been previously hypothesized), then it is possible that the metabolism of progesterone to allopregnanolone (or other neurosteroids) exhibits individual variability, that is, secondary to genetic differences in neurosteroidogenic enzymes, factors related to health status preinjury, medical complications secondary to polytrauma, medication interactions, and/or multiple other contributing etiologies. In addition, neurosteroids are expressed in a regionally specific manner, and $GABA_A$ receptor-mediated inhibition by specific neurosteroids such as allopregnanolone also appears to vary by brain region. Injury heterogeneity in which multiple brain regions may be impacted differently following TBI could thus be particularly confounding for neurosteroid intervention studies. In addition, duration of treatment may have been suboptimal.

Biomarker Studies

Biomarker investigations in clinical populations focusing on neurosteroids in TBI are currently somewhat limited. Neurosteroids are highly relevant to the stress response and increase acutely following a number of stressors in animal models.[66–68] Allopregnanolone, for example, increases acutely following multiple stressors in animal models, but then appears to be involved in a negative feedback mechanism that diminishes CRF, ACTH, and corticosterone release in rodents.[69–71] These actions have been postulated to constitute an endogenous autoregulatory mechanism that returns an organism to a homeostatic state after stress.[72] In contrast to acute stress, chronic stress (such as prolonged social isolation in rodents) appears to decrease allopregnanolone levels.[73,74] In addition, the neurosteroid DHEA increases markedly in plasma following ACTH stimulation in humans.[75] To the extent that multiple neurosteroids interact with the hypothalamic–pituitary–adrenal axis, it is possible that neurosteroid regulation may be impacted by pituitary injury following TBI.

Pituitary hypofunction appears to be a relatively common occurrence following TBI, with a reported prevalence of approximately 15%–68%, as recently reviewed by Sundaram et al.[76,77–87] The likelihood of developing an endocrine disorder following TBI is not necessarily solely related to TBI severity, however,[78,88] and pituitary hypofunction has been observed in 10%–25% of patients who sustained a mild TBI (reviewed by Sundaram et al.[76]). Growth hormone deficiency was also observed in 4 of 16 (25%) of veterans with a history of mild combat related TBI.[89] Endocrine dysfunction may also occur months to years postinjury.[77,78,80,82,84,87] Many efforts in the area of pituitary hypofunction and TBI to date have focused on growth hormone (GH), luteinizing hormone (LH), follicle-stimulating hormone (FSH), adrenocorticotropic hormone (ACTH), cortisol, and thyroid hormone, and prolactin disruption.

Pituitary hypofunction also appears to be a possible consequence of blast-related TBI specifically, and its detection and treatment could be important to Iraq and Afghanistan veterans (given high rates of blast-related TBI exposures in this cohort). Although data are currently limited and confined to small studies, one report observed that 6 of 19 soldiers (32%) with a history of blast-related moderate to severe TBI who had served in the military in the United Kingdom had evidence of pituitary dysfunction,[90] and another investigation reported that 11 of 26 (42%) U.S. veterans with a history of blast-related TBI demonstrated abnormal hormone levels in one or more pituitary-related endocrine systems.[91] To our knowledge, neurosteroid alterations following blast-related TBI have not yet been reported. We next present pilot results from a recent investigation focusing on this area in Iraq and Afghanistan era veterans.

NEUROSTEROIDS AS BIOMARKER CANDIDATES: PILOT INVESTIGATIONS IN IRAQ AND AFGHANISTAN ERA VETERANS WITH A HISTORY OF BLAST-RELATED TBI

Neurosteroid levels in serum samples were quantified in 55 male veterans who had sustained a blast-related TBI in Iraq or Afghanistan (with or without loss of consciousness), and in 55 male control veterans who had been deployed to Iraq or Afghanistan but who had not sustained a blast-related TBI (matched for age, time of blood draw, and smoking status). Nonparametric statistical comparisons between groups were conducted for each neurosteroid (pregnanolone, pregnenolone, androsterone, allopregnanolone). Neurosteroid levels in sera were determined by a highly sensitive and specific gas chromatography/mass spectrometry (GC/MS) method, as previously described,[45,92–94] with modifications. One milliliter of serum was extracted three times in ethyl acetate prior to high-performance liquid chromatography purification utilizing tetrahydrofuran, ethanol, and hexane in the mobile phase. A subset of samples (30%) was injected in duplicate; mean intra-assay coefficients of variation in these duplicate subsets were less than 6% for all four neurosteroids. The limit of detection for this mass spectrometry method was 1–2 picograms for each neurosteroid.

Pregnanolone (the 5β-stereoisomer of allopregnanolone, also a positive modulator of $GABA_A$ receptors) was significantly reduced in veterans who had sustained a blast-related TBI in Iraq or Afghanistan (median level = 485.6 pg/mL) compared to veterans who had no TBI history (median level = 649.5 pg/mL; Mann-Whitney, $p = 0.0008$, $n = 55$ per group; Figure 7.1a). Androsterone (another GABAergic neurosteroid) was also significantly reduced in veterans who had sustained a blast-related TBI (median level = 315.2 pg/mL) compared to veterans with no history of TBI (median level = 400.0 pg/mL; Mann-Whitney, $p = 0.001$, $n = 55$ per group; Figure 7.1b). Pregnenolone also tended to be decreased in veterans who had sustained a blast-related TBI (median level = 584.8 pg/mL) compared to veterans with a history of deployment to Iraq or Afghanistan but who had no blast-related TBI history (median level = 735.8 pg/mL), but this finding was a "trend" only and not statistically significant (Mann-Whitney, $p = 0.077$, $n = 55$ per group; Figure 7.1c). Allopregnanolone levels were not significantly different between groups (Mann-Whitney, $p > 0.05$; $n = 55$ per group). Neurosteroid levels in veterans with a history of blast-related TBI and co-occurring post-traumatic stress disorder (PTSD) were not significantly different compared to neurosteroid levels in veterans with a history of blast-related TBI and no concurrent PTSD (Mann-Whitney, $p > 0.05$ for each neurosteroid). These data thus suggest that specific neurosteroids (pregnanolone, androsterone, pregnenolone) may be dysregulated following blast-related TBI, and that supplementation with neurosteroids could be a promising pharmacological intervention strategy to ameliorate deficits in these molecules. As endocrinopathies frequently result in irritability, poor concentration, depressive symptoms, and affective lability, among

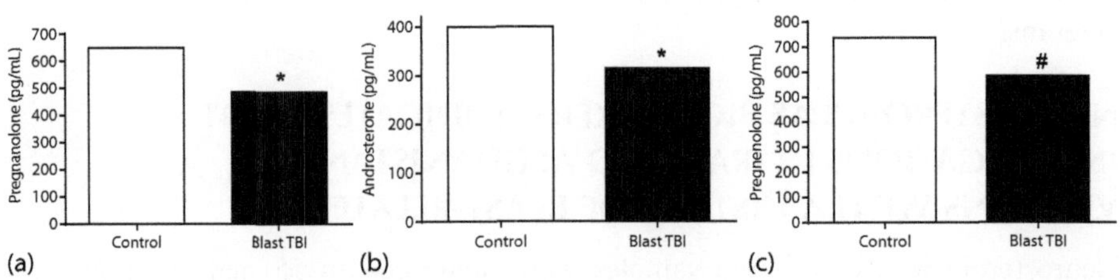

FIGURE 7.1 (a) Pregnanolone is significantly reduced in Iraq and Afghanistan era veterans with a history of blast-related TBI compared to control veterans with a history of deployment but no TBI exposure. Mann-Whitney, $^*p = 0.0008$; $n = 55$ per group. (b) Androsterone is significantly reduced in Iraq and Afghanistan era veterans with a history of blast-related TBI compared to control veterans with a history of deployment but no TBI exposure. Mann-Whitney, $^*p = 0.001$; $n = 55$ per group. (c) Pregnenolone tends to be reduced in Iraq and Afghanistan era veterans with a history of blast-related TBI compared to control veterans with a history of deployment but no TBI exposure. Mann-Whitney, $\#p = 0.077$; $n = 55$ per group.

other manifestations, it is logical to hypothesize that neurosteroid reductions (possibly secondary to pituitary hypofunction post-TBI) may contribute to cognitive and behavioral symptoms following TBI. This possibility is supported by evidence that a number of neurosteroids are decreased and potentially dysregulated in depression[95] and PTSD.[96] Results are preliminary, however, and this pilot study will require replication in larger cohorts. The extent to which neurosteroid reductions following blast-related TBI may predispose individuals to the development of neuropsychiatric sequelae remains to be elucidated.

NEUROSTEROIDS AS THERAPEUTICS: PILOT PROOF-OF-CONCEPT CONTROLLED TRIAL IN IRAQ AND AFGHANISTAN ERA VETERANS WITH A HISTORY OF MILD TBI

To examine the potential utility of neurosteroids as therapeutic interventions in TBI, we conducted a pilot proof-of-concept randomized controlled trial with adjunctive pregnenolone in Iraq and Afghanistan era veterans with a history of mild TBI. The aforementioned preliminary data showing that pregnenolone tends to be reduced in Iraq and Afghanistan veterans who have a history of blast-related TBI ("trend" level finding), and that pregnanolone (a GABAergic neurosteroid) is significantly reduced following blast-related TBI, strengthen the rationale for the use of this neurosteroid as a TBI treatment, and potentially represents a biomarker-informed therapeutic strategy. We have previously determined that pregnenolone administration results in significant increases in downstream pregnanolone and allopregnanolone levels (in addition to expected increases in levels of pregnenolone itself).[45] It is thus possible that pregnenolone may have promise as a precursor loading strategy to achieve downstream elevations in pregnanolone and allopregnanolone, two GABAergic neurosteroids. Since endocrine dysregulation may be an underrecognized consequence of TBI, it is logical to consider this pathophysiological element in the development of novel treatment options. Allopregnanolone has shown neuroprotective actions in preclinical TBI models, as described earlier. In addition, pregnenolone and its sulfated derivative (pregnenolone sulfate) enhance learning and memory in rodents.[54,97–105] There are thus multiple lines of evidence that support the potential utility of pregnenolone as a therapeutic candidate in TBI.

Iraq and Afghanistan era veterans with a history of mild TBI were randomized to adjunctive pregnenolone or placebo for 8 weeks in this pilot randomized controlled trial (following a 2-week placebo lead-in phase for all participants). Participants met criteria for mild TBI according to the World Health Organization Task Force guidelines,[106] with the exception of the Glasgow Coma Scale score (which was not generally available for these patients); these criteria are also consistent with the CDC Report to Congress (September 2003) and the American Congress of Rehabilitation Medicine (ACRM) definition of mild TBI (1993). In this exploratory study, endpoints included a cognitive battery (Brief Assessment of Cognition composite z-score),[107,108] the Clinician-Administered PTSD Scale (CAPS), a resilience measure (Connor-Davidson Resilience Scale [CD-RISC]), and other assessments; rating scales were administered at randomization, 4 weeks, and 8 weeks. The study took place at the Durham Veterans Affairs Medical Center (Durham, North

Carolina; NCT00623506; FDA IND #78,270) and was approved by the Durham VA Medical Center's Institutional Review Board and conducted in accordance with the Declaration of Helsinki. Study visits occurred every 2 weeks (6 study visits total); electrocardiograms (ECGs) were performed at baseline and completion of the study; clinical chemistry panels, liver function testing, and complete blood counts were done at the screening visit, randomization visit, week 4, and week 8; and cholesterol levels, thyroid stimulating hormone levels, and prolactin levels were done at baseline and completion. Pregnenolone dosing consisted of the following regimen for participants randomized to this intervention: 50 mg twice daily × 2 weeks, followed by 150 mg twice daily × 2 weeks, followed by 250 mg twice daily × 4 weeks. Inclusion criteria included the following: Participants who had served in the U.S. military since September 11, 2001 (Iraq and Afghanistan era) and who were 18 to 55 years old, of any ethnic group, either sex. Patients must have been able to participate fully in the informed consent process and have had a history of mild TBI since September 2001, with no anticipated need to alter any concomitant medications for the 10-week duration of the study. Key exclusion criteria included: serious unstable medical illness, history of seizures, history of cerebrovascular accident, prostate, uterine, or breast cancer, use of oral contraceptives or other hormonal supplementation such as estrogen, current active suicidal and/or homicidal ideation, intent or plan, current *DSM-IV* diagnosis of bipolar disorder, schizophrenia or other psychotic disorder, cognitive disorder due to a general medical condition other than TBI, female patients who are pregnant or breast-feeding, and known allergy to study medication.

Of 30 patients randomized, 22 (73%) completed at least 4 weeks of treatment (11 participants per group). Mean changes in composite BACS cognitive scores were not significantly different between groups posttreatment (change in z-score = 0.61 in the pregnenolone group vs. 0.80 in the placebo group; baseline z-scores were –0.75 and –0.55, respectively, for the pregnenolone and placebo group prior to randomization). Total CAPS scores were also not significantly different between groups posttreatment, decreasing by a mean of 8.5 points in the pregnenolone group and 7.3 points in the placebo group (mean baseline CAPS scores were 43.1 and 38.6 in the pregnenolone and placebo groups, respectively, prior to randomization). A potential clinical signal emerged among "Cluster D" symptoms of the CAPS, however, which include sleep difficulties, trouble concentrating, irritability or anger outbursts, increased startle response, and hypervigilance. Interestingly, these symptoms overlap considerably with those of postconcussive syndrome. While changes in "Cluster D" symptoms did not differ significantly between groups, alterations were in the hypothesized direction (Figure 7.2). Perhaps more important, CAPS Cluster D symptom reductions demonstrated significant inverse correlations with pregnenolone ($r_s = -0.82$; $p = 0.011$), allopregnanolone ($r_s = -0.82$; $p = 0.011$), and pregnanolone ($r_s = -0.82$; $p = 0.011$) changes posttreatment (i.e., the greater the increases in these neurosteroids posttreatment with pregnenolone, the greater the reduction in Cluster D symptoms at study completion; Figure 7.3). In addition, increases in serum pregnenolone (Spearman $r = -0.85$; $p = 0.006$) and allopregnanolone ($r_s = -0.85$; $p = 0.006$) posttreatment were also significantly inversely correlated with reductions in total CAPS scores in the group randomized to the pregnenolone intervention. Pregnanolone increases posttreatment with pregnenolone also predicted improvements in resilience (CD-RISC;

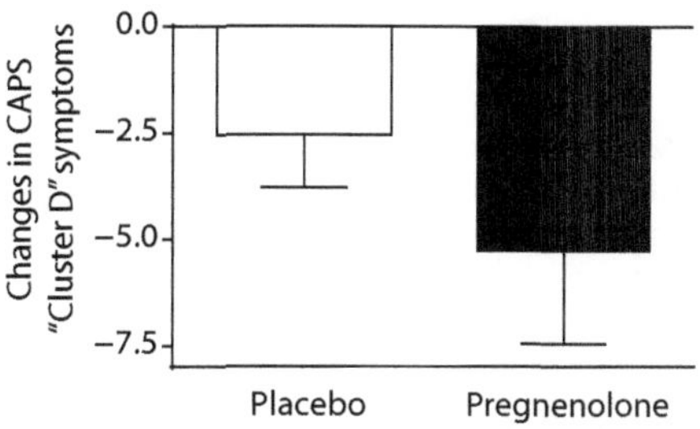

FIGURE 7.2 Changes in CAPS "Cluster D" symptoms were not significantly different between groups ($p > 0.05$), but improvements were in the hypothesized direction in this symptom domain in participants randomized to pregnenolone; $n = 11$ per group.

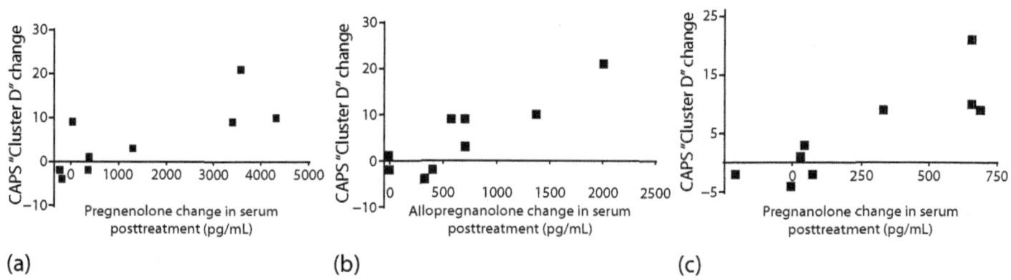

FIGURE 7.3 (a) Pregnenolone increases that follow treatment with this intervention appear to predict PTSD "Cluster D" symptom improvement (Spearman $r = -0.82$, $p = 0.011$, $n = 9$ participants randomized to pregnenolone who completed the entire 8-week study). (b) Allopregnanolone increases that follow treatment with pregnenolone appear to predict PTSD "Cluster D" symptom improvement (Spearman $r = -0.82$, $p = 0.011$, $n = 9$ participants randomized to pregnenolone who completed the entire 8-week study). (c) Pregnanolone increases that follow treatment with pregnenolone appear to predict PTSD "Cluster D" symptom improvement (Spearman $r = -0.82$, $p = 0.011$, $n = 9$ participants randomized to pregnenolone who completed the entire 8-week study).

$r = 0.80$, $p = 0.014$). Pregnenolone administration significantly increased pregnenolone and allopregnanolone levels posttreatment, and tended to increase pregnanolone levels also (Figure 7.4); these results are very consistent with previously reported neurosteroid changes following treatment with pregnenolone.[45,109] There were no serious adverse events in either group; pregnenolone was well tolerated. Side effects that presented to a greater extent in the pregnenolone group compared to the placebo group included headache (reported by 4 participants in the pregnenolone group vs. 1 participant in the placebo group; severity rated as "mild" in all instances) and excitation/agitation (reported by 2 participants in the pregnenolone group vs. no participants in the placebo group; severity rated as "mild" in both instances). There were no clinically significant changes in ECG parameters, clinical chemistry assessments, liver function tests, complete blood count, or weight in either group posttreatment. Cholesterol levels were significantly decreased posttreatment in the pregnenolone

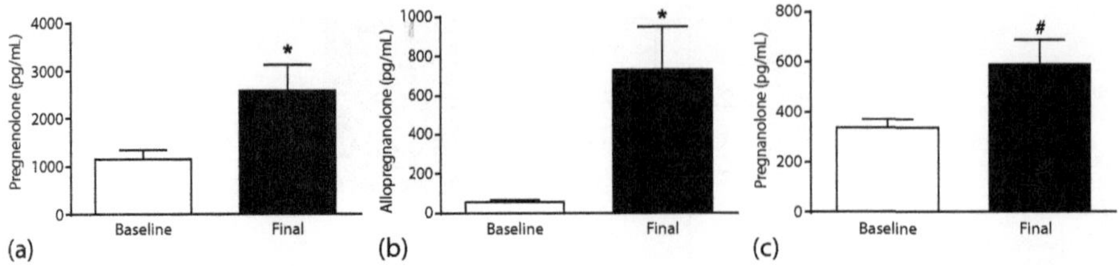

FIGURE 7.4 (a) Pregnenolone levels in sera are significantly increased posttreatment with this neurosteroid (mean 2591 pg/mL) compared to baseline sera levels (mean 1151 pg/mL); paired t-test p = 0.045; n = 9 per group. (b) Allopregnanolone levels in sera are significantly increased posttreatment with pregnenolone (mean 731.1 pg/mL) compared to baseline sera levels (mean 55.53 pg/mL); paired t-test p = 0.015; n = 9 per group. (c) Pregnanolone levels in sera tend to be increased posttreatment with pregnenolone (mean 590.0 pg/mL) compared to baseline sera levels (mean 337.8 pg/mL); paired t-test p = 0.057; n = 9 per group.

group compared to the placebo group (p = 0.035), but these were nonfasting assessments. There were no changes in TSH or prolactin levels in either group.

These pilot results are very preliminary, but it is possible that neurosteroids may have utility as candidate biomarkers for therapeutic response (since neurosteroid increases posttreatment were correlated with clinical improvements). These data may also provide early biomarker information regarding optimal therapeutic concentrations of pregnenolone and other neurosteroids, as well as potential parameters for participant selection in future studies. Investigations in larger cohorts will be required to test these possibilities. Should pregnenolone be a useful pharmacological candidate for TBI, contributions to its therapeutic efficacy may involve multiple mechanisms that encompass more than one neurosteroid, given pregnenolone's metabolism profile and pleiotropic actions. As the course of TBI involves physiological processes that span inflammatory responses, excitotoxicity, neurodegeneration, apoptosis, myelination, and other CNS events, a pharmacological intervention such as pregnenolone that impacts a number of these parameters and targets multiple mechanisms (either by itself and/or via its metabolism to downstream GABAergic neurosteroids such as pregnanolone and allopregnanolone) may be a logical therapeutic candidate.

SUMMARY AND FUTURE DIRECTIONS

A number of neurosteroids demonstrate pleiotropic actions that may be highly relevant to the pathophysiology and treatment of TBI. The area of neurosteroid therapeutics is still in its earliest stages, however, and few clinical studies are currently available. Larger clinical trials that have been conducted to date have focused on acute moderate and/or severe TBI utilizing progesterone, with two Phase III trials recently reporting negative results. It is possible that characterizing progesterone levels (and progesterone metabolite and precursor levels) at baseline and posttreatment could potentially elucidate subsets of individuals who may have been more or less likely to respond to a progesterone intervention in the setting of acute TBI. Genetic and epigenetic information with regard to neurosteroidogenic enzyme function could also be useful in "personalizing" future strategies and refining patient selection for neurosteroid clinical trials.

Clinical studies focusing on neurosteroids other than progesterone for TBI are currently limited, either as acute or chronic TBI interventions. A Phase II clinical trial utilizing intravenous allopregnanolone is currently underway for acute TBI (NCT01673828); allopregnanolone may be a particularly promising intervention, as recent data suggest that it was effective in severe refractory status epilepticus in a small number of patients (Globe Newswire Report/SAGE-547).[110] A larger randomized controlled trial in Iraq and Afghanistan era veterans utilizing pregnenolone for chronic mild TBI that builds upon the aforementioned pilot data is also enrolling (NCT01336413). Tailoring therapeutics to individual patients and their specific neurobiological alterations following TBI will be critical. Along these lines, neurosteroid interventions could potentially be efficacious in patients with relative deficits in these molecules (either at baseline or following TBI), whether secondary to pituitary hypofunction, dysregulation in neurosteroidogenic enzymes (involving decreased synthesis and/or increased metabolism of specific neurosteroids), or other mechanistic etiologies. It will also be important to establish optimal dosing strategies for neurosteroids, and to characterize windows of therapeutic response that may predict maximal efficacy of neurosteroid interventions.

With regard to biomarker efforts, data are currently somewhat limited in the neurosteroid area vis-à-vis TBI. Pilot biomarker efforts suggest that specific neurosteroids may be decreased chronically following blast-related TBI in Iraq and Afghanistan era veterans, potentially strengthening the rationale for the utility of neurosteroids as pharmacological interventions; these neurosteroid reductions have included pregnenolone, pregnanolone, and androsterone among veterans with a history of blast-related TBI, as presented earlier. Given the complexity of biosynthetic pathways leading to neurosteroid synthesis and metabolism, and likely individual variation in neurosteroidogenic enzyme activity (that is genetically and/or epigenetically mediated, or modulated by multiple other variables), additional research will be required to comprehensively assess the expression of these molecules as possible biological signatures for TBI; such biomarker data could then effectively inform and refine future neurosteroid treatment strategies.

ACKNOWLEDGMENTS

Dr. Naylor is supported by the Department of Veterans Affairs Rehabilitation Research and Development Career Development Award (1IK2RX000908). Dr. Acheson had a VA Career Development Award (IK2BX001267). This work was supported by a VA Career Development Transition Award (CEM), VA Career Development-II Awards (CEM and JCN), and the VA Mid-Atlantic MIRECC (PI John Fairbank, PhD). The VA Mid-Atlantic MIRECC Workgroup consists of the following members: Jean C. Beckham, Mira Brancu, Patrick S. Calhoun, Eric B. Elbogen, John A. Fairbank, Robin A. Hurley, Gregory McCarthy, Scott D. McDonald, John Mason, Marinell Miller-Mumford, Scott D. Moore, Treven C. Pickett, Kristy A. Straights-Troster, and Katherine H. Taber. We thank Gillian Parke for her excellent technical assistance. The views expressed in this manuscript are those of the authors and do not necessarily represent the views of the Department of Veterans Affairs.

DISCLOSURES

Pending patents on the use of neurosteroids and derivatives in CNS disorders and for lowering cholesterol; no patents issued; no licensing in place. VA 208 waiver in place.

REFERENCES

1. Diaz-Arrastia, R. et al. 2014. Pharmacotherapy of traumatic brain injury: State of the science and the road forward: Report of the Department of Defense Neurotrauma Pharmacology Workgroup. *J Neurotrauma* 31:135–158.
2. Chen, G. et al. 2007. Progesterone prevents traumatic brain injury-induced intestinal nuclear factor kappa B activation and proinflammatory cytokines expression in male rats. *Mediators Inflamm* 2007:93431.
3. Cutler, S.M. et al. 2007. Progesterone improves acute recovery after traumatic brain injury in the aged rat. *J Neurotrauma* 24:1475–1486.
4. He, J. et al. 2004. Progesterone and allopregnanolone reduce inflammatory cytokines after traumatic brain injury. *Exp Neurol* 189:404–412.
5. Pan, D.S. et al. 2007. Inhibitory effect of progesterone on inflammatory factors after experimental traumatic brain injury. *Biomed Environ Sci* 20:432–438.
6. Sarkaki, A.R. et al. 2013. Time- and dose-dependent neuroprotective effects of sex steroid hormones on inflammatory cytokines after a traumatic brain injury. *J Neurotrauma* 30:47–54.
7. Cutler, S.M. et al. 2006. Slow-release and injected progesterone treatments enhance acute recovery after traumatic brain injury. *Pharmacol Biochem Behav* 84:420–428.
8. Galani, R., Hoffman, S.W., Stein, D.G. 2001. Effects of the duration of progesterone treatment on the resolution of cerebral edema induced by cortical contusions in rats. *Restor Neurol Neurosci* 18:161–166.
9. Grossman, K.J., Goss, C.W., Stein, D.G. 2004. Effects of progesterone on the inflammatory response to brain injury in the rat. *Brain Res* 1008:29–39.
10. Guo, Q. et al. 2006. Progesterone administration modulates AQP4 expression and edema after traumatic brain injury in male rats. *Exp Neurol* 198:469–478.
11. Kasturi, B.S., Stein, D.G. 2009. Progesterone decreases cortical and sub-cortical edema in young and aged ovariectomized rats with brain injury. *Restor Neurol Neurosci* 27:265–275.
12. Robertson, C.L. et al. 2006. Physiologic progesterone reduces mitochondrial dysfunction and hippocampal cell loss after traumatic brain injury in female rats. *Exp Neurol* 197:235–243.
13. Roof, R.L. et al. 1996. Progesterone rapidly decreases brain edema: Treatment delayed up to 24 hours is still effective. *Exp Neurol* 138:246–251.
14. Roof, R.L., Duvdevani, R., Stein, D.G. 1992. Progesterone treatment attenuates brain edema following contusion injury in male and female rats. *Restor Neurol Neurosci* 4:425–427.
15. Shahrokhi, N. et al. 2010. Effect of sex steroid hormones on brain edema, intracranial pressure, and neurologic outcomes after traumatic brain injury. *Can J Physiol Pharmacol* 88:414–421.
16. Grossman, K.J., Goss, C.W., Stein, D.G. 2011. Sickness behaviors following medial frontal cortical contusions in male rats. *Behav Brain Res* 217:202–208.
17. Jones, N.C. et al. 2005. The neuroprotective effect of progesterone after traumatic brain injury in male mice is independent of both the inflammatory response and growth factor expression. *Eur J Neurosci* 21:1547–1554.

18. Peterson, T.C. et al. 2012. A comparison of the effects of nicotinamide and progesterone on functional recovery of cognitive behavior following cortical contusion injury in the rat. *J Neurotrauma* 29:2823–2830.
19. O'Connor, C.A. et al. 2007. Effects of progesterone on neurologic and morphologic outcome following diffuse traumatic brain injury in rats. *Exp Neurol* 205:145–153.
20. Djebaili, M. et al. 2005. The neurosteroids progesterone and allopregnanolone reduce cell death, gliosis, and functional deficits after traumatic brain injury in rats. *J Neurotrauma* 22:106–118.
21. Djebaili, M., Hoffman, S.W., Stein, D.G. 2004. Allopregnanolone and progesterone decrease cell death and cognitive deficits after a contusion of the rat pre-frontal cortex. *Neuroscience* 123:349–359.
22. Majewska, M.D. et al. 1986. Steroid hormone metabolites are barbiturate-like modulators of the GABA receptor. *Science* 232:1004–1007.
23. Morrow, A.L. et al. 1990. Characterization of steroid interactions with gamma-aminobutyric acid receptor-gated chloride ion channels: Evidence for multiple steroid recognition sites. *Mol Pharmacol* 37:263–270.
24. Morrow, A.L., Suzdak, P.D., Paul, S.M. 1987. Steroid hormone metabolites potentiate GABA receptor-mediated chloride ion flux with nanomolar potency. *Eur J Pharmacol* 142:483–485.
25. Belelli, D., Bolger, M.B., Gee, K.W. 1989. Anticonvulsant profile of the progesterone metabolite 5 alpha-pregnan-3 alpha-ol-20-one. *Eur J Pharmacol* 166:325–329.
26. Bitran, D., Hilvers, R.J., Kellogg, C.K. 1991. Anxiolytic effects of 3 alpha-hydroxy-5 alpha[beta]-pregnan-20-one: Endogenous metabolites of progesterone that are active at the GABAA receptor. *Brain Res* 561:157–161.
27. Hogskilde, S. et al. 1988. Anticonvulsive properties of pregnanolone emulsion compared with althesin and thiopentone in mice. *Br J Anaesth* 61:462–467.
28. Kokate, T.G. et al. 1998. Lack of anticonvulsant tolerance to the neuroactive steroid pregnanolone in mice. *J Pharmacol Exp Ther* 287:553–558.
29. Wieland, S. et al. 1991. Anxiolytic activity of the progesterone metabolite 5 alpha-pregnan-3 alpha-o1-20-one. *Brain Res* 565:263–268.
30. Charalampopoulos, I. et al. 2006. Neurosteroids as endogenous inhibitors of neuronal cell apoptosis in aging. *Ann NY Acad Sci* 1088:139–152.
31. Charalampopoulos, I. et al. 2004. Dehydroepiandrosterone and allopregnanolone protect sympathoadrenal medulla cells against apoptosis via antiapoptotic Bcl-2 proteins. *Proc Natl Acad Sci U S A* 101:8209–8214.
32. Xilouri, M., Papazafiri, P. 2006. Anti-apoptotic effects of allopregnanolone on P19 neurons. *Eur J Neurosci* 23:43–54.
33. Wang, J.M. et al. 2005. The neurosteroid allopregnanolone promotes proliferation of rodent and human neural progenitor cells and regulates cell-cycle gene and protein expression. *J Neurosci* 25:4706–4718.
34. Ahmad, I. et al. 2005. Allopregnanolone treatment, both as a single injection or repetitively, delays demyelination and enhances survival of Niemann-Pick C mice. *J Neurosci Res* 82:811–821.
35. Griffin, L.D. et al. 2004. Niemann-Pick type C disease involves disrupted neurosteroidogenesis and responds to allopregnanolone. *Nat Med* 10:704–711.
36. Mellon, S.H., Gong, W., Schonemann, M.D. 2008. Endogenous and synthetic neurosteroids in treatment of Niemann-Pick Type C disease. *Brain Res Rev* 57:410–420.
37. Sayeed, I. et al. 2006. Allopregnanolone, a progesterone metabolite, is more effective than progesterone in reducing cortical infarct volume after transient middle cerebral artery occlusion. *Ann Emerg Med* 47:381–389.
38. Sayeed, I., Stein, D.G. 2009. Progesterone as a neuroprotective factor in traumatic and ischemic brain injury. *Prog Brain Res* 175:219–237.

39. Ciriza, I., Azcoitia, I., Garcia-Segura, L.M. 2004. Reduced progesterone metabolites protect rat hippocampal neurones from kainic acid excitotoxicity in vivo. *J Neuroendocrinol* 16:58–63.
40. Ciriza, I. et al. 2006. Reduced metabolites mediate neuroprotective effects of progesterone in the adult rat hippocampus. The synthetic progestin medroxyprogesterone acetate (Provera) is not neuroprotective. *J Neurobiol* 66:916–928.
41. Ishihara, Y. et al. 2013. Allopregnanolone-mediated protective effects of progesterone on tributyltin-induced neuronal injury in rat hippocampal slices. *J Steroid Biochem Mol Biol* 135:1–6.
42. Gerak, L.R. et al. 2004. Effects of pregnanolone alone and in combination with other positive GABAA modulators on complex behavior in rats. *Psychopharmacology* 173:195–202.
43. Majewska, M.D. 1992. Neurosteroids: Endogenous bimodal modulators of the GABAA receptor. Mechanism of action and physiological significance. *Prog Neurobiol* 38:379–395.
44. Turner, D.M. et al. 1989. Steroid anesthetics and naturally occurring analogs modulate the gamma-aminobutyric acid receptor complex at a site distinct from barbiturates. *J Pharmacol Exp Ther* 248:960–966.
45. Sripada, R.K. et al. 2014. The neurosteroids allopregnanolone and dehydroepiandrosterone modulate resting-state amygdala connectivity. *Hum Brain Mapp* 35:3249–3261.
46. Hawkinson, J.E. et al. 1994. Correlation of neuroactive steroid modulation of [35S] t-butylbicyclophosphorothionate and [3H]flunitrazepam binding and gamma-aminobutyric acidA receptor function. *Mol Pharmacol* 46:977–985.
47. Peters, J.A. et al. 1988. Modulation of the GABAA receptor by depressant barbiturates and pregnane steroids. *Br J Pharmacol* 94:1257–1269.
48. Kaminski, R.M. et al. 2005. Anticonvulsant activity of androsterone and etiocholanolone. *Epilepsia* 46:819–827.
49. Milman, A. et al. 2008. DHEAS repeated treatment improves cognitive and behavioral deficits after mild traumatic brain injury. *Eur Neuropsychopharmacol* 18:181–187.
50. Juhasz-Vedres, G. et al. 2006. Dehydroepiandrosterone sulfate is neuroprotective when administered either before or after injury in a focal cortical cold lesion model. *Endocrinology* 147:683–686.
51. Hoffman, S.W. et al. 2003. The delayed administration of dehydroepiandrosterone sulfate improves recovery of function after traumatic brain injury in rats. *J Neurotrauma* 20:859–870.
52. Imamura, M., Prasad, C. 1998. Modulation of GABA-gated chloride ion influx in the brain by dehydroepiandrosterone and its metabolites. *Biochem Biophys Res Commun* 243:771–775.
53. Park-Chung, M. et al. 1999. Sulfated and unsulfated steroids modulate gamma-aminobutyric acidA receptor function through distinct sites. *Brain Res* 830:72–87.
54. Flood, J.F., Morley, J.E., Roberts, E. 1992. Memory-enhancing effects in male mice of pregnenolone and steroids metabolically derived from it. *Proc Natl Acad Sci U S A* 89:1567–1571.
55. Koenig, H.L. et al. 1995. Progesterone synthesis and myelin formation by Schwann cells. *Science* 268:1500–1503.
56. Zhu, T.S., Glaser, M. 2008. Regulatory role of cytochrome P450scc and pregnenolone in myelination by rat Schwann cells. *Mol Cell Biochem* 313:79–89.
57. Hsu, H.J. et al. 2006. Pregnenolone stabilizes microtubules and promotes zebrafish embryonic cell movement. *Nature* 439:480–483.
58. Fontaine-Lenoir, V. et al. 2006. Microtubule-associated protein 2 (MAP2) is a neurosteroid receptor. *Proc Natl Acad Sci U S A* 103:4711–4716.
59. Garcia-Estrada, J. et al. 1999. Dehydroepiandrosterone, pregnenolone and sex steroids down-regulate reactive astroglia in the male rat brain after a penetrating brain injury. *Int J Dev Neurosci* 17:145–151.

60. Guennoun, R. et al. 2015. Progesterone and allopregnanolone in the central nervous system: Response to injury and implication for neuroprotection. *J Steroid Biochem Mol Biol* 146C:48–61.
61. Meffre, D. et al. 2007. Steroid profiling in brain and plasma of male and pseudopregnant female rats after traumatic brain injury: Analysis by gas chromatography/mass spectrometry. *Endocrinology* 148:2505–2517.
62. Wright, D.W. et al. 2007. ProTECT: A randomized clinical trial of progesterone for acute traumatic brain injury. *Ann Emerg Med* 49:391–402, 402.e1–e2.
63. Xiao, G. et al. 2008. Improved outcomes from the administration of progesterone for patients with acute severe traumatic brain injury: A randomized controlled trial. *Crit Care* 12:R61.
64. Wright, D.W. et al. 2014. Very early administration of progesterone for acute traumatic brain injury. *N Engl J Med* 371:2457–2466.
65. Skolnick, B.E. et al. 2014. A clinical trial of progesterone for severe traumatic brain injury. *N Engl J Med* 371:2467–2476.
66. Barbaccia, M.L. et al. 1996. Time-dependent changes in rat brain neuroactive steroid concentrations and GABAA receptor function after acute stress. *Neuroendocrinology* 63:166–172.
67. Barbaccia, M.L. et al. 1997. The effects of inhibitors of GABAergic transmission and stress on brain and plasma allopregnanolone concentrations. *Br J Pharmacol* 120:1582–1588.
68. Purdy, R.H. et al. 1991. Stress-induced elevations of gamma-aminobutyric acid type A receptor-active steroids in the rat brain. *Proc Natl Acad Sci U S A* 88:4553–4557.
69. Guo, A.L. et al. 1995. Evidence for a role of neurosteroids in modulation of diurnal changes and acute stress-induced corticosterone secretion in rats. *Gynecol Endocrinol* 9:1–7.
70. Patchev, V.K. et al. 1996. The neurosteroid tetrahydroprogesterone attenuates the endocrine response to stress and exerts glucocorticoid-like effects on vasopressin gene transcription in the rat hypothalamus. *Neuropsychopharmacology* 15:533–540.
71. Patchev, V.K. et al. 1994. The neurosteroid tetrahydroprogesterone counteracts corticotropin-releasing hormone-induced anxiety and alters the release and gene expression of corticotropin-releasing hormone in the rat hypothalamus. *Neuroscience* 62:265–271.
72. Morrow, A.L. et al. 1995. Neuroactive steroid modulators of the stress response. *Ann NY Acad Sci* 771:257–272.
73. Pinna, G. et al. 2003. In socially isolated mice, the reversal of brain allopregnanolone down-regulation mediates the anti-aggressive action of fluoxetine. *Proc Natl Acad Sci U S A* 100:2035–2040.
74. Pisu, M.G. et al. 2013. Altered stress responsiveness and hypothalamic-pituitary-adrenal axis function in male rat offspring of socially isolated parents. *J Neurochem* 126:493–502.
75. Rasmusson, A.M. et al. 2004. An increased capacity for adrenal DHEA release is associated with decreased avoidance and negative mood symptoms in women with PTSD. *Neuropsychopharmacology* 29:1546–1557.
76. Sundaram, N.K., Geer, E.B., Greenwald, B.D. 2013. The impact of traumatic brain injury on pituitary function. *Endocrinol Metab Clin North Am* 42:565–583.
77. Agha, A. et al. 2005. The natural history of post-traumatic hypopituitarism: Implications for assessment and treatment. *Am J Med* 118:1416.
78. Agha, A. et al. 2004. Anterior pituitary dysfunction in survivors of traumatic brain injury. *J Clin Endocrinol Metab* 89:4929–4936.
79. Aimaretti, G. et al. 2005. Residual pituitary function after brain injury-induced hypopituitarism: A prospective 12-month study. *J Clin Endocrinol Metab* 90:6085–6092.
80. Benvenga, S. et al. 2000. Clinical review 113: Hypopituitarism secondary to head trauma. *J Clin Endocrinol Metab* 85:1353–1361.

81. Lenzen, J. et al. 1993. Function tests on the neuroendocrine hypothalamo-pituitary system following acute midbrain syndrome, with special reference to computertomographical and magnetic resonance imaging results. *Neurosurg Rev* 16:183–187.
82. Lieberman, S.A. et al. 2001. Prevalence of neuroendocrine dysfunction in patients recovering from traumatic brain injury. *J Clin Endocrinol Metab* 86:2752–2756.
83. Masel, B.S., Urban, R.J. 2015. Chronic endocrinopathies in traumatic brain injury disease. *J Neurotrauma*.
84. Schneider, H.J. et al. 2007. Hypothalamopituitary dysfunction following traumatic brain injury and aneurysmal subarachnoid hemorrhage: A systematic review. *JAMA* 298:1429–1438.
85. Schneider, H.J. et al. 2006. Prevalence of anterior pituitary insufficiency 3 and 12 months after traumatic brain injury. *Eur J Endocrinol* 154:259–265.
86. Tanriverdi, F. et al. 2013. A five-year prospective investigation of anterior pituitary function after traumatic brain injury: Is hypopituitarism long-term after head trauma associated with autoimmunity? *J Neurotrauma* 30:1426–1433.
87. Tanriverdi, F. et al. 2006. High risk of hypopituitarism after traumatic brain injury: A prospective investigation of anterior pituitary function in the acute phase and 12 months after trauma. *J Clin Endocrinol Metab* 91:2105–2111.
88. Lauzier, F. et al. 2014. Clinical outcomes, predictors, and prevalence of anterior pituitary disorders following traumatic brain injury: A systematic review. *Crit Care Med* 42:712–721.
89. Ioachimescu, A.G. et al. 2014. Growth hormone deficiency after mild combat-related traumatic brain injury. *Pituitary* 18(4):535–541.
90. Baxter, D. et al. 2013. Pituitary dysfunction after blast traumatic brain injury: The UK BIOSAP study. *Ann Neurol* 74:527–536.
91. Wilkinson, C.W. et al. 2012. High prevalence of chronic pituitary and target-organ hormone abnormalities after blast-related mild traumatic brain injury. *Front Neurol* 3:11.
92. Kilts, J.D. et al. 2010. Neurosteroids and self-reported pain in veterans who served in the U.S. Military after September 11, 2001. *Pain Med* 11:1469–1476.
93. Marx, C.E. et al. 2006. Neuroactive steroids, negative affect, and nicotine dependence severity in male smokers. *Psychopharmacology* 186:462–472.
94. Marx, C.E. et al. 2006. The neurosteroid allopregnanolone is reduced in prefrontal cortex in Alzheimer's disease. *Biol Psychiatry* 60:1287–1294.
95. Uzunova, V. et al. 1998. Increase in the cerebrospinal fluid content of neurosteroids in patients with unipolar major depression who are receiving fluoxetine or fluvoxamine. *Proc Natl Acad Sci U S A* 95:3239–3244.
96. Rasmusson, A.M. et al. 2006. Decreased cerebrospinal fluid allopregnanolone levels in women with posttraumatic stress disorder. *Biol Psychiatry* 60:704–713.
97. Akwa, Y. et al. 2001. The synthetic enantiomer of pregnenolone sulfate is very active on memory in rats and mice, even more so than its physiological neurosteroid counterpart: Distinct mechanisms? *Proc Natl Acad Sci U S A* 98:14033–14037.
98. Darnaudery, M. et al. 2002. The neurosteroid pregnenolone sulfate infused into the medial septum nucleus increases hippocampal acetylcholine and spatial memory in rats. *Brain Res* 951:237–242.
99. Flood, J.F., Morley, J.E., Roberts, E. 1995. Pregnenolone sulfate enhances post-training memory processes when injected in very low doses into limbic system structures: The amygdala is by far the most sensitive. *Proc Natl Acad Sci U S A* 92:10806–10810.
100. Ladurelle, N. et al. 2000. Prolonged intracerebroventricular infusion of neurosteroids affects cognitive performances in the mouse. *Brain Res* 858:371–379.
101. Mayo, W. et al. 1993. Infusion of neurosteroids into the nucleus basalis magnocellularis affects cognitive processes in the rat. *Brain Res* 607:324–328.

102. Meziane, H. et al. 1996. The neurosteroid pregnenolone sulfate reduces learning deficits induced by scopolamine and has promnestic effects in mice performing an appetitive learning task. *Psychopharmacology* 126:323–330.
103. Pallares, M. et al. 1998. The neurosteroid pregnenolone sulfate infused into the nucleus basalis increases both acetylcholine release in the frontal cortex or amygdala and spatial memory. *Neuroscience* 87:551–558.
104. Vallee, M. et al. 1997. Neurosteroids: Deficient cognitive performance in aged rats depends on low pregnenolone sulfate levels in the hippocampus. *Proc Natl Acad Sci U S A* 94:14865–14870.
105. Vallee, M. et al. 2001. Steroid structure and pharmacological properties determine the anti-amnesic effects of pregnenolone sulphate in the passive avoidance task in rats. *Eur J Neurosci* 14:2003–2010.
106. Holm, L. et al. 2005. Summary of the WHO Collaborating Centre for Neurotrauma Task Force on Mild Traumatic Brain Injury. *J Rehabil Med* 37:137–141.
107. Keefe, R.S. et al. 2004. The Brief Assessment of Cognition in Schizophrenia: Reliability, sensitivity, and comparison with a standard neurocognitive battery. *Schizophr Res* 68:283–297.
108. Keefe, R.S. et al. 2008. Norms and standardization of the Brief Assessment of Cognition in Schizophrenia (BACS). *Schizophr Res* 102:108–115.
109. Marx, C.E. et al. 2009. Proof-of-concept trial with the neurosteroid pregnenolone targeting cognitive and negative symptoms in schizophrenia. *Neuropsychopharmacology* 34:1885–1903.
110. Broomall, E. et al. 2014. Pediatric super-refractory status epilepticus treated with allopregnanolone. *Ann Neurol* 76:911–915.

[illegible]

[illegible]

[illegible] 2001. [illegible]

8 Neuroplasticity after Traumatic Brain Injury

YouRong Sophie Su, Anand Veeravagu, and Gerald Grant

CONTENTS

Traumatic brain injury (TBI) is a challenging disease process, both to treat and investigate. Broadly speaking, TBI consists of structural injuries or physiologic changes in brain function secondary to external forces.[1] Such injuries may result in cell death, gliotic scar formation, and/or damage from reactive oxygen species and inflammation.[1]

Prior TBI population studies revealed that the prevalence of TBI in adults over 18 was 8.5%.[2] In 2010, 2.5 million emergency department visits, hospitalizations, and deaths were associated with TBI with data from the Centers for Disease Control and Prevention[3] suggesting that TBI was related to 30% of mortalities. Pediatric TBI, while largely understudied, is also an important concern, as it can reach an annual incidence of 100,000–200,000 with children between the ages of 0 and 4 years having the highest percentage of incidence and mortality.[4,5] The economic impact of brain injury is difficult to determine when considering compensation for work loss, quality of life, rehabilitation, and home services in addition to medical costs. Nonetheless, estimates of total lifetime costs range from $147 billion for fatal TBI to $18 billion for nonhospitalized TBI.[2]

Although brain injuries are a significant portion of trauma, the severity of TBI ranges from mild, defined as a momentary change in consciousness, to severe, which involves sustained periods of unconsciousness and/or amnesia. Fortunately, more than 85% of TBI that is medically treated is considered mild and most patients are able to recover from their injuries.[2]

NEUROPLASTICITY

The central nervous system (CNS) retains an innovative ability to recover and adapt secondary compensatory mechanisms to injury. The basis of recovery stems from neuroplasticity, defined as the ability for neuronal circuits to make adaptive changes on both a structural and functional level, ranging from molecular, synaptic, and cellular changes to more global network changes. The adult brain was traditionally thought to be stagnant, with neuroplasticity confined to cortical development.

Now, however, neuronal plasticity occurs after an injury in a sequence of three phases.[6] Immediately after injury, cell death occurs along with decrease in cortical inhibitory pathways for 1 to 2 days that is thought to recruit or unmask new and secondary neuronal networks.[7] Eventually, the activity of cortical pathways shift from inhibitory to excitatory followed by neuronal proliferation and synaptogenesis. Both neuronal and nonneuronal cells (i.e., endothelial progenitors, glial cells, and inflammatory cells) are recruited to replace the damaged cells, facilitate gliotic scar tissue, and revascularize.[6] Weeks after injury, new synaptic markers and axonal sprouting are upregulated,[8] allowing for remodeling and cortical changes for recovery. Chronic changes have been studied in several mouse injury models, although outcomes vary and are affected by the age of the mouse at time of injury.[4] Preliminary work does suggest that long-lasting morphologic changes occur in the hippocampus after TBI, including growth of cell soma and recruitment of neurons to the hippocampus.[4,9]

Direct evidence for neurogenesis and plasticity have existed for decades, since the discovery of labeling agents such as BrdU, ^{3}H-thymidine, and 14C,[10–12] allowing for the direct visualization of cell division and turnover. Mapping studies, using these same agents on nonhuman primates after injury, demonstrated that the original injured cortical region in monkeys after behavioral recovery would localize to an adjacent territory.[13] Studies in the somatosensory and motor cortex, in particular, have implicated these regions to be capable and receptive to neuroplasticity, although little work has looked at the entire cortical circuitry. Early evidence for neuroplasticity was determined in the healthy brain in the somatosensory and motor cortex. Both somatosensory and motor maps of the body surface could be retrained so that adjacent body parts come to represent a larger cortical region.[14,15] Clark et al.[16] sutured two adjacent digits together in primates and found that, over time, the receptive fields of certain neurons, especially at the junction of the two fingers, would span across both digits. Nudo[14] trained primates to move multiple joints in a reach-and-retrieval task that similarly increased cortical representations of adjacent muscles. In the injured brain, with work focusing on stroke-related injuries, researchers have found that the outcome of neuroplasticity ultimately depends on postinjury behaviors. Furthermore, imaging studies using diffusion tensor imaging have found that plasticity can occur in areas that were initially spared from stroke-related damage, such as changes in the arcuate fasciculus that occur secondary to chronic Broca's aphasia.[17] Plasticity furthermore correlates with changes in the functional performance of patients. Fraser et al.,[15] in looking at corticobulbar excitability in stroke patients suffering from dysphagia, found that stimulus to the corticobulbar nucleus prompted reorganization of the cortex, as gauged by functional MRI (fMRI), and yielded improvements in swallowing. Evidence for whether neuroplasticity occurs on a more global, cortical level is limited as most studies have focused on either the somatosensory or motor cortex through electrophysiology or fMRI. Nonetheless, work by Schlaug et al.[17] in patients with dysphagia raises the possibility that the entire cortex may be indirectly shaped by neuroplastic changes, although more imaging studies will be needed. These studies set the groundwork that neuroplasticity exists for certain regions of the cortex and occurs throughout life and that injuries, such as stroke or trauma, serve as stimuli to prompt further regenerative events.

Studies on the response of the pediatric brain to injury have yielded interesting findings and differing views on the effect of age on recovery after injury and its functional effect. One view, first proposed by Margaret Kennard, later to be named the Kennard principle, proposed that the developing brain is capable of more significant reorganization and recovery after injury.[18] Furthermore, the younger brain, in contrast to the elderly brain, is less likely to develop progressive cognitive decline, and the ongoing development may in actuality promote recovery.[19] The opposing perspective, however, sees the developing brain as more vulnerable to damage given that it is undergoing significant growth and circuitry formation during critical periods, which may lead to more severe or more permanent physiologic changes in the presence of injury. Animal work by Casella et al.[4] in juvenile (postnatal day 17) and immature (postnatal day 7) rats who underwent focal TBI with contusion revealed that the age of the mice at time of injury affects the plasticity and recovery of the brain postinjury. In their work, the researchers find that juvenile rats have both memory and learning deficits in the Morris water maze that last until postinjury day 17 (PID 17). Juvenile mice have longer-lasting somatic and emotional dysfunction, up until PID 60, as tested by behaviors displaying anxiety and sensorimotor function.[4] The poor functional outcomes in juvenile mice were found to correlate with anatomical changes specifically in the hippocampus, including increases in the soma size, dendritic length and branching points in cells of the dentate gyrus after injury. Soma size of cells in CA3 also increased, whereas the dendritic length and branch points of cells in CA1 decreased after injury. Interestingly, immature mice were found to have no morphological changes in cells of the dentate gyrus, CA1, or CA3 after injury.[4] Casella reports his work as evidence for dependency on not only age, but also the region of the brain in regards to response after injury. Unfortunately, his work does not correlate the morphological findings in PND7 mice with any cognitive, motor, or sensorimotor outcomes. Other work in mice report that outcomes after age-related injuries is associated more with the particular stage of cortical development at a certain age rather than the age itself.[18]

In human studies, contradictory evidence exists regarding the effect of age on the response of children after injury. The consistency across the studies, however, suggests that the type of injury plays a large factor in the final response. Several studies support the theory that younger age is a protective factor. Berger et al.[20] reviewed their series of 37 children under 17 years of age with traumatic brain injury and found that the children had better functional recovery than adults. Another group shows that patients injured at a younger age (less than 26 years) were less disabled compared to older patients (greater than 40 years) despite having more severe injuries.[19] In contrast, other studies report that children less than 4 years old have a worse motor and cognitive outcome compared with older children who suffer from TBI.[21] More recent work found that younger age had more adverse impacts on language in work done by Levin et al.[22] who tested word generation, repetition, receptive vocabulary, narration, and recall in both young and older children with varying degrees of close head injuries (CHI). Young children with severe CHI took longer to recover word fluency when compared to same-aged children with mild CHI and older children with any degree of head injury. This finding is thought to be explained by the disruption of white matter development and tracks by the diffuse axonal injury that typically occurs secondary to severe trauma.[22] Furthermore, it is thought that older age allows for not only more neuronal tracks to be incorporated into the appropriate circuitry, but also normal brain development, thus improving overall functions.[22] Still other studies seem to suggest that age is not a critical factor in determining functional outcomes and capacity for recovery, as was the case for Schuett and Zihl[23] in their study of age-related effects on visual field disorders. Both older and younger patients had no difference in the severity of impairments, functional outcomes, and response to treatments. Although studies have yielded varying results, they did establish the use of neuroimaging in tracking and investigating neuroplasticity.

FUNCTIONAL IMAGING OF TRAUMATIC BRAIN INJURY AND NEUROPLASTICITY

Until very recently, noninvasive neuroimaging had limited power to detect white matter structural changes. The development of techniques such as positron emission tomography (PET), functional MRI (fMRI), diffusion tensor imaging (DTI), and transcranial magnetic stimulation (TMS) have changed the detection of response after brain injury (see Table 8.1).

Positron Emission Tomography and Functional MRI

PET and fMRI are two techniques that do not assess neuronal activity directly; rather, they use vascular and metabolic changes, respectively, as indications of neuronal activity. PET is a more invasive imaging technique that involves either inhalation or injection of radioactive tracers that accumulate in activated brain regions. PET relies on the premise that cerebral blood flow increases to regions of neuronal activity. Signal changes are then mapped onto a standard MRI scan of the brain to allow for anatomical correlation. While PET offers high spatial resolution up to 5–10 mm, it has poor temporal resolution given the time needed to record blood flow.[24]

Similarly, fMRI operates on the assumption that neuronal activity increases oxygen consumption and glucose metabolism. A particular sequence called blood oxygenation level dependent (BOLD) is sensitive to the presence of deoxyhemoglobin in the blood, which distorts the magnetic fields and uses the ratio of deoxy- to

TABLE 8.1
Comparison of Various Imaging Modalities for Neuroplasticity

Imaging Modality	Measurement of Neuronal Activity	Advantages	Limitations
PET	Indirect; measures cerebral blood flow	Good spatial resolution of 5–10 mm	Invasive; inhale or inject radioactive tracers; poor temporal resolution
fMRI	Indirect; measures oxygen consumption and glucose metabolism or BOLD signal	No radiation; highest spatial resolution of 1–6 mm; good temporal resolution	Data affected by multiple individual factors, such as medication, blood flow, cerebral dominance
DTI	Direct; measures microstructural white matter changes using water diffusion	Greater sensitivity of white matter injury compared to MRI/CT; detects diffuse axonal injury	Accuracy restricted to single white matter tracts
TMS	Uses magnetic pulses to suppress or enhance cortical excitation	Noninvasive; both stimulates and measures response; able to compare before and after treatment	Currently restricted to motor cortex; limited efficacy

oxyhemoglobin to create signals. BOLD fMRI allows for multiple acquisitions that offer useful temporal resolution to detect differences between brain regions.[24,25] Preliminary fMRI studies in patients with TBI show differential activation patterns; for example, in a patient with right temporoparietal contusion now experiencing dyscalculia and reading disability, fMRI reveals considerable left hemisphere activation and minimal right hemisphere activation in contrast to the bilateral activation detected in neurologically intact patients.[26] Unfortunately, given that PET and fMRI depend on blood flow and metabolic activities, the readout can be affected by age and cerebrovascular diseases such as atherosclerosis.

Diffusion Tensor Imaging

Diffusion tensor imaging (DTI) has high sensitivity for microscopic injury and is increasingly used to detect earlier signs of injury. DTI analyzes the microstructure of white matter based on vector maps created from diffusion patterns of water molecules. Algorithms that analyze properties of water diffusion can determine fiber tracts[27] and can confer information on fiber orientation and damage that are not detectable through conventional MRI.[28,29] Preliminary work indicates that DTI can detect microscopic injury in moderate to severe TBI, whereas imaging on patients with mild TBI have thus far not shown any significant difference compared to neurologically intact patients.[29,30] Unfortunately, there are limited studies on patients with mild TBI, making it difficult to determine the effect of time lapse after injury on the lack of changes. Nonetheless, DTI does provide information on the amount and severity of brain injury with promising findings on the structural disorganization associated with diffuse axonal injury.[26,30] Diffuse axonal injury has never been directly detected but rather inferred at a later time point when white matter injury leads to degenerative changes contributing to ventriculomegaly.[26] As such, there is interest in using DTI to not only quantify the degree of white matter injury, but also prognosticate severity based on the abnormalities discovered. Future work will also focus on the effects of interventions on the trajectories and connections of white matter tracts.[26]

Transcranial Magnetic Stimulation

Transcranial magnetic stimulation (TMS) uses magnetic fields and electrical currents to stimulate cortical regions of the brain in a noninvasive manner. Primarily used to trigger brain plasticity in the motor system, TMS involves applying a current over the scalp corresponding to a motor region that then triggers an electrographic response in its target muscles called motor evoked potentials (MEP).[31] Comparing MEPs before and after injury or across injured and uninjured hemispheres determines residual and changing cortical function. Preliminary work thus far demonstrates that cortical maps change in response to injury through two mechanisms: (1) the region of excitation for a cortical map will either enlarge or shrink and (2) the region corresponding to a cortical map may migrate to adjacent regions.[32,33] The applicability of TMS in studying plasticity was shown by Liepert et al., who imaged patients trained in new fine motor skills using both hands and feet and found shifts in cortical representations for the muscle groups. The shifts in cortical representations detected by TMS recapitulated the early mapping studies done by Glees and Cole and was thought to be secondary to cortical modulations. TMS can also detect asymmetry between hemispheres in motor map stimulation after a hemispheric stroke that is likely related to disuse on the injured side and compensatory changes on the uninjured side.

Imaging modalities such as fMRI, PET, and DTI are exciting for their potential use in monitoring ongoing neuroplasticity. DTI, however, is limited to monitoring only single white matter tracks as its sensitivity is decreased with the presence of multiple white matter tracks intersecting or degenerating in complex or injured regions. While long-term neuroplasticity and improvement has not been fully monitored after injury, stimulation and training seems to promote neural changes that are long lasting, suggesting that neuroplasticity should generally be a chronic process. One group reviewed fMRI across multiple studies to determine that poststroke treatment promoted cortical changes in the motor region beyond the recovery plateau typically seen after stroke.[34] Limited longitudinal studies are available on the presence of chronic neuroplasticity in other domains, such as the sensory cortex or language.

PROMISING THERAPEUTICS AFTER TBI

Traumatic brain injury causes both direct damage, through shear injury of neurons and blood vessels, and indirect damage from secondary ischemia, edema, or inflammation. Through destruction of the blood–brain barrier (BBB), TBI allows immune cells to enter the injured region to activate inflammatory responses. TBI also activates microglia and astrocytes to release inflammatory cytokines, chemokines, and prostaglandins that further increase the permeability of the BBB.[35]

Acute treatment algorithms for TBI include minimizing intracranial pressure and optimizing cerebral perfusion pressure to limit secondary damage. Long-term therapies focus on improving motor, cognitive, and behavioral outcomes. However, therapies for TBI and intrinsic repair mechanisms in the brain are often constrained by the extent and severity of injury, age of the patient, prior or polytrauma, and time lapse to medical management. Thus, newer therapies target prevention of secondary sequelae to enhance neuroprotection. In particular, mounting evidence for neuroplasticity and neural regeneration in the adult CNS has encouraged the development of pharmacologic therapies to enhance the regenerative process. Many of the pharmacological therapies described below are still in their early stages of investigation, but target processes such as neurogenesis, inflammation, angiogenesis, and synaptic remodeling and formation (see Table 8.2).

Stem Cells

Animal studies have shown promising results in the use of stem cells to ameliorate the sequelae of TBI. Because stem cells are capable of self-renewal and differentiation into multiple cell types, exogenous stem cell transplantation into injured brain can counteract multiple damaging mechanisms, from direct neuronal loss to secondary inflammatory sequelae and even provide trophic factors for a nurturing microenvironment.[36] By compensating for many aspects of recovery, stem cells are protective even for chronic phases of recovery. Types of stem cells under investigation for TBI include neural stem cells (NSCs), bone-marrow-derived mesenchymal stem cells (BM-MSCs), and umbilical cord-derived mesenchymal stem cells (UC-MSCs).

The existence of NSCs were first discovered from adult mice striatal tissue,[37] but have since been isolated from diverse parts of the adult brain, including the cortex, subventricular zone, and ventricular zone.[12,38] NSCs can differentiate into functional neurons, astrocytes, and oligodendrocytes and integrate into existing neuronal circuitry.[39] The argument for using adult neural stem cells rather than embryonic and other multipotent stem cells is related to possible tumorigenicity with the latter cells. In contrast, others argue that NSCs is limited in that it provides only the neural cells and trophic factors, while ignoring the role of the surrounding microenvironment, vasculature, and immune system on repair. However, perhaps because of their ability to respond to and secrete trophic factors, NSCs have an advantage over other types of stem cells in migrating to regions of injury. NSCs express cell adhesion proteins, integrins, and chemokine receptors that hone onto inflammatory regions of the brain.[38] Once at the site of injury, NSCs can confer certain functional benefits. In mice subjected to controlled cortical impact (CCI) injuries, transplanted NSCs can return motor but not cognitive function.[40,41] The benefits of NSCs are likely derived from increased expression of neurotrophic factors and release of chemokines. Neurotrophic factors, such as brain-derived neurotrophic factor (BDNF), nerve growth factor (NGF), and fibroblast growth factor (FGF) promote cell survival,

TABLE 8.2
Summary of Promising New Therapies

Treatment	Mechanism of Benefit	Types of Studies	Severity of TBI	Advantages	Disadvantages
Neural stem cells (NSCs)	Differentiate into neurons, astrocytes, oligodendrocytes, integrate into neural networks	Mouse	Varied models	Secretes and responds to trophic factors; committed neural lineage	Potential tumorigenicity; difficult to preserve cells; ignores tumor microenvironment
Bone-marrow derived mesenchymal stem cells (BM-MSCs)	Unclear, thought to alter immune system and ameliorate inflammatory response after injury; affects neurogenesis and angiogenesis	Mouse and human	Varied models	Low immunogenicity; affects microenvironment; easily obtained, rapid proliferative rate	Benefit is confounded by patient's age, injury, delivery of stem cells
Umbilical cord-derived mesenchymal stem cells (UC-MSCs)	Same as BM-MSCs	Mouse and human (limited)	Varied models of TBI	Easily obtained; proliferative rate greater than BM-MSCs; immunomodulatory	Unclear genetic or chromosomal makeup of cells
Antioxidant therapy	Prevent formation of reactive oxygen species that cause neuronal damage or death	Human (phase II and III) trials with limited clinical benefit	Severe	Good safety profile	Large molecules so may have limited BBB permeability

(*Continued*)

TABLE 8.2 (*Continued*)
Summary of Promising New Therapies

Treatment	Mechanism of Benefit	Types of Studies	Severity of TBI	Advantages	Disadvantages
Cyclosporin A (CsA)	Inhibits permeability of mitochondrial transition pore and inhibits calcineurin	Mouse and human (phase II)	Mild, moderate, and severe TBI	Reduces axonal injury and lesion size	No long-term study; outcomes in mouse studies depend on injury model; varying benefits in motor outcomes, no improvements in cognitive outcomes
Erythropoietin (EPO)	Unclear mechanism; cytokine involved in apoptotic, oxidative, and inflammatory reactions	Mouse and human (phase II and III trials)	Mild, moderate, and severe TBI	Neuroprotective	Concerned about effects on angiogenesis; may have limited time window to confer benefit
Progesterone	Unclear mechanism; steroid likely involved in inflammatory pathway	Mouse and human (phase II, III, and RCTs)	Moderate to severe	Neuroprotective; can be metabolized to other steroids	Can be metabolized and involved in several pathways
High mobility group box-1 neutralizing antibody (mAB-HMGB1)	Reduces blood–brain barrier permeability and reduces inflammatory proteins	Mouse	Moderate	BBB protective	Mechanism unclear; long-term outcomes unclear

growth, and differentiation through activation of signaling pathways, whereas chemokines help to modulate the inflammatory response.[38] The utility of NSCs unfortunately may be limited by the severity of TBI, improving function only in cases of mild TBI[42] and for younger populations experiencing TBI.[43] In mice subjected to CCI, Shindo et al.[42] found that transplanted NSCs survived in mild TBI injuries in contrast to severe TBI thought to be due to differences in the surrounding microenvironment prompting differential expression of neurotrophic factors. Furthermore, intrinsic NSCs, detectable in TBI-injured brain, is present in higher quantities and have increased survival rates in younger mice compared to older mice, and also thought to be related to increased expression of neurotrophic factors and decreased expression of proinflammatory cytokines such as tumor necrosis factor α (TNF-α) and interleukins (IL)-1/1β.[43]

Mesenchymal stem cells (MSCs) are multipotent nonhematopoietic cells that have been shown to have a protective role in TBI. While the exact protective mechanism of transplanted MSCs still needs to be elucidated and likely multifactorial, initial work suggests that the promise of MSCs lies with modulating the immune system and altering the inflammatory responses commonly triggered by TBI. As part of the immune response to TBI, astrocytes become reactive and trigger a glial scar response surrounding the brain injury. The glial scar serves to limit the excitotoxicity that occurs with neuronal death and will also recruit microglia and macrophages to digest dead neurons. However, the presence of the glial scar itself often inhibits synaptic reformation and repair. Transplanted MSCs modulate this response by migrating to sites of injury and decreasing the thickness of the glial scar, allowing for enhanced regeneration and improved functional recovery.[44] Furthermore, MSCs differentiate into neurons, glial cells, and vascular endothelial cells, and secrete factors and cytokines to promote neurogenesis and angiogenesis.[36,45] The ability of MSCs to influence the surrounding microenvironment and their low immunogenicity when allogenic also confer advantages over other types of stem cells.[35,45] In addition, MSCs are easily obtained and will rapidly proliferate ex vivo. MSCs derived from bone marrow (BM-MSCs) and umbilical cord blood (UCB-MSCs) have both been investigated. Functional recovery improves with either intravenous or intra-arterial infusion of BM-MSCs as does intracisternal MSCs.[46,47] Tian et al.[48] also found that a subpopulation of patients with TBI had functional improvements after transplantation with BM-MSCs through lumbar puncture. Outcomes, however, can be confounded by a number of factors, such as age of patients, time after injury, mode of delivery, and MSC isolation and culture techniques. Nonetheless, the benefits of MSCs are conferred through BDNF and NGF[49] as well as differential expressions of pro-inflammatory cytokines depending on the time period after TBI. MSCs initially decrease levels of IL-6 in the acute period, but subsequently upregulate IL-6 levels to promote revascularization and scar formation.[35]

Despite the success with cells from the bone marrow, umbilical cord blood (UCB) is desirable for multiple reasons: (1) UCB offers a rich source of multiple types of stem cells, from hematopoietic and mesenchymal stem cells to embryonic stem cells; (2) is easily and readily obtained compared to bone marrow aspiration, and (3) ethically accepted.[49] UC-MSCs also have more proliferative activity than BM-MSCs when initially cultured, which may confer greater benefits in transplantation to an injured brain.[50,51] UC-MSCs have been investigated in both ischemic and traumatic brain injuries. Zanier et al.[44] show that that UC-MSCs behave similarly to BM-MSCs; when placed intracerebroventricularly, UC-MSCs survive at a high rate, migrate to the injury site, and secrete BDNF, which then alters the response of microglia and macrophages to inflammation to decrease the size of the brain scar. In mice with CCI injuries, UC-MSCs were able to clinically improve sensorimotor functions.[52] Limited studies with UC-MSCs exist in humans, but in patients with distant TBI, UC-MSCs improved motor scores and functional independence measures.[51]

Antioxidant Therapy

Reactive oxygen species (ROS) are a common source of damage secondary to ischemic-related injury. Ischemia induces excitotoxicity when neurons release glutamate that then sets off cascades for free radical production. Free radical production after TBI has been shown to induce NSC degeneration and death,[53,54] preventing the neuronal regeneration needed for ultimate repair. ROS also interfere with autoregulatory mechanisms in the vasculature and induce lipid peroxidation, which damages neuronal membranes. As such, antioxidant therapies seek to inhibit the formation of ROS, neutralize the ROS, or antagonize the enzymes that act upon the ROS. As such, antioxidant therapies seek to inhibit the formation of ROS, neutralize the ROS, or antagonize the enzymes that act upon the ROS. The presence of antioxidative enzymes decrease with age, leading to more radical-induced damage and cell death when TBI occurs among the elderly. This gives antioxidative agents a special niche in functional recovery for the elderly.[55] Examples of ROS scavenging compounds include polyethylene glycol-conjugated superoxide dismutase (PEG-SOD), the 2-methylaminochroman compound, U-83836E, and edaravone. PEG-SOD prevents post-traumatic microvascular dysfunction by isolating the $O_2^{\bullet}$ radicals responsible for the damage.[56] Although phase II trials studying PEG-SOD showed initial promise, subsequent phase III studies did not produce any significant clinical benefit, related to either poor penetration through the blood–brain barrier or degree of TBI studied.[57] U-83836E, considered to be the most effective lipid peroxidation inhibitor due to high affinity for membrane phospholipids, has reduced lipid peroxidation and protein nitration and preserved mitochondrial function in mouse injury models.[56] Clinical trials on U-83836E, however, are still needed to demonstrate clinical efficacy. Edavarone, also known as 3-methyl-1-phenyl-pyrazoline-5-one, has the ability to penetrate through the BBB and has already shown remarkable neuroprotective effects in ischemic mouse models and patients with stroke. In TBI, edavarone attenuates ischemic damage through interactions with O_2^- and $^{\bullet}OH$, which reduce brain edema. Edavarone also has the ability to block apoptotic pathways through inhibition of cytochrome c and caspase-3 and upregulation of phosphatidylinositol 3-kinase-Akt pathway. Combined, these effects prevent neuronal and glial death and allow for NSCs to appear and survive at the site of TBI.[54] Functionally, edavarone administration allowed mice with TBI to perform the Morris water maze faster than control mice, suggesting its potential in ameliorating TBI.[58]

Cyclosporin A (CsA) preserves mitochondrial function by inhibiting permeability of the transition pore and reducing the amount of reactive oxygen species.[59]

Combined with its ability to inhibit calcineurin, CsA reduces the amount of axonal injury and size of lesion after TBI.[60] Functionally, CsA improves motor outcomes as gauged by the Morris water maze when tested in mice after lateral fluid percussion injuries (FPI).[61] However, depending on the type of mouse injury model studied, CsA can yield conflicting results. In the CCI injury model, CsA had no effect on cognitive outcomes.[62] In humans, initial clinical trials comparing CsA to a placebo in adults with severe TBI have been promising; CsA both improves Glasgow Coma Scores (GCS) at 6 months and increases mean arterial pressure (MAP) and cerebral perfusion pressure (CPP).[63] Unfortunately, many studies with CsA have small sample sizes and lack long-term follow-up for evaluation of toxicity.

Pharmacologic Treatment

Erythropoietin (EPO) is a cytokine, known for its role in erythropoiesis, but also capable of counteracting a multitude of apoptotic, oxidative, and inflammatory reactions.[64] EPO came under study when mice without EPO receptor were found to have worse outcomes after CCI compared to wild-type mice.[65] EPO confers benefits across multiple mouse injury models, such as CCI, FPI, impact acceleration, and combined injuries.[64] For example, carbamylated EPO, which does not affect hematocrit, reduces lesion size in CCI and promotes both neurogenesis and angiogenesis.[66] Work with EPO in human TBI has mostly focused on determining timing and dosing, but preliminary data reveals that EPO decreases hospital mortality.[67,68]

The steroid progesterone, enriched in the brain, is another agent with multiple mechanisms for neuroprotection. Its metabolic derivatives and action on $GABA_A$ receptors produce an anti-inflammatory state by reducing brain edema, apoptosis, and neuronal cell death.[69–71] Limited randomized controlled trials currently exist on the effects of progesterone in TBI, but reveal that progesterone decreases the mortality rate, compared to placebo, following acute TBI and also increases scores measuring functional outcomes (Wright et al., 2007).[72]

As part of the mechanical damage to brain tissue and the ensuing inflammatory response, the blood–brain barrier often becomes disrupted. Prior work demonstrated that a nonhistone chromatin DNA-binding protein, called high mobility group box-1 (HMGB1), is released from damaged cells particularly in ischemic regions, which sets off the inflammatory events responsible for BBB disruption.[73] As such, therapies such as a neutralizing monoclonal antibody against HMGB1 (mAB-HMGB1) have been investigated, discovering that mAB-HMGB1 is able to reduce the extent of brain injury and edema from fluid percussion by reducing the extent of BBB permeability.[74,75] Furthermore, anti-HMGB1 reduces the amount of inflammatory proteins expressed, thus limiting the degree of secondary insults.[74]

CONCLUSION

The recovery process after traumatic brain injury is long, but with emerging evidence for neuroplasticity, the prospects for recovery are no longer so grim. The exact mechanism remains unknown, however, many hypotheses are currently being investigated. Many potential therapeutic opportunities are being explored to target known changes with neuroplasticity, from differential gene expression and cellular proliferation, to the upregulation of synaptic proteins and junctions for new network connections, to the modulation of inflammatory reactions and the recruitment of immune cells to limit the size and volume of damage. Future therapies may find benefit in targeting multiple mechanisms of recovery and as such, stem cell therapies or a combination of different pharmacologic therapies are of utmost interest and currently under heavy investigation.

REFERENCES

1. Murphy, T.H., Corbett, D. 2009. Plasticity during stroke recovery: From synapse to behaviour. *Nat Rev Neurosci* 10(12):861–872.
2. Orman, J.A.L. et al. 2011. Epidemiology. In *Textbook of Traumatic Brain Injury*, eds. J.M. Silver, T.W. McAllister, and S.C. Yudofsky, 1–22. Virginia: American Psychiatric Publishing.
3. Center for Disease Control and Prevention. 2014. Traumatic Brain Injury in the United States: Fact Sheet. Available at http://www.cdc.gov/traumaticbraininjury/get_the_facts.html.
4. Casella, E.M. et al. 2014. Traumatic brain injury alters long-term hippocampal neuronal morphology in juvenile, but not immature, rats. Official Journal of the International Society for Pediatric Neurosurgery. *Childs Nerv Syst* 30(8):1333–1342.
5. Kraus, J.F., Fife, D., Conroy, C. 1987. Pediatric brain injuries: The nature, clinical course, and early outcomes in a defined United States population. *Pediatrics* 79(4): 501–507.
6. Burda, J.E., Sofroniew, M.V. 2014. Reactive gliosis and the multicellular response to CNS damage and disease. *Neuron* 81(2):229–248.
7. Nahmani, M., Turrigiano, G.G. 2014. Adult cortical plasticity following injury: Recapitulation of critical period mechanisms? *Neurosci* 283:4–16.
8. Carmichael, S.T. 2003. Plasticity of cortical projections after stroke. The Neuroscientist: A Review Journal bringing Neurobiology. *Neurol Psychiatry* 9(1):64–75.
9. Card, J.P. et al. 2005. Plastic reorganization of hippocampal and neocortical circuitry in experimental traumatic brain injury in the immature rat. *J Neurotrauma* 22(9):989–1002.
10. Altman, J., Das, G.D. 1965. Autoradiographic and histological evidence of postnatal hippocampal neurogenesis in rats. *J Comp Neurol* 124(3):319–335.
11. Corotto, F.S., Henegar, J.A., Maruniak, J.A. 1993. Neurogenesis persists in the subependymal layer of the adult mouse brain. *Neurosci Lett* 149(2):111–114.
12. Taupin, P., Gage, F.H. 2002. Adult neurogenesis and neural stem cells of the central nervous system in mammals. *J Neurosci Res* 69(6):745–749.
13. Glees, P., Cole, J. 1949. The reappearance of coordinated movements of the hand after lesions in the hand area of the motor cortex of the rhesus monkey. *J Physiol* 108(1):33.
14. Nudo, RJ. 2011. Neural bases of recovery after brain injury. *J Commun Disord* 44(5):515–520.
15. Fraser, C. et al. 2002. Driving plasticity in human adult motor cortex is associated with improved motor function after brain injury. *Neuron* 34(5):831–840.
16. Clark, S.A. et al. 1988. Receptive fields in the body-surface map in adult cortex defined by temporally correlated inputs. *Nature* 332(6163):444–445.
17. Schlaug, G., Marchina, S., Norton, A. 2009. Evidence for plasticity in white-matter tracts of patients with chronic Broca's aphasia undergoing intense intonation-based speech therapy. *Ann N Y Acad Sci* 1169:385–394.
18. Kolb, B. 2014. Brain plasticity and behavior during development. In *Recovery after Traumatic Brain Injury*, ed. B.P. Uzzell, 199–212. Psychology Press.

19. Marquez de la Plata, C.D. et al. 2008. Impact of age on long-term recovery from traumatic brain injury. *Arch Phys Med Rehabil* 89(5):896–903.
20. Berger, M.S. et al. 1985. Outcomes from severe head injury in children and adolescents. *J Neurosurg* 62(2):194–199.
21. Luerssen, T.G., Klauber, M.R., Marshall, L.F. 1988. Outcome from head injury related to patient's age: A longitudinal prospective study of adult and pediatric head injury. *J Neurosurg* 68(3):409–416.
22. Levin, H.S. et al. 2001. Word fluency in relation to severity of closed head injury, associated frontal brain lesions, and age at injury in children. *Neuropsychologia* 39(2):122–131.
23. Schuett, S., Zihl, J. 2013. Does age matter? Age and rehabilitation of visual field disorders after brain injury. *Cortex* 49(4):1001–1012.
24. Belanger, H.G., Vanderploeg, R.D., Curtiss, G., Warden, D.L. 2007. Recent neuroimaging techniques in mild traumatic brain injury. *J Neuropsych Clin N* 19(1):5–20.
25. Seifert, F., Maihöfner, C. 2011. Functional and structural imaging of pain-induced neuroplasticity. *Curr Opin Anaesthesiol* 24(5):515–523.
26. Levin, H.S. 2003. Neuroplasticity following non-penetrating traumatic brain injury. *Brain Injury* 17(8):665–674.
27. Suskauer, S.J., Huisman, T.A.G.M. 2009. Neuroimaging in pediatric traumatic brain injury: Current and future predictors of functional outcome. *Dev Dis Res Rev* 15(2):117–123.
28. Rugg-Gunn, F.J., Symms, M.R., Barker, G.J., Greenwood, R., Duncan, J.S. 2001. Diffusion imaging shows abnormalities after blunt head trauma when conventional magnetic resonance imaging is normal. *J Neurol Neurosurg Psychiatr* 70(4):530–533.
29. Arfanakis, K., Haughton, V.M., Carew, J.D., Rogers, B.P., Dempsey, R.J., Meyerand, M.E. 2002. Diffusion tensor MR imaging in diffuse axonal injury. *Am J Neuroradiol* 23(5):794–802.
30. Inglese, M., Makani, S., Johnson, G., Cohen, B.A., Silver, J.A., Gonen, O., Grossman, R.I. 2005. Diffuse axonal injury in mild traumatic brain injury: A diffusion tensor imaging study. *J Neurosurg* 103(2):298–303.
31. Rossini, P.M., Pauri, F. 2000. Neuromagnetic integrated methods tracking human brain mechanisms of sensorimotor areas 'plastic' reorganisation. *Brain Res Rev* 33(2–3):131–154.
32. Bütefisch, C.M., Kleiser, R., Seitz, R.J. 2006. Post-lesional cerebral reorganisation: Evidence from functional neuroimaging and transcranial magnetic stimulation. *J Physiol* 99(4–6):437–454.
33. Nudo, R.J. 2013. Recovery after brain injury: Mechanisms and principles. *Front Hum Neurosci* 7:887.
34. Hodics, T., Cohen, L.G., Cramer, S.C. 2006. Functional imaging of intervention effects in stroke motor rehabilitation. *Arch Phys Med Rehabil* 87(12 Suppl 2):S36–S42.
35. Galindo, L.T. et al. 2011. Mesenchymal stem cell therapy modulates the inflammatory response in experimental traumatic brain injury. *Neurol Res Int* 2011:564089.
36. Sun, T., Ma, Q.H. 2013. Repairing neural injuries using human umbilical cord blood. *Mol Neurobiol* 47(3):938–945.
37. Reynolds, B.A., Weiss, S. 1992. Generation of neurons and astrocytes from isolated cells of the adult mammalian central nervous system. *Science* 255(5052):1707–1710.
38. English, D. et al. 2013. Neural stem cells-tends and advances. *J Cell Biochem* 114(4):764–772.
39. Song, H., Stevens, C.F., Gage, F.H. 2002. Neural stem cells from adult hippocampus develop essential properties of functional CNS neurons. *Nat Neurosci* 5(5):438–445.
40. Ma, H., Yu, B., Kong, L., Zhang, Y., Shi, Y. 2011. Transplantation of neural stem cells enhances expression of synaptic protein and promotes functional recovery in a rat model of traumatic brain injury. *Mol Med* 4(5):849–856.

41. Riess, P. et al. 2002. Transplanted neural stem cells survive, differentiate, and improve neurological motor function after experiemental traumatic brain injury. *Neurosurgery* 51(4):1043–1052.
42. Shindo, T. et al. 2006. Differences in the neuronal stem cells survival, neuronal differentiation, and neurological improvement after transplantation of neural stem cells between mild and severe experimental traumatic brain injury. *J Med Invest* 53(1–2): 42–51.
43. Itoh, T. et al. 2012. Appearance of neural stem cells around the damaged area following traumatic brain injury in aged rats. *J Neural Transm* 120:361–374.
44. Zanier, E.R. et al. 2011. Human umbilical cord blood mesenchymal stem cells protect mice brain after trauma. *Crit Care Med* 39(11):2501–2510.
45. Parr, A.M., Tator, C.H., Keating, A. 2007. Bone marrow-derived mesenchymal stromal cells for the repair of central nervous system injury. *Bone Marrow Transplant* 40(7):609–619.
46. Li, Y., Chopp, M. 2009. Marrow stromal cells transformation in stroke and traumatic brain injury. *Neurosci Lett* 456:120–123.
47. Hu, D.Z., Zhou, L.F., Zhu, J.H. 2004. Marrow stromal cells administrated intracisternally to rats after traumatic brain injury migrate into the brain and improve neurological function. *Chin Med J (Engl)* 117:1576–1578.
48. Tian, C. et al. 2013. Autologous bone marrow mesenchymal stem cell therapy in the subacute stage of traumatic brain injury by lumbar puncture. *Exp Clin Transplant* 11(2):176–181.
49. Hu, D.Z. et al. 2005. Upregulated gene expression of local brain-derived neurotrophic factor and nerve growth factor after intracisternal administration of marrow stromal cells in rats with traumatic brain injury. *Chin J Traumatol* 8(1):23–26.
50. Sun, T., Ma, Q. 2013. Repairing neural injuries using human umbilical cord blood. *Mol Neurobiol* 47(3):938–945.
51. Wang, S., Cheng, H., Dai, G., Wang, X., Hua, R., Liu, X., Wang, P., Chen, G., Yue, W., An, Y. 2013. Umbilical cord mesenchymal stem cell transplantation significantly improves neurological function in patients with sequelae of traumatic brain injury. *Brain Res* 1532:76–84.
52. Gornicka-Pawlak, E. et al. 2011. Systemic treatment of focal brain injury in the rat by human umbilical cord blood cells being at different levels of neural commitment. *Acta Neurobiol Exp (Wars)* 71:46–64.
53. Itoh, T. et al. 2005. Isolation of neural stem cells from damaged rat cerebral cortex after traumatic brain injury. *Neuroreport* 16(15):1687–1691.
54. Itoh, T. et al. 2009. The novel free radical scavenger, edaravone, increases neural stem cell number around the area of damage following rat traumatic brain injury. *Neurotox Res* 16(4):378–389.
55. Itoh, T., Imano, M., Nishida, S., Tsubaki, M., Nakayama, T., Mizuguchi, N., Yamanaka, S., Tabuchi, M., Munakata, H., Hashimoto, S., Ito, A., Satou, T. 2013. Appearance of neural stem cells around the damaged area following traumatic brain injury in aged rats. *J Neural Transm* 120(3):361–374.
56. Hall, E.D., Vaishnav, R.A., Mustafa, A.G. 2010. Antioxidant therapies for traumatic brain injury. *Neurotherap: Journal Am Soc Exp NeuroTherap* 7(1):51–61.
57. Muizelaar, J.P., Kupiec, J.W., Rapp, L.A. 1995. PEG-SOD after head injury. *J Neurosurg* 83(5):942.
58. Itoh, T. et al. 2010. Edaravone protects against apoptotic neuronal cell death and improves cerebral function after traumatic brain injury in rats. *Neurochem Res* 35(2):348–355.
59. Hansson, M.J. et al. 2003. Powerful cyclosporine inhibition of calcium-induced permeability transition in brain mitochrondria. *Brain Res* 960(1–2):99–111.

60. Sullivan, P.G., Thompson, M., Scheff, S.W. 2000. Continuous infusion of Cyclosporin A postinjury significantly ameliorates cortical damage following traumatic brain injury. *Exp Neurol* 161(2):631–637.
61. Alessandri, B. et al. 2002. Cyclosporin A improves brain tissue oxygen consumption and learning/memory performance after lateral fluid percussion injury in rats. *J Neurotrauma* 19(7):829–841.
62. Gijtenbeek, J.M., van den Bent, M.J., Vecht, C.J. 1999. Cyclosporine neurotoxicity: A review. *J Neurol* 246(5):339–346.
63. Mazzeo, A.T. et al. 2009. Safety and tolerability of Cyclosporin A in severe traumatic brain injury patients: Results from a prospective randomized trial. *J Neurotrauma* 26(12):2195–2206.
64. Diaz-Arrastia, R., Kochanek, P.M., Bergold, P., Kenney, K., Marx, C.E., Grimes, J.B., Yince Loh, L.T.C., Adam, G.E., Oskvig, D., Curley, K.C., Salzer, W. 2014. Pharmacotherapy of traumatic brain injury: State of the science and the road forward: Report of the Department of Defense Neurotrauma Pharmacology Workgroup. *J Neurotrauma* 31(2):135–158.
65. Xiong, Y., Lu, D., Qu, C., Goussev, A., Schallert, T., Mahmood, A., Chopp, M. 2008. Effects of erythropoietin on reducing brain damage and improving functional outcome after traumatic brain injury in mice. *J Neurosurg* 109(3):510–521.
66. Xiong, Y. et al. 2011. Effects of posttraumatic carbamylated erythropoietin therapy on reducing lesion volume and hippocampal cell loss, enhancing angiogenesis and neurogenesis, and improving functional outcome in rats following traumatic brain injury. *J Neurosurg* 114(2):549–559.
67. Corwin, H.L. et al. 2002. Efficacy of recombinant human erythropoietin in critically ill patients: A randomized controlled trial. *J Am Med Assoc* 288(22):2827–2835.
68. Corwin, H.L. et al. 2007. Efficacy and safety of epoetin alfa in critically ill patients. *N Engl J Med* 357(10):965–976.
69. Galani, R., Hoffman, S.W., Stein, D.G. 2001. Effects of the duration of progesterone treatment on the resolution of cerebral edema induced by cortical contusions in rats. *Restor Neurol Neurosci* 18(4):161–166.
70. Grossman, K.J., Goss, C.W., Stein, D.G. 2004. Effects of progesterone on the inflammatory response to brain injury in the rat. *Brain Res* 1008(1):29–39.
71. Pan, D-S., Liu, W-G., Yang, X-F., Cao, F. 2007. Inhibitory effect of progesterone on inflammatory factors after experimental traumatic brain injury. *Biomed Environ Sci* 20(5):432–438.
72. Xiao, G. et al. 2008. Improved outcomes from the administration of progesterone for patients with acute severe traumatic brain injury: A randomized controlled trial. *Crit Care* 12:R61.
73. Qiu, J. et al. 2008. Early release of HMGB-1 from neurons after the onset of brain ischemia. *J Cereb Blood Flow Metab* 28(5):927–938.
74. Okuma, Y., Liu, K., Wake, H., Zhang, J., Maruo, T., Date, I., Yoshino, T., Ohtsuka, A., Otani, N., Tomura, S., Shima, K., Yamamoto, Y., Yamamoto, H., Takahashi, H.K., Mori, S., Nishibori, M. 2012. Anti-high mobility group box-1 antibody therapy for traumatic brain injury. *Ann Neurol* 72(3):373–384.
75. Zhang, J. et al. 2011. Anti-high mobility group box-1 monoclonal antibody protects the blood-brain barrier from ischemia-induced disruption in rats. *Stroke* 42(5):1420–1428.
76. Lee, M.W. et al. 2010. Stem and progenitor cells in human umbilical cord blood. *Int J Hematol* 92(1):45–51.

9 Genetic Influences in Traumatic Brain Injury

Ellen R. Bennett, Karin Reuter-Rice, and Daniel T. Laskowitz

CONTENTS

INTRODUCTION

Traumatic brain injury (TBI) is a leading cause of death and disability worldwide, affecting all ages and demographics. In the United States alone, approximately 1.7 million new cases are reported yearly,[1–3] resulting in death in roughly 5% of individuals, long-term disability in greater than 40%, and 25% of affected adults unable to return to work 1 year following the injury.[4] Symptoms associated with TBI can appear immediately following injury or days to weeks later, and result in wide-ranging physical and psychological deficits including motor impairment, epilepsy, personality change, and memory impairment. TBI is classified into three categories designated mild, moderate, and severe, based on the severity of injury using the Glasgow Coma Score (GCS) and other clinical measures.[5,6] These tools assess whether the individual was unconscious and duration, length of amnesia, resulting cognitive, behavioral or physical disability, and subsequent recovery. Mild TBI (mTBI) is the most common subtype of TBI, with estimates ranging from 1.6 million to 3.8 million annually among U.S. athletes alone. Despite its designation, a *mild* TBI should not be viewed as an inconsequential injury, as some mTBIs can result in prolonged cognitive, emotional, and functional disabilities, significantly impacting quality of life.

Predicting outcome following TBI is challenging, and cannot be made based solely on clinical presentation and radiological findings since patients with comparable injuries may have variable outcomes. The injury itself can be viewed as occurring in two distinct phases, a primary phase and secondary phase. The *primary phase* occurs at impact from the mechanical forces of the injury which can disrupt the brain parenchyma and integrity of the blood–brain barrier (BBB). This is followed by a systemic and neuroinflammatory response or *secondary phase*, mediated by peripheral immune cells and activation of resident neural cells, triggering the release of molecular mediators such as cytokines, growth factors, and adhesion molecules, and activation of a complex network of pathways. Secondary injury can develop over a period of hours to days and months following the primary injury. Some of these pathways are involved in reparative processes,[7,8] whereas others contribute to metabolic dysregulation that may result in secondary brain injury. Apoptosis of neurons and glia contribute to the overall pathology of TBI, and neurons undergoing apoptosis have been identified within contusions in the acute post-traumatic phase, as well as in regions remote from the site of injury in the days and weeks following trauma.[9]

The genes involved in TBI can be roughly categorized into those that influence the extent of the injury (e.g., pro- and anti-inflammatory cytokines) and those that effect repair and plasticity (e.g., neurotrophic genes). An additional category of genes that should be considered are those that effect pre- and postinjury cognitive and neurobehavioral capacity (e.g., catecholamine genes).[10] A growing body of literature has attributed a role for genetic factors in the interindividual variability observed in TBI, and in predicting functional and cognitive outcome following brain injury.[10–14] These variations are a result of alterations in the DNA sequence within a given gene and are referred to as genetic polymorphisms. Polymorphisms can arise from insertions or deletions of short lengths of DNA within a particular gene, interfering with the normal function of the gene, or at a single nucleotide (G, A, T, or C). When a single nucleotide is responsible for the modification in the DNA, it is referred to as a single nucleotide polymorphism or SNP. SNPs are the most common type of genetic variation, occurring once every 100–300 nucleotides, amounting to approximately 10 million in the human genome. A SNP can reside within the coding sequence of a gene where it may alter the amino acid composition of a protein, or within a noncoding region of a gene, such as a promoter or intron, where it may influence expression of the gene and protein production. The nomenclature for SNPs can be confusing since an individual SNP may be represented in several different ways in the literature. A common depiction can be illustrated with the SNP –174 G/C in the interleukin-6 (IL-6) gene; here, the "174" denotes the nucleotide number at which the variation occurs, the "–" designates that it occurs *upstream* of the transcription start site (designated +1) in the noncoding region of the gene, and the G/C refers to the nucleotide change at that position, in this case within the promoter region. For the SNP +3953 C/T in the interleukin-1 beta (IL1B) gene, the "+" denotes that the variation occurs *downstream* from the transcription start site at nucleotide position 3953. The most prevalent variation is commonly referred to as "allele 1" or "major allele". When a SNP is officially registered in a public database maintained by the U.S. National Center for Biotechnology Information, it is assigned a unique identifier referred to as an *rs* number (e.g., rs1800795 identifies IL6 SNP –174 G/C).

To date, numerous genes have been implicated in the pathophysiology and outcome following moderate to severe TBI. More recently, considerable attention has focused on genes associated with mild and repetitive mTBIs, notably among combat veterans and professional athletes.[15–17] Although inheriting a single "good" or "bad" allele of a specific gene may predispose an individual to better or worse outcome following injury, it is becoming increasingly apparent that recovery from TBI is polygenic in nature, involving the interaction of numerous genes from multiple pathways. Moreover, one must also consider the role of epigenetic mechanisms in disease and injury,[18–20] processes that can effect gene expression without altering the DNA sequence (e.g., DNA methylation, chromatin modifications).

The purpose of this chapter is to provide a current overview of genetic polymorphisms associated with recovery and outcome following acute TBI in an adult population. It is not intended to serve as an in-depth study of the individual genes and possible mechanisms of action; several reviews exist in the literature that address this in greater detail.

CYTOKINE POLYMORPHISMS AND TRAUMATIC BRAIN INJURY (TBI) OUTCOME

BACKGROUND

The mechanisms mediating the host's response to brain injury are complex and not fully understood. Following TBI, the integrity of the BBB is compromised, accompanied by a state of oxidative stress and an increase in expression of cell adhesion molecules on the brain endothelium, which in turn promotes the influx of peripheral inflammatory cells into the injured brain parenchyma.[21] Coupled with this, the brain undergoes gliosis, a state in which resident cells within the central nervous system

(CNS), such as astrocytes and microglia, become activated and secrete inflammatory mediators and other cellular mediators.[22,23] As a result, the initial neuroinflammatory response following injury can promote secondary neuronal death and the development of cerebral edema. Cytokines are among the major mediators during the initial phase of brain injury, where they can both exacerbate the injury early, yet paradoxically may contribute to recovery and repair at later stages.[21,24,25] Experimental models of closed head injury show that tumor necrosis factor α (TNFα), a proinflammatory cytokine, is upregulated immediately following TBI, peaking within hours, and returning to normal baseline levels within 24 hours of injury.[26] Preclinical studies using knockout animals have demonstrated that TNFα deficiency is beneficial early after experimental brain injury but deleterious over the long term, suggesting that TNFα plays both a neuroprotective and neurotoxic role.[27] The interleukin-1 (IL-1) family consists of two proinflammatory cytokines, IL-1α and IL-1β, and a receptor antagonist, IL-1RA. IL-1 acts together with TNFα to increase inflammation and pyrexia following injury. In experimental models of concussion, IL-1α and IL-1β increase within hours of injury, with IL-1α spiking initially and IL-1β exhibiting a more gradual increase, remaining elevated for days following the injury.[28,29] In addition, IL-1β levels correspond to the level of injury severity, with greater levels of IL-1β mRNA observed in more severe injuries.[28] The production of IL-6 is stimulated by TNFα and IL-1 and secreted by microglial during the acute phase of TBI.[30] Winter et al. reported that increased levels of IL-6 correlated with improved outcome following TBI, suggesting that IL-6 has neuroprotective properties following acute brain injury.[7] An additional biological mediator in TBI is transforming growth factor-β (TGF-β).[31] TGF-β is induced by the presence of cytokines yet forms a negative feedback system where it exerts a predominantly anti-inflammatory role during the acute phase of brain injury, inhibiting the production of IL-1, IL-6, and TNFα. TGF-β peaks within 24 hours in clinical studies of severe TBI (sTBI)[32] and can promote tissue repair, decrease brain edema, and reduce brain lesions following ischemia,[24,25,33] although excessive TGF-β expression can hinder repair mechanisms in the brain and predispose to infections.[33,34]

The inflammatory modulators described above represent only a fraction of the molecules involved in this complex neuroinflammatory response following TBI. The following section will focus on genetic polymorphisms within these neuroinflammatory mediators that have been associated with TBI.

TNFα Polymorphisms

TNFα plays a role in mediating neuronal death in the acute phase of brain injury as well as promoting neuronal repair and plasticity in the long term.[35,36] The gene encoding human TNFα is ~3 kb and is located within the major histocompatibility complex (MHC) on chromosome 6.[37] Herrmann et al. screened the entire coding region of TNFα as well as over 1 kb of upstream regulatory sequences and identified five polymorphisms, four of which reside within regulatory sequences.[38] Two of these are SNPs located at nucleotide position –238 (rs361525) and –308 (rs1800629) and are G/A substitutions; the G-allele, sometimes referred to as allele 1, is the more common allele, occurring at a frequency of ~90% in the population as a whole (1000 genomes[39]).

These SNPs (–238 A/G and –308 A/G) have been extensively studied and associated with outcome from a number of inflammatory conditions including rheumatoid arthritis (RA), psoriasis, Crohn's disease, cerebral malaria, septic shock, and meningococcal disease.[40] By contrast, very few studies have been reported on TNFα SNPs and TBI. In 2013, Waters et al. looked for an association between the TNFα –238 and –308 SNPs and clinical outcome at 6 months (GOS) in 1096 sTBI patients, 937 of which were included in the genetic analysis.[12] They observed no association between the –238 A/G polymorphism and outcome; however, when the data was dichotomized into unfavorable (death, vegetative survival, or severe disability) and favorable (moderate disability or good recovery) outcome, individuals with the –308 A-allele (allele 2) were more likely to have unfavorable outcome at 6 months compared to noncarriers (39% vs. 31%). Although mortality was similar for those with a –308 A-allele relative to those without (13% vs. 12%), among survivors, those harboring the –308 A-allele were more likely to have a severe disability (25% vs. 18%) and less likely to have a favorable outcome (37% vs. 43%). The authors argue that association of the –308 A-allele with poor outcome following TBI is biologically plausible since this SNP resides within the promoter region of the gene, and the A-allele (allele 2) has been associated with a higher expression of TNFα.[41,42] They reason that patients with the –308 A polymorphism have increased expression of TNFα following TBI and as a result have worse outcome. It is tempting to suggest that the low frequency rate of the –308 A-allele in the general population (~10%) is selectively advantageous, however, given the multifaceted nature of the neuroinflammatory response and the temporal regulation of its many roles within the brain, cerebrospinal fluid (CSF), and plasma, the relationship is likely more complex. Waters et al. propose that further studies measuring TNFα levels in serum, CSF, and the brain, will help elucidate the hypothesis. Other studies have examined TNFα SNPs in other neuroinflammatory conditions such as Alzheimer's disease (AD)[43] and stroke[44]; however, to our knowledge, this represents the only reported study on TNFα polymorphism and TBI.

IL-1 Polymorphisms

The IL-1 family consists of two proinflammatory cytokines, IL-1α and IL-1β, and a receptor antagonist, IL-1RA, which are located on chromosome 2, and encoded by the genes IL-1A, IL-1B, and IL-1RN, respectively.[45] The IL-1RA modulates the immune response by binding to the IL-1 receptors, which in turn inhibit the actions of IL-1α and IL-1β. In the CNS, microglial are the primary producers of IL-1, and following TBI there is a rapid increase in IL-1 production,[46] which has been associated with injury to hippocampal neurons.[47] A number of genetic polymorphisms have been identified within IL-1 that influence gene expression and are associated with AD and stroke.[48,49] These include: IL-1A –899 C/T (rs1800587), IL-1B –31 C/T (rs1143627), IL-1B –511 G/T (rs16944), and IL-1B +3953 C/T (rs1143634). An additional polymorphism is associated with IL-1RN and can result in five different alleles with a variable copy number (2-6) of an 86-bp repeat within intron 2.[50,51] The most prevalent repeats, a 4-repeat and 2-repeat, are designated as IL1RN*1 and IL1RN*2, respectively.[51] Numerous groups have looked for an association between the IL-1 polymorphisms and outcome after TBI (Table 9.1). In 2005, Uzan et al.

TABLE 9.1
Gene Association Studies in Traumatic Brain Injury

Author	Gene/Allele	TBI (m, mild; M, moderate; s, severe)	n	Evaluation	Post-TBI Testing	Findings
Waters et al.[12]	TNFα –238 G/A, –308 G/A; IL-6 –174 G/C, –572 G/C, –597 G/A; IL-1B –31 C/T, –511 G/T, +3953 C/T and for TGFB –509 C/T, –800 G/A	s	1096	GOS	6 months	Only TNFα –308 A (allele 2) carriers show unfavorable outcome compared to noncarriers
Hadjigeorgiou et al.[53]	IL-1RN*2; IL-1B –511 G/T	m, M, s	151	GOS	6 months	IL-1RN*2 associated with increase likelihood of cerebral hemorrhage after TBI; no association observed with IL-1B
Tanriverdi et al.[54]	IL-1A –889 C/T	m, M, s	71	GOS	6 months	No association observed following TBI
Uzan et al.[52]	IL-1B –511 G/T, +3933 C/T	m, M, s	69	GOS	6 months	Unfavorable outcome associated with IL-1B +3953 T and –511 G allele
Dardiotis et al.[55]	IL-1A –889 C/T	m, M, s	215	GOS	6 months	No association observed with favorable vs. unfavorable outcome
Minambres et al.[62]	IL-6 –174 G/C	s	62	GOS	6 months	No correlation between –174 C and outcome
Dalla Libera et al.[63]	IL-6 –174 G/C	s	77	Survival time/ICU discharge; GOS at SICU discharge	ICU discharge 1–53 days	GG genotype more frequent in survivors; –174 G allele more frequent among survivors

(Continued)

TABLE 9.1 (*Continued*)
Gene Association Studies in Traumatic Brain Injury

Author	Gene/Allele	TBI (m, mild; M, moderate; s, severe)	n	Evaluation	Post-TBI Testing	Findings
Zhou et al.[70]	APOE4	m, M, s	2527	Meta-analysis	6 months	No association with initial TBI severity but associated with increased risk of poor long-term outcome at 6 months after injury
Ponsford et al.[127]	APOE4	m, M, s	648	GOSE	Mean 1.9 years	Greater susceptibility in older females for worse GOSE scores, and E4 may be associated with poorer long-term outcome, but not injury severity
Lendon et al.[138]	APOE4; APOE –219 G/T, –419 A/T, –427 C/T	m, M, s	92	GOS	6 months	–291 T associated with poor recovery in TT carriers; no association found with –419 A/T or –427 C/T and outcome
Jiang et al.[139]	APOE4; APOE –219 G/T, –419 A/T, –427 C/T	m, M, s	110	Clinical deterioration or stabilization	<7 days	After adjustment for patients' age, injury severity and injury mechanism, both –49IAA promoter polymorphism and E4 allele were risk factors for poor outcome
Terrell et al.[149]	Tau $^{tyr}53^{Pro}$, $^{his}47^{tyr}$, APOE4; APOE –219 G/T	m	195	History of multiple concussions	Reported history	Tau $^{tyr}53^{pro}$ TT was weakly associated with increase concussion risk; no association with E4 and history of concussion

(*Continued*)

TABLE 9.1 (*Continued*)
Gene Association Studies in Traumatic Brain Injury

Author	Gene/Allele	TBI (m, mild; M, moderate; s, severe)	n	Evaluation	Post-TBI Testing	Findings
Robertson et al.[179]	NOS3 –786 T/C	s	51	CBF, mortality internal carotid artery flow volume	12, 48 hours	C-allele associated with reduced CBF following TBI
Lipsky et al.[180]	COMT Val158Met	m, M, s	131	WCST, TMT, COWA, Stroop Color and Word Test	1 year maximum	Val allele was associated with perseveration suggesting problem with cognitive flexibility
McAllister et al.[184]	COMT Val158Met	m	39	WRAT-3, WAIS-III, CVLT, CPT	Mean 38days	Val allele associated with response latency
Willmott et al.[185]	COMT Val158Met	M, s	223	GOSE; measures of cognitive function	GOSE	No association with Val158Met and cognitive function
Ariza et al.[189]	ACE I/D	M, s	73	AVLT, CFT, GOSE, TMT, Grooved peg, COWAT, NRS-R	6 months	Individuals with D-allele performed worse cognitively
McAllister et al.[210]	BDNF Val66Met, rs11030102, rs11030107, rs12273363	m, M	75	CVLT, SRTRT, WRAT-3, WAIS-III	1 month	Met-allele associated with slower processing speed among TBI and controls rs11030102, rs11030107, and rs12273363 associated with memory measures within the mTBI
Martinez-Lucas et al.[218]	TP53 Argp53Pro	s	90	GOS	Discharge from ICU 6 months	Arg/Arg associated with bad outcome at GOS-0 but not at GOS-6

(*Continued*)

TABLE 9.1 (*Continued*)
Gene Association Studies in Traumatic Brain Injury

Author	Gene/Allele	TBI (m, mild; M, moderate; s, severe)	n	Evaluation	Post-TBI Testing	Findings
Sarnaik et al.[221]	PARP-1 SNP A/T (rs3219119); SNP A/G (rs2271347)	s	191	GOS	6 months	SNP rs3219119 A/A genotype associated with favorable outcome at 6 months; no association with SNP (rs2271347)
Johnson et al.[224]	MME dinucleotide GT repeat polymorphism	s	81	Presence of Aβ		Subjects with >41 GT repeats associated with Aβ accumulation following TBI compared to Aβ nonaccumulators
McDevitt et al.[227]	NEFH (rs165602) A/C	m	48	History of multiple concussions, duration of symptoms, return to play	Reported history	No association between rs165602 and occurrence or severity of concussion
Chan et al.[232]	5-HTT S/L genotype	m, M, s	75	Hamilton Depression Rating Score	Within 12 months of TBI	No association with S/L alleles and depression scores
Bales et al.[235]	PPP3CC SNP A/G (rs2443504)	N/A	277	GOS	3, 6, and 12 months	A/G SNP predictor of susceptibility and recovery; AA genotype associated with initial GCS<6 and worse GOS

(*Continued*)

TABLE 9.1 (*Continued*)
Gene Association Studies in Traumatic Brain Injury

Author	Gene/Allele	TBI (m, mild; M, moderate; s, severe)	n	Evaluation	Post-TBI Testing	Findings
McAllister et al.[184,237]	ANKK1 Taq1A A/T SNP (rs1800497)	m, M	39; 93	WRAT-3, WAIS-III, CVLT, CPT	1 month	T-allele of Taq1A associated with poorer episodic memory and response latency
Wagner et al.[243]	KIBRA T/C SNP (rs17070145)	s	129	GOS, DRS, CVLT, SRT, FIM, NRS	6, 12 months	T-allele associated with poorer performance on episodic memory at 6 and 12 months
Bulstrode et al.[258]	Mitochondrial haplogroup H, J, T, U APOE4	m, M, s	880	GOS	6 months	K variant associated with better outcome; possession of K or T lessened the negative effect of aging on outcome following TBI; no association with APOE4 alone but individuals with APOE4 and K had better outcome
Conley[252]	ND3 A10398G, T195C	s	336	GOS, NRS, DRS, mortality, lactate/pyruvate ratio in CSF	1 year	A10398 allele (H haplogroup) associated with slower recovery; 10398 G allele (K/J haplogroup) associated with faster recovery; T195 allele associated with worse outcome in females and mitochondria dysfunction in females
Brichtová and Kozák[276]	APOE4	m, M, s (pediatric)	70	GOS	1 year	E4 carriers more likely to have severe unfavorable outcome

(*Continued*)

TABLE 9.1 (*Continued*)
Gene Association Studies in Traumatic Brain Injury

Author	Gene/Allele	TBI (m, mild; M, moderate; s, severe)	n	Evaluation	Post-TBI Testing	Findings
Moran et al.[277]	APOE2, E3, E4	m (pediatric)	99	CVLT, VMI, CANTAB, WASI, WRAT-3, PCS, HBI	2 weeks, 3 and 12 months	APOE4 allele was associated with poor global outcome relative to individuals with E2/E3 or E3/E3 genotype
Lo et al.[278]	APOE4, E3/E3	s (pediatric)	65	CPP, consciousness at discharge, GOS	6 months	E4 carriers with poor outcome had lower CPP insult levels than noncarriers; E3/E3 patients had better recovery compared to E3 heterozygotes

Note: Glasgow Outcome Scale (GOS); Glasgow Outcome Scale Extended (GOSE); cerebral blood flow (CBF); Glasgow Coma Score (GCS); cerebral spinal fluid (CSF); surgical ICU (SICU); Disability Rating Scale (DRS); cerebral perfusion pressure (CPP); Disability Rating Scale (DRS); Neuro-Behavioral Rating Scale (NRS); California Learning Verbal Test (CVLT); Cambridge Neuropsychological Test Automated Battery (CANTAB); Visual Motor Integration (VMI); Wechsler Abbreviated Scale of Intelligence (WASI); Wide Range Achievement Test (WRAT-3); Post Concussive Symptom Interview (PCS); Health Behavioral Inventory (HBI); Functional Independence Measure (FIM); Selective Reminding Test (SRT); Continuous Performance Test (CPT); Auditory Verbal Learning Test (AVLT); Complex Figure Test (CFT); Controlled Oral Word Association Test (COWAT); Trail Making Test (TMT); Wisconsin Card Scoring Test (WCST).

looked in a small cohort of 69 mild to severe TBI patients and reported that the minor alleles (allele 2) of IL-1B –511 (G-allele) and +3953 (T-allele) were significantly associated with 6-month unfavorable outcome.[52] In contrast, Hadjigeorgiou et al. detected no association between the IL-1B –511 G-allele and outcome in a population of 151 TBI patients, but did observe an association with the IL1RN*2 allele and increased likelihood of cerebral hemorrhages at 6 months after TBI.[53] Two additional groups observed no association with IL-1A –889 and 6-month outcome in cohorts of 71 and 215 patients[54,55] with mild to severe TBI. In a report published by Waters and colleagues in 2013, 11 cytokine polymorphisms, including IL-1A –899 C/T, IL-1B –31 C/T, IL-1B –511 G/T, and IL-1B +3953 C/T, were analyzed for an association with 6-month outcome in a population of 1096 sTBI patients.[12] In their initial multi-SNP screen, none of the SNPs reached statistical significance at the 0.05 level with 6-month outcome, although they did observe several associations between these SNPs and secondary complications. For example, possession of a copy of the minor T-allele of IL-1A –899 was weakly associated with seizures ($p = 0.049$) as well as raised intracranial pressure (ICP; $p = 0.01$). The IL-1B +3953 T-allele was similarly associated with raised ICP ($p = 0.027$), whereas the less common C-allele of IL-1B –31 was associated with a reduced likelihood of experiencing a serious infection following TBI.

IL-6 Polymorphisms

The human IL-6 gene is a relatively small gene located on chromosome 7. Three common DNA variants identified in the regulatory region of the gene include IL-6 –174 G/C (rs1800795), –572 G/C (rs1800796), and –597 G/A (rs1800797). Together, these three SNPs have been shown to influence IL-6 transcription through a complex interaction, and the G alleles at these loci have been associated with increased IL-6 expression.[56] Numerous studies examining IL-6 expression in the context of disease and brain injury have been reported in the literature.[57–59] The G alleles of the –597 and –174 polymorphisms have been associated with higher levels of circulating IL-6 in healthy female subjects.[60] In other clinical studies, the IL-6 –174 GG genotype is associated with higher IL-6 protein and mRNA levels in the brains of patients with brain arteriovenous malformations relative to the GC and CC genotypes.[61] Several additional groups have suggested that CSF and parenchymal IL-6 levels may correlate with improved outcome following TBI.[7,30] In a small study conducted in 2003, Minambres et al.[62] examined jugular vein and systemic arterial IL-6 levels in 62 patients with sTBI or hemorrhagic acute brain injury, genotyping the IL-6 –174 promoter polymorphism. Their results revealed that the –174 C allele was significantly associated with lower concentrations of IL-6 but not with high or low responders and 6-month outcome. Furthermore, they detected a significant relationship between severity of injury and transcranial IL-6 gradient at admission. More recently, Dalla Libera[63] tested whether the –174 C/G SNP is associated with primary short-term outcome (death or ICU discharge) in 77 male patients in Brazil with sTBI. The frequencies of the IL-6 –174 C and G-alleles among the 77 patients were 26% and 74%, respectively. When the group was divided into survivor and nonsurvivors, the GG genotype was significantly more frequent among the survivor group compared to nonsurvivors (67% vs. 41%; $p = 0.038$), and the G-allele was more prevalent in survivors compared to nonsurviving patients (81% vs. 65%; $p = 0.031$), suggesting that the G-allele has significant impact on short-term outcome following sTBI. In contrast, others have reported no significant association between the IL-6 promoter polymorphisms –174 C/G, –572 G/C, and –597 G/A and outcome following sTBI[12]; however, in a recent prospective study of 3255 college athletes, Terrell and associates observed a significant association between IL-6 –572 C/C genotype and concussion risk (personal communication, in press).

TGF-β Polymorphisms

TGF-β is a potent suppressive cytokine that plays a pivotal role in maintaining a balanced host response during inflammation and inflammatory conditions.[64,65] TNFB1, the gene encoding human TGF-β, is located on human chromosome 19. Two SNPs identified within TNFB1 reside within its promoter region at nucleotide positions –800 (G/A substitution; rs1800468) and –509 (C/T substitution; rs1800469).[66] Clinical studies have demonstrated that the –509 T-allele is associated with enhanced TGF-β plasma levels,[67] and it has been postulated to be a risk factor for sporadic AD, cystic fibrosis, and other diseases.[68] In vitro studies suggest that presence of a C at nucleotide position –500 reduces gene expression by interfering with transcription factor binding.[69] Few clinical studies have examined TGF-β polymorphisms in the context of TBI. To date, the only reported study is that of Waters et al., who genotyped 11 cytokine SNPs (including TNFB SNPs –800 G/A and –509 C/T) in a large cohort ($n = 1096$) of sTBI patients, and observed no association between these two SNPs and secondary complications following injury, nor with outcome at 6 months post-TBI.[12] Consistent with a lack of association between TGF-β and outcome, Csuka et al. observed no longitudinal differences in serum and CSF TGF-β levels in a small population ($n = 28$) of sTBI patients.[32]

APOLIPOPROTEIN E

Background

Apolipoprotein E (apoE, protein; APOE, gene) is the most studied gene with respect to outcome after neurotrauma.[70] Although originally studied for its role in cholesterol metabolism,[71] apoE is the major apolipoprotein produced in the central nervous system (CNS) where it is synthesized by astrocytes and microglia, and in neurons under conditions of stress.[72] Numerous functions have been attributed to apoE including promoting repair and growth of neurons, maintenance of synaptodendritic connections,[73] and mediating brain inflammatory responses.[74–78] Interest in the neurobiology of apoE intensified following its identification as a susceptibility gene associated with familial and sporadic Alzheimer's disease (AD)[79,80] and later shown to be associated with neuroinflammatory disorders including multiple sclerosis,[81] recovery following cardiopulmonary bypass,[82] intracerebral hemorrhage[83–85] and TBI.[70] The human APOE gene resides on chromosome 19 and is a relatively small gene, comprised of approximately 3.7 kb consisting of 4 exons separated by 3 introns, with additional

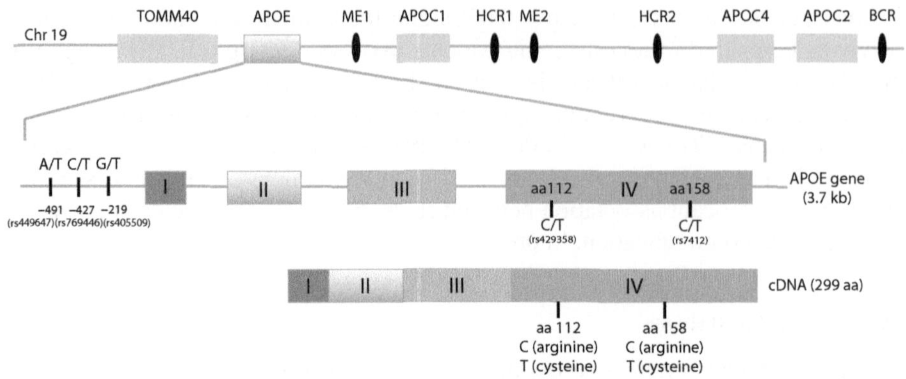

FIGURE 9.1 Schematic of APOE locus on chromosome 19 showing regional genes (TOMM40, APOC1, APOC4, and APOC2), putative multienhancer regulatory elements (ME1 and ME2), brain control region (BCR), and hepatic enhancers (HCR1 and HCR2). The APOE gene is approximately 3.7 kb and is comprised of 4 exons (I, II, III, and IV) and encodes a 299 amino acid (aa) protein. Amino acid residues 112 and 158 are shown within exon IV with corresponding SNPs (C/T) and resulting arginine/cysteine substitutions. These amino acid substitutions define the APOE alleles (APOE2, E3, or E4). Additional SNPs (–491, –427, and –219) located upstream of APOE in promoter regulatory sequences are indicated with corresponding nucleotide changes (G, A, T, or C). The rs number corresponding to each polymorphism is shown below each SNP.

upstream and downstream regulatory regions (Figure 9.1). Human apoE is a 34 kDa, 299 amino acid protein, and has three common isoforms, designated apoE2, E3, and E4, which differ by single amino acid interchanges (cysteine and arginine) at residues 112 and 158 (Figure 9.1) (E2,Cys112-Cys158; E3,Cys112-Arg158; E4Arg112-Arg158[71]). The E2/E3/E4 haplotype is defined by two nonsynonymous SNPs within exon 4 resulting in three distinct alleles (E2, E3, and E4) with six possible genotypes (APOE2/E2, E2/E3, E2/E4, E3/E3, E3/E4, and E4/E4).[86] The allele frequencies in Caucasians is 7% (E2), 78% (E3), and 15% (E4); however, these gene allele frequencies vary significantly among different geographical or ethnic groups.[87] Two SNPs consisting of C to T nucleotide changes at amino acid residue 112 (rs429358) and 158 (rs7412), result in an arginine (C) or cysteine (T) substitution (Figure 9.1). Additional SNPs within and surrounding APOE have been identified and studied in brain injury and disease.[88–92] Three of these, –219 G/T (rs405509), –427 rs769446 T/C, and –491 A/T (rs449647), reside within regulatory sequences of the gene where they can potentially influence APOE gene expression.

APOE Coding Polymorphisms and TBI

Before focusing on specific APOE genetic polymorphisms studied in the context of TBI, it will be helpful to briefly review several independent lines of evidence that first alluded to a possible connection between APOE, AD, and TBI. First, before an association was made between APOE and AD, case control and epidemiologic studies performed by numerous groups showed that prior head injury was a risk factor for developing AD later in life.[93–96] Second, boxers who developed chronic traumatic encephalopathy (CTE) due to repetitive blows to the head throughout their boxing careers were shown to be at risk for developing AD and, moreover, shared similar brain pathology at autopsy to that of AD brains.[97,98] Third, when the brains of sTBI patients were examined postmortem, 30% were shown to have deposition of amyloid beta (Aβ), a key constituent of plaques associated with AD.[99–101] Interestingly, accumulation of Aβ plaques following both acute and mild TBI was subsequently confirmed by others and shown to occur within hours of injury.[102–105] Last, the APOE genotype has been shown to influence Aβ accumulation in senile plaques of AD brains, with more abundant levels observed among E4 carriers relative to noncarriers (reviewed by Kanekiyo et al.[106]). Taken together, it seemed reasonable to hypothesize an association between APOE genotype and severity and outcome following TBI.

In a key study conducted by Mayeux and associates, they observed that individuals with an APOE4 allele and history of TBI had a 10-fold increase risk of developing AD relative to a twofold increase of risk for those with APOE4 in the absence of injury, whereas TBI in the absence of APOE4 did not increase risk.[107] Although the study was limited in size and relied on patient interview to establish a history of head injury with loss of consciousness, the results nonetheless suggested that APOE4 may act synergistically with TBI to increase the risk of AD. Shortly thereafter, Teasdale and coworkers conducted a prospective study of head injured patients admitted to the neurosurgical unit, grouping them according to the Glasgow Coma Score (GCS 3–8, n = 22; GCS 9–12, n = 19; and GCS 13–15, n = 48), and followed them for 6 months.[73] They observed that patients with an APOE4 allele were more than twice as likely to have an unfavorable outcome at 6 months following injury relative to those without. Moreover, homozygotes for APOE4 were more disabled relative to heterozygotes, suggesting an APOE4 dose effect. These initial investigations were followed by a myriad of other studies examining APOE in TBI, some of which reported that TBI patients with an APOE4 allele experienced longer hospital stays and loss of consciousness,[108] higher fatality rates and length of coma,[109–111] increase hematoma volume,[112] poorer memory outcome,[113] a higher risk of post-traumatic seizures,[114] and slower recovery rate.[115] By contrast, other studies described little or no evidence supporting an association between the two.[116–123] The conflicting results may in part be attributed to inconsistencies in severity of injury, outcome methodology, and limitations in sample size and statistical power, thereby limiting the accuracy of results. To better delineate these discrepancies, in 2008 Zhou and colleagues conducted a meta-analysis designed to determine whether the presence of APOE4 contributes to initial injury severity and/or poor outcome following TBI.[70] One hundred studies conducted from 1993 to 2007 were identified and of those, 14 cohort studies were selected, totaling 2527 participants comprising mild, moderate, and severe TBI, 736 with and 1791 without the APOE4 allele (Table 9.1). From this meta-analysis, the authors concluded that the presence of the APOE4 allele is not associated with the initial severity of brain injury but is associated with a slight (1.36) increased risk of poor long-term outcome at 6 months following injury. It is noteworthy that within this analysis is a later prospective study reported by Teasdale's group,[124] designed to confirm their earlier results. Approximately 1000 TBI patients (GCS 3–8, n = 267;

GCS 9–12, n = 178; and GCS 13–15, n = 513) were assessed for outcome at 6 months following injury using the Glasgow Outcome Scale (GOS), a measure of gross functional outcome often used in the TBI population.[125,126] In agreement with others, they observed no association between APOE4 and severity of injury, however, in contrast to their previous results, they observed no overall association between APOE genotype and unfavorable outcome at 6 months. Interestingly, they did observe an effect between age and APOE4 on outcome, which was most pronounced in children (age <15) and young adults (16–30), concluding that the influence of APOE4 on unfavorable outcome is greater among younger TBI patients. In a more recent study of 648 patients with mild to severe TBI, Ponsford and colleagues[127] detected an association between APOE4 and poorer long-term outcome between 1 and 5 years after injury, proposing that the detrimental effects of APOE4 become more apparent over time. In addition, poorer outcome was more pronounced in APOE4 positive females over 55 years of age compared to men, inferring a protective role for estrogen in APOE4 females. In a smaller study of sTBI patients (n = 48), it was suggested that patients with APOE4 may be more predisposed to brain cellular damage.[128]

APOE Promoter Polymorphisms and TBI

In addition to the SNPs that reside within the protein coding region of the gene and determine APOE genotype (rs429358 and rs7412), numerous other SNPs have been detected in and around the APOE gene.[129] Three SNPs identified within the promoter region of the gene, –219 G/T (rs405509), –427 T/C (rs769446), and –491 A/T (rs449647),[89,90,130,131] have been studied in the context of head injury (Figure 9.1). This region of DNA, located immediately upstream of the gene, serves as a binding site for the transcriptional machinery to initiate transcription and gene expression. As a result, single nucleotide changes within this region can alter transcriptional activity in a cell-specific manner. For example, the –219 G/T and –491 A/T polymorphisms have been shown to influence APOE expression in hepatocytes, by either increasing (–219 G) or decreasing (–491 T) mRNA expression, presumably by altering the binding of specific cellular factors to the promoter, whereas the –427 C/T appears to have no effect.[89] In 1988, Lambert et al. compared the effect of the same promoter polymorphisms in 573 AD cases relative to 509 controls.[131] Interestingly, the risk of AD was increased in individuals with –219 T (also referred to as the Th1/E47cs mutation) and decreased in those carrying the –491 T-allele, whereas the –427 C/T had no effect. Moreover, the relative mRNA expression level of the E4 allele in brain tissue from AD subjects (n = 49) correlated according to which promoter polymorphism was present, with –219 T augmenting expression and –491 T reducing mRNA levels. A subsequent study by Laws et al. demonstrated that the –491 A allele imparted a dose-dependent effect on apoE levels in the frontal cortex.[132] A later meta-analysis investigation led by Lambert et al. examining 1732 AD subjects and 1926 controls, showed that in addition to APOE4 acting as a risk factor for AD, the –491 and –219 polymorphisms serve as additional risk factors, with age accentuating the effect of the –219 TT genotype.[133] Additional studies have similarly implicated an association with the –219 and –491 polymorphisms and AD.[88,91,134–136] Other groups examining the effect of these promoter polymorphisms in plasma report that the –219 T allele is associated with decreased plasma apoE expression in a dose-dependent fashion, whereas the –427 and –491 have no effect.[137] Berkis et al. studied the effect of the –491 A/T SNP in CSF and observed that carriers of an AA genotype have higher levels of CSF apoE levels relative to individuals with an AT or TT genotype.[90] This is consistent with the observations of Laws et al. who reported that a –491 AA genotype was associated with increased apoE expression in the brain and plasma.[132]

Fewer studies have been conducted investigating an association between the APOE promoter polymorphisms and TBI. In one study, Lendon and coworkers looked for an association between –219 G/T, –427 T/C, and –491 A/T SNPs and 6-month outcome in 92 patients with moderate to severe TBI.[138] Although no association was detected between the –427 and –491 promoter polymorphisms and poorer outcome, an association was observed between –219 TT individuals and poor recovery. In addition, the –219 G/T frequency was comparable between the TBI group and controls, suggesting that this polymorphism may influence poorer outcome following TBI. In a subsequent study, Jiang et al.[139] looked at the relationship between APOE4 and the three promoter polymorphisms in clinical deterioration within the acute phase (<7 days) of TBI in a Chinese population (n = 100). Patients were monitored in the intensive care unit, and clinical deterioration was judged by one of the following criteria: decrease GCS relative to admission, increase in hematoma volume, or delayed hematoma detected by serial CT scanning. Their results indicated that –491 AA acted synergistically with APOE4 to confer an increased risk of clinical deterioration following TBI. By contrast, they observed no association with –219 G/T or –427 T/C and clinical deterioration. It should be noted that the allele frequencies of genes can vary considerably between different populations, and as a result one must be cautious when making comparisons between diverse ethnic and geographical populations. For example, the frequency rate of the –491 T-allele promoter polymorphism of APOE is 17% and 3% in European and Asian populations, respectively. Similarly, the C-allele of APOE rs429358, which accounts for the arginine (C) to cysteine (T) substitution at amino acid 112, occurs at a frequency of 16% (European) and 9% (Asian).[87]

APOE, Concussion, and CTE

An increased awareness of the dangers and long-term effects associated with concussions, particularly among professional and college athletes, has recently received considerable attention. In the United States alone, it is estimated that approximately 300,000 sports-related concussions occur annually and that the true incidence is even higher due to many going unreported.[140–143] Recent reports in the media of high-profile professional football players struggling with the deleterious consequences associated with concussion and those being diagnosed with chronic traumatic encephalopathy (CTE), as well as the growing numbers of young athletes involved in contact sports, underscore the importance of increasing the awareness of concussion risks and developing strategies to reduce its devastating long-term effects. CTE, previously referred to as *punch drunk syndrome* or dementia pugilistica, is a neurodegenerative disorder believed to result from repetitive brain trauma, with or without symptomatic concussion, similar to that experienced among athletes involved in contact sports such as boxing, football, and hockey.[144,145] It is postulated that these repeated blows to the head set off a cascade of neurodegenerative changes that can present clinically years or decades after the sports career ends, causing progressive neurological and cognitive impairment as well as behavioral and mood disturbances. As early as 1997, Jordan et al. reported that boxers with the APOE4 allele suffered from CTE,[146] and in 2000, Kutner and colleagues reported that older professional football players with an APOE4 allele scored lower on cognitive tests than those without.[147] More recently, in a literature review of 47 cases of neuropathologically verified CTE, 46 of which were of athletes, APOE genotyping was reported in 10 cases and of these, half carried at least one APOE4 allele.[148] Given that the percentage of APOE4 carriers in the population is significantly lower, the authors suggest that APOE4 is a risk factor for development of CTE. Several other groups have considered APOE4 as a possible risk factor for concussion among college athletes and have reported conflicting results.[149–151] For example, Terrell et al. and Kristman and associates observed no association between the APOE4 allele and prior concussion.[149,150] By comparison, Tierney et al.[151] report an association with APOE4 and history of concussion among athletes that carry a combination of an APOE2 and E4 allele together with the APOE promoter –219 T allele, but not in athletes with the APOE4 variant alone. Terrell et al. similarly observed a correlation between the APOE –219 TT genotype and increased risk for history of concussion. However, in a more recent prospective study of 3255 college athletes, this group found no association.[152] In addition, in this latter study the APOE4 allele was significantly associated with reduced concussion risk. As intriguing as these observations are, they should be approached with caution given differences in study design and the population size. It is of interest that the 2013 "Summary of Evidence-Based Guideline Update: Evaluation and Management of Concussion in Sports" listed APOE4 genotype and concussion exposure as a risk factor for chronic neurobehavioral impairment.[153]

MICROTUBULE-ASSOCIATED PROTEIN TAU AND TBI

Tau, a member of a large family of microtubule (MT)-associated binding proteins, stabilizes the assembly of MT in neurites and is abundantly expressed in axons of neurons in both the CNS and peripheral nervous system (PNS), and to lower levels in oligodendrocytes and astrocytes.[154–158] Tau is phosphorylated at multiple sites and hyperphosphorylation of tau has been linked to the formation of neurofibrillary tangles (NFTs) associated with AD.[159–162] In addition to its association with AD and CTE,[163] pathological tau protein inclusions have been associated with numerous neurodegenerative disorders, collectively referred to as tauopathies. These include frontotemporal lobar degeneration and progressive supranuclear palsy (PSP) (reviewed in Pittman et al.[164]). Human tau is encoded by the MAPT gene located on chromosome 17 and is a large gene comprised of ~150 kb and 15 exons. Alternative splicing of exons 2, 3, and 10 of the tau gene yields six major isoforms in the healthy adult human brain,[165] however, several tauopathies have been associated with aberrant splicing of exon 10, presumably causing an imbalance of specific tau isoform ratios.[164] In addition to neurodegenerative diseases, tau has also been associated with acute brain injury. Elevated levels of tau protein have been detected in human CSF following acute ischemic stroke[166,167] as well as in severe TBI, with higher levels correlating with severity of brain damage and 1-year outcome.[168,169] Several studies conducted among amateur boxers have shown elevated CSF tau levels following matches, which correlate with the number and severity of head blows, returning to normal levels after rest.[170,171] In a recent study conducted among Swedish professional hockey players (n = 47), concussed players had increased levels of plasma tau compared to preseason values, with the highest levels detected immediately following concussion and decreasing during rehabilitation.[172] Furthermore, tau levels detected at 1 hour following concussion predicted the length of time for symptoms to resolve and the players return to unrestricted competition. Despite the studies examining CSF and plasma tau protein levels following acute brain injury, very few studies have evaluated the association between tau genetic polymorphisms and TBI. In fact to date, only one group has reported on tau polymorphisms and acute head injury.[149] In 2008, Terrell et al. performed tau genetic testing ($^{Ser}53^{Pro}$ and $^{His}47^{Tyr}$ in exon 6) in 195 college football and soccer athletes and observed the TT genotype of tau $^{Ser}53^{Pro}$ (rs10445337) to be weakly associated with increased concussion risk. More recently, the same analysis was repeated in a larger cohort of college athletes (n = 3218), and no significant association was observed between either of the tau SNPs and concussion incidence; however, a trend between self-reported prior concussion and the tau $^{Ser}53^{Pro}$ TT genotype was observed (T. Terrell, 2014, personal communication). Given the paucity of studies investigating tau genetic polymorphisms and TBI, further exploration is warranted.

OTHER GENETIC ASSOCIATIONS AND TBI

NOS3

Endothelial nitric oxide synthase (NOS3) has been postulated to play a role in maintaining cerebral blood flow following TBI.[173] The gene for human NOS3 is located on chromosome 7 and several polymorphisms identified (–786 T/C, 894 G/T, and a 27 bp insertion/deletion variant, VNTR) have been examined for a possible link with cerebrovascular disease. One of these, –786 T/C (rs2070744), is located within the promoter and has been shown to influence nitric oxide (NO) mRNA levels[174–176] and to be associated with cerebral vasospasm.[177,178] In 2011, Robertson et al. reported an association with the NOS3 –786 C-allele and reduced cerebral blood flow (CBF) among a population of 51 sTBI patients.[179]

COMT

The catechol-O-methyltransferase (COMT) gene, located on chromosome 22, encodes for a protein that is believed to functionally modulate dopamine (DA) neurons, and thereby likely influence executive function. A decrease in cortical DA is thought to result in poorer executive function and vice versa.[180] A common SNP (rs4680) within COMT results in a valine (Val) to methionine substitution at amino acid 158.[181] The Val variant of the COMT Val158Met genotype is associated with high COMT enzymatic activity and lower levels of endogenous DA, whereas COMT

Met is associated with low activity and higher levels of DA.[182] Given the evidence that deficits in executive function are observed in patients suffering from TBI,[183] Lipsky and associates hypothesized that individuals with the Met polymorphism, accompanied by less active enzyme activity and increased DA levels, would perform better on tests of executive function than those with the COMT Val variant.[180] When they compared measures of executive function among a population (n = 113) of active duty and retired army veterans who had sustained TBI and controls, they observed an association between COMT Val158Met genotype and frontal-executive function at 1 year following TBI; specifically, the Val variant was associated with perseveration, suggestive of problems with cognitive flexibility. This result was in agreement with a previous study[184] that demonstrated an association between the Val allele and response latency following mild TBI (n = 39). Contrary to these reports, a more recent report by Willmott and associates of 223 participants with moderate-to-severe TBI, observed no significant interactions between COMT Val158Met genotype and cognitive outcome measures following TBI.[185] They suggest that the differences in outcome measures employed in the two studies could account for the observed disparity, speculating that methods used in the earlier study may have been more selective at detecting catecholamine-mediated cognitive function compared to the more-standard clinical neuropsychological testing used in their current study (Table 9.1).

ACE

The gene for angiotensin-converting enzyme (ACE) resides on human chromosome 17 and codes for an enzyme that regulates blood pressure (reviewed in Sayed-Tabatabaei et al.[186]). More than 160 ACE polymorphisms have been identified, the majority of which are SNPs. One ACE genetic variant commonly studied, the ACE I/D polymorphism, involves the presence of an insertion (I-allele) or deletion (D-allele) of a 287 bp DNA sequence within intron 16. The ACE D/D genotype has been associated with increase ACE activity and higher circulating levels[187] and has been extensively investigated in studies of hypertension.[186,188] The ACE I/D polymorphism was recently explored in a population of moderate to severe TBI patients (n = 73).[189] Individuals with the D-allele performed worse on tests of cognitive performance compared to those with the I-allele. Although the mechanism by which ACE influences TBI outcome is unknown, the authors postulate that it may involve disturbances in cerebral blood flow or cerebrovascular autoregulation, which may influence post-traumatic cerebral hypoperfusion.

BDNF

Additional genes with functional polymorphic alleles have been examined in the context of TBI. One of these, human brain-derived growth factor (BDNF), resides on chromosome 11 and is a major regulator in synaptic connections, plasticity, and in neuronal survival and growth.[190–192] In addition, it has been implicated in facilitating episodic and working memory, as well as executive function.[193–196] Many SNPs have been identified in the BDNF gene, one of which is a G to A transition at nucleotide

+196 (rs6265), resulting in a valine to methionine (Val66Met) substitution that alters the regulated secretion and neuroplastic function of BDNF.[193,197–199] In clinical studies, the Met-allele has been associated with impaired cognitive function in healthy individuals,[193,200,201] stroke patients,[202] and psychiatric patients.[203–205] Interestingly, several recent reports of Vietnam War combat veterans who had sustained focal penetrating TBI found the Met-allele to be protective for general intelligence.[196,206,207] An additional study detected no association with the BDNF Met-allele and recovery of consciousness and cognitive functions in patients in a vegetative state following TBI.[208] Given the abundance of BDNF in the hippocampus and its role in memory, the vulnerability of the hippocampus to TBI[209] and attention and processing problems observed following injury, McAllister and colleagues hypothesized that BDNF polymorphisms would influence cognition 1 month after mild to moderate TBI.[210] In a study of 75 patients with mild to moderate TBI and 38 healthy controls, an association of the Met-allele with slower processing speed (cognition) among the entire group of subjects was detected.[210] Furthermore, several additional BDNF SNPs (rs11030102, rs11030107, and rs12273363) not previously studied were associated with memory measures within the mTBI group after adjusting for the Met-allele. These results support the hypothesis that polymorphisms in BDNF influence cognitive performance following mTBI.

TP53

The tumor protein 53 gene (TP53) resides on chromosome 17 and encodes p53, a crucial regulator of the cell cycle and apoptosis.[211] In animal models of closed head injury and ischemia, changes in p53 expression have been reported[212,213] and apoptotic neurons have been positively identified in autopsied human brain tissue following TBI.[214] The most prevalent polymorphism in p53 is a G/T polymorphism within codon 72 (rs1042522) resulting in an arginine (Arg) to proline (Pro) substitution.[215–217] It has been suggested that homozygous arginine variants are more effective inducers of apoptosis than proline variants.[216] To determine the relationship between the Argp53Pro polymorphism and outcome following TBI, Martinez-Lucas et al. genotyped 90 sTBI patients and measured outcome at discharge from the surgical ICU (GOS-0) and at 6 months (GOS-6) postinjury.[218] They reported that the homozygous Arg/Arg genotype was associated with an almost threefold greater risk of having a bad outcome at surgical ICU discharge (GOS-0), and served as an independent risk factor for bad outcome in TBI. A similar association was not observed with GOS-6 nor was it associated with mortality.[218]

PARP-1

The PARP-1 (poly[ADP-ribose]polymerase-1) gene, located on human chromosome 1, encodes a ubiquitously expressed enzyme that functions in the cellular response to stress and DNA damage.[219,220] Excessive activity of PARP-1 is thought to worsen brain injury via NAD+ depletion and energy failure. In a retrospective study, Sarnaik et al.[221] reported that an A/T SNP (rs3219119) associated with the automodification and catalytic domains of the PARP-1, was an independent predictor of neurological outcome. Specifically, in a population of 191 patients with sTBI, individuals with the AA-genotype had a favorable GOS at 6 months. A second SNP, rs2271347, involving an A to G substitution in the promoter region of the gene, did not correlate with outcome but did correlate with poly-ADP-ribose levels in the CSF.

MME

Neprilysin is a zinc-dependent metalloprotease encoded by the MME (membrane metallo-endopeptidase) gene located on chromosome 3. The protease is expressed in a variety of tissues, degrades Aβ protein, and is thought to be the rate-limiting step in Aβ degradation.[222] Experiments performed in neprilysin-knockout mice have provided evidence for the association of the neprilysin with the AD process.[223] TBI has been shown to induce the rapid formation of AD-like Aβ plaques in about 30% of patients although the mechanism behind this selective plaque formation is unclear.[224] A dinucleotide GT repeat polymorphism has been identified in the upstream regulatory region of MME and may be involved in regulating neprilysin expression in neurons.[225] Johnson et al. investigated the potential association between amyloid deposition acutely after TBI and the MME polymorphism in 81 patients.[224] They observed that the length of the MME GT repeats was longer (>41 repeats) in patients that accumulated Aβ following TBI compared to Aβ nonaccumulators. In addition, the presence of 22 repeats in at least one allele was independently associated with Aβ deposition. The authors suggest that these findings provide a potential genetic screening test for individuals at high risk for TBI.

NEFH

The neurofilament heavy (NEFH) gene resides on chromosome 22 and codes for the heavy protein subunit of neurofilament, which combines with the medium and light subunits to make up neuronal cytoskeleton. There have been several reports in the literature suggesting that mutations in NEFH may play a role in neurodegenerative diseases.[226] One study has been reported in the literature studying the effect of NEFH genetic polymorphisms and TBI. McDevitt and colleagues genotyped 48 college athletes for an A to C polymorphism of NEFH (rs165602) looking for an association between genotype and the frequency and severity of previous self-reported concussion.[227] This polymorphism results in an A to C missense SNP that changes a glutamic acid to an alanine at amino acid residue 805. The authors observed no significant association between rs165602 and the occurrence or severity of concussion (mTBI) among the athletes.

5-HTT

The serotonin transporter (5-HTT) gene (solute carrier family 6 member; SLC6A4) is thought to be involved in the pathophysiology of major depression.[228] The 5-HTT gene resides on chromosome 17 and is modulated by the serotonin transporter gene linked polymorphic region (5-HTTLPR). The variant involves the insertion or deletion of a 44 bp DNA sequence resulting in a short (S) allele or long (L) allele.[229,230] In addition to the S/L variations, an A/G SNP (rs2553) has been identified in this region resulting in additional allelic combinations: L(A) or L(G); and S(A) or S(G).[231] It is generally accepted that one or two copies of the S-allele confer a greater susceptibility at developing major depression in response to stressful life events.[232] Because major depression is a common neurobehavioral complication following TBI, Chan et al. investigated the role of the S/L polymorphisms in the risk of depression secondary to TBI.[232] The population included 75 patients who had sustained a TBI within the past 12 months and met the criteria for mood disorder, and 99 controls with TBI but no mood disorder. The majority of the TBI population fell within an initial GCS of 13–15 (86%), whereas 10.1% were GCS 9–12 and 3.8% GCS <9. They observed no difference in S/L genotype frequencies between the depressed and control groups. Furthermore, for depressed individuals, there was no association between depression scores (Hamilton Depression Rating Score) and these polymorphisms.

PPP3CC

The protein phosphatase 3 catalytic subunit gamma isozyme (PPP3CC) gene is located on chromosome 8 in humans and encodes a regulatory subunit for calcineurin, a calcium-dependent protein phosphatase. Calcineurin is thought to play a role in the regulation of dopaminergic signal transduction and synaptic plasticity,[233] and has been implicated in neurological disease.[234] In a preliminary study of 277 TBI patients reported by Bales et al. in 2011,[235] an A/G polymorphism (rs2443504) in PPP3CC was found to be a predictor of susceptibility and recovery in TBI, using initial clinical severity (GCS) and recovery (GOS) at 3, 6, and 12 months post-TBI as measures. When individuals with an AA genotype were compared to those homozygous for the G-allele, the results were as follows: chance of an initial GCS <6, 54.2% (AA) vs. 26.6% (GG); 3-month severe GOS, 59.3% (AA) vs. 31.3% (GG); 6-month severe GOC, 53.1% (AA) vs. 32.4% (GG); and 12-month severe GOS, 56.2% (AA) vs. 33.3% (AA). Although these results are compelling, the report was presented in abstract form and lacks a detailed description of the patient population studied.

ANKK1

The ankyrin repeat and kinase domain containing 1 (ANKK1) gene, located on human chromosome 11, is a member of the serine/threonine protein kinase family. ANKK1 contains an A/T SNP referred to as the Taq1A polymorphism (rs1800497) and was originally thought to be the dopamine receptor D2 (DRD2) gene.[236,237] The T-allele is associated with a 40% reduction in expression of D2 receptors.[238] McAllister et al.[184] reported an association with the Taq1A T-allele and poorer episodic memory and response latency in a combined group of subjects that included 39 mTBI subjects and controls.[184] The T-allele mTBI carriers had slower response latencies compared to T-allele positive controls. These observations were confirmed when the same group extended their study into a larger population (n = 93) of mild and moderate TBI subjects.[237]

KIBRA

KIBRA, kidney and brain expressed protein, represents one of the first genes to be associated with episodic memory. In humans, KIBRA resides on chromosome 5 and encodes a protein involved in synaptic plasticity and transmission, long-term-potentiation, and signal transduction, which, in the brain, is predominantly expressed in the temporal lobe and hippocampus.[239–241] The T-allele of a common T to C polymorphism (rs17070145) identified in KIBRA, has been associated with better performance on episodic memory measures in healthy adults compared to noncarriers.[242] However, in elderly adults with memory complaints, the reverse is true.[242] Given this observation, Wagner et al. predicted that TBI patients harboring a T-allele would perform less well cognitively following injury.[243] To this end, they conducted a study of 129 patients with sTBI and reported an association between the rs17070145 T-allele and poorer performance on episodic memory at 6 and 12 months, supporting their hypothesis.[243]

MITOCHONDRIAL GENES

Background

Mitochondria are the cytoplasmic organelles whose primary function is to generate the cell's energy in the form of ATP. In addition to being the site for oxidative-phosphorylation and electron transport, they are involved in a number of cellular processes including maintenance of the cell cycle, differentiation, cell signaling, calcium storage, and apoptosis (reviewed in McBride et al.[244]). Mitochondria are distributed along the microtubules in cells and can vary significantly in number and location, often residing in subcellular regions with high metabolic demands. They exhibit a high degree of plasticity and interconnectivity, forming complex networks inside the cell with the cytoskeleton. The location and mobility of mitochondria in neurons has been well studied, and they are most prominent at sites of intense activity such as synapses and active growth cones in developing neurons.[245] Recent evidence suggests that disturbance of this finely balanced mitochondrial network plays a pivotal role in the pathophysiology of a number of diseases including neurodegenerative disorders (reviewed in Burte et al.[246]). Although some of these diseases are associated with mutations within nuclear DNA that encode mitochondrial proteins, over 150 point mutations and DNA rearrangements have been identified in human mitochondrial (mt) DNA and associated with a variety of conditions (reviewed in Dimauro and Davidzon[247]). Two examples of neurological diseases associated with mutations in mtDNA include mitochondrial encephalopathy, lactic acidosis, stroke-like episodes (MELAS) and myoclonus epilepsy and ragged-red fibers (MERRF).[247]

Human mtDNA is a double-stranded, circular molecule comprised of 16,569 nucleotides that code for 37 genes, including 2 ribosomal (r) RNA genes, 22 transfer (t) RNA genes, and 13 structural genes critical to the function of the respiratory chain/oxidative phosphorylation system.[248] Although nuclear genes encode many mitochondrial proteins including some necessary for cellular respiration, proteins encoded specifically by mtDNA, such as NADH dehydrogenase, are required for a functional electron transport chain and oxidative phosphorylation. Mitochondrial DNA is maternally inherited and each cell contains hundreds to thousands of mtDNA molecules that replicate independently and distribute randomly in daughter cells during cell division. The production of oxygen-free radicals within the mitochondria can cause somatic mutations that occur in some but not all mtDNA molecules, which can then accumulate in postmitotic tissue; the clinical manifestation of a pathogenic mtDNA mutation is largely determined by the relative proportion of mutant to normal genomes. Unlike nuclear DNA, mtDNA lacks protective histones and DNA repair enzymes, and as a result cells accumulate a progressively larger burden of mtDNA damage as they age.[249] During cell division, the ratio of mutant to normal molecules in daughter cells can change; this phenomena may partially explain why the disease phenotype of certain patients with mtDNA disorders may vary with age.

Over millennia, a substantial number of mtDNA mutations or SNPs have accumulated, the combination of which define mitochondrial haplotypes, commonly referred to as "haplogroups."[250] Mitochondrial haplogroups have been used to map genetic differences among populations and tend to be continent specific. Nine major haplogroups identified among individuals of European descent include H, T, U, V, W, X, J, Y, and K, with haplogroup H representing the majority (40%) of any European population.[251] Specific SNPs are associated with particular haplogroups. For example, SNP A10398G is a mitochondrial SNP located within the NADH-ubiquinone oxireductase subunit-3 locus (ND3)[252]; the A10398 allele is found in haplogroup H whereas the 10398G allele is associated with haplogroups K and J. Mitochondrial haplogroups have been reported as factors for degenerative pathologies and neurodegenerative conditions including Parkinson's disease (PD), AD, Huntington's disease (HD), and motor neuron disease.[253]

Mitochondrial Haplogroups and TBI

Mitochondrial dysfunction following TBI has been well documented in the literature in both animal[254–256] and human studies[252,257,258] and is characterized by a decrease in ATP production, an increase in lactate accumulation, and uncontrolled regulation of intracellular calcium prior to neuronal loss.[259,260] Despite these observations, the mitochondrial genome has received little attention with respect to TBI. In a recent article, Bulstrode and colleagues[258] examined the relationship between mitochondrial genomic variants and 6-month outcome following TBI in a large cohort of 880 patients (mild to severe TBI) admitted to hospitals in Scotland. As expected within a European population, approximately 40% of the patients had the mitochondrial haplogroup H. Interestingly, the authors observed a significant predictive effect of mtDNA genotype with outcome at 6-month (GOS) when the haplogroups were examined individually, with patients possessing the K variant exhibiting a better outcome than those without (odds ratio = 1.64%, confidence interval = 1.08–2.51, $p = 0.02$). By contrast, no significant association was observed within haplogroups H, J, T, and U on their own. In a recent preclinical model of TBI, Gilmer et al. showed that mitochondrial dysfunction increases in an age-related manner.[256] Interestingly, when mitochondrial dysfunction was examined in the current human TBI population,

Bulstrode et al. reported that possession of either the haplogroup K or T significantly lessened the negative effect of aging on outcome ($p = 0.017$ and $p = 0.015$, respectively).[258] Furthermore, these findings corroborate epidemiological studies showing an overrepresentation of haplogroup K in longevity and underrepresentation of haplogroups K and T in neurodegenerative conditions.[261–263] Several lines of evidence have suggested that abnormalities in oxidative metabolism play a role in AD,[264,265] and APOE has been shown to protect neuronal cell lines from oxidative stress in an allele-specific fashion (E2>E3>E4).[266] To this end, Bulstrode et al. looked for an interaction between mtDNA haplogroups and APOE polymorphisms in their TBI cohort but failed to detect an association between APOE genotype and outcome. They did, however, observe that patients with an APOE4 allele had a better outcome following TBI if they also possessed the mtDNA K haplogroup,[258] supporting the observations of others from an earlier study of 213 TBI patients.[267] It is noteworthy that enzymatic components of the electron transport chain encoded by haplotype K are less tightly coupled relative to other mitochondrial variants, resulting in a reduction of reactive oxygen species, key mediators of secondary cell damage following TBI.[262,268,269]

Conley et al. recently investigated the impact of several mtDNA SNPs on outcomes in 336 sTBI patients up to one year following injury.[252] In this study, all TBI patients had external ventricular drains, and measures included the GOS, Neurobehavioral Rating Scale (NRS), Disability Rating Scale (DRS), in-hospital mortality or length of stay, and lactate:pyruvate ratio from CSF. They observed that the A10398G SNP within ND3 was associated with DRS at 6 and 12 months. Specifically, the A10398 allele (associated with haplogroup H) and 10398G (haplogroups K and J) were associated with slower and faster recovery, respectively. A second SNP, T195C, located within a noncoding region of mtDNA, was associated with outcome depending on sex, with the T195 allele associating with poorer outcome in females ($p = 0.03$). The same allele was similarly associated with mitochondrial dysfunction in females ($p = 0.01$). These results suggest that mitochondrial variation may be particularly relevant to recovery among female patients. It is noteworthy that a sex-based association favoring protection in females has also been observed with mitochondrial SNP 10398G in PD.[270] Altogether, these studies are intriguing and suggest a relationship between mtDNA haplogroups, aging, and the pathophysiology of TBI, as well as a possible interaction of APOE pathways with genetically regulated mitochondrial functions in acute injury.

GENETICS OF PEDIATRIC TBI

TBI in pediatric patients differs significantly from that of adults in multiple aspects and as a result, the pediatric brain should not be viewed simply as a "small adult brain."[271] Several of these distinctions include anatomical and functional developmental differences, enhanced apoptotic-induced cell death, increased risk of post-traumatic seizures, and age-specific differences in cerebral blood flow and metabolism.[271–273] Surprisingly, despite the observation that the vast majority of TBIs occur in children and young adults, there is a paucity of genetic association studies specific to pediatric TBI. In fact, a 2012 comprehensive review of genetic association studies and pediatric TBI identified only a handful of vetted studies specific to pediatric TBI, each of which investigated an association between APOE4 and outcome.[274] One of these evaluated an association between APOE4 and post-traumatic cerebral swelling in 165 postmortem cases of children aged 2–19 years (median 13 years) who survived 1 hour to 5 months (median 3 days) after injury.[275] No significant relationship was detected with possession of an APOE4 allele and brain swelling, however, the authors acknowledge that a bias toward worst possible outcome had been introduced because the patient cohort was selected by death. As cited earlier in this chapter, Teasdale and coworkers observed that possession of an APOE4 allele in children (n = 215) less than 16 years was associated with a 3.06 (95% confidence interval; 1.22–7.65) greater odds of an unfavorable outcome at 6 months following TBI.[124] An additional study of 70 children ranging in age from 1 month to 17 years, conducted in the Czech Republic, reported that presence of an APOE4 allele was associated with poorer global outcome at 12 months following TBI.[276] In 2009, Moran and colleagues evaluated whether APOE genotype could predict outcome in children following mild TBI.[277] The study included 99 children (8–15 years) who presented to the emergency room and were subsequently diagnosed with concussion according to the following criteria: loss of consciousness (LOC) less than 30 minutes, GCS score of 13 or 14, or two or more acute symptoms of concussion. Outcomes were assessed at presentation and again at 2 weeks, 3 months, and 12 months postinjury employing a battery of measures to assess memory, development, neuropsychological functioning, general intelligence, and academic achievement. These authors concluded that the APOE4 allele was associated with poor global outcome relative to individuals with an E2/E3 or E3/E3 genotype. A fourth study assessed the association of APOE with cerebral perfusion pressure (CPP) in 65 critically ill children with TBI using modified GCS, global outcome, and CPP as outcome measures, concluding that individuals with an APOE4 allele were less tolerant to increases in CPP.[278] Conversely, an earlier study of 71 TBI children requiring inpatient rehabilitation reported that the APOE4 allele (n = 4) was associated with improved outcome at discharge, however, the E4 frequency (4%) was well below that observed in the population at large (15%).[279] Despite these differences, a statistical analysis of the combined data of the aforementioned studies revealed a 2.44 (95% confidence interval: 1.25–4.80) greater odds of having poor outcome in children who are carriers of APOE4 compared to noncarriers.[280] A more comprehensive review of the literature conducted in May 2014, examining genes associated with pediatric TBI, similarly identified APOE4 as being the lone gene studied in the arena of pediatric TBI.[281]

CONCLUSION

The study of genes that influence outcome following TBI is still in its early stages. Evidence has been rapidly accumulating in the literature, particularly in the last decade, identifying genetic polymorphisms associated with the pathophysiology and outcome following TBI. The genes associated with these polymorphisms play multifactorial roles in regulating the brain's response to injury and include both anti- and pro-inflammatory cytokines, DNA repair enzymes, signaling molecules, and neurotrophins. The importance of mitochondria in the pathophysiology of brain injury has been well documented, although the mitochondrial genome has received relatively little attention in the setting of TBI. How these diverse genes interact with one another following TBI remains to be elucidated and requires extensive research. Furthermore, an additional significant area of research not addressed in this chapter that must be considered in the context of TBI is that of epigenetic mechanisms (DNA methylation, histone modifications and nucleosome remodeling, RNA editing) that modulate gene expression and cellular activity. The majority of TBI-genetic association studies reported in the literature have been conducted in adult populations, despite the observation that the vast majority of TBIs occur in children and young adults. Clearly, additional genetic association studies are essential to facilitate both the prediction of outcome and clinical management in the pediatric TBI population.

ACKNOWLEDGMENT

We would like to thank Austin M. Cook for his help in constructing Figure 9.1 and Table 9.1.

REFERENCES

1. Faul, M., Coronado, V. 2015. Epidemiology of traumatic brain injury. *Handb Clin Neurol* 127:3–13.
2. Coronado, V.G. et al. 2011. Surveillance for traumatic brain injury-related deaths—United States, 1997–2007. *MMWR Surveill Summ* 60(5):1–32.
3. Corrigan, J.D., Selassie, A.W., Orman, J.A. 2010. The epidemiology of traumatic brain injury. *J Head Trauma Rehabil* 25(2):72–80.
4. Rutland-Brown, W. et al. 2006. Incidence of traumatic brain injury in the United States, 2003. *J Head Trauma Rehabil* 21(6):544–548.
5. Teasdale, G., Jennett, B. 1974. Assessment of coma and impaired consciousness. A practical scale. *Lancet* 2(7872):81–84.
6. Management of Concussion/mTBI Working Group. 2009. VA/DoD Clinical Practice Guideline for Management of Concussion/Mild Traumatic Brain Injury. *J Rehabil Res Dev* 46(6):CP1–CP68.
7. Winter, C.D. et al. 2004. Raised parenchymal interleukin-6 levels correlate with improved outcome after traumatic brain injury. *Brain* 127(Pt 2):315–320.
8. Schmidt, O.I. et al. 2004. Tumor necrosis factor-mediated inhibition of interleukin-18 in the brain: A clinical and experimental study in head-injured patients and in a murine model of closed head injury. *J Neuroinflammation* 1(1):13.
9. Raghupathi, R., Graham, D.I., McIntosh, T.K. 2000. Apoptosis after traumatic brain injury. *J Neurotrauma* 17(10):927–938.
10. McAllister, T.W. 2010. Genetic factors modulating outcome after neurotrauma. *PM R* 2(12 Suppl 2):S241–S252.
11. Weaver, S.M. et al. 2014. Genetic polymorphisms and traumatic brain injury: The contribution of individual differences to recovery. *Brain Imaging Behav* 8(3):420–434.
12. Waters, R.J. et al. 2013. Cytokine gene polymorphisms and outcome after traumatic brain injury. *J Neurotrauma* 30(20):1710–1716.
13. Wilson, M., Montgomery, H. 2007. Impact of genetic factors on outcome from brain injury. *Br J Anaesth* 99(1):43–48.
14. Davidson, J., Cusimano, M.D., Bendena, W.G. 2014. Post-traumatic brain injury: Genetic susceptibility to outcome. *Neuroscientist*.

15. Dashnaw, M.L., Petraglia, A.L., Bailes, J.E. 2012. An overview of the basic science of concussion and subconcussion: Where we are and where we are going. *Neurosurg Focus* 33(6):E5: 1–9.
16. McKee, A.C. et al. 2014. The neuropathology of sport. *Acta Neuropathol* 127(1):29–51.
17. Rigg, J.L., Mooney, S.R. 2011. Concussions and the military: Issues specific to service members. *PM R* 3(10 Suppl 2):S380–S386.
18. Graves, B.T., Munro, C.L. 2013. Epigenetics in critical illness: A new frontier. *Nurs Res Pract* 2013:503686.
19. Qureshi, I.A., Mehler, M.F. 2010. Emerging role of epigenetics in stroke: Part 1: DNA methylation and chromatin modifications. *Arch Neurol* 67(11):1316–1322.
20. Liyanage, V.R. et al. 2014. DNA modifications: Function and applications in normal and disease States. *Biology (Basel)* 3(4):670–723.
21. Chodobski, A., Zink, B.J., Szmydynger-Chodobska, J. 2011. Blood-brain barrier pathophysiology in traumatic brain injury. *Transl Stroke Res* 2(4):492–516.
22. Singh, R.B. et al. 2012. Metabolic syndrome: A brain disease. *Can J Physiol Pharmacol* 90(9):1171–1183.
23. Galic, M.A., Riazi, K., Pittman, Q.J. 2012. Cytokines and brain excitability. *Front Neuroendocrinol* 33(1):116–125.
24. Kadhim, H.J., Duchateau, J., Sebire, G. 2008. Cytokines and brain injury: Invited review. *J Intensive Care Med* 23(4):236–249.
25. Patterson, Z.R., Holahan, M.R. 2012. Understanding the neuroinflammatory response following concussion to develop treatment strategies. *Front Cell Neurosci* 6:58.
26. Shohami, E. et al. 1997. Oxidative stress in closed-head injury: Brain antioxidant capacity as an indicator of functional outcome. *J Cereb Blood Flow Metab* 17(10):1007–1019.
27. Scherbel, U. et al. 1999. Differential acute and chronic responses of tumor necrosis factor-deficient mice to experimental brain injury. *Proc Natl Acad Sci U S A* 96(15):8721–8726.
28. Holmin, S. et al. 1997. Delayed cytokine expression in rat brain following experimental contusion. *J Neurosurg* 86(3):493–504.
29. Dalgard, C.L. et al. 2012. The cytokine temporal profile in rat cortex after controlled cortical impact. *Front Mol Neurosci* 5:6.
30. Singhal, A. et al. 2002. Association between cerebrospinal fluid interleukin-6 concentrations and outcome after severe human traumatic brain injury. *J Neurotrauma* 19(8):929–937.
31. Ziebell, J.M., Morganti-Kossmann, M.C. 2010. Involvement of pro- and anti-inflammatory cytokines and chemokines in the pathophysiology of traumatic brain injury. *Neurotherapeutics* 7(1):22–30.
32. Csuka, E. et al. 1999. IL-10 levels in cerebrospinal fluid and serum of patients with severe traumatic brain injury: Relationship to IL-6, TNF-alpha, TGF-beta1 and blood-brain barrier function. *J Neuroimmunol* 101(2):211–221.
33. Wahl, S.M. et al. 1992. Transforming growth factor-beta in synovial fluids modulates Fc gamma RII (CD16) expression on mononuclear phagocytes. *J Immunol* 148(2):485–490.
34. Lowrance, J.H. et al. 1994. Spontaneous elaboration of transforming growth factor beta suppresses host defense against bacterial infection in autoimmune MRL/lpr mice. *J Exp Med* 180(5):1693–1703.
35. Oshima, T. et al. 2009. TNF-alpha contributes to axonal sprouting and functional recovery following traumatic brain injury. *Brain Res* 1290:102–110.
36. Bermpohl, D. et al. 2006. Traumatic brain injury in mice deficient in Bid: Effects on histopathology and functional outcome. *J Cereb Blood Flow Metab* 26(5):625–633.
37. Nedwin, G.E. et al. 1985. Effect of interleukin 2, interferon-gamma, and mitogens on the production of tumor necrosis factors alpha and beta. *J Immunol* 135(4):2492–2497.

38. Herrmann, S.M. et al. 1998. Polymorphisms of the tumour necrosis factor-alpha gene, coronary heart disease and obesity. *Eur J Clin Invest* 28(1):59–66.
39. Cariaso, M., Lennon, G. 2012. SNPedia: A wiki supporting personal genome annotation, interpretation and analysis. *Nucleic Acids Res* 40(Database issue):D1308–D1312.
40. Wong, M. et al. 2008. TNFalpha blockade in human diseases: Mechanisms and future directions. *Clin Immunol* 126(2):121–136.
41. Wilson, J.X. 1997. Antioxidant defense of the brain: A role for astrocytes. *Can J Physiol Pharmacol* 75(10–11):1149–1163.
42. Kroeger, K.M., Carville, K.S., Abraham, L.J. 1997. The -308 tumor necrosis factor-alpha promoter polymorphism effects transcription. *Mol Immunol* 34(5):391–399.
43. Culpan, D. et al. 2003. Tumour necrosis factor-alpha gene polymorphisms and Alzheimer's disease. *Neurosci Lett* 350(1):61–65.
44. Harcos, P. et al. 2006. Decreased frequency of the TNF2 allele of TNF-alpha -308 promoter polymorphism is associated with lacunar infarction. *Cytokine* 33(2):100–105.
45. Nicklin, M.J., Weith, A., Duff, G.W. 1994. A physical map of the region encompassing the human interleukin-1 alpha, interleukin-1 beta, and interleukin-1 receptor antagonist genes. *Genomics* 19(2):382–384.
46. Holmin, S., Hojeberg, B. 2004. In situ detection of intracerebral cytokine expression after human brain contusion. *Neurosci Lett* 369(2):108–114.
47. Lu, K.T. et al. 2005. Effect of interleukin-1 on traumatic brain injury-induced damage to hippocampal neurons. *J Neurotrauma* 22(8):885–895.
48. Nicoll, J.A. et al. 2000. Association of interleukin-1 gene polymorphisms with Alzheimer's disease. *Ann Neurol* 47(3):365–368.
49. Um, J.Y. et al. 2003. Association of interleukin-1 alpha gene polymorphism with cerebral infarction. *Brain Res Mol Brain Res* 115(1):50–54.
50. Tarlow, J.K. et al. 1993. Polymorphism in human IL-1 receptor antagonist gene intron 2 is caused by variable numbers of an 86-bp tandem repeat. *Hum Genet* 91(4):403–404.
51. Blakemore, A.I. et al. 1996. Interleukin-1 receptor antagonist allele (IL1RN*2) associated with nephropathy in diabetes mellitus. *Hum Genet* 97(3):369–374.
52. Uzan, M. et al. 2005. Association between interleukin-1 beta (IL-1beta) gene polymorphism and outcome after head injury: An early report. *Acta Neurochir (Wien)* 147(7):715–720; discussion 720.
53. Hadjigeorgiou, G.M. et al. 2005. IL-1RN and IL-1B gene polymorphisms and cerebral hemorrhagic events after traumatic brain injury. *Neurology* 65(7):1077–1082.
54. Tanriverdi, T. et al. 2006. Lack of association between the IL1A gene (-889) polymorphism and outcome after head injury. *Surg Neurol* 65(1):7–10; discussion 10.
55. Dardiotis, E. et al. 2006. Re: Lack of association between the IL1A gene (-889) polymorphism and outcome after head injury. Tanriverdi T et al. *Surg Neurol* 2006;65:7–10; discussion 10. *Surg Neurol* 66(3):334–335.
56. Terry, C.F., Loukaci, V., Green, F.R. 2000. Cooperative influence of genetic polymorphisms on interleukin 6 transcriptional regulation. *J Biol Chem* 275(24):18138–18144.
57. Fishman, D. et al. 1998. The effect of novel polymorphisms in the interleukin-6 (IL-6) gene on IL-6 transcription and plasma IL-6 levels, and an association with systemic-onset juvenile chronic arthritis. *J Clin Invest* 102(7):1369–1376.
58. Rivera-Chavez, F.A. et al. 2003. Interleukin-6 promoter haplotypes and interleukin-6 cytokine responses. *Shock* 20(3):218–223.
59. Brull, D.J. et al. 2001. Interleukin-6 gene -174g>c and -572g>c promoter polymorphisms are strong predictors of plasma interleukin-6 levels after coronary artery bypass surgery. *Arterioscler Thromb Vasc Biol* 21(9):1458–1463.
60. Villuendas, G. et al. 2002. The -597 G-->A and -174 G-->C polymorphisms in the promoter of the IL-6 gene are associated with hyperandrogenism. *J Clin Endocrinol Metab* 87(3):1134–1141.

61. Chen, Y. et al. 2006. Interleukin-6 involvement in brain arteriovenous malformations. *Ann Neurol* 59(1):72–80.
62. Minambres, E. et al. 2003. Correlation between transcranial interleukin-6 gradient and outcome in patients with acute brain injury. *Crit Care Med* 31(3):933–938.
63. Dalla Libera, A.L. et al. 2011. IL-6 polymorphism associated with fatal outcome in patients with severe traumatic brain injury. *Brain Inj* 25(4):365–369.
64. Letterio, J.J., Roberts, A.B. 1998. Regulation of immune responses by TGF-beta. *Annu Rev Immunol* 16:137–161.
65. Massague, J. et al. 1990. TGF-beta receptors and TGF-beta binding proteoglycans: Recent progress in identifying their functional properties. *Ann NY Acad Sci* 593: 59–72.
66. Derynck, R. et al. 1987. Intron-exon structure of the human transforming growth factor-beta precursor gene. *Nucleic Acids Res* 15(7):3188–3189.
67. Grainger, D.J. et al. 1999. Genetic control of the circulating concentration of transforming growth factor type beta1. *Hum Mol Genet* 8(1):93–97.
68. Shah, R. et al. 2006. Allelic diversity in the TGFB1 regulatory region: Characterization of novel functional single nucleotide polymorphisms. *Hum Genet* 119(1–2):61–74.
69. Shah, R., Hurley, C.K., Posch, P.E. 2006. A molecular mechanism for the differential regulation of TGF-beta1 expression due to the common SNP -509C-T (c. -1347C > T). *Hum Genet* 120(4):461–469.
70. Zhou, W. et al. 2008. Meta-analysis of APOE4 allele and outcome after traumatic brain injury. *J Neurotrauma* 25(4):279–290.
71. Weisgraber, K.H. 1994. Apolipoprotein E: Structure-function relationships. *Adv Protein Chem* 45:249–302.
72. Xu, Q. et al. 2006. Profile and regulation of apolipoprotein E (ApoE) expression in the CNS in mice with targeting of green fluorescent protein gene to the ApoE locus. *J Neurosci* 26(19):4985–4994.
73. Teasdale, G.M. et al. 1997. Association of apolipoprotein E polymorphism with outcome after head injury. *Lancet* 350(9084):1069–1071.
74. Laskowitz, D.T., Horsburgh, K., Roses, A.D. 1998. Apolipoprotein E and the CNS response to injury. *J Cereb Blood Flow Metab* 18(5):465–471.
75. Laskowitz, D.T. et al. 2006. Apolipoprotein E-derived peptides reduce CNS inflammation: Implications for therapy of neurological disease. *Acta Neurol Scand Suppl* 185:15–20.
76. Lynch, J.R. et al. 2005. A novel therapeutic derived from apolipoprotein E reduces brain inflammation and improves outcome after closed head injury. *Exp Neurol* 192(1):109–116.
77. Wang, H. et al. 2013. Apolipoprotein E mimetic peptides improve outcome after focal ischemia. *Exp Neurol* 241:67–74.
78. Laskowitz, D.T. et al. 2010. Traumatic brain injury exacerbates neurodegenerative pathology: Improvement with an apolipoprotein E-based therapeutic. *J Neurotrauma* 27(11):1983–1995.
79. Strittmatter, W.J. et al. 1993. Apolipoprotein E: High-avidity binding to beta-amyloid and increased frequency of type 4 allele in late-onset familial Alzheimer disease. *Proc Natl Acad Sci U S A* 90(5):1977–1981.
80. Saunders, A.M. et al. 1993. Association of apolipoprotein E allele epsilon 4 with late-onset familial and sporadic Alzheimer's disease. *Neurology* 43(8):1467–1472.
81. Fazekas, F. et al. 2006. The impact of our genes: Consequences of the apolipoprotein E polymorphism in Alzheimer disease and multiple sclerosis. *J Neurol Sci* 245(1–2):35–39.
82. Newman, M.F. et al. 1995. Predictors of cognitive decline after cardiac operation. *Ann Thorac Surg* 59(5):1326–1330.

83. Alberts, M.J. et al. 1995. ApoE genotype and survival from intracerebral haemorrhage. *Lancet* 346(8974):575.
84. McCarron, M.O. et al. 2003. Effect of apolipoprotein E genotype on in-hospital mortality following intracerebral haemorrhage. *Acta Neurol Scand* 107(2):106–109.
85. James, M.L. et al. 2009. Apolipoprotein E modifies neurological outcome by affecting cerebral edema but not hematoma size after intracerebral hemorrhage in humans. *J Stroke Cerebrovasc Dis* 18(2):144–149.
86. Hixson, J.E., Vernier, D.T. 1990. Restriction isotyping of human apolipoprotein E by gene amplification and cleavage with HhaI. *J Lipid Res* 31(3):545–548.
87. Abecasis, G.R. et al. 2010. A map of human genome variation from population-scale sequencing. *Nature* 467(7319):1061–1073.
88. Bullido, M.J. et al. 1998. A polymorphism in the regulatory region of APOE associated with risk for Alzheimer's dementia. *Nat Genet* 18(1):69–71.
89. Artiga, M.J. et al. 1998. Allelic polymorphisms in the transcriptional regulatory region of apolipoprotein E gene. *FEBS Lett* 421(2):105–108.
90. Bekris, L.M. et al. 2008. Multiple SNPs within and surrounding the apolipoprotein E gene influence cerebrospinal fluid apolipoprotein E protein levels. *J Alzheimers Dis* 13(3):255–266.
91. Lescai, F. et al. 2011. An APOE haplotype associated with decreased epsilon4 expression increases the risk of late onset Alzheimer's disease. *J Alzheimers Dis* 24(2): 235–245.
92. Bizzarro, A. et al. 2009. The complex interaction between APOE promoter and AD: An Italian case-control study. *Eur J Hum Genet* 17(7):938–945.
93. Mortimer, J.A. et al. 1991. Head trauma as a risk factor for Alzheimer's disease: A collaborative re-analysis of case-control studies. EURODEM Risk Factors Research Group. *Int J Epidemiol* 20(Suppl 2):S28–S35.
94. Mortimer, J.A. et al. 1985. Head injury as a risk factor for Alzheimer's disease. *Neurology* 35(2):264–267.
95. Graves, A.B. et al. 1990. The association between head trauma and Alzheimer's disease. *Am J Epidemiol* 131(3):491–501.
96. Mayeux, R. et al. 1993. Genetic susceptibility and head injury as risk factors for Alzheimer's disease among community-dwelling elderly persons and their first-degree relatives. *Ann Neurol* 33(5):494–501.
97. Roberts, G.W., Allsop, D., Bruton, C. 1990. The occult aftermath of boxing. *J Neurol Neurosurg Psychiatry* 53(5):373–378.
98. Roberts, G.W. 1988. Immunocytochemistry of neurofibrillary tangles in dementia pugilistica and Alzheimer's disease: Evidence for common genesis. *Lancet* 2(8626–8627):1456–1458.
99. Roberts, G.W. et al. 1991. Beta A4 amyloid protein deposition in brain after head trauma. *Lancet* 338(8780):1422–1423.
100. Roberts, G.W. et al. 1994. Beta amyloid protein deposition in the brain after severe head injury: Implications for the pathogenesis of Alzheimer's disease. *J Neurol Neurosurg Psychiatry* 57(4):419–425.
101. Nicoll, J.A., Roberts, G.W., Graham, D.I. 1995. Apolipoprotein E epsilon 4 allele is associated with deposition of amyloid beta-protein following head injury. *Nat Med* 1(2):135–137.
102. Ikonomovic, M.D. et al. 2004. Alzheimer's pathology in human temporal cortex surgically excised after severe brain injury. *Exp Neurol* 190(1):192–203.
103. Blumbergs, P.C. et al. 1994. Staining of amyloid precursor protein to study axonal damage in mild head injury. *Lancet* 344(8929):1055–1056.
104. McKenzie, K.J. et al. 1996. Is beta-APP a marker of axonal damage in short-surviving head injury? *Acta Neuropathol* 92(6):608–613.

105. Johnson, V.E., Stewart, W., Smith, D.H. 2012. Widespread tau and amyloid-beta pathology many years after a single traumatic brain injury in humans. *Brain Pathol* 22(2):142–149.
106. Kanekiyo, T., Xu, H., Bu, G. 2014. ApoE and Abeta in Alzheimer's disease: Accidental encounters or partners? *Neuron* 81(4):740–754.
107. Mayeux, R. et al. 1995. Synergistic effects of traumatic head injury and apolipoprotein-epsilon 4 in patients with Alzheimer's disease. *Neurology* 45(3 Pt 1):555–557.
108. Chiang, M.F., Chang, J.G., Hu, C.J. 2003. Association between apolipoprotein E genotype and outcome of traumatic brain injury. *Acta Neurochir (Wien)* 145(8):649–653; discussion 653–654.
109. Friedman, G. et al. 1999. Apolipoprotein E-epsilon4 genotype predicts a poor outcome in survivors of traumatic brain injury. *Neurology* 52(2):244–248.
110. Sorbi, S. et al. 1995. ApoE as a prognostic factor for post-traumatic coma. *Nat Med* 1(9):852.
111. Lichtman, S.W. et al. 2000. Apolipoprotein E and functional recovery from brain injury following postacute rehabilitation. *Neurology* 55(10):1536–1539.
112. Liaquat, I. et al. 2002. Effect of apolipoprotein E genotype on hematoma volume after trauma. *J Neurosurg* 96(1):90–96.
113. Crawford, F.C. et al. 2002. APOE genotype influences acquisition and recall following traumatic brain injury. *Neurology* 58(7):1115–1118.
114. Diaz-Arrastia, R. et al. 2003. Increased risk of late posttraumatic seizures associated with inheritance of APOE epsilon4 allele. *Arch Neurol* 60(6):818–822.
115. Alexander, S. et al. 2007. Apolipoprotein E4 allele presence and functional outcome after severe traumatic brain injury. *J Neurotrauma* 24(5):790–797.
116. Liberman, J.N. et al. 2002. Apolipoprotein E epsilon 4 and short-term recovery from predominantly mild brain injury. *Neurology* 58(7):1038–1044.
117. Millar, K. et al. 2003. Long term neuropsychological outcome after head injury: Relation to APOE genotype. *J Neurol Neurosurg Psychiatry* 74(8):1047–1052.
118. Sundstrom, A. et al. 2004. APOE influences on neuropsychological function after mild head injury: Within-person comparisons. *Neurology* 62(11):1963–1966.
119. Chamelian, L., Reis, M., Feinstein, A. 2004. Six-month recovery from mild to moderate traumatic brain injury: The role of APOE-epsilon4 allele. *Brain* 127(Pt 12):2621–2628.
120. Nathoo, N. et al. 2003. Genetic vulnerability following traumatic brain injury: The role of apolipoprotein E. *Mol Pathol* 56(3):132–136.
121. Isoniemi, H. et al. 2006. Outcome of traumatic brain injury after three decades—Relationship to ApoE genotype. *J Neurotrauma* 23(11):1600–1608.
122. Han, S.D. et al. 2007. Verbal paired-associate learning by APOE genotype in non-demented older adults: fMRI evidence of a right hemispheric compensatory response. *Neurobiol Aging* 28(2):238–247.
123. Han, S.D. et al. 2007. Apolipoprotein E and traumatic brain injury in a military population: Evidence of a neuropsychological compensatory mechanism? *J Neurol Neurosurg Psychiatry* 78(10):1103–1108.
124. Teasdale, G.M., Murray, G.D., Nicoll, J.A. 2005. The association between APOE epsilon4, age and outcome after head injury: A prospective cohort study. *Brain* 128(Pt 11):2556–2561.
125. Jennett, B., Bond, M. 1975. Assessment of outcome after severe brain damage. *Lancet* 1(7905):480–484.
126. Wilson, J.T., Pettigrew, L.E., Teasdale, G.M. 1998. Structured interviews for the Glasgow Outcome Scale and the extended Glasgow Outcome Scale: Guidelines for their use. *J Neurotrauma* 15(8):573–585.
127. Ponsford, J. et al. 2011. The association between apolipoprotein E and traumatic brain injury severity and functional outcome in a rehabilitation sample. *J Neurotrauma* 28(9):1683–1692.

128. Olivecrona, Z., Koskinen, L.O. 2012. The release of S-100B and NSE in severe traumatic head injury is associated with APOE epsilon4. *Acta Neurochir (Wien)* 154(4):675–680; discussion 680.
129. Bekris, L.M., Lutz, F., Yu, C.E. 2012. Functional analysis of APOE locus genetic variation implicates regional enhancers in the regulation of both TOMM40 and APOE. *J Hum Genet* 57(1):18–25.
130. Lambert, J.C. et al. 1998. Pronounced impact of Th1/E47cs mutation compared with -491 AT mutation on neural APOE gene expression and risk of developing Alzheimer's disease. *Hum Mol Genet* 7(9):1511–1516.
131. Lambert, J.C. et al. 1998. A new polymorphism in the APOE promoter associated with risk of developing Alzheimer's disease. *Hum Mol Genet* 7(3):533–540.
132. Laws, S.M. et al. 2002. Variation at the APOE -491 promoter locus is associated with altered brain levels of apolipoprotein E. *Mol Psychiatry* 7(8):886–890.
133. Lambert, J.C. et al. 2002. Contribution of APOE promoter polymorphisms to Alzheimer's disease risk. *Neurology* 59(1):59–66.
134. Casadei, V.M. et al. 1999. APOE-491 promoter polymorphism is a risk factor for late-onset Alzheimer's disease. *Neurology* 53(8):1888–1889.
135. Bray, N.J. et al. 2004. Allelic expression of APOE in human brain: Effects of epsilon status and promoter haplotypes. *Hum Mol Genet* 13(22):2885–2892.
136. Parker, G.R. et al. 2005. Apolipoprotein gene E4 allele promoter polymorphisms as risk factors for Alzheimer's disease. *Psychiatr Genet* 15(4):271–275.
137. Lambert, J.C. et al. 2000. Independent association of an APOE gene promoter polymorphism with increased risk of myocardial infarction and decreased APOE plasma concentrations—The ECTIM study. *Hum Mol Genet* 9(1):57–61.
138. Lendon, C.L. et al. 2003. Genetic variation of the APOE promoter and outcome after head injury. *Neurology* 61(5):683–685.
139. Jiang, Y. et al. 2007. Correlation between APOE -491AA promoter in epsilon4 carriers and clinical deterioration in early stage of traumatic brain injury. *J Neurotrauma* 24(12):1802–1810.
140. Thurman, D.J., Branche, C.M., Sniezek, J.E. 1998. The epidemiology of sports-related traumatic brain injuries in the United States: Recent developments. *J Head Trauma Rehabil* 13(2):1–8.
141. Guskiewicz, K.M. et al. 2000. Epidemiology of concussion in collegiate and high school football players. *Am J Sports Med* 28(5):643–650.
142. Langlois, J.A., Rutland-Brown, W., Wald, M.M. 2006. The epidemiology and impact of traumatic brain injury: A brief overview. *J Head Trauma Rehabil* 21(5):375–378.
143. McCrea, M. et al. 2004. Unreported concussion in high school football players: Implications for prevention. *Clin J Sport Med* 14(1):13–17.
144. Blennow, K., Hardy, J., Zetterberg, H. 2012. The neuropathology and neurobiology of traumatic brain injury. *Neuron* 76(5):886–899.
145. Lakhan, S.E., Kirchgessner, A. 2012. Chronic traumatic encephalopathy: The dangers of getting "dinged." *Springerplus* 1:2.
146. Jordan, B.D. et al. 1997. Apolipoprotein E epsilon4 associated with chronic traumatic brain injury in boxing. *JAMA* 278(2):136–140.
147. Kutner, K.C. et al. 2000. Lower cognitive performance of older football players possessing apolipoprotein E epsilon4. *Neurosurgery* 47(3):651–657; discussion 657–658.
148. McKee, A.C. et al. 2009. Chronic traumatic encephalopathy in athletes: Progressive tauopathy after repetitive head injury. *J Neuropathol Exp Neurol* 68(7):709–735.
149. Terrell, T.R. et al. 2008. APOE, APOE promoter, and Tau genotypes and risk for concussion in college athletes. *Clin J Sport Med* 18(1):10–17.
150. Kristman, V.L. et al. 2008. Does the apolipoprotein epsilon 4 allele predispose varsity athletes to concussion? A prospective cohort study. *Clin J Sport Med* 18(4):322–328.

151. Tierney, R.T. et al. 2010. Apolipoprotein E genotype and concussion in college athletes. *Clin J Sport Med* 20(6):464–468.
152. Terrell, T. et al. 2014. Multi center cohort study on association of genotypes with prospective sports concussion: Methods, lessons learned, and recommendations. *J Sports Med Phys Fitness.*
153. Giza, C.C. et al. 2013. Summary of evidence-based guideline update: Evaluation and management of concussion in sports: Report of the Guideline Development Subcommittee of the American Academy of Neurology. *Neurology* 80(24):2250–2257.
154. Witman, G.B. et al. 1976. Tubulin requires tau for growth onto microtubule initiating sites. *Proc Natl Acad Sci U S A* 73(11):4070–4074.
155. Binder, L.I., Frankfurter, A., Rebhun, L.I. 1985. The distribution of tau in the mammalian central nervous system. *J Cell Biol* 101(4):1371–1378.
156. Gu, Y., Oyama, F., Ihara, Y. 1996. Tau is widely expressed in rat tissues. *J Neurochem* 67(3):1235–1244.
157. LoPresti, P. et al. 1995. Functional implications for the microtubule-associated protein tau: Localization in oligodendrocytes. *Proc Natl Acad Sci U S A* 92(22):10369–10373.
158. Dehmelt, L., Halpain, S. 2005. The MAP2/Tau family of microtubule-associated proteins. *Genome Biol* 6(1):204.
159. Iqbal, K. et al. 1975. Chemical pathology of neurofibrils. Neurofibrillary tangles of Alzheimer's presenile-senile dementia. *J Histochem Cytochem* 23(7):563–569.
160. Grundke-Iqbal, I. et al. 1986. Abnormal phosphorylation of the microtubule-associated protein tau (tau) in Alzheimer cytoskeletal pathology. *Proc Natl Acad Sci U S A* 83(13):4913–4917.
161. Ihara, Y. et al. 1986. Phosphorylated tau protein is integrated into paired helical filaments in Alzheimer's disease. *J Biochem* 99(6):1807–1810.
162. Bancher, C. et al. 1989. Accumulation of abnormally phosphorylated tau precedes the formation of neurofibrillary tangles in Alzheimer's disease. *Brain Res* 477(1–2):90–99.
163. Smith, D.H., Johnson, V.E., Stewart, W. 2013. Chronic neuropathologies of single and repetitive TBI: Substrates of dementia? *Nat Rev Neurol* 9(4):211–221.
164. Pittman, A.M., Fung, H.C., de Silva, R. 2006. Untangling the tau gene association with neurodegenerative disorders. *Hum Mol Genet* 15(Spec No. 2):R188–R195.
165. Goedert, M. et al. 1989. Multiple isoforms of human microtubule-associated protein tau: Sequences and localization in neurofibrillary tangles of Alzheimer's disease. *Neuron* 3(4):519–526.
166. Hesse, C. et al. 2001. Transient increase in total tau but not phospho-tau in human cerebrospinal fluid after acute stroke. *Neurosci Lett* 297(3):187–190.
167. Hesse, C. et al. 2000. Cerebrospinal fluid markers for Alzheimer's disease evaluated after acute ischemic stroke. *J Alzheimers Dis* 2(3–4):199–206.
168. Ost, M. et al. 2006. Initial CSF total tau correlates with 1-year outcome in patients with traumatic brain injury. *Neurology* 67(9):1600–1604.
169. Franz, G. et al. 2003. Amyloid beta 1-42 and tau in cerebrospinal fluid after severe traumatic brain injury. *Neurology* 60(9):1457–1461.
170. Zetterberg, H. et al. 2006. Neurochemical aftermath of amateur boxing. *Arch Neurol* 63(9):1277–1280.
171. Neselius, S. et al. 2014. Neurological assessment and its relationship to CSF biomarkers in amateur boxers. *PLoS One* 9(6):e99870.
172. Shahim, P. et al. 2014. Blood biomarkers for brain injury in concussed professional ice hockey players. *JAMA Neurol* 71(6):684–692.
173. Buchanan, J.E., Phillis, J.W. 1993. The role of nitric oxide in the regulation of cerebral blood flow. *Brain Res* 610(2):248–255.
174. Asif, A.R. et al. 2009. T-786C polymorphism of the NOS-3 gene and the endothelial cell response to fluid shear stress—A proteome analysis. *J Proteome Res* 8(6):3161–3168.

175. Rossi, G.P. et al. 2003. The T-786C and Glu298Asp polymorphisms of the endothelial nitric oxide gene affect the forearm blood flow responses of Caucasian hypertensive patients. *J Am Coll Cardiol* 41(6):938–945.
176. Rossi, G.P. et al. 2003. The T-786C endothelial nitric oxide synthase genotype is a novel risk factor for coronary artery disease in Caucasian patients of the GENICA study. *J Am Coll Cardiol* 41(6):930–937.
177. Nasreen, S. et al. 2002. T-786C polymorphism in endothelial NO synthase gene affects cerebral circulation in smokers: Possible gene-environmental interaction. *Arterioscler Thromb Vasc Biol* 22(4):605–610.
178. Ko, N.U. et al. 2008. Endothelial nitric oxide synthase polymorphism (-786T->C) and increased risk of angiographic vasospasm after aneurysmal subarachnoid hemorrhage. *Stroke* 39(4):1103–1108.
179. Robertson, C.S. et al. 2011. Variants of the endothelial nitric oxide gene and cerebral blood flow after severe traumatic brain injury. *J Neurotrauma* 28(5):727–737.
180. Lipsky, R.H. et al. 2005. Association of COMT Val158Met genotype with executive functioning following traumatic brain injury. *J Neuropsychiatry Clin Neurosci* 17(4):465–471.
181. Grossman, M.H. et al. 1992. Electrophoretic analysis of low and high activity forms of catechol-O-methyltransferase in human erythrocytes. *Life Sci* 50(7):473–480.
182. Weinshilboum, R.M., Otterness, D.M., Szumlanski, C.L. 1999. Methylation pharmacogenetics: Catechol O-methyltransferase, thiopurine methyltransferase, and histamine N-methyltransferase. *Annu Rev Pharmacol Toxicol* 39:19–52.
183. Moghaddam, B. 2002. Stress activation of glutamate neurotransmission in the prefrontal cortex: Implications for dopamine-associated psychiatric disorders. *Biol Psychiatry* 51(10):775–787.
184. McAllister, T.W. et al. 2005. Effect of the dopamine D2 receptor T allele on response latency after mild traumatic brain injury. *Am J Psychiatry* 162(9):1749–1751.
185. Willmott, C. et al. 2014. COMT Val158Met and cognitive and functional outcomes after traumatic brain injury. *J Neurotrauma* 31(17):1507–1514.
186. Sayed-Tabatabaei, F.A. et al. 2006. ACE polymorphisms. *Circ Res* 98(9):1123–1133.
187. Rigat, B. et al. 1990. An insertion/deletion polymorphism in the angiotensin I-converting enzyme gene accounting for half the variance of serum enzyme levels. *J Clin Invest* 86(4):1343–1346.
188. Tiret, L. et al. 1992. Evidence, from combined segregation and linkage analysis, that a variant of the angiotensin I-converting enzyme (ACE) gene controls plasma ACE levels. *Am J Hum Genet* 51(1):197–205.
189. Ariza, M. et al. 2006. Influence of Angiotensin-converting enzyme polymorphism on neuropsychological subacute performance in moderate and severe traumatic brain injury. *J Neuropsychiatry Clin Neurosci* 18(1):39–44.
190. McAllister, A.K., Lo, D.C., Katz, L.C. 1995. Neurotrophins regulate dendritic growth in developing visual cortex. *Neuron* 15(4):791–803.
191. Lu, B. 2003. Pro-region of neurotrophins: Role in synaptic modulation. *Neuron* 39(5):735–738.
192. Zuccato, C. et al. 2001. Loss of huntingtin-mediated BDNF gene transcription in Huntington's disease. *Science* 293(5529):493–498.
193. Egan, M.F. et al. 2003. The BDNF val66met polymorphism affects activity-dependent secretion of BDNF and human memory and hippocampal function. *Cell* 112(2):257–269.
194. Lu, B., Gottschalk, W. 2000. Modulation of hippocampal synaptic transmission and plasticity by neurotrophins. *Prog Brain Res* 128:231–241.
195. Poo, M.M. 2001. Neurotrophins as synaptic modulators. *Nat Rev Neurosci* 2(1):24–32.

196. Krueger, F. et al. 2011. The role of the Met66 brain-derived neurotrophic factor allele in the recovery of executive functioning after combat-related traumatic brain injury. *J Neurosci* 31(2):598–606.
197. Lipsky, R.H., Marini, A.M. 2007. Brain-derived neurotrophic factor in neuronal survival and behavior-related plasticity. *Ann NY Acad Sci* 1122:130–143.
198. Chen, Z.Y. et al. 2004. Variant brain-derived neurotrophic factor (BDNF) (Met66) alters the intracellular trafficking and activity-dependent secretion of wild-type BDNF in neurosecretory cells and cortical neurons. *J Neurosci* 24(18):4401–4411.
199. Chen, Z.Y. et al. 2006. Genetic variant BDNF (Val66Met) polymorphism alters anxiety-related behavior. *Science* 314(5796):140–143.
200. Bath, K.G., Lee, F.S. 2006. Variant BDNF (Val66Met) impact on brain structure and function. *Cogn Affect Behav Neurosci* 6(1):79–85.
201. Hariri, A.R. et al. 2003. Brain-derived neurotrophic factor val66met polymorphism affects human memory-related hippocampal activity and predicts memory performance. *J Neurosci* 23(17):6690–6694.
202. Siironen, J. et al. 2007. The Met allele of the BDNF Val66Met polymorphism predicts poor outcome among survivors of aneurysmal subarachnoid hemorrhage. *Stroke* 38(10):2858–2860.
203. Kunugi, H. et al. 2001. A novel polymorphism of the brain-derived neurotrophic factor (BDNF) gene associated with late-onset Alzheimer's disease. *Mol Psychiatry* 6(1):83–86.
204. Riemenschneider, M. et al. 2002. A polymorphism of the brain-derived neurotrophic factor (BDNF) is associated with Alzheimer's disease in patients lacking the Apolipoprotein E epsilon4 allele. *Mol Psychiatry* 7(7):782–785.
205. Ventriglia, M. et al. 2002. Association between the BDNF 196 A/G polymorphism and sporadic Alzheimer's disease. *Mol Psychiatry* 7(2):136–137.
206. Neves-Pereira, M. et al. 2002. The brain-derived neurotrophic factor gene confers susceptibility to bipolar disorder: Evidence from a family-based association study. *Am J Hum Genet* 71(3):651–655.
207. Barbey, A.K. et al. 2014. Preservation of general intelligence following traumatic brain injury: Contributions of the Met66 brain-derived neurotrophic factor. *PLoS One* 9(2):e88733.
208. Bagnato, S. et al. 2012. Brain-derived neurotrophic factor (Val66Met) polymorphism does not influence recovery from a post-traumatic vegetative state: A blinded retrospective multi-centric study. *J Neurotrauma* 29(11):2050–2059.
209. Farkas, O., Povlishock, J.T. 2007. Cellular and subcellular change evoked by diffuse traumatic brain injury: A complex web of change extending far beyond focal damage. *Prog Brain Res* 161:43–59.
210. McAllister, T.W. et al. 2012. Polymorphisms in the brain-derived neurotrophic factor gene influence memory and processing speed one month after brain injury. *J Neurotrauma* 29(6):1111–1118.
211. Levine, A.J., Momand, J., Finlay, C.A. 1991. The p53 tumour suppressor gene. *Nature* 351(6326):453–456.
212. Lu, J. et al. 2000. Changes in apoptosis-related protein (p53, Bax, Bcl-2 and Fos) expression with DNA fragmentation in the central nervous system in rats after closed head injury. *Neurosci Lett* 290(2):89–92.
213. Chopp, M., Li, Y., Jiang, N. 1999. Increase in apoptosis and concomitant reduction of ischemic lesion volume and evidence for synaptogenesis after transient focal cerebral ischemia in rat treated with staurosporine. *Brain Res* 828(1–2):197–201.
214. Clark, R.S. et al. 1999. Increases in Bcl-2 and cleavage of caspase-1 and caspase-3 in human brain after head injury. *FASEB J* 13(8):813–821.
215. Ara, S. et al. 1990. Codon 72 polymorphism of the TP53 gene. *Nucleic Acids Res* 18(16):4961.

216. Dumont, P. et al. 2003. The codon 72 polymorphic variants of p53 have markedly different apoptotic potential. *Nat Genet* 33(3):357–365.
217. Harris, N. et al. 1986. Molecular basis for heterogeneity of the human p53 protein. *Mol Cell Biol* 6(12):4650–4656.
218. Martinez-Lucas, P. et al. 2005. Relationship between the Arg72Pro polymorphism of p53 and outcome for patients with traumatic brain injury. *Intensive Care Med* 31(9):1168–1173.
219. Ueda, K., Hayaishi, O. 1985. ADP-ribosylation. *Annu Rev Biochem* 54:73–100.
220. Virag, L., Szabo, C. 2002. The therapeutic potential of poly(ADP-ribose) polymerase inhibitors. *Pharmacol Rev* 54(3):375–429.
221. Sarnaik, A.A. et al. 2010. Influence of PARP-1 polymorphisms in patients after traumatic brain injury. *J Neurotrauma* 27(3):465–471.
222. Iwata, N. et al. 2000. Identification of the major Abeta1-42-degrading catabolic pathway in brain parenchyma: Suppression leads to biochemical and pathological deposition. *Nat Med* 6(2):143–150.
223. Madani, R. et al. 2006. Lack of neprilysin suffices to generate murine amyloid-like deposits in the brain and behavioral deficit in vivo. *J Neurosci Res* 84(8):1871–1878.
224. Johnson, V.E. et al. 2009. A neprilysin polymorphism and amyloid-beta plaques after traumatic brain injury. *J Neurotrauma* 26(8):1197–1202.
225. Iwata, N. et al. 2001. Metabolic regulation of brain Abeta by neprilysin. *Science* 292(5521):1550–1552.
226. Al-Chalabi, A. et al. 1999. Deletions of the heavy neurofilament subunit tail in amyotrophic lateral sclerosis. *Hum Mol Genet* 8(2):157–164.
227. McDevitt, J.K. et al. 2011. Neuronal structural protein polymorphism and concussion in college athletes. *Brain Inj* 25(11):1108–1113.
228. Trivedi, M.H. et al. 2006. Evaluation of outcomes with citalopram for depression using measurement-based care in STAR*D: Implications for clinical practice. *Am J Psychiatry* 163(1):28–40.
229. Heils, A. et al. 1996. Allelic variation of human serotonin transporter gene expression. *J Neurochem* 66(6):2621–2624.
230. Collier, D.A. et al. 1996. A novel functional polymorphism within the promoter of the serotonin transporter gene: Possible role in susceptibility to affective disorders. *Mol Psychiatry* 1(6):453–460.
231. Nakamura, M. et al. 2000. The human serotonin transporter gene linked polymorphism (5-HTTLPR) shows ten novel allelic variants. *Mol Psychiatry* 5(1):32–38.
232. Chan, F. et al. 2008. The serotonin transporter polymorphisms and major depression following traumatic brain injury. *Brain Inj* 22(6):471–479.
233. Gerber, D.J. et al. 2003. Evidence for association of schizophrenia with genetic variation in the 8p21.3 gene, PPP3CC, encoding the calcineurin gamma subunit. *Proc Natl Acad Sci U S A* 100(15):8993–8998.
234. Horiuchi, Y. et al. 2007. Support for association of the PPP3CC gene with schizophrenia. *Mol Psychiatry* 12(10):891–893.
235. Bales, J.W. et al. 2011. Association between the PPP3CC gene, coding for the calcineurin gamma catalytic subunit, and severity of traumatic brain injury in humans. *J Neurotrauma* 28:A-1-134.
236. Neville, M.J., Johnstone, E.C., Walton, R.T. 2004. Identification and characterization of ANKK1: A novel kinase gene closely linked to DRD2 on chromosome band 11q23.1. *Hum Mutat* 23(6):540–545.
237. McAllister, T.W. et al. 2008. Single nucleotide polymorphisms in ANKK1 and the dopamine D2 receptor gene affect cognitive outcome shortly after traumatic brain injury: A replication and extension study. *Brain Inj* 22(9):705–714.

238. Thompson, J. et al. 1997. D2 dopamine receptor gene (DRD2) Taql A polymorphism: Reduced dopamine D2 receptor binding in the human striatum associated with the A1 allele. *Pharmacogenetics* 7(6):479–484.
239. Papassotiropoulos, A. et al. 2006. Common Kibra alleles are associated with human memory performance. *Science* 314(5798):475–478.
240. Makuch, L. et al. 2011. Regulation of AMPA receptor function by the human memory-associated gene KIBRA. *Neuron* 71(6):1022–1029.
241. Schneider, A. et al. 2010. KIBRA: A new gateway to learning and memory? *Front Aging Neurosci* 2:4.
242. Schaper, K. et al. 2008. KIBRA gene variants are associated with episodic memory in healthy elderly. *Neurobiol Aging* 29(7):1123–1125.
243. Wagner, A.K. et al. 2012. Association of KIBRA rs17070145 polymorphism and episodic memory in individuals with severe TBI. *Brain Inj* 26(13–14):1658–1669.
244. McBride, H.M., Neuspiel, M., Wasiak, S. 2006. Mitochondria: More than just a powerhouse. *Curr Biol* 16(14):R551–R560.
245. Mironov, S.L. 2007. ADP regulates movements of mitochondria in neurons. *Biophys J* 92(8):2944–2952.
246. Burte, F. et al. 2015. Disturbed mitochondrial dynamics and neurodegenerative disorders. *Nat Rev Neurol* 11(1):11–24.
247. Dimauro, S., Davidzon, G. 2005. Mitochondrial DNA and disease. *Ann Med* 37(3):222–232.
248. Anderson, S. et al. 1981. Sequence and organization of the human mitochondrial genome. *Nature* 290(5806):457–465.
249. Wallace, D.C. 2005. A mitochondrial paradigm of metabolic and degenerative diseases, aging, and cancer: A dawn for evolutionary medicine. *Annu Rev Genet* 39:359–407.
250. Wallace, D.C. 1994. Mitochondrial DNA sequence variation in human evolution and disease. *Proc Natl Acad Sci U S A* 91(19):8739–8746.
251. Torroni, A. et al. 1996. Classification of European mtDNAs from an analysis of three European populations. *Genetics* 144(4):1835–1850.
252. Conley, Y.P. et al. 2014. Mitochondrial polymorphisms impact outcomes after severe traumatic brain injury. *J Neurotrauma* 31(1):34–41.
253. Lehman, E.J. et al. 2012. Neurodegenerative causes of death among retired National Football League players. *Neurology* 79(19):1970–1974.
254. Sullivan, P.G. et al. 1998. Traumatic brain injury alters synaptic homeostasis: Implications for impaired mitochondrial and transport function. *J Neurotrauma* 15(10):789–798.
255. Vink, R., Golding, E.M., Headrick, J.P. 1994. Bioenergetic analysis of oxidative metabolism following traumatic brain injury in rats. *J Neurotrauma* 11(3):265–274.
256. Gilmer, L.K. et al. 2010. Age-related mitochondrial changes after traumatic brain injury. *J Neurotrauma* 27(5):939–950.
257. Verweij, B.H. et al. 2000. Impaired cerebral mitochondrial function after traumatic brain injury in humans. *J Neurosurg* 93(5):815–820.
258. Bulstrode, H. et al. 2014. Mitochondrial DNA and traumatic brain injury. *Ann Neurol* 75(2):186–195.
259. Verweij, B.H. et al. 2000. Improvement in mitochondrial dysfunction as a new surrogate efficiency measure for preclinical trials: Dose-response and time-window profiles for administration of the calcium channel blocker Ziconotide in experimental brain injury. *J Neurosurg* 93(5):829–834.
260. Singh, I.N. et al. 2006. Time course of post-traumatic mitochondrial oxidative damage and dysfunction in a mouse model of focal traumatic brain injury: Implications for neuroprotective therapy. *J Cereb Blood Flow Metab* 26(11):1407–1418.

261. Pyle, A. et al. 2005. Mitochondrial DNA haplogroup cluster UKJT reduces the risk of PD. *Ann Neurol* 57(4):564–567.
262. Wallace, D.C. 2013. A mitochondrial bioenergetic etiology of disease. *J Clin Invest* 123(4):1405–1412.
263. Khusnutdinova, E. et al. 2008. A mitochondrial etiology of neurodegenerative diseases: Evidence from Parkinson's disease. *Ann N Y Acad Sci* 1147:1–20.
264. Gibson, G.E., Sheu, K.F., Blass, J.P. 1998. Abnormalities of mitochondrial enzymes in Alzheimer disease. *J Neural Transm* 105(8–9):855–870.
265. Shoffner, J.M. 1997. Oxidative phosphorylation defects and Alzheimer's disease. *Neurogenetics* 1(1):13–19.
266. Miyata, M., Smith, J.D. 1996. Apolipoprotein E allele-specific antioxidant activity and effects on cytotoxicity by oxidative insults and beta-amyloid peptides. *Nat Genet* 14(1):55–61.
267. Carrieri, G. et al. 2001. Mitochondrial DNA haplogroups and APOE4 allele are non-independent variables in sporadic Alzheimer's disease. *Hum Genet* 108(3):194–198.
268. Lifshitz, J. et al. 2004. Mitochondrial damage and dysfunction in traumatic brain injury. *Mitochondrion* 4(5–6):705–713.
269. Tavazzi, B. et al. 2007. Temporal window of metabolic brain vulnerability to concussions: Oxidative and nitrosative stresses—Part II. *Neurosurgery* 61(2):390–395; discussion 395–396.
270. van der Walt, J.M. et al. 2003. Mitochondrial polymorphisms significantly reduce the risk of Parkinson disease. *Am J Hum Gene* 72(4):804–811.
271. Giza, C.C., Mink, R.B., Madikians, A. 2007. Pediatric traumatic brain injury: Not just little adults. *Curr Opin Crit Care* 13(2):143–152.
272. Pinto, P.S. et al. 2012. The unique features of traumatic brain injury in children. Review of the characteristics of the pediatric skull and brain, mechanisms of trauma, patterns of injury, complications and their imaging findings—Part 1. *J Neuroimaging* 22(2):e1–e17.
273. Pinto, P.S. et al. 2012. The unique features of traumatic brain injury in children. Review of the characteristics of the pediatric skull and brain, mechanisms of trauma, patterns of injury, complications, and their imaging findings—Part 2. *J Neuroimaging* 22(2):e18–e41.
274. Kurowski, B.G. et al. 2013. Association of parent ratings of executive function with global- and setting-specific behavioral impairment after adolescent traumatic brain injury. *Arch Phys Med Rehabil* 94(3):543–550.
275. Quinn, T.J. et al. 2004. There is no evidence of an association in children and teenagers between the apolipoprotein E epsilon4 allele and post-traumatic brain swelling. *Neuropathol Appl Neurobiol* 30(6):569–575.
276. Brichtová, E., Kozák, L. 2008. Apolipoprotein E genotype and traumatic brain injury in children—Association with neurological outcome. *Childs Nerv Syst* 24(3):349–356.
277. Moran, L.M. et al. 2009. Apolipoprotein E4 as a predictor of outcomes in pediatric mild traumatic brain injury. *J Neurotrauma* 26(9):1489–1495.
278. Lo, T.Y. et al. 2009. Modulating effect of apolipoprotein E polymorphisms on secondary brain insult and outcome after childhood brain trauma. *Childs Nerv Syst* 25(1):47–54.
279. Blackman, J.A., Worley, G., Strittmatter, W.J. 2005. Apolipoprotein E and brain injury: Implications for children. *Dev Med Child Neurol* 47(1):64–70.
280. Kurowski, B., Martin, L.J., Wade, S.L. 2012. Genetics and outcomes after traumatic brain injury (TBI): What do we know about pediatric TBI? *J Pediatr Rehabil Med* 5(3):217–231.
281. Reuter-Rice, K., Eads, J.K., Berndt, S.B., Bennett, E. 2015. State of the science of pediatric traumatic brain injury: Biomarkers and gene association studies. *Rev Nursing Res* 33, Chapter 6.

10 Modeling Post-Traumatic Epilepsy for Therapy Development

Giulia Curia, Clifford L. Eastman, John W. Miller, and Raimondo D'Ambrosio

CONTENTS

INTRODUCTION

Epilepsy is the most prevalent serious neurological disorder, afflicting almost 1% of the population worldwide.[1,2] It is a heterogeneous disorder, comprising numerous syndromes with a wide range of etiologies,[3] that is defined by the manifestation of chronic spontaneous recurrent seizures (CSRSs). An epileptic seizure, in turn, is defined by the International League Against Epilepsy (ILAE) as "transient occurrence of signs and/or symptoms due to abnormal excessive or synchronous neuronal activity in the brain."[4,5]

Post-traumatic epilepsy (PTE) arises after mechanical damage to the brain and is diagnosed when spontaneous seizures are observed at least a week after brain injury. PTE is the most prevalent acquired epilepsy in young adults and accounts for 5% of epilepsies overall.[6,7] There are currently no cures for PTE and no means to prevent the disorder in those at risk.[8] Available treatments of PTE are symptomatic, and at least 40% of patients have seizures that cannot be controlled with any of the available drugs.[9,10] This dire situation requires rethinking the development and use of animal models for the development of therapies for PTE.[11,12] In this chapter we will introduce the problem and discuss several topics crucial for modeling PTE for therapy development.

What to Model

Human PTE results from a traumatic brain injury (TBI), which may result from traffic accident (~47%), fall (~30%), recreation or sport accident (14%), and assault (7%).[13] The complexity and diversity of PTE parallels that of TBI, which includes both focal and diffuse components due to acceleration/deceleration injuries, blunt force closed head injuries with or without skull fracture, contusions, and penetrating brain injuries. Acceleration/deceleration injuries induce diffuse damage, while predominantly focal injuries induce surface contusion and laceration that can be accompanied by skull fracture or hematoma. Thus, TBI may affect different epileptogenic areas of the human brain.[14] Although subcortical structures may be involved when the impact is severe, surface focal injuries are often restricted to the vicinity of the mechanical impact and preferentially affect the frontal and temporal lobes, possibly due to the particular shape of the human skull.[15] Thus, in view of the well documented high epileptogenicity of brain contusions in both humans and experimental animals,[16–19] it is not surprising that human PTE presents most often as frontal and temporal lobe epilepsies, with or without secondary generalization.[20–28] In addition, the frontal and temporal lobes are intrinsically highly epileptogenic brain regions, and experiments with frontal, parietal, and occipital FPI demonstrate the much higher epileptogenicity of the rat frontal neocortex.[15]

The probability that patients will develop PTE varies greatly depending on the type and severity of TBI, and on such risk factors as intracerebral hemorrhage, subdural hematoma, dural penetration, depressed skull fracture, multilobar injury, retained metal fragments, early (<1 week) post-traumatic seizures, and loss of consciousness or post-traumatic amnesia >24 hours.[7,13,29–32]

The rate of epileptogenesis after head injury is also very variable.[33–36] Patients may become epileptic within days after the injury or many years later.[30,31,37–39] The reason for this large variability in epileptogenesis is not well understood. While experimental work clearly shows that the severity of the injury directly affects the speed of epileptogenesis, other physiological and genetic factors must also contribute.[15]

The frequency of seizure in human PTE also varies widely and has not been well characterized, because head injury patients do not undergo prospective electroencephalography (EEG) monitoring and PTE patients do not routinely undergo invasive electrocorticographic (ECoG) monitoring. Also, PTE patients may present with localized or spreading focal seizures, termed simple partial seizures (SPSs), complex partial seizures (CPSs), or secondarily generalized seizures (SGSs) in the older ILAE classification.[40] Focal seizures can have very diverse ictal behaviors, ranging from undetectable to tonic clonic convulsions, depending on the location of the epileptic focus and the pattern and extent of the seizures' spread.[14,24,33,35,41–46] When these seizures have subtle or no detectable (subclinical) behavioral correlates, PTE may be missed or misdiagnosed.[28,47–53] Human SPSs, especially those producing subjective symptoms without motor output, are often unreported or misdiagnosed.[54,55] CPSs, which can manifest with a wide variety of convulsive or nonconvulsive behaviors and subjective experiences,[56–58] can easily be misdiagnosed or confused with psychiatric disorders.[48–50] In addition, seizures with limited electrical spread that are clearly evident in ECoG often elude detection by standard scalp EEG.[49,50,55,59–61] When patients are invasively monitored, the differences between electrographically detected seizures and seizures detected by other means can differ greatly.[62] Thus, epidemiological data based on self-reported seizures underestimate nonconvulsive SPSs and CPSs,[52] and seizures that are not severe enough to lead the patient to seek the advice of a physician are never counted.

In summary, there is no canonical PTE syndrome to target, and efforts to model PTE should focus on mimicking epileptogenic head injuries that closely correspond to head injuries commonly resulting in PTE in humans and adapting them to specific experimental aims.

Animal Models of Post-Traumatic Epilepsy

Existing animal models of PTE include: (1) models developed to investigate particular mechanisms that have been postulated to contribute to PTE, (2) models based on chemoconvulsant or electrical stimulation-induced status epilepticus (SE), and (3) etiologically realistic models in which epilepsy is induced after an experimental head injury. The first class includes the cortical undercut model[63] and acute intraparenchymal injection of $FeCl_3$.[64] The cortical undercut reproduces trauma-induced hemorrhage and partial deafferentiation,[63] while the $FeCl_3$ model mimics features of intracerebral hemorrhage. However, while several aspects of TBI, such as intraparenchymal hemorrhage, ischemia, and hypoxia, can induce epilepsy in experimental models,[65–69] neither clinical nor experimental studies have determined which, if any, of theses factors is actually responsible for genesis and progression of human PTE. Thus, until the roles of the hypothesized mechanisms are better understood, mechanism-based models may not be prudent choices for therapy development. Since human PTE often involves the temporal lobes, existing SE-based models of temporal lobe epilepsy have been suggested for the study of PTE. In these widely used models, SE is induced by systemic or intracerebral administration of pilocarpine or kainate, or electrical stimulation of the amygdala or hippocampus. Although convenient to implement and extensively studied, SE-based models have limitations for PTE translational research: (1) the epileptogenic mechanisms involved are not well known and may not necessarily be clinically relevant for human PTE,[70–72] and (2) it is generally difficult to determine whether experimental preventive treatments for epilepsy modify the disease or merely the initiating insult.[73,74]

Etiologically realistic PTE models reproduce a mechanical injury that induces TBI in humans. While animal models have been developed to reproduce both the focal and the diffuse injuries found in human TBI, predominantly diffuse injuries have not yet been studied with a focus on epilepsy and will not be discussed further. There are three main experimental models that reproduce predominantly focal contusive brain injuries. The weight-drop model,[75,76] which has not yet been demonstrated to induce epilepsy, the controlled cortical impact (CCI) model, and fluid percussion injury (FPI) model. In the CCI model, a rigid impactor delivers a calibrated injury to the intact dura.[77–79] Several laboratories have now reported the CCI-induced development of CSRSs in rodents with behavioral correlates that range from behavioral arrest to tonic–clonic convulsions.[80–83] However, the reported incidence (20%–50%) and frequency (0.2–0.3 seizures/day) of CSRSs is low, the latency to seizure onset is long (>75 days),[83] and the progression of PTE has yet to be well characterized. In the FPI model, a calibrated fluid pulse is delivered to the intact dura through an open craniotomy. FPI reproduces most of the pathophysiological features of human contusive closed head injury,[84] and it has become one of the best-studied rodent PTE models.[85–89] However, FPI is a complex injury, and the character of FPI-induced PTE varies greatly depending on the details of the injury protocol.[15] A recent study[90] in which FPI was administered with a picospritzer, rather than a conventional FPI apparatus,[84] failed to induce PTE despite injury parameters (pulse duration and pressure) claimed to be nominally similar to those used routinely to induce PTE with conventional FPI. Fortunately, this sensitivity of the FPI model to the details of implementation also provides opportunities for optimization, and we have found that FPI can be effectively optimized for therapy development.

MODEL OPTIMIZATION FOR THERAPY DEVELOPMENT

An ideal model for therapy development would provide accurate prediction of the effect of investigational interventions in the modeled human population with high throughput and minimal cost (Table 10.1). Given our poor understanding of the mechanisms mediating human post-traumatic epileptogenesis, it is expected that the best prediction would be obtained with models, such as the FPI and CCI models, in which the epileptogenic stimulus is very similar (i.e., realistic) to the corresponding one in humans and, thus, likely to engage the same mechanisms. High throughput requires rapid and reliable epileptogenesis and syndrome assessment, and costs are determined by the number of subjects required to reach statistically valid conclusions and the length of time required for epileptogenesis, treatment, and the collection and analysis of video/ECoG data. We have invested considerable effort in optimizing the throughput and minimizing the number of subjects required for translational studies using the FPI model. Since human data clearly show that the incidence and severity of PTE vary with the character and severity of the initiating injury, the parameters of experimental injury can be manipulated to increase the incidence and frequency of CSRSs without threat to model validity.

The well-studied lateral FPI model induced using the classic cylinder-piston device was adapted by McIntosh for TBI/neuroprotection studies in the rat.[84] It induces PTE, but epileptogenesis is reported to be slow (several months), seizure frequencies are low (<1/day), and fewer than 50% of injured animals are reported to develop epilepsy by 7–12 months postinjury.[87,88] In such a model, experiments to assess antiseizure or antiepileptogenic interventions would require 6–9 months with large numbers of animals recorded for long periods of time and would be prohibitively expensive (Figure 10.1). However, we have found that the rate of epileptogenesis and the incidence

TABLE 10.1
Desirable Features of a PTE Animal Model for Therapy Development

Etiologically realistic brain injury matching a specific and common population of head injury patients[a]
Electroclinical syndrome that is similar to a human PTE[a]
Electroclinical syndrome that is difficult to treat[b] and has some of the risk factors for drug resistance epilepsy in humans:

- Frequent seizures
- Induced before adulthood
- Associated with dual pathology
- Focal without secondary generalization

High incidence of epilepsy[c]
High seizure frequency[c]
Absence of extreme seizure clustering[c]
Low intersubject variability in seizure frequency[c]
Rapid epileptogenesis[c]
A localized epileptic focus[d]

[a] Therapy development requires models that maximize the chance that the treatment will be effective in humans. To this end, the initiating etiology and the induced PTE should closely match the human population modeled.
[b] The indicated risk factors for drug resistance are from French.[91]
[c] Model features that increase throughput and decrease variability and cost are indicated.
[d] Knowing the location of the focus allows for targeted focal treatments.

and frequency of seizures all vary depending upon the location of injury and the amplitude and duration of the pressure pulse.[15] The overall incidence and frequency of seizures was lower after milder (2.0 atm) than more severe (3.4 atm) trauma, and a rostral injury delivered over the frontal neocortex resulted in more rapid and reliable epileptogenesis than injury delivered at more caudal sites, as in the classic FPI model. The parameters of FPI are implementation-dependent and should be interpreted very cautiously. While a 3.4-atm-pressure wave delivered through a rostral parasagittal 3 mm craniotomy with our methods reliably induces PTE, similar pressure and pulse durations are most likely not suitable for a different device with different mechanical properties or different methods. For example, it is known that the mere presence of an air bubble between the dura of the animal and the pressure transducer can dramatically affect the severity of the injury. Effective parameters must be separately identified for any novel implementation and validated. In addition, the use of volatile anesthetics and mechanical ventilation permits better control of the model: (1) it controls post-traumatic apnea, which induces epileptogenic hypoxia; and (2) it reduces post-traumatic mortality (~10%), relative to the classic lateral FPI (30%–35%), which permits severely injured animals to enter the pool of PTE subjects. The FPI model we have developed (Figure 10.2) results in over 90% incidence of CSRSs with a mean frequency of about 2 seizures/hour, by 4–5 weeks postinjury. This permits completion of seizure-control and seizure-prevention studies

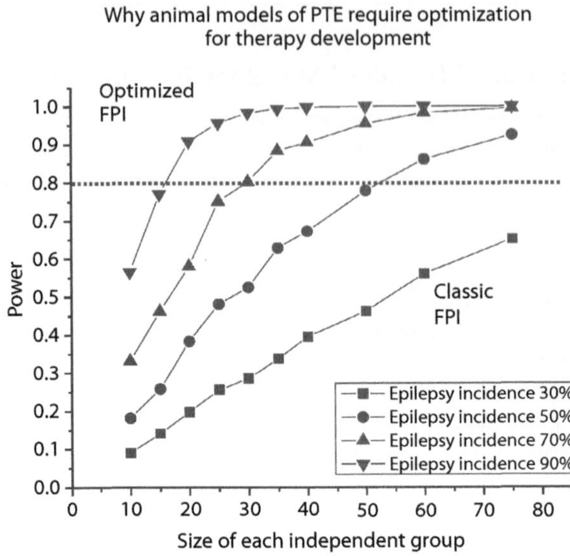

FIGURE 10.1 The statistical power to detect the prevention of epileptic seizures after head injury critically depends on the animal model's seizure incidence. This Monte Carlo simulation of antiepileptogenesis studies is based on 10^4 simulated experiments per data point. In each experiment, samples were randomly drawn with replacement from a control population and from a treated population with a simulated 50% decrease in the incidence of seizure. The untreated control and treated samples were compared using a one-tailed Fisher exact test (fisher.test; Revolution R Enterprise 5.0.1). Statistical power was computed as the proportion of comparisons that yielded probabilities less than 0.05. These data demonstrate that practical experiments to identify antiepileptogenic effects cannot be conducted using models that induce epilepsy in 30%–50% of subjects. To provide 80% power to detect a 50% decrease in the incidence of seizures, such experiments would require preparation, housing, monitoring, and analysis of ECoG data of more than 100 subjects (2 groups of 50 subjects). In contrast, 80% power to detect a 50% decrease in the incidence of seizures can be obtained with 2 groups of just 16 animals when experimental injury induces detectable seizures in 90% of subjects.

in less than 6 weeks, and manageable numbers of animals recorded for less than 72 hours per experimental condition provide adequate statistical power to detect robust and clinically interesting treatment effects.[92–94]

This FPI model, which is adapted for epilepsy therapy development (etdFPI), offers additional advantages that could facilitate advances in target identification and therapy development. First, the focus that reliably develops in the perilesional cortex is readily accessible for experimental manipulation. This feature was critical for the demonstration that mild focal cooling of the epileptic focus prevents the development of post-traumatic epileptic seizures.[94] Also, the PTE syndrome that develops after etdFPI prominently includes nonconvulsive neocortical seizures,[94,95] which have been largely overlooked in most preclinical studies and which have proven difficult to treat with available antiseizure drugs (ASDs).[92,93] Like human PTE, FPI-induced

FPI implementation	Classic for TBI/ neuroprotection studies	Adapted for epilepsy therapy development
Purpose	Massive injury with focal and diffuse components	Focal neocortical injury, with minimal diffuse component
Rat size at time of injury	~400 g	~120 g
Craniotomy size	Large (5 mm dia)	Small (3 mm dia)
Craniotomy location	Parietal parasagittal	Frontal parasagittal
Pressure pulse duration	Long (22–25 ms) to maximize impulse and diffuse injury	Short (8–10 ms) to minimize impulse and diffuse injury
Anesthesia	Injected anesthetics: no resuscitation by ventilation	Ventilated anesthetics: severly injured animals are rescued by ventilation
Post-traumatic apnea	Uncontrolled (10 s–1.5 min)	Controlled (10 s) to decrease injury variability due to variable global hypoxia
Mortality	High (~35%)	Low (~10%)
Reported incidence of chronic seizures	Low (30%–50% at >6 months postinjury)	High (80%–100% at >4 weeks postinjury)
Reported incidence of focal neocortical seizures	N.A.	High
Reported incidence of generalized seizures	Low	No tonic–clonic convulsions
Reported frequency of chronic seizures	Low (up to 1.2 per day; mean ~0.3/day)	High (up to 35 per hour; mean ~2/hour)
Reported duration of chronic seizures	Range 6 seconds to 676 seconds*	Range 1 second to 615 seconds**

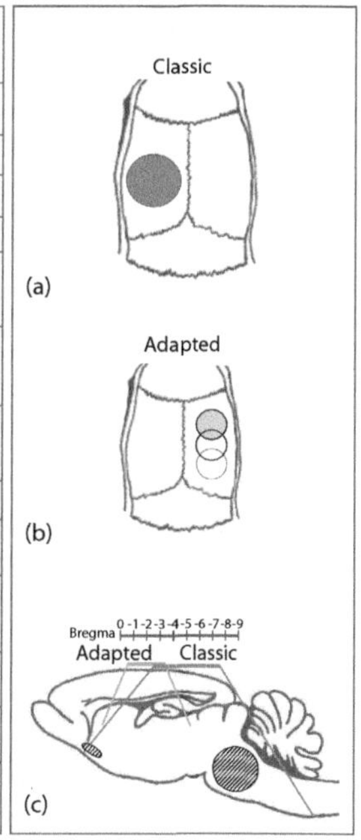

FIGURE 10.2 Optimization of the FPI model to achieve the high seizure frequency and incidence needed for therapy development. In the classic FPI model developed for neuroprotection studies, a long fluid pulse is delivered through a large craniotomy centered on the parietal cortex, resulting in both focal and diffuse injury, the latter of which can affect brainstem areas involved in seizure generalization. Injury results in high mortality, particularly when post-traumatic apnea is uncontrolled. Epileptogenesis is reported to be slow, and up to 50% of injured animals are reported to develop convulsive seizures with frequencies up to 1.2/day. To produce a more focal injury with less impact on the brainstem, we deliver a short fluid pulse through a smaller craniotomy, and we have found that a more rostral injury induces rapid epileptogenesis and frequent nonconvulsive neocortical seizures in over 90% of subjects. With post-traumatic apnea limited to 10 s, mortality is just 10%. Right panel: (a) Top view of the skull of a 400 g rat showing the placement of the craniotomy (filled circle) in the classic FPI model. (b) Top view of the skull of a 120 g rat showing the placement of the craniotomy (filled circle) in the adapted FPI model. Open circles show sites of more caudal craniotomies, at which FPI resulted in slower epileptogenesis and less frequent seizures. (c) Schematic of a sagittal section through a rat brain illustrating placement and extent of injury after classic (black) and adapted (gray) FPI. Thatched areas depict the area tempestas and the brainstem regions involved in secondary generalization of seizures. *These studies defined seizures as events longer than 5 seconds.[87,88] **These studies are based on the ILAE definition of epileptic seizure.[95–97]

PTE includes frontal lobe seizures that are associated with behavioral arrest, with or without mild automatisms, and which are electrically and behaviorally consistent with those seen in humans.[47,48,50,57,98–108] The rare parieto-occipital neocortical seizures induced by etdFPI were never associated with detectable behavioral change but might have been associated with subjective visual phenomena that cannot be assessed in animals.[15] Frontal and limbic seizures with loss of body posture became progressively more common in rats over time[86,95] and had semiology reminiscent of human CPSs. Many of these seizures consist of stereotyped crouching, sometimes with facial or body automatisms.

Although the latency of PTE after etdFPI is shorter, and the incidence of PTE and the frequency of post-traumatic epileptic seizures is much higher than the *averages* reported for humans, the differences are more apparent than real. First, human data show a very large variability in both speed of epileptogenesis and seizure frequency. Many head injury patients develop epilepsy within 1–2 weeks, just as most animals receiving etdFPI. Second, all estimates of the incidence of epilepsy after head injury are based on aggregated outcomes of diverse injuries that may differ widely in epileptogenicity depending on the character and severity of the initiating injury. Even when stratified by a measure of severity, head injury patients differ widely in the type, number, and location(s) of their injuries, and in the presence of risk factors for PTE. Patients are also typically heterogeneous in age, gender, genetic background, and comorbid conditions that may affect their susceptibility to PTE.[104,105] In contrast, the etdFPI protocol is standardized and empirically designed to induce rapid and robust epileptogenesis, and it is administered to rats of identical age and gender that vary much less in genetic background and antecedent comorbidities. In addition, the experimental animals are much more closely and comprehensively monitored than PTE patient cohorts. They are continuously monitored, prospectively, using sensitive video/ECoG, while human head injury patients are not typically monitored prospectively and seizures are most often monitored by self-report, which underestimates the incidence and frequency of post-traumatic CSRSs (see Section "What to Model"). Thus, the model cannot be compared to average epidemiological findings, and its features and performance can be optimized so long as the selected injury protocol induces epilepsy and realistically reproduces the pathophysiology of a clinically important human epileptogenic injury.

ASSESSING EPILEPSY TREATMENTS WITH FLUID PERCUSSION INJURY-INDUCED PTE

The etdFPI-PTE model has been deployed in studies of both antiseizure and antiepileptogenic treatments.[92–94] Results from these studies suggest that (1) the pharmacological sensitivity of etdFPI-induced epilepsy differs from other models of evoked and spontaneous seizures, (2) etdFPI-induced PTE may have value as a model of pharmacoresistant epilepsy, and (3) the etdFPI-PTE model is capable of identifying antiepileptogenic interventions.

The limited data available on the pharmacological properties of the SE- and etdFPI-based epilepsy models suggest important differences. Several conventional ASDs (carbamazepine, phenobarbital, phenytoin, and topiramate) have been demonstrated to be effective against convulsive seizures in the chronic phase after SE,[106–108] as has the investigational ASD, carisbamate (CRS).[109] In contrast, etdFPI-induced CSRSs were poorly controlled by carbamazepine, CRS, and valproate (in a majority of animals), but well controlled by halothane and, in a subset of animals, by valproate.[92,93] The effect of valproate on CSRSs in responders developed progressively over a week of exposure and outlasted the exposure period, suggesting a mechanism distinct from the one mediating its antiseizure action in acute models.[92] Thus, etdFPI-induced PTE appears less responsive to conventionally developed ASDs than older models. Indeed, the etdFPI-PTE model incorporates most known risk factors for pharmacoresistant epilepsy (Table 10.1), and its poor response to conventional ASDs suggests it may have value as a model of pharmacoresistant epilepsy. This suggestion is supported by studies of CRS—a promising investigational ASD that was withdrawn from the regulatory process for epilepsy treatment after disappointing performance as an adjunctive treatment in pharmacoresistant patients. In stark to its poor performance against etdFPI-induced PTE,[93] CRS has demonstrated broad efficacy in evoked seizure models and was reported to perform better than topiramate against convulsive seizures in a SE-based epilepsy model.[109,110] However, these disparate results may both be consistent with the clinical trial experience. Halford et al.[111] reported that CRS, at the highest dose tested, resulted in 17%, 12%, and 29% reductions (over placebo levels) in the frequency of SPSs, CPSs, and SGSs, respectively. Thus, while the very modest effects of CRS on partial seizures without generalization are consistent with its effect on nonconvulsive seizures in the etdFPI-PTE model, it may have had a larger effect on SGSs, which were relatively rare in the study population. Together, these data support efforts to determine the scope and mechanisms of treatment resistance in etdFPI rats.

The etdFPI-PTE model has also demonstrated the capacity to distinguish effective and ineffective antiepileptogenic interventions. In a blind, randomized and adequately powered study, a 2-week treatment with CRS begun 15 min after injury, had no significant effect on the development of CSRSs.[93] Although the study was not powered to detect less than an 85% reduction in seizure frequency compared to controls, the nearly identical seizure frequencies in control and treated rats at both 4 and 12 weeks after injury suggests a much weaker effect. In contrast, 5 weeks of mild focal cooling of the perilesional neocortex demonstrated prevention of about 99% of CSRSs by the end of treatment in a similar blind and randomized study.[94] This effect persisted for the duration of the study, which extended over 10 weeks after treatment termination.

TECHNICAL AND METHODOLOGICAL ISSUES IN PTE THERAPY DEVELOPMENT

Definition of Epilepsy and of Seizures

In addition to optimizing an experimental injury to induce a high incidence and frequency of epileptic seizures, the experimentally induced seizures must also be reliably and sensitively detected for an epilepsy model to be practical and useful. Overlooking seizures poses the same threat to a cost-efficient study as low seizure frequency and incidence. Seizure detection is straightforward when attention is restricted to Racine scale 4–5 convulsive seizures that roughly correspond to secondarily generalized convulsive seizures in humans. However, detection of the nonconvulsive seizures that predominate in etiologically realistic epilepsy models, and roughly correspond to human SPSs and CPSs without secondary generalization, is more challenging. Both electrographic and video monitoring are essential for the characterization of these seizures, which may have subtle behavioral correlates, and criteria must be established for their reliable and reproducible identification. To ensure that the same phenomena are being studied in animals and humans, these criteria should closely resemble those that guide seizure identification in human video/EEG monitoring. Currently, there is no consensus on how to identify and classify seizures in animals, and definitions of experimental seizures are varied, arbitrary, and often far more restrictive than those used for humans. The definition of human clinical epileptic seizures accepted by the ILAE[4,5] is equally applicable to seizures in other species, and its use in preclinical studies should help align clinical and preclinical study endpoints and improve bench to bedside translation.[96,97]

As in human PTE, the epileptic seizures observed in etiologically realistic PTE models come in a wide range of durations and clinical manifestations, and there is no scientific reason to systematically exclude any class of events from preclinical investigation. However, many investigators have required that seizures have minimum durations or specific types of behavioral correlates, which may vary widely among laboratories. Such criteria were introduced because they (1) simplify the distinction between ictal and interictal events, (2) simplify exclusion of both the typically short age-dependent idiopathic seizures that are common in many rodent strains, and brief focal seizures that may be induced by cortical damage inflicted unintentionally during electrode implantation, and (3) reduce false positives in both visual and algorithm-based automated seizure detection. However, application of such arbitrary criteria can exclude large numbers of seizures, introducing large errors in seizure frequency and epilepsy incidence. The duration of clinical seizures in etdFPI rats ranges widely from 1 second to several minutes (See figure 4I in Ref. 94; Figure 10.2), and shows a clear progression over time postinjury regardless of the location of injury.[15] We have previously shown how altering the definition of epileptic seizures can change estimates of seizure incidence and frequency, as well as the latent period (figure 9 in Ref. 95). In addition, an investigational ASD that increased the frequency of seizure but shortened their duration or prevented generalization could be misidentified as an effective antiepileptogenic or antiseizure agent. More important, such criteria can result in exclusion of specific classes of seizure from the study. For example, an investigation that systematically overlooked nonconvulsive seizures could seriously overestimate the clinical effectiveness of a treatment for patients who, like those enrolled in most phase III clinical trials, suffer predominantly from SPSs and CPSs with infrequent or absent secondary convulsions.

Detection of Epileptic Seizures

The clinical manifestations of a seizure depend on its site of origin and the area and extent of its spread. Thus, both animals and humans with PTE may experience seizures with electrographic changes on EEG that lack detectable behavioral accompaniment. Human SPSs and CPSs can be associated not only with convulsive events but also with nonmotor symptoms, including somatosensory, special sensory, autonomic, or affective components.[54] We have observed that brief focal seizures with very limited spread in human motor cortex areas were associated with a mild stereotyped ictal behavior (eye blinking or muscle twitching), while similar events located outside of the motor cortex were not associated with visible behavioral output.[95] Similarly, in FPI rat, we observed that frontal cortex focal seizures are associated to behavioral change (e.g., motor arrest with or without facial automatisms), while focal seizures in the occipital cortex are never associated with obvious behavioral changes.[15] Therefore, seizures are most accurately diagnosed on the basis of both behavioral and electrographic data.

Simultaneous epidural ECoG and scalp EEG recordings in etdFPI-injured rats have shown that nonspreading focal cortical seizures were not detected by scalp EEG, which also failed to detect ~62% of the spreading seizures that were detected by the epidural electrodes. These data suggest that scalp EEG only picked up the activity when it had spread to a sufficiently large volume of cortex.[95] Similar results have been obtained in humans: brief epileptic events that were detected by invasive ECoG were missed by scalp EEG,[59,60,95] and invasive monitoring often shows many times the frequency of seizures that regular scalp EEG detects.[53,62] Thus, assessment of ictal activity is better performed with ECoG.

In animals, as in humans, multiple ECoG electrodes are required to better appreciate focal seizures with limited spread. When the location of an injury-induced epileptic focus is predictable (as it is after etdFPI), strategic montage and analysis can be used to maximize seizure detection using a reasonable number of electrodes. In the rat, we routinely monitor invasively with a five-electrode montage that directly samples the perilesional cortex that reliably generates an epileptic focus after FPI. This guarantees the detection of most seizures. Because many of the focal neocortical seizures that develop after etdFPI are brief and fail to spread, a montage including a perilesional electrode is critical for accurate assessment of FPI-induced PTE[95] and also to test treatments.[94]

Long-term continuous recording is often considered important to prove experimental subjects to be seizure-free. However, clinical seizures can be both focal and local, and ictal behaviors can be mild or internal. Thus, because it is not possible to comprehensively map the electrical activity of a whole brain with an indefinitely large numbers of electrodes, no subject can ever be proven to be seizure free, even with continuous observation using the most sensitive methods. Indeed, seizure freedom can only be proven for generalized convulsive seizures whose diffuse spread ensures both motor output and electrical detection. In general, the experimenter can only report the number of seizures detected in a specified interval using a particular recording montage, and the effects of interventions must be assessed in terms of treatment-related changes in the numbers and/or duration of seizures. While prolonged recording may be needed for models that exhibit low seizure frequency, treatment effects on frequent absence seizures in WAG/Rij or GAERS rats are routinely assessed on the basis of a few hours of recording (e.g., Refs. 112 and 113). Nonparametric power analyses show that clinically interesting reductions of 50% or more in seizure frequency after etdFPI can be detected (80% power; $\alpha = 0.05$) in groups of eight rats recorded for just 48–72 hours before and during treatment.[92,93]

Thus, the use of sensitive ECoG recording in an optimized acquired epilepsy model permits assessment of antiseizure effects in small groups of subjects recorded for manageable periods of time.

Avoiding Experimental Bias

Experimental outcomes can be biased by comparisons of nonequivalent experimental groups or by biases in investigator judgment. Individual experimental animals may differ in age, gender, weight, genetics, fine details of experimental treatment, responses to injury and treatment, and in innumerable other factors that may or may not affect experimental outcomes. Many factors that are known to affect outcomes (e.g., strain, age, and gender) can be explicitly controlled. However, outcomes may also be affected by factors that are unknown, not measured or not controllable. For example, despite the carefully controlled injury parameters, the duration of post-traumatic apnea, and the age and gender of rats (Figure 10.2), seizure frequency and rate of progression of epileptogenesis still vary greatly among animals.[15] The possibility that experimental outcomes could be biased by unbalanced allocation of subjects to experimental groups can be minimized by *formal* randomization procedures. Such procedures also protect against another form of bias. Unconstrained and seemingly random subject selection could easily introduce bias if the investigator selected the easiest animal to grab (low activity level) or the first in a cohort to exhibit seizure. Formal randomization measures have recently been recommended for improving the translation of preclinical data in both the epilepsy and TBI fields.[114,115]

Experimenter bias in data interpretation can be minimized either by blind manual assessment of experimental data or by using an exclusively algorithmic approach. A recent study determined that when human video EEG monitoring data were analyzed by clinicians blinded to all other clinical data, they might arrive at different conclusions about whether seizures are epileptic in origin.[116] For epilepsy studies, blinding entails manual review of video and/or ECoG data by expert personnel who are unaware of the identity of the data. In our previous preclinical studies on carisbamate[93] and cooling[94] all ECoG data were analyzed manually by trained investigators blinded to the treatment group and collection date of data files. Such blinding is not difficult to do and only requires a third person, not involved in the analysis, to rename the data files with a coding software.

Unbiased analysis of ECoG data from etiologically realistic models of PTE could be greatly expedited by improvements in automated ECoG analysis. Currently, expert manual analysis is the only approach to detect all seizures in ECoG, including the focal nonconvulsive ones, with high sensitivity and specificity. This approach, while reliable, requires extensive training and takes qualified investigators about 1.5 hours to analyze 24 hours of continuous rat ECoG. An ECoG-based antiepileptogenesis study to examine a single compound with 20 animals recorded for a week would require one person about 6 weeks just for the primary analysis of the data. Systems capable of detecting both convulsive and nonconvulsive seizures with acceptable sensitivity and specificity to fully automate the process do not yet exist. Significant effort should be put into their development to improve therapy development.

Using the Right Control Groups

Sham Injury and Electrode Damage

Studies of FPI-induced PTE often compare data obtained from injured animals to data obtained from sham-injured animals given a craniotomy but no FPI. While such shams provide a useful comparison group, they may not be regarded as normal. Cole and colleagues have demonstrated that the sham surgery itself caused brain injury distinct from the impact injury.[117] In their study, rats that received anesthesia only were compared to rats with craniotomies carefully performed either by manual trephine or by electric drill. Despite their care to avoid damage to the dura, both drill and trephine caused morphological, behavioral, and biochemical changes consistent with TBI. The demonstration that the craniotomy itself, without the fluid percussion, can cause significant proinflammatory, morphological, and behavioral damage, suggests that the routine use of this type of sham controls may lead to confusion in interpretation of conventional experimental brain injury models. Therefore, great care should be taken in selecting the appropriate models and controls for comparing post-traumatic changes in the brain and for evaluating the efficacy of potential treatments.

We routinely take several precautions to ensure that epileptiform ECoG events we examine are induced by FPI and not by compression or frictional heating of the cortex during drilling.[94,95] We cool the skull and drill bit with room-temperature sterile saline during drilling and take care to not deform the skull or the dura with the drill bit. The depth of epidural electrodes is carefully adjusted to avoid brain compression. In addition, glial fibrillary acidic protein (GFAP) immunostaining can be performed in FPI and sham-injured animals to assess astroglial reactivity beneath the epidural electrodes, and animals exhibiting foci of GFAP immunoreactivity (see figure 1 in Ref. 95) can be excluded from study. While the relationship between the subtle neuropathology induced by suboptimal electrode implantation and the generation of epileptic discharges has not yet been fully characterized, we have never observed focal discharges in control animals that were not associated with foci of glial reactivity beneath the electrode at which they were detected. Similar concerns apply even more urgently to the use of depth electrodes, which penetrate the meninges and brain parenchyma, and locally disrupt the blood–brain barrier. Chronic depth-electrode implantation likely induces inflammation[118] and prolonged implantation of depth electrodes has been shown to facilitate kindling, worsen evoked seizures, and increase the severity of electrically induced SE.[119–121] Thus, depth recordings may have the potential to significantly affect the outcome of acquired epilepsy studies.

Impact of Rat Strain, Age, and Gender on Idiopathic Seizures

In studies of acquired epilepsies, it is desirable that control animals be seizure-free and that seizures in injured animals be solely attributable to the experimental injury. Idiopathic epilepsy is common in rats, however, with incidence and onset depending on strain, age, and gender.[122–125] Since acquired epilepsy studies frequently involve recording from animals well beyond 6 months of age, many investigators have reported brief epileptiform events in control animals. These potentially confounding events can be minimized by using younger animals of a strain with low incidence and/or late onset of idiopathic seizures. Willoughby and MacKenzie[122] reported Sprague Dawley rats to have the lowest rate of idiopathic discharges among eight strains examined. More recently, Pearce et al.[125] reported the appearance of idiopathic seizures in female Sprague Dawley rats and in males older than 6 months, but not in younger male Sprague Dawley rats. This is consistent with our experience. Most of our studies utilized male Sprague Dawley rats younger than 6 months of age, and seizures were not observed in undamaged controls.[86] Idiopathic seizures were observed in undamaged control rats in one study in which recordings were obtained from rats older than 6 months. These events, typically 2–10 seconds long and bilateral at onset, were characterized by a sharp-wave pattern and were readily distinguished from PTE because they were significantly larger in amplitude in the parietal-occipital cortex (see figure 1D in Ref. 86). This posterior dominance was similar to that previously reported in idiopathic seizures recorded from Wistar rats using a similar bilateral fronto-occipital multielectrode montage.[122,126] We observed these events in about 33% of the rats that were recorded through 7–8 months of age and they represented just 3.6% of the cortical discharges recorded from FPI rats at 27–28 weeks postinjury. Thus, young male Sprague Dawley rats can provide seizure-free controls for acquired epilepsy studies. In acquired epilepsy studies that must extend past the age at which idiopathic seizures appear, such seizures can be identified and excluded from analysis if a montage is employed that permits spatial characterization of the cortical discharge.[86] A recent study[90] claimed to detect SWDs in two-thirds of young male Sprague Dawley rats from Harlan Laboratories, but it is unclear whether this colony of rats is genetically more prone to SWDs because the detection was based on an unvalidated automated algorithm with unknown specificity. Since behaviorally salient genetic differences have been documented in Sprague Dawley rats from different colonies,[127] it may be prudent to use animals from colonies known to have a low incidence of idiopathic epilepsy. Pearce et al.[125] used rats from Charles River in Wilmington, Massachusetts, and our rats have predominantly been shipped from Charles River in Hollister, California.

CONCLUSIONS

Limitations and diminishing returns of conventional methods of epilepsy drug development and the absence of treatments to prevent acquired epileptogenesis have spurred interest in epilepsy models featuring etiologically realistic epilepsy syndromes. An ideal model would engage mechanisms that contribute to human epilepsy, exhibit pathology consistent with human epilepsy, produce CSRSs, and neurobehavioral impairment consistent with human epilepsy, and yet be practical to use in translational studies. Models for translational research must deliver adequate statistical power to detect treatment effects, and throughput issues are an additional concern. The etdFPI-PTE model has benefited from optimizations that make it suitable for therapy development. Indeed, it has already contributed to novel insights on investigational treatments. This demonstrates the feasibility and value of adapting etiologically realistic models to the development of novel treatments for the corresponding human epileptic syndromes.

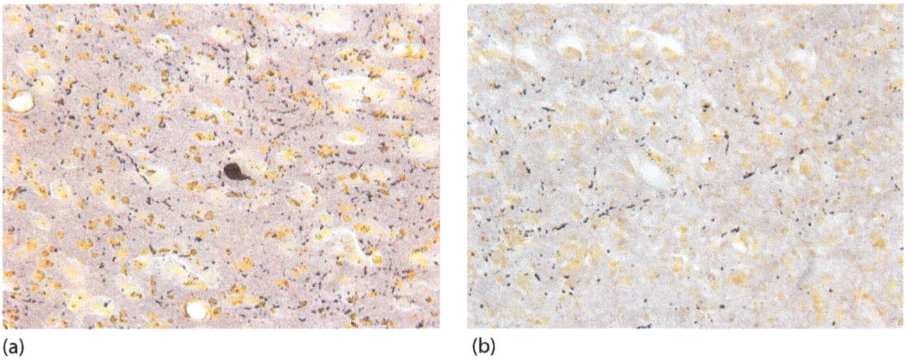

FIGURE 3.1 (a) Silver-stained tissue of juvenile rat brain tissue at 40× following impact-acceleration TBI revealing axonal retraction bulb at center. (b) Silver-stained tissue of juvenile rat brain tissue at 40× following impact-acceleration TBI revealing linear degenerating axonal tracts. (Image courtesy of Vassilis Koliatsos, MD/Leyan Xu, MD.)

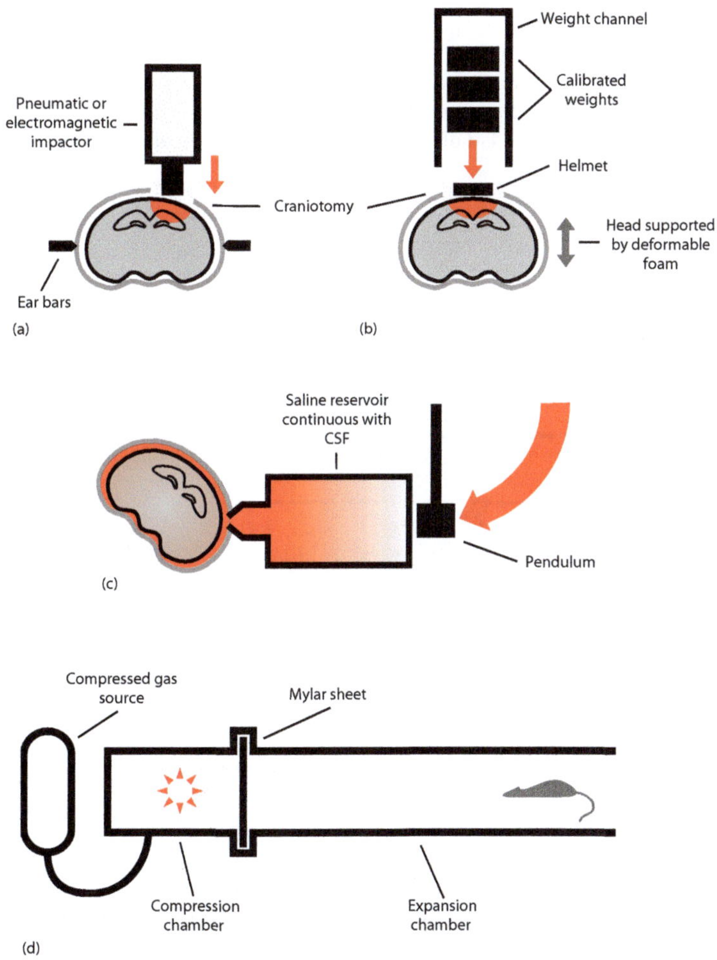

FIGURE 3.2 Major experimental models of traumatic brain injury used in simulating DAI in lissencephalic animals: (a) controlled cortical impact, (b) impact acceleration, (c) fluid percussion, (d) compressed gas blast tube.

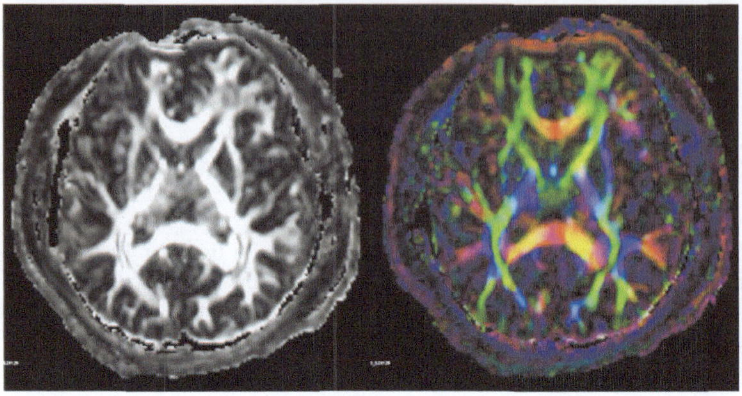

FIGURE 3.5 Diffusion tensor imaging images of relative fractional anisotropy in an 8-year-old patient after TBI from being struck by a motor vehicle. The left grayscale image reflects FA magnitude and the right image incorporates direction data, with green labeling identifying anterior-posterior tensor alignment, red identifying the left-right direction, and blue corresponding to the cephalad-caudad direction. Note decreased FA magnitude in the R frontal white matter.

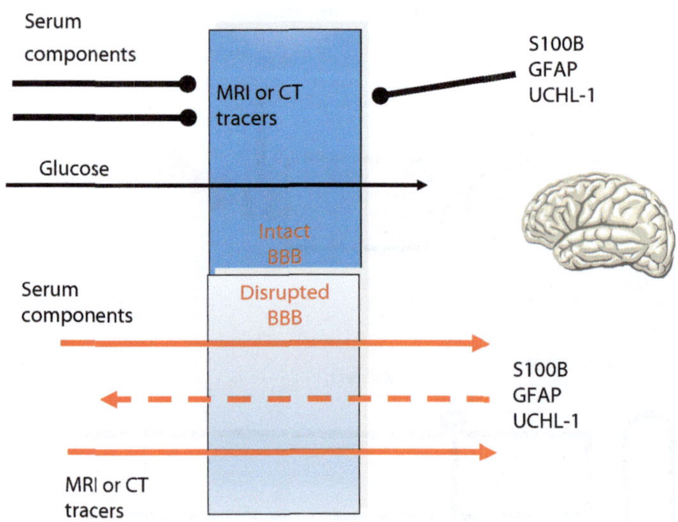

FIGURE 12.1 Under conditions of an intact BBB, serum extravasation of brain markers and leakage of radiological tracers is prevented by the tight junction. When the BBB is disrupted, both radiologic and protein markers cross the endothelial layer.

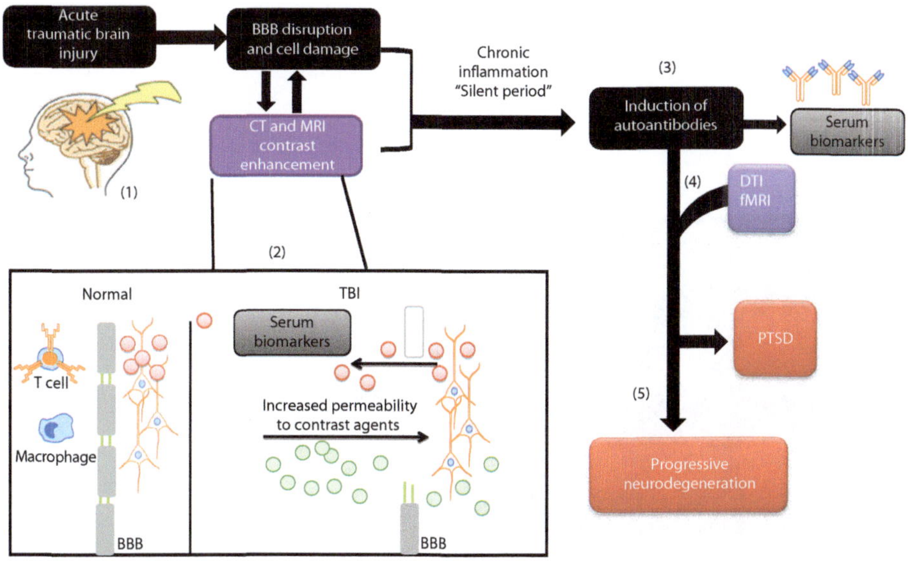

FIGURE 12.2 Schematic of research hypothesis: (1) Single or repetitive TBI triggers CNS damage, mechanical BBB disruption (BBBD) and inflammation. (2) Release of pro-inflammatory cytokines including IL-1β, TNF-α, and IL-6 induces activation of peripheral immune cells and activation of resident microglia. Increased BBB permeability due to cytokine release, apoptosis, and mechanical BBBD facilitates extravasation of leukocytes into the parenchyma. (3) Factors such as genetic background (e.g., HLA), injury severity, history of infection, or excessive immune activation (e.g., due to poor clearance of debris) predispose an individual to chronic inflammation and autoantibody induction. These autoantibodies may serve as biomarkers for increased Alzheimer's disease risk. (4) CNS autoreactive antibodies circulate in the periphery and contribute to the development of Alzheimer's disease in conjunction with other risk factors that increase inflammation, BBB permeability, etc. (5) Identification of biomarker autoantibodies paves the way for early interventions including immunosuppressive therapy or cognitive therapy.

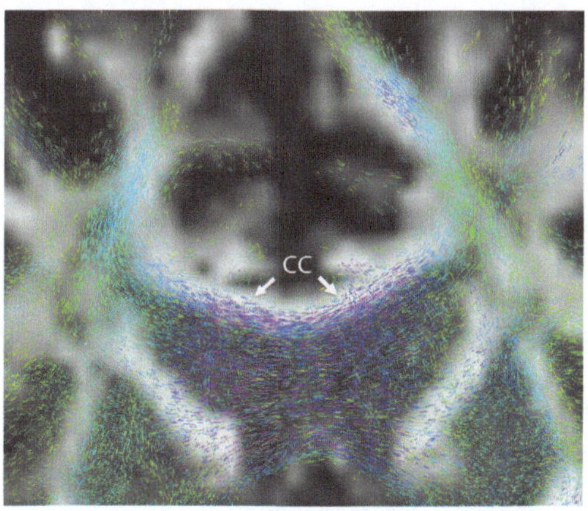

FIGURE 13.2 Diffusion tensor imaging makes it possible to trace the preferred diffusion direction of water molecules. On a coronal view, the higher directionality of diffusion can be seen within highly organized brain structures such as the corpus callosum (CC). The blue color corresponds to high FA values, which translates to high directionality of diffusion.

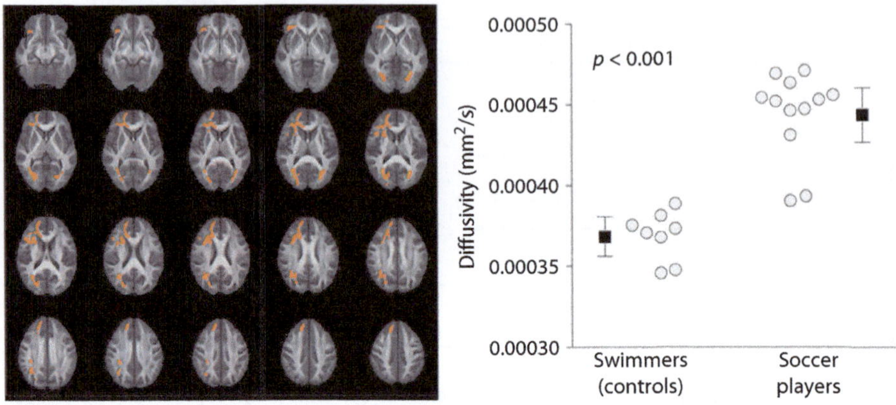

FIGURE 13.3 Changes in radial diffusivity in asymptomatic soccer players frequently exposed to subconcussive impacts to their heads. Soccer players showed increased radial diffusivity when compared to age- and gender-matched swimmers using tract-based spatial statistics (TBSS, FSL 4.03, FMRIB, UK). (From Koerte I.K. et al., *JAMA* 308:1859–1861, 2012. With permission.)

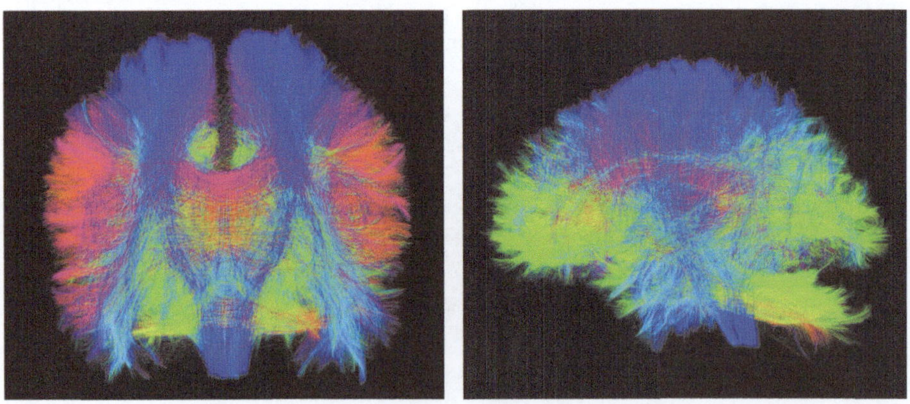

FIGURE 13.4 Multitensor algorithms allow for a more precise differentiation between the end of a fiber tract (FA <0.15) and crossing fiber tracts by calculating multiple tensors within each voxel. A full brain tractography, color coded for the main orientation along the xyz-axis (z-axis, blue; x-axis, red; y-axis, green) is shown.

ACKNOWLEDGMENTS

This work was supported by the Italian Ministry of Education, University and Research ("Rientro Cervelli" 17DZE8RZEA to GC), by the National Institutes of Health (NS076570 to RD), and by Citizens United for Research in Epilepsy (Prevention of Acquired Epilepsies Award to RD).

The authors declare that the research was conducted in the absence of any commercial or financial relationships that could be construed as potential conflicts of interest.

REFERENCES

1. Sander, J.W. 2003. The epidemiology of epilepsy revisited. *Curr Opin Neurol* 16(2):165–170.
2. Duncan, J.S. et al. 2006. Adult epilepsy. *Lancet* 367(9516):1087–1100.
3. Bhalla, D. et al. 2011. Etiologies of epilepsy: A comprehensive review. *Expert Rev Neurother* 11(6):861–876.
4. Fisher, R.S. et al. 2005. Epileptic seizures and epilepsy: Definitions proposed by the International League Against Epilepsy (ILAE) and the International Bureau for Epilepsy (IBE). *Epilepsia* 46(4):470–472.
5. Fisher, R.S. et al. 2014. ILAE Official Report: A practical clinical definition of epilepsy. *Epilepsia* 55(4):475–482.
6. Hauser, W.A., Annegers, J.F., Kurland, L.T. 1991. Prevalence of epilepsy in Rochester, Minnesota: 1940–1980. *Epilepsia* 32(4):429–445.
7. Garga, N., Lowenstein, D.H. 2006. Posttraumatic epilepsy: A major problem in desperate need of major advances. *Epilepsy Curr* 6(1):1–5.
8. Temkin, N.R. 2009. Preventing and treating posttraumatic seizures: The human experience. *Epilepsia* 50(Suppl 2):10–13.
9. Semah, F. et al. 1998. Is the underlying cause of epilepsy a major prognostic factor for recurrence? *Neurology* 51(5):1256–1262.
10. Lu, Y., Yu, W., Wang, X. 2009. Efficacy of topiramate in adult patients with symptomatic epilepsy: An open-label, long-term, retrospective observation. *CNS Drugs* 23(4):351–359.
11. Löscher, W. et al. 2013. New avenues for anti-epileptic drug discovery and development. *Nat Rev Drug Discov* 12(10):757–776.
12. Weaver, D.F., Pohlmann-Eden, B. 2013. Pharmacoresistant epilepsy: Unmet needs in solving the puzzle(s). *Epilepsia* 54(Suppl 2):80–85.
13. Annegers, J.F., Coan, S.P. 2000. The risk of epilepsy after traumatic brain injury. *Seizure* 9:453–457.
14. Becker, D.P. et al. 1979. Head injuries—Panel 3. *Arch Neurol* 36:750–758.
15. Curia, G. et al. 2011. Impact of injury location and severity on posttraumatic epilepsy in the rat: Role of frontal neocortex. *Cereb Cortex* 21:1574–1592.
16. Messori, A. et al. 2005. Predicting posttraumatic epilepsy with MRI: Prospective longitudinal morphologic study in adults. *Epilepsia* 46(9):1472–1481.
17. Stewart, T.H. et al. 2010. Chronic dysfunction of astrocytic inwardly rectifying K+ channels specific to the neocortical epileptic focus after fluid percussion injury in the rat. *J Neurophysiol* 104(6):3345–3360.
18. Friedman, A., Heinemann, U. 2012. Role of blood-brain barrier dysfunction in epileptogenesis. In *Jasper's Basic Mechanisms of the Epilepsies*, 4th ed., eds. J.L. Noebels, M. Avoli, M.A. Rogawski, R.W. Olsen, A.V. Delgado-Escueta, 353–361. Bethesda, MD: National Center for Biotechnology Information.
19. Vezzani, A., Friedman, A., Dingledine, R.J. 2013. The role of inflammation in epileptogenesis. *Neuropharmacology* 69:16–24.

20. Jennett, W.B. 1961. Late epilepsy after blunt head injuries: A clinical study based on 282 cases of traumatic epilepsy. *Ann R Coll Surg Engl* 29:370–384.
21. Ludwig, B., Marsan, C.A., Buren, J.V. 1975. Cerebral seizures of probable orbitofrontal origin. *Epilepsia* 16:141–158.
22. Williamson, P.D. 1992. Frontal lobe seizures. Problems of diagnosis and classification. *Adv Neurol* 57:289–309.
23. Marks, D.A. et al. 1995. Seizure localization and pathology following head injury in patients with uncontrolled epilepsy. *Neurology* 45:2051–2057.
24. Wohlrab, G., Schmitt, B., Boltshauser, E. 1997. Benign focal epileptiform discharges in children after severe head trauma: Prognostic value and clinical course. *Epilepsia* 38(3):275–278.
25. Diaz-Arrastia, R. et al. 2000. Neurophysiologic and neuroradiologic features of intractable epilepsy after traumatic brain injury in adults. *Arch Neurol* 57:1611–1616.
26. Diaz-Arrastia, R. et al. 2009. Post-traumatic epilepsy: The endophenotypes of a human model of epileptogenesis. *Epilepsia* 50(S2):14–20.
27. Hudak, A.M. et al. 2004. Evaluation of seizure-like episodes in survivors of moderate and severe traumatic brain injury. *J Head Trauma Rehabil* 19(4):290–295.
28. Gupta, P. et al. 2014. Subtypes of post-traumatic epilepsy: Clinical, electrophysiologic, and imaging features. *J Neurotrauma* 31(16):1439–1443.
29. Frey, L.C. 2003. Epidemiology of posttraumatic epilepsy: A critical review. *Epilepsia* 44(S10):11–17.
30. Agrawal, A. et al. 2006. Post-traumatic epilepsy: An overview. *Clin Neurol Neurosurg* 108:433–439.
31. Eftekhar, B. et al. 2009. Prognostic factors in the persistence of posttraumatic epilepsy after penetrating head injuries sustained in war. *J Neurosurg* 110:319–326.
32. Yeh, C.C. et al. 2013. Risk of epilepsy after traumatic brain injury: A retrospective population-based cohort study. *J Neurol Neurosurg Psychiatry* 84(4):441–445.
33. Caveness, W.F. 1963. Onset and cessation of fits following craniocerebral trauma. *J Neurosurg* 20:570–583.
34. Annegers, J.F. et al. 1998. A population-based study of seizures after traumatic brain injuries. *N Engl J Med* 338:20–24.
35. Jabbari, B. et al. 1986. Clinical and radiological correlates of EEG in the late phase of head injury: A study of 515 Vietnam veterans. *Electroenceph Clin Neurphysiol* 64:285–293.
36. Christensen, J. et al. 2009. Long-term risk of epilepsy after traumatic brain injury in children and young adults: A population-based cohort study. *Lancet* 373:1105–1110.
37. Walker, A.E. 1967. The significance of posttraumatic epilepsy. *Conn Med* 31(2):109–114.
38. Caveness, W.F. et al. 1979. The nature of post-traumatic epilepsy. *J Neurosurg* 50(5):545–553.
39. Aarabi, B. et al. 2000. Prognostic factors in the occurrence of posttraumatic epilepsy after penetrating head injury suffered during military service. *Neurosurg Focus* 8(1):1–6.
40. Muro, V.M., Connolly, M.B. 2014. Classifying epileptic seizures and epilepsies. In *Epilepsy*, eds. J.W. Miller, H.P. Goodkin, 10–14. Hoboken, NJ: Wiley Blackwell.
41. Caveness, W.F., Liss, H.R. 1961. Incidence of post-traumatic epilepsy. *Epilepsia* 2:123–129.
42. Salazar, A.M. et al. 1985. Epilepsy after penetrating head injury. I. Clinical correlates: A report of the Vietnam head injury study. *Neurology* 35:1406–1414.
43. Salazar, A.M., Schwab, K., Grafman, J.H. 1995. Penetrating injuries in the Vietnam war: Traumatic unconsciousness, epilepsy, and psychosocial outcome. *Neurosurg Clin North Am* 6(4):715–726.
44. Pohlmann-Eden, B., Bruckmeir, J. 1997. Predictors and dynamics of posttraumatic epilepsy. *Acta Neurol Scand* 95(5):257–262.

45. Weiss, G.H., Caveness, W.F. 1972. Prognostic factors in the persistence of posttraumatic epilepsy. *J Neurosurg* 37:164–169.
46. Walker, A.E., Blumer, D. 1989. The fate of World War II veterans with posttraumatic seizures. *Arch Neurol* 46:23–26.
47. Geier, S. et al. 1977. The seizures of frontal lobe epilepsy. A study of clinical manifestations. *Neurology* 27:951–958.
48. Williamson, P.D. et al. 1985. Complex partial seizures of frontal lobe origin. *Ann Neurol* 18:497–504.
49. Williamson, P.D. et al. 1985. Complex partial status epilepticus: A depth-electrode study. *Ann Neurol* 18:647–654.
50. Williamson, P.D., Spencer, S.S. 1986. Clinical and EEG features of complex partial seizures of extratemporal origin. *Epilepsia* 27(S2):S46–S63.
51. Williamson, P.D. et al. 1992. Occipital lobe epilepsy: Clinical characteristics, seizure spread patterns, and results of surgery. *Ann Neurol* 31:3–13.
52. Kerling, F. et al. 2006. When do patients forget their seizures? An electroclinical study. *Epilepsy Behav* 9(2):281–285.
53. Dichter, M.A. 2009. Posttraumatic epilepsy: The challenge of translating discoveries in the laboratory to pathways to a cure. *Epilepsia* 50(S2):41–45.
54. Devinsky, O. et al. 1988. Clinical and electrographic features of simple partial seizures. *Neurology* 38(9):1347–1352.
55. Devinsky, O. et al. 1989. Electroencephalographic studies of simple partial seizures with subdural electrode recordings. *Neurology* 39:527–533.
56. Penfield, W., Jasper, H.H. 1954. *Epilepsy and the Functional Anatomy of the Human Brain*. Boston: Little, Brown.
57. Swartz, B.E. 1992. Pseudo-absence seizures. A frontal lobe phenomenon. *J Epilepsy* 5:80–93.
58. Leppik, I.E. 1997. The classification of seizures. In *Contemporary Diagnosis and Management of the Patient with Epilepsy*, ed. I.E. Leppik, 8–15. Newton, PA: Handbooks in Health Care Co.
59. Cukiert, A. et al. 2001. Results of surgery in patients with refractory extratemporal epilepsy with normal or nonlocalizing magnetic resonance findings investigated with subdural grids. *Epilepsia* 42(7):889–894.
60. Binnie, C.D., Stefan, H. 2003. The EEG in epilepsy. In *Clinical Neurophysiology. Vol 2: EEG, Paediatric Neurophysiology, Special Techniques and Applications*, eds. C. Binnie, R. Cooper, F. Mauguière, J. Osselton, P. Prior, B. Tedman, 268–303. Amsterdam, the Netherlands: Elsevier.
61. Worrell, G.A. et al. 2008. High-frequency oscillations in human temporal lobe: Simultaneous microwire and clinical macroelectrode recordings. *Brain* 131(Pt 4): 928–937.
62. Cook, M.J. et al. 2013. Prediction of seizure likelihood with a long-term, implanted seizure advisory system in patients with drug-resistant epilepsy: A first-in-man study. *Lancet Neurol* 12(6):563–571.
63. Graber, K.D., Prince, D.A. 2006. Chronic partial cortical isolation. In *Models of Seizures and Epilepsy*, eds. A. Pitkanen, P. Schwartzkroin, S. Moshe, 477–493. San Diego, CA: Elsevier Academic Press.
64. Ueda, Y., Triggs, W.J., Willmore, L.J. 2006. Head trauma: Hemorrhage-iron deposition. In *Models of Seizures and Epilepsy*, eds. A. Pitkanen, P.A. Schwartzkroin, S.L. Moshé, 495–500. San Diego, CA: Elsevier Academic Press.
65. Lee, K.R. et al. 1997. Seizures induced by intracerebral injection of thrombin: A model of intracerebral hemorrhage. *J Neurosurg* 87:73–78.
66. Seiffert, E. et al. 2004. Lasting blood-brain barrier disruption induces epileptic focus in the rat somatosensory cortex. *J Neurosci* 24(36):7829–7836.

67. Epsztein, J. et al. 2008. Late-onset epileptogenesis and seizures genesis: Lessons from models of cerebral ischemia. *Neuroscientist* 14(1):78–90.
68. Maggio, N. et al. 2008. Thrombin induces long-term potentiation of reactivity to afferent stimulation and facilitates epileptic seizures in rat hippocampal slices: Toward understanding the functional consequences of cerebrovascular insults. *J Neurosci* 28(3):732–736.
69. Kadam, S.D. et al. 2010. Continuous electroenephalographic monitoring with radiotelemetry in a rat model of perinatal hypoxia-ischemia reveals progressive post-stroke epilepsy. *J Neurosci* 30(1):404–415.
70. Curia, G. et al. 2014. Pathophysiogenesis of mesial temporal lobe epilepsy: Is prevention of damage antiepileptogenic? *Curr Med Chem* 21(6):663–688.
71. Pitkänen, A., Bolkvadze, T., Immonen, R. 2011. Anti-epileptogenesis in rodent post-traumatic epilepsy models. *Neurosci Lett* 497(3):163–171.
72. Schmidt, D. 2012. Is antiepileptogenesis a realistic goal in clinical trials? Concerns and new horizons. *Epileptic Disord* 14(2):105–113.
73. Löscher, W., Brandt, C. 2010. Prevention or modification of epileptogenesis after brain insults: Experimental approaches and translational research. *Pharmacol Rev* 62(4):668–700.
74. Sloviter, R.S. 2011. Progress on the issue of excitotoxic injury modification vs. real neuroprotection; Implications for post-traumatic epilepsy. *Neuropharmacology* 61(5–6):1048–1050.
75. Feeney, D.M. et al. 1981. Response to cortical injury. I. Methodology and local effects of contusions in the rat. *Brain Res* 211:67–77.
76. Dail, W.G. et al. 1981. Responses to cortical injury: II. Widespread depression of the activity of an enzyme in cortex remote from a focal injury. *Brain Res* 211:29–89.
77. Lighthall, J.W. 1988. Controlled cortical impact: A new experimental brain injury model. *J Neurotrauma* 5:1–15.
78. Dixon, C.E. et al. 1991. A controlled cortical impact model of traumatic brain injury in the rat. *J Neurosci Methods* 39:253–262.
79. Smith, D.H. et al. 1995. A model of parasagittal controlled cortical impact in the mouse: Cognitive and histopathologic effects. *J Neurotrauma* 12(2):169–178.
80. Hunt, R.F., Scheff, S.W., Smith, B.N. 2009. Posttraumatic epilepsy after controlled cortical impact injury in mice. *Exp Neurol* 215(2):243–252.
81. Statler, K.D. et al. 2009. A potential model of pediatric posttraumatic epilepsy. *Epilepsy Res* 86(2–3):221–223.
82. Bolkvadze, T., Pitkänen, A. 2012. Development of post-traumatic epilepsy after controlled cortical impact and lateral fluid-percussion-induced brain injury in the mouse. *J Neurotrauma* 29(5):789–812.
83. Guo, D. et al. 2013. Rapamycin attenuates the development of posttraumatic epilepsy in a mouse model of traumatic brain injury. *PLoS One* 8(5):e64078.
84. Thompson, H.J. et al. 2005. Lateral fluid percussion brain injury: A 15-year review and evaluation. *J Neurotrauma* 22(1):42–75.
85. D'Ambrosio, R. et al. 2004. Post-traumatic epilepsy following fluid percussion injury in the rat. *Brain* 127:304–314.
86. D'Ambrosio, R. et al. 2005. Progression from frontal-parietal to mesial-temporal epilepsy after fluid percussion injury in the rat. *Brain* 128:174–188.
87. Kharatishvili, I. et al. 2006. A model of posttraumatic epilepsy induced by lateral fluid-percussion brain injury in rats. *Neuroscience* 140(2):685–697.
88. Shultz, S.R. et al. 2013. Can structural or functional changes following traumatic brain injury in the rat predict epileptic outcome? *Epilepsia* 54(7):1240–1250.
89. Goodrich, G.S. et al. 2013. Cefriaxone treatment after traumatic barin injury restores expression of the glutamate transporter, GLT-1, reduces regional gliosis, and reduces post-traumatic seizures in the rat. *J Neurotrauma* 15:30(16):1434–1441.

90. Rodgers, K.M.1. et al. 2015. Progressive, seizure-like, spike-wave discharges are common in both injured and uninjured Sprague Dawley rats: Implications for the fluid percussion injury model of post-traumatic epilepsy. *J Neurosci.* 35(24):9194–9204.
91. French, J. 2007. Refractory epilepsy: Clinical overview. *Epilepsia* 48(Suppl 1):3–7.
92. Eastman, C.L. et al. 2010. ECoG studies of valproate, carbamazepine and halothane in frontal-lobe epilepsy induced by head injury in the rat. *Exp Neurol* 224:369–388.
93. Eastman, C.L. et al. 2011. Antiepileptic and antiepileptogenic performance of carisbamate after head injury in the rat: Blind and randomized studies. *J Pharmacol Exp Ther* 336:779–790.
94. D'Ambrosio, R. et al. 2013. Mild passive focal cooling prevents epileptic seizures after head injury in rats. *Ann Neurol* 73:199–209.
95. D'Ambrosio, R. et al. 2009. Functional definition of seizures provides new insights into posttraumatic epileptogenesis. *Brain* 132:2805–2821.
96. D'Ambrosio, R., Miller, J.W. 2010. What is an epileptic seizure? Unifying definitions in clinical practice and animal research to develop novel treatments. *Epilepsy Curr* 10(3):61–66.
97. D'Ambrosio, R., Miller, J.W. 2010. Point on "What is an Epileptic Seizure?" *Epilepsy Curr* 10(4):90.
98. Bancaud, J., Talairach, J. 1992. Clinical semiology of frontal lobe seizures. *Adv Neurol* 57:3–58.
99. Lüders, H.O. et al. 1992. A negative motor response elicited by electrical stimulation of the human frontal cortex. *Adv Neurol* 57:149–157.
100. Lüders, H.O. et al. 1995. Cortical electrical stimulation in humans. The negative motor areas. *Adv Neurol* 67:115–129.
101. Salanova, V. et al. 1995. Frontal lobe seizures: Electroclinical syndromes. *Epilepsia* 36(1):16–24.
102. So, N.K. 1998. Mesial frontal epilepsy. *Epilepsia* 39(S4):S49–S61.
103. Ikeda, A. et al. 2009. Negative motor seizure arising from the negative motor area: Is it ictal apraxia? *Epilepsia* 50(9):2072–2084.
104. Ferguson, P.L. et al. 2010. A population-based study of risk of epilepsy after hospitalization for traumatic brain injury. *Epilepsia* 51(5):891–898.
105. Maegele, M. et al. 2005. Characterization of a new rat model of experimental combined neurotrauma. *Shock* 23(5):476–481.
106. Leite, J.P., Cavalheiro, E.A. 1995. Effects of conventional antiepileptic drugs in a model of spontaneous recurrent seizures in rats. *Epilepsy Res* 20(2):93–104.
107. Grabenstatter, H.L. et al. 2005. Use of chronic epilepsy models in antiepileptic drug discovery: The effect of topiramate on spontaneous motor seizures in rats with kainate-induced epilepsy. *Epilepsia* 46(1):8–14.
108. Grabenstatter, H.L., Clark, S., Dudek, F.E. 2007. Anticonvulsant effects of carbamazepine on spontaneous seizures in rats with kainate-induced epilepsy: Comparison of intraperitoneal injections with drug-in-food protocols. *Epilepsia* 48(12):2287–2295.
109. Grabenstatter, H.L., Dudek, F.E. 2008. A new potential AED, carisbamate, substantially reduces spontaneous motor seizures in rats with kainate-induced epilepsy. *Epilepsia* 49(10):1787–1794.
110. Bialer, M. et al. 2010. Progress report on new antiepileptic drugs: A summary of the Tenth Eilat Conference (EILAT X). *Epilepsy Res* 92(2–3):89–124.
111. Halford, J.J. et al. 2011. A randomized, double-blind, placebo-controlled study of the efficacy, safety, and tolerability of adjunctive carisbamate treatment in patients with partial-onset seizures. *Epilepsia* 52(4):816–825.
112. Russo, E. et al. 2010. Comparison of the antiepileptogenic effects of an early long-term treatment with ethosuximide or levetiracetam in a genetic animal model of absence epilepsy. *Epilepsia* 51(8):1560–1569.

113. Dezsi, G. et al. 2013. Ethosuximide reduces epileptogenesis and behavioral comorbidity in the GAERS model of genetic generalized epilepsy. *Epilepsia* 54(4):635–643.
114. Loane, D.J., Faden, A.I. 2010. Neuroprotection for traumatic brain injury: Translational challenges and emerging therapeutic strategies. *Trends Pharmacol Sci* 31(12):596–604.
115. Galanopoulou, A.S. et al. 2013. Epilepsy therapy development: Technical and methodologic issues in studies with animal models. *Epilepsia* 54(Suppl 4):13–23.
116. Benbadis, S.R. et al. 2009. Interrater reliability of EEG-video monitoring. *Neurology* 73:843–846.
117. Cole, J.T. et al. 2011. Craniotomy: True sham for traumatic brain injury, or a sham of a sham? *J Neurotrauma* 28(3):359–369.
118. Holguin, A. et al. 2007. Characterization of the temporo-spatial effects of chronic bilateral intrahippocampal cannulae on interleukin-1beta. *J Neurosci Methods* 161(2): 265–272.
119. Blackwood, D.H., Martin, M.J., McQueen, J.K. 1982. Enhanced rate of kindling after prolonged electrode implantation into the amygdala of rats. *J Neurosci Methods* 5(4):343–348.
120. Löscher, W. et al. 1995. Does prolonged implantation of depth electrodes predispose the brain to kindling? *Brain Res* 697:197–204.
121. Bankstahl, J.P., Brandt, C., Löscher, W. 2014. Prolonged depth electrode implantation in the limbic system increases the severity of status epilepticus in rats. *Epilepsy Res* 108(4):802–805.
122. Willoughby, J.O., Mackenzie, L. 1992. Nonconvulsive electrocorticographic paroxysms (absence epilepsy) in rat strains. *Lab Anim Sci* 42(6):551–554.
123. van Luijtelaar, E.L. et al. 2002. Rat models of genetic absence epilepsy: What do EEG spike-wave discharges tell us about drug effects? *Methods Find Exp Clin Pharmacol* 24(Suppl D):65–70.
124. Shaw, F.Z. 2004. Is spontaneous high-voltage rhythmic spike discharge in Long Evans rats an absence-like seizure activity? *J Neurophysiol* 91(1):63–77.
125. Pearce, P.S. et al. 2014. Spike-wave discharges in adult Sprague Dawley rats and their implications for animal models of temporal lobe epilepsy. *Epilepsy Behav* 32:121–131.
126. Aporti, F. et al. 1986. Age-dependent spontaneous EEG bursts in rats. Effects of brain phosphatidylseline. *Neurobiol Aging* 7(2):115–120.
127. Fitzpatrick, C.J. et al. 2013. Variation in the form of Pavlovian conditioned approach behavior among outbred male Sprague Dawley rats from different vendors and colonies: Sign-tracking vs. goal-tracking. *PLoS One* 8(10):e75042.

11 Translational Principles of Neuroprotective and Neurorestorative Therapy Testing in Animal Models of Traumatic Brain Injury

Edward D. Hall

CONTENTS

INTRODUCTION

It is estimated that there are in excess of 1.5 million cases of traumatic brain injury (TBI) each year in the United States. Of these, 1.2 million seek medical care. Out of this number, approximately 90% of TBIs are "mild" in severity (Glasgow Coma Score = 13–15) although some require brief hospitalization, and the neurological and psychological consequences are often life changing. The other 10% of TBIs result in either severe (Glasgow Coma Score = 3–8) or moderate (Glasgow Coma Score = 9–12) acute sequelae and typically require intensive medical treatment and extended recovery periods. Although TBI can victimize active individuals at any age, most occur in young adults in the second and third decades of life. Moreover, the majority of TBI patients now survive their neurological insults due to improvements in emergency, neurological intensive care, and surgical treatments. Nevertheless, the need for intensive rehabilitation and the reality of prolonged disability exacts a significant toll on the individual, his or her family, and society. Effective therapies to either attenuate post-TBI secondary damage to brain tissue, thereby preserving the anatomic substrate needed for recovery of neurological function, or that enhance the functional capacity of the surviving brain tissue, represents one of the most pressing unmet needs in medicine to improve the outlook for those with TBI by enabling higher levels of independence and productivity and a reduction of the costs of long-term care to the injured individual, their families, and society.

The goal of this chapter is to discuss the importance of preclinical evaluation of potential therapies for TBI in animal models that mimic the human disorder as a prelude to the translation of these into clinical trials. The focus will be on a brief description of the various rodent (mouse and rat) TBI models, what subtypes of TBI they mainly model, and the presentation of basic principles of preclinical therapeutic testing. For a more detailed discussion of the specifics of particular TBI models, the reader is referred to the several excellent in-depth reviews on TBI modeling that have been published during the last decade.[1–8] The discussion of how to test therapies in TBI, or TBI-relevant models, will be conveyed primarily in relation to pharmacological therapies. However, many, if not all of the principles that define a thorough preclinical evaluation of drugs in animal models are in fact equally applicable to the application of therapeutic hypothermia as well as gene and cellular transplant therapies.

GOALS OF DRUG THERAPIES FOR ACUTE TRAUMATIC BRAIN INJURY (TBI)

NEUROPROTECTION

Much of the opportunity for pharmacological intervention to preserve neurological function after acute TBI is based on the fact that most of the vascular and/or neurodegeneration that follows these injuries is not due to the primary mechanical (i.e., shearing of blood vessels, and nerve cells) insults, but rather to secondary molecular and pathophysiological injury events set in motion by the primary injury. Indeed, during the first minutes, hours, and days following injury, a neurodegenerative process is initiated that is proportional to the magnitude of the initial insult. Nevertheless, the initial anatomical continuity of much of the injured brain tissue in the majority of cases, together with our present knowledge of many of the factors involved in the secondary injury process, has lead to the notion that pharmacological treatments that interrupt the secondary cascade, if applied early, could improve brain tissue survival (preservation of axons, nerve cell bodies and synaptic connections), and thus preserve the necessary anatomic substrates for functional recovery to take place. The pharmacological strategy of interfering with acute (first 72 hours) and subacute (72 hours to 14 days) post-TBI neuronal, axonal, and synaptic loss is referred to as *neuroprotection*.

In particular, a key determinant in neurological recovery after TBI concerns the secondary loss of axons and consequently the lost connection between neuronal cell bodies, the distal nerve terminals that make synaptic connections with downstream neurons. This often widespread loss of axons in the injured brain is referred to as *diffuse axonal injury* (DAI) or *traumatic axonal injury* (TAI), much of which is due to a potentially interruptible process involving localized disruptions of axonal ionic balance; intra-axonal accumulation of calcium (Ca^{++}), which is exacerbated by free radical-induced oxidative damage to Ca^{++} homeostatic mechanisms; mitochondrial failure, and the triggering of Ca^{++}-activated calpain-mediated proteolysis of axonal neurofilaments.[9–14] Accordingly, various compounds that attenuate these secondary injury processes have been documented to reduce axonal loss. However, it should be realized that a significant factor in influencing the extent of neural injury is a decrease in brain microvascular perfusion (i.e., secondary ischemia). When this occurs, the result is an exacerbation of the injury process due to superimposed brain tissue ischemic hypoxia. Moreover, deficiencies in brain cerebral blood flow (CBF) can be aggravated by systemic hypotension and hypoxia, which are well documented causes of outcome worsening after TBI. Thus, it is important to note that secondary injury involves both parenchymal and microvascular events, and that some neuroprotective compounds can work by protecting brain parenchymal neurons and glia directly or indirectly by improving cerebral perfusion pressure (CPP), CBF, and microvascular oxygen delivery that is required for maintaining cellular respiration and tissue viability.

Neurorestoration

The second approach to the treatment of TBI involves attempting to restore lost neurological function once the extent of the traumatic or ischemic damage to the brain and the associated neurological deficits has stabilized. Until a decade ago, it was firmly believed that once the brain (or spinal cord) was damaged by the secondary injury process, there was little, if any, capability for regeneration of axons and formation of new synapses to take the place of the lost neural elements. However, over the last several years, it has been discovered that the brain is indeed capable of significant structural and functional repair, plasticity and regeneration that can be pharmacologically enhanced. Approaches for accomplishing this include reawakening the growth potential of the surviving neurons or antagonizing the multiple inhibitory factors that have been discovered whose activity is aimed at inhibiting axonal growth and synaptogenesis. Alternatively, cellular replacement may be achievable in certain brain regions that possess nascent neural stem cells. It is increasingly apparent that these endogenous stem cell populations in the injured brain might be pharmacologically stimulated to divide, migrate, and differentiate into neuronal or glial precursor cell types, and ultimately neurons and remyelinating oligodendroglia, respectively. To enable this process, the molecular mechanisms that control neurogenesis and gliogenesis are rapidly being revealed providing targets for pharmacological neurorestoration of function. Several pharmacological mechanisms have been identified that can be targeted to try to enhance the function and/or structural plasticity of neuronal pathways that survive the ravages of post-traumatic secondary injury.[15–18]

PREVIOUS TBI CLINICAL TRIALS OF NEUROPROTECTIVE AGENTS AND WHAT WE LEARNED RELEVANT TO THE NEEDS FOR PRECLINICAL DRUG EVALUATION IN ANIMAL MODELS

In the early 1980s, several pharmaceutical companies became attracted to the idea of discovering neuroprotective drugs for the acute treatment of TBI and stroke. As a result, several compounds were discovered that were entered into development with some making their way into large, double-blind, multicenter phase III clinical trials in TBI. These efforts, which actually began in the late 1970s and continued into the early 2000s, were primarily directed at three general pharmacological mechanistic strategies to interrupt secondary injury processes: (1) reduction of intracellular calcium overload (L-type calcium channel blockers), (2) inhibition of glutamate-mediated excitotoxicity (i.e., glutamate receptor antagonists), and (3) interruption of reactive oxygen-mediated damage (i.e., free radical scavengers/antioxidants).

CALCIUM CHANNEL BLOCKER NIMODIPINE

Accumulation of intracellular calcium is a major player in secondary injury after TBI. One of the mechanisms for the postinsult calcium overload involves depolarization-induced entry via voltage-dependent L-type channels. Accordingly, the first neuroprotective approach to be tested in Phase III clinical trials in TBI was the competitive L-type calcium channel blocker nimodipine, which was entered into clinical trials beginning in the late 1970s. In two different Phase III multicenter TBI (moderate and severe) trials,[19] no overall benefit was revealed. However, retrospective analysis of the TBI trials revealed that nimodipine may have improved outcome in patients with traumatic subarachnoid hemorrhage (tSAH).[19] This is not an insignificant finding since about half of all patients with severe TBI have tSAH as part of their post-traumatic pathology, which carries with it a largely vascular pathophysiology including severe blood–brain barrier (BBB) compromise, loss of microvascular autoregulation and spasm of major cerebral arteries that restricts CPP and CBF. Furthermore, nimodipine has been shown to produce a slight but significant increase in survival in aneurysmal SAH patients[20] and was approved by the U.S. Food and Drug Administration, and in most countries for the treatment of this condition. While its neuroprotective efficacy is quite limited, nimodipine represents the first agent to be approved for neuroprotective use even though much of its effect is probably mediated via protection of the microvasculature and vasodilation-mediated improvements in CBF. However, due to its dose-related vasodilatory action, the compound must be used with care since it can lower arterial and cerebral perfusion pressures, which can potentially exacerbate post-TBI secondary brain injury. In any event, we now know that prolonged opening of the voltage-dependent L-type calcium channel is only one of the mechanisms for post-traumatic intracellular calcium overload, and probably much less important than the others including glutamate receptor overstimulation-induced "excitotoxicity" and oxidative damage-impairment of the multiple intracellular mechanisms responsible for maintaining calcium homeostasis.[21]

Glutamate Receptor Antagonists

In order to inhibit post-TBI glutamate-mediated excitotoxicity, multiple glutamate receptor antagonists were taken into Phase II and III trials including the competitive N-methyl-D-aspartate (NMDA) receptor antagonists selfotel (CGS 19755) and aptiganel (CNS 1102) that block the binding of glutamate to its receptor complex recognition site, eliprodil that blocks the polyamine site, and CP-101,606 (traxoprodil) that blocks the NR2B subunit on the NMDA receptor complex.[22] None of these produced a statistically significant improvement in neurological recovery in TBI patients, although traxoprodil came close ($p < 0.1$) to producing a significant improvement in favorable outcome in a relatively large (400 severe TBI patients) phase II clinical trial.[23] In retrospect, there was precious little literature generated concerning the testing of the competitive NMDA antagonists in TBI animal models before the aforementioned glutamate receptor antagonists were placed into clinical development. Much of the rationale for trying them in clinical TBI was based upon their efficacy in ischemic stroke models and the common notion that whatever was neuroprotective in a stroke model would also be effective in TBI paradigms. More important, whether they were tested in stroke or TBI models or both, the published work involved very early postinjury drug administration times of usually less than one hour. There was little systematic definition of the therapeutic window (i.e., how long after TBI could the treatment be delayed and still produce neuroprotective efficacy), but what was done showed that neuroprotective efficacy was lost during the first hour after injury. Nevertheless, the apparently short therapeutic window for NMDA antagonists in animal models did not dissuade companies from conducting trials of NMDA receptor blockers allowing treatment initiation as much as 8 hours after TBI. This was due to the prevailing assumption among TBI (and stroke) researchers and clinicians, that whatever the therapeutic window might be in rodent animal models, it would probably be considerably longer in humans. In other words, since mice and rats are hypermetabolic compared to humans, the progression of post-TBI secondary injury and neurodegeneration would be more rapid and would need to be treated with a neuroprotective agent quicker than in humans. However, the only evidence that supports this assumption is from microdialysis studies of glutamate release in injured rat brain showing that the massive injured-triggered release of glutamate only last for minutes, whereas in human TBI patients, the duration of elevated glutamate release may go on for 6 hours to several days.[24] In contrast to the older microdialysis studies of glutamate-release kinetics in TBI models, more recent results in the rat central fluid percussion model using real-time continuous microelectrode monitoring of extracellular glutamate shows that the injury-induced increase in glutamate levels in injured rat cerebral cortex in fact persists for days,[25,26] as it does in human TBI. Therefore, if the duration of glutamate release in both rodents and humans is prolonged for days, then why would rodent TBI and stroke models only show a NMDA receptor antagonist neuroprotective therapeutic window of less than an hour?

Following the failures of the NMDA antagonist clinical trials conducted in TBI patients, Shohami and colleagues[27] revealed in their mouse weight-drop TBI model that the short therapeutic window for these compounds may be due to the fact that the initially glutamate overstimulated NMDA receptors in the injured brain actually enter into a hypofunctional state by an hour after TBI, and that this loss of function persists for days. Consequently, blocking the receptors after the first hour with the NMDA receptor blocker is no longer useful since the NMDA receptor is already in a nonfunctional state.[27] These investigators additionally showed that what was needed in regard to NMDA receptor modulation was to administer NMDA to reactivate the NMDA receptors beginning at 24 or 48 hours after TBI and that this lead to an improvement in 14-day cognitive performance. Subsequent experiments have shown that this can be more safely achieved by administering a weaker NMDA receptor agonist or a partial agonist such as D-cycloserine,[28] which is actually in early clinical trials in TBI patients as a neurorestorative approach to improve long-term post-TBI recovery. In summary, while few, if any, investigators would question the important role of glutamate NMDA receptor activation in the initiation of excitotoxic brain damage after TBI, the efficacy of NMDA receptor blocking as a neuroprotective strategy is most likely clinically impractical due to a short therapeutic window, whereas what may be more useful is to pharmacologically reactivate brain NMDA receptors in the postacute recovery phase after TBI.

Free Radical Scavengers/Antioxidants

In the case of efforts to interrupt reactive oxygen damage, the polyethylene conjugated form of the superoxide radical scavenger Cu/Zn superoxide dismutase (PEG-SOD) was evaluated in trials conducted in moderate and severe TBI patients. Although it showed a positive trend in an initial, small phase II trial,[29] subsequent phase III trials failed to show any enhancement of neurological recovery.[22]

A much bigger development program was undertaken with the 21-aminosteroid lipid peroxidation inhibitor tirilazad. Tirilazad was extensively evaluated in animal models of TBI as well as SCI, ischemic stroke, and subarachnoid hemorrhage, and shown to exert a variety of neuro- and vasoprotective effects.[30,31] Based upon these preclinical studies, clinical trials of tirilazad were conducted in TBI,[22,32] SAH,[33] ischemic stroke[34,35] and SCI.[36] In TBI, an initial North American trial of 1100 patients, comparing tirilazad treatment with a placebo for 5 days initiated with 4 hours postinjury, ended with such a confounding randomization imbalance that no meaningful efficacy analysis could be extracted. In contrast, a European phase III trial was successfully completed, but it failed to show an overall effect in moderate and severely injured patients. However, post-hoc analysis revealed that the compound significantly improved survival in both moderately and severely injured male patients with traumatic SAH.[32] This beneficial effect in the tSAH subgroup, which represents about half of severe TBIs, was not surprising in that the drug had previously been shown to improve recovery and survival in a Phase III trial in aneurysmal SAH patients.[33] Interestingly, this effect in tSAH and aneurysmal SAH was mainly apparent in male patients. This gender difference was found to be partially due to a faster rate of tirilazad metabolism in females. Nevertheless, subsequent female-only trials with higher tirilazad doses that were calculated to duplicate the exposure levels in males did not reveal the same level of efficacy as seen in male patients, although beneficial effects were apparent in the more severe SAH females.[37,38] Thus, the issue of gender differences in neuroprotective drug responsiveness clouds the interpretation of tirilazad's neuroprotective efficacy.

Learnings

This brief history of neuroprotective drug discovery and development over the past 20–25 years could be fairly characterized as a series of often high profile and expensive failures. Although these have largely dampened the enthusiasm of the pharmaceutical industry for this therapeutic area, much has been learned from them that could, and should, serve as a roadmap for future efforts aimed at pharmacological neuroprotection and improved neurological recovery after TBI. Postmortem analyses of mistakes made in TBI drug development have been published, and a careful reading of them reveals a host of shortcomings in past preclinical testing of candidate neuroprotective agents and in clinical trial design and conduct that need to be addressed in the future.[22] A summary is provided in Table 11.1.

First, the discovery of the first generation of neuroprotective agents, which included glutamate receptor antagonists, calcium channel blockers, and free radical scavengers/antioxidants, occurred prior to the elucidation of an adequate understanding of the intricacies of the targeted secondary injury mechanisms. In each case, there was inadequate knowledge of the time course and interrelationships of these events; their therapeutic windows for effective treatment intervention; and how these were either similar or different between species, injury models, genders, and between animals and humans. In the case of reactive oxygen mechanisms, our knowledge of the key ROS species and their sources or cellular origins and targets was insufficient to guide the design of optimum antioxidant neuroprotective compounds. Second, the preclinical efficacy testing of compounds was often naively deficient and failed to consider the pathological complexity of human TBI and the need to test potential neuroprotective agents in multiple models that might define whether a particular agent might be broadly used across the full TBI spectrum or for a specific TBI phenotype as discussed in the following section. Third, a major problem was that the clinical trials were poorly designed in regard to a failure to allow the preclinical TBI animal models results (e.g., therapeutic window) to inform the clinical trial design, inclusion of all types of TBI pathological phenotypes to be in the trial, and lack of standardization of acute treatment guidelines and neurorehabilitation protocols across the patient enrollment sites.

TABLE 11.1
Reasons for Past Failures in Neuroprotective Drug Discovery and Development

Inadequate Understanding of Secondary Injury Mechanisms
Lack of definition of time course of glutamate receptor functional changes
Lack of definition of the sources and spatial and temporal characteristics of reactive oxygen generation; inability to rationally determine therapeutic window and optimum treatment duration
Lack of understanding of the interrelationship of secondary injury mechanisms
Focus on secondary injury mechanisms with short therapeutic windows; need to identify and target injury mechanisms with longer therapeutic windows
Lack of understanding of the relative therapeutic windows in animal models and humans; is the time course of secondary injury in mice, rats, and men similar
Inadequate Preclinical Testing
Lack of testing in multiple models
Failure to compare efficacy in male and female animals
Incomplete dose-response and definition of therapeutic plasma levels
Incomplete definition of therapeutic window
Lack of definition of pharmacokinetics, timing of needed maintenance dosing, and optimum treatment duration
Poor Clinical Trial Design
Gross mismatch between preclinical and clinical testing
Imprecise/insensitive outcome scales (e.g., Glasgow Outcome Scale)
Lumping of all kinds of TBIs
Lack of identification and a priori plan to analyze pathophysiological subgroups (e.g., tSAH)
Lack of biomarker to follow the progression of the pathophysiology and monitor mechanistic drug effects
Lack of standardization of acute TBI treatment guidelines and neurorehabilitation protocols between sites

VARIABILITY IN TBI PATHOLOGY AND PATHOPHYSIOLOGY: RATIONALE FOR TESTING NEUROPROTECTIVE OR NEURORESTORATIVE DRUGS IN MULTIPLE TBI MODELS

As eluded to at the beginning of this chapter, the traditional clinical assessment tool for expressing the severity of TBI in patients is the Glasgow Coma Scale (GCS) score, so named because of its development in the mid-1970s by two prominent TBI-focused neurosurgeons in Glasgow, Scotland, Brian Jennett and Graham Teasdale.[39,40] The scale involves scoring of three responses to either a painful stimulus or to command: eye opening, verbal response, and basic motor response. Eye opening is worth a maximum of 4 points: 1 for no opening in response to pain or to command, 2 for eye opening in response to pain, 3 for eye opening in response to a verbal command, and 4 points for spontaneous eye opening. Verbal response is worth a maximum of 5 points: 1 for no response, 2 for a response to sounds, 3 for a nonsense response, 4 for a confused response, and 5 for the ability to carry on a conversation with the examiner. The basic motor response in response to a pain stimulus is worth a maximum of 6 points: 1 for no response, 2 for a generalizing response to the pain, 3 for arm extension, 4 for arm flexion, 5 for a localizing response to the pain (ability to identify where the pain is emanating from), and 6 if a motor response occurs on command. The total GCS ranges from a low score of 3, which would be a person who has suffered a TBI and is in a deep coma with no eye opening, verbal, or motor response, to a full score of 15, which is a patient who is has full normal responses. The scoring range is typically divided into three severity categories: 3–8 for someone who is said to have suffered a "severe" TBI, 9–12 for a "moderate" TBI, and 13–15 for a "mild" TBI. Someone who is at 8 or below is characterized as being comatose in contrast to someone who has improved to a 9 who is considered to be emerging from a coma. It should also be remembered that while scores will increase as the TBI patient improves his or her neurological status, it is also not uncommon for the GCS to decrease from an initially moderate or even mild level down to the severe level. Clinicians who treat acute TBIs refer to patients who are initially responsive when first admitted to the emergency department, but subsequently slips progressively into a coma as falling into a category of TBI patients who "talk and deteriorate." Such patients demonstrate the influence of progressive post-traumatic secondary pathophysiological changes (e.g., brain swelling, decreased cerebral blood flow, hemorrhage expansion) that are taking place during the first minutes or hours after TBI.

To illustrate the variability in TBI-induced pathology, Figure 11.1[41] displays the computerized tomography (CT) scans of six acutely severe TBI patients who all have a GCS <8. Yet, each patient reflects differing post-traumatic pathological manifestations caused by their TBIs. The top left CT shows an epidural hematoma (EDH), which by itself can lead to a good outcome if it is promptly surgically evacuated and additional hemorrhage stopped. However, because they are generally caused by traumatic meningeal arterial hemorrhage as a result they can expand rapidly raising intracranial pressure (ICP) and patient death due to tentorial herniation and resulting brainstem compression. A patient who sustains a TBI and presents in the emergency room able to talk, but with a rapidly developing EDH is perhaps the most likely to talk and deteriorate. The bottom left CT scan shows a subdural hematoma (SDH), which is from ruptured cerebral veins that may more slowly increase in their dimension over time. If small, they may be missed on the first CT and thus not surgically removed during the acute post-TBI period. However, if it grows slowly it will begin to increase ICP and cause neurological deterioration. In both of these patients, the CT reveals the hematoma being large enough to press the midline structures toward the opposite side of the brain.

The middle top CT in Figure 11.1 is from a TBI patient who has manifested a focal brain contusion as well as an SDH on the left side of the CT. The middle bottom CT is from a TBI patient with a traumatically induced subarachnoid hemorrhage (SAH) along with an SDH on the left side of the CT and a smaller intraventricular hematoma (IVH) on the right side of the CT. TBI patients with hemorrhagic contusions, SAHs, and IVHs, if they are large, typically have the worst prognosis because these pathologies have a good chance of producing blood–brain barrier compromise and rapidly evolving vasogenic brain edema, and acute and chronic arterial vasospasm

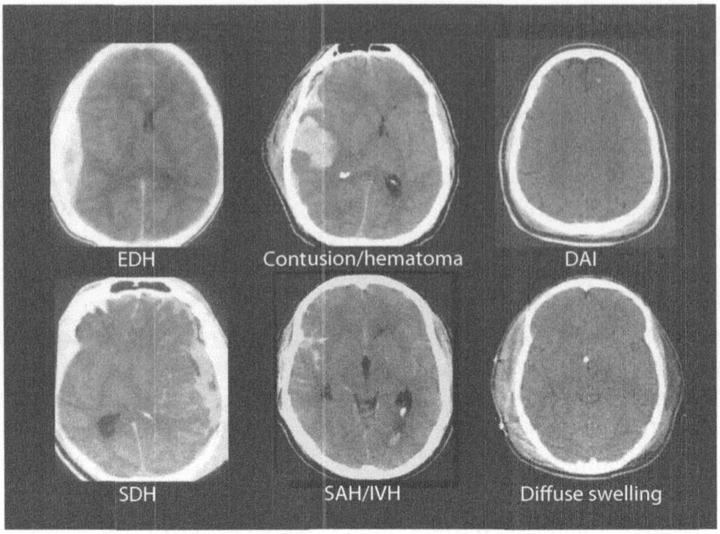

FIGURE 11.1 Heterogeneity of severe traumatic brain injury (TBI). Computed tomography (CT) scans of six different patients with severe TBI, defined as a Glasgow Coma Scale score of <8, highlighting the significant heterogeneity of pathological findings. CT scans represent patients with epidural hematomas (EDH), contusions and parenchymal hematomas (contusion/hematoma), diffuse axonal injury (DAI), subdural hematoma (SDH), subarachnoid hemorrhage and intraventricular hemorrhage (SAH/IVH), and diffuse brain swelling (diffuse swelling). (Reproduced with permission from K.E. Saatman et al., *J. Neurotrauma* 25: 719–738, 2008.)

compromising cerebral blood flow (CBF) and brain oxygen delivery. On the bottom right of Figure 11.1 is a CT from a patient with no apparent hemorrhagic manifestations but has massive diffuse swelling as judged from the inability to discern visible cortical gyri and sulci or ventricles since the swelling has closed them off. Finally, in the top right CT is a TBI patient in which the cortical structure is largely intact, the midline structures are not shifted, and no hemorrhagic pathologies are visible. This type of patient who was verified to have a TBI and is displaying neurological sequelae as a result is referred to as having diffuse axonal injury (DAI). However, each of the patients in Figure 11.1 will have DAI along with their other pathologies.

To further illustrate the variability in TBI pathologies and pathophysiologies, Tables 11.2 and 11.3 list the post-traumatic pathologies seen in a recent series of 17 patients (13 males, 4 females) who presented at the University of Kentucky Medical Center with initial post-TBI GCS scores between 4 and 10. The most frequent pathology seen in 7/17 patients (41.2%) was traumatic SAH. However, what is apparent from the tables is that the typical severe or moderate TBI patient can have multiple post-traumatic pathologies. In other words, the pathology and associated pathophysiology of individual TBIs is highly variable even though their admission GCS values are all in the same range.

Similarly, a prominent group of TBI experts from the Departments of Neuropathology and Neurosurgery associated with the University of Glasgow Institute

TABLE 11.2
Injury Types Seen in 17 Individual TBI Patients

Patient	M/F	Postresuscitation GCS	SAH	DAI	EDH	SDH	Cortical Contusion	IPH	IVH	Skull Fracture
1	F	10	✓							
2	M	9	✓	✓						
3	M	9			✓					✓
4	M	6	✓							
5	M	6		✓						
6	F	6				✓	✓			✓
7	F	7		✓			✓			
8	M	8		✓			✓			
9	F	6				✓		✓	✓	
10	M	6	✓				✓	✓		
11	M	7	✓			✓				
12	M	4		✓						
13	M	7	✓					✓		✓
14	M	6			✓					
15	M	7	✓					✓		
16	M	9					✓	✓	✓	
17	M	5				✓				✓

Note: Patients presented at the University of Kentucky Medical Center between June 2010 and November 2012. Thirteen patients were severe (GCS = 4–8) and four moderate (GCS = 9–10). DAI = diffuse axonal injury; EDH = epidural hematoma; IPH = intraparenchymal hemorrhage; IVH = intraventricular hemorrhage. M/F = male or female; SAH = traumatic subarachnoid hemorrhage; SDH = subdural hematoma.

TABLE 11.3
Summary of TBI Characteristics Seen in 17 Moderate or Severe TBI Patients

Patient Characteristics (N = 17)	**Values (Range)**
Age (Years)	32 (18–63)
Male (%)	77%
Postresuscitation GCS	7 (4–10)
Initial Intracranial pressure (ICP; mm Hg)	13.6 (2–23)
Injury Type	**N (%)**
Traumatic subarachnoid hemorrhage (SAH)	7 (41.2%)
Diffuse axonal injury (DAI)	5 (29.4%)
Epidural hematoma (EDH)	2 (11.8%)
Subdural hematoma (SDH)	4 (23.5%)
Cortical contusion	5 (29.4%)
Intraparenchymal hemorrhage (IPH)	5 (29.4%)
Intraventricular hemorrhage (IVH)	2 (11.8%)
Skull fracture	4 (23.5%)
Extracranial injury	6 (35.3%)

Note: Patients were admitted to the University of Kentucky Medical Center between June 2010 and November 2012.

of Neurological Sciences published in 2011 a summary of neuropathological findings in a group of 85 TBI victims treated at their center who survived at least a month after their TBIs but subsequently died. As assessed by the Glasgow Outcome Scale (GOS),[42] the 85 at the time of their deaths were either in a vegetative state (35), severely disabled (30), or moderately disabled (20). Neuropathological assessment of their brains revealed that 84% had cerebral contusions, 58% had DAI, 67% had ischemic brain damage, 68% had bilateral enlargement of the ventricles indicative of widespread brain atrophy reflecting extensive neurodegeneration, 52% had thalamic damage, and 55% showed evidence of raised ICP. Moreover, 41% had undergone surgical evacuation of an intracranial hematoma (either EDH, SDH, IVH, IPH, or SAH). Of these variable pathologies, DAI, raised ICP, thalamic damage, ventricular enlargement, and ischemic brain damage were associated with the worse outcomes. In particular, diffuse or multifocal primary axonal injury or secondary ischemic damage was associated with the most severely impaired outcomes after a TBI. These statistics from Adams et al.,[42] as well as those in Tables 11.2 and 11.3 from the University of Kentucky, underscore the complexity and variability of post-TBI secondary injury-associated neuropathologies. Accordingly, this reality has necessitated the development of multiple preclinical TBI paradigms in order to model the different TBI phenotypes and more importantly for exploring the efficacy of neuroprotective and neurorestorative treatments for each injury type.

TRAUMATIC BRAIN INJURY AND TRAUMATIC BRAIN INJURY-RELEVANT MODELS

As pointed out earlier in this chapter, several excellent and usually comprehensive reviews on animal TBI models have been published in the past 15 years.[1–8,43] However, one the most recent reviews that is aimed at not only classifying the types of models, but discussing their pathological and pathophysiological characteristics as those that relate to particular TBI phenotypes and their appropriateness for preclinical neuroprotective and neurorestorative drug testing is adopted by the author for the present discussion.[4] However, differing from the prior TBI animal model reviews, the current presentation additionally includes TBI-relevant models that while not involving an actual mechanical brain trauma are neurological injury paradigms that have relevance to particular TBI pathologies. These include SAHs, epidural hematomas, subdural hematomas, and ischemic focal stroke models that replicate some of the major sequelae of human TBI. The TBI and TBI relevant models are listed in Table 11.4.

TRAUMATIC BRAIN INJURY MODELS

In vivo TBI models include six basic types: diffuse, focal, mixed diffuse/focal, complex (e.g., diffuse or focal TBI complicated by either a hypotensive, hypoxic, or peripheral traumatic insult such as a tibial fracture), penetrating, and blast.

Diffuse TBI Models

There are four general types of diffuse TBI models. The first of these is the rat fluid percussion TBI paradigm that involves the application of a transient hydraulic pressure pulse onto the exposed dura mater over the midline of the brain, which is referred to as the *central fluid percussion* model. This model was first developed in the cat[44] as well as pigs, dogs, and sheep, but later modified for use in rats[45,46] and mice.[4] It produces a widespread (i.e., "diffuse") pattern of cellular and axonal neurodegeneration as recently shown using the de Olmos silver-staining method.[47] The second is the rat impact-acceleration injury model developed by Anthony Marmarou at the Medical College of Virginia and hence often referred to as the *Marmarou model*. It involves a 0.5 or 1.0 kg weight-drop impact onto a helmet cemented onto the exposed skull. The acceleration component of the model is achieved by the sudden downward movement of the impacted head into a foam pad.[48,49] The third is the mouse weight-drop concussion paradigm first developed by the current author[50] and used extensively for neuroprotective drug testing.[51–55] It involves dropping a 50 or 100 g weight at variable distance onto the exposed head of the hand-restrained mouse, which produces a diffuse brain injury along with either contracoup or lateral contusions in approximately a third of the mice.[54] Although not widely used, it has been adopted by other laboratories for neuroprotective drug screening.[55–58] More recently, other laboratories have developed mouse and rat weight-drop diffuse TBI models employing single or repetitive weight drop or pneumatic piston-induced injuries to the closed skull.[1,59–63] The final category of diffuse TBI model used in the pig, cat, rabbit, or sheep, or primates involves induction of a rapid rotational acceleration of the constrained head that is particularly useful for induction of diffuse axonal injury.[64]

TABLE 11.4
In Vivo TBI or TBI Relevant Models Employable for Discovery of Neuroprotective and Neurorestorative Agents

Traumatic Brain Injury Models

Diffuse TBI

Rat or mouse central fluid percussion
Rat or mouse impact acceleration
Rat or mouse closed skull weight drop, single or repetitive
Pig, cat, rabbit, sheep, or nonhuman primate inertial acceleration (nonimpact)

Focal TBI

Rat or mouse controlled cortical impact (CCI)
Rat open skull weight drop
Rat bifrontal contusion

Mixed Diffuse/Focal

Rat or mouse lateral fluid percussion

Complex

Rat or mouse diffuse of focal complicated by secondary insult (hypotensive, hypoxic, or traumatic insult)

Penetrating

Rat penetrating ballistic-like brain injury (PBBI)

Blast

Rat blast tube

Traumatic Brain Injury Relevant Models

Subarachnoid Hemorrhage

Rabbit, cat, or dog intracisterna magna injection of autologous blood
Rat intracranial injection of autologous blood via dorsolateral cranial burr hole
Monkey SAH via surgical placement of autologous blood clot around base of MCA

Intracerebral/Intraparenchymal Hemorrhage

Rat striatal ICH

Subdural Hematoma

Rat intracranial injection of autologous blood via dorsolateral cranial burr hole (same as SAH model)

Focal Ischemia

Rat or mouse temporary middle cerebral artery occlusion (MCAO)-microclip or intraluminal suture for 30 min–2 hours
Rat or mouse permanent MCAO-electrocoagulation or intraluminal suture

Focal TBI Models

For the induction of focal TBIs, there is the widely employed controlled cortical impact (CCI) model first developed by Lighthall and colleagues for use in ferrets to produce a contusion TBI[65] that has since been adapted for use in rats[66] and mice.[67,68] It involves the infliction of a contusion injury through a small craniotomy via a pneumatically driven piston. The magnitude of the injury is generally varied by the depth of the cortical indentation (usually 0.5–1.0 mm in mice; 1.0–2.2 mm in rats). The CCI paradigm is a model of TBI-induced brain contusions. However, recent studies from the current author's laboratory using the de Olmos silver-staining technique have demonstrated that the subsequent neurodegeneration is not as focal as generally thought.[69,70] Widespread axonal degeneration is seen in not only the cortex, but also in the hippocampus and thalamus. Moreover, anterograde and retrograde axonal degeneration is also seen in the contralateral hemisphere, previously thought to be uninjured and thus incorrectly considered to serve as a contralateral control for the opposite injured hemisphere. Open skull weight-drop focal injury models have also been utilized in rats and mice.[4] Last, a bilateral focal contusion model has also been used by Stein and colleagues.[71]

Mixed Diffuse/Focal TBI Models

The widely employed lateral fluid percussion TBI model developed in the late 1980s for use in rats[72] represents a model that combines a cortical contusion in the deeper layers of the cerebral cortex underlying the fluid percussion site as well as more diffuse axonal degeneration in the ipsilateral as well as the contralateral hemisphere. Thus, it models the TBI mixed TBI phenotype seen in patients who have a cortical contusion as well as diffuse axonal injury. Together with the CCI model, the lateral fluid percussion TBI model is among the most widely used in experimental TBI research.

Complex TBI Models

The complex TBI models simply involve the use of one of the aforementioned TBI paradigms and exacerbating the pathophysiology of the mechanical TBI, whether it be a diffuse or focal type, by combining the injury with an episode of hypoxia or hemorrhagic hypotension. This is based upon the firmly established finding that if the acute TBI patient sustains a period of reduced oxygen delivery to the injured brain by suppressed respiration or by a reduction of blood volume due to traumatic hemorrhage or damage to the brainstem, cardiovascular, or respiratory control centers, that their outcome after their TBI will likely be worsened. Thus, combining the Marmarou impact acceleration, lateral fluid percussion, or the controlled cortical impact models with a defined episode of experimentally induced hypoxia (e.g., PaO_2 <60 mm Hg) or hemorrhagic hypotension (e.g., systolic blood pressure <90 mm Hg or CPP <60 mm Hg) will predictably worsen the brain neurochemical and histological damage and exacerbate motor and cognitive recovery. Accordingly, this complexity provides a real-world relevant model of TBI complicated by a systemic insult that provides a larger hurdle for testing the efficacy of neuroprotective compounds. Similarly, a handful of investigators have begun to examine the combination of a polytraumatic insult involving a TBI complicated by a tibial fracture.[73]

Penetrating Ballistic-Like TBI Model

In the early days of TBI research, a few laboratories took on the task of developing models of penetrating TBI generally involving shooting a bullet into the brain of an anesthetized animal. However, shooting animals in the head was one of the lightning rods for attacks by animal welfare groups. Moreover, such models were difficult to control. However, since ballistic-induced penetrating TBI sustained by warfighters or by civilians who sustain criminal attacks, the need to model penetrating ballistic-like injuries is a significant need. Accordingly, Tortella and coworkers at the Walter Reed Army Institute of Research have recently developed and extensively characterized a highly controlled, nonbullet, penetrating ballistic-like brain injury (PBBI) model in rats.[74–79] Furthermore, the same group has successfully tested a neuroprotective anti-inflammatory compound, NNZ-2566, that is currently in early human development.[80]

Blast-Induced TBI Models

Over the past 13 years, since the beginning of the War on Terror, blast-induced TBI, it has become a growing focus in contemporary TBI research. Most of the blast models published to date involve the use of various "blast tube" devices to direct either the sudden triggering of a small amount of explosive or the sudden release of compressed air at the closed end of the blast tube to inflict a blast-induced TBI in a restrained rat placed near the opening at the opposite end of the tube. The pathology and pathophysiology of blast TBI is being widely studied by a number of laboratories as discussed in several recent papers and reviews.[81–87]

Traumatic Brain Injury Relevant Models

The chapter also includes a limited presentation of hemorrhagic and ischemic stroke models, because human TBIs, mainly on the more severe end of the spectrum, often include brain hemorrhagic or ischemic–hypoxic insults. This is due to post-traumatic vasogenic and cellular edema that causes a rise in intracranial pressure (ICP), which in turn decreases cerebral perfusion pressure (CPP) as explained by the simple equation: CPP = Mean arterial blood pressure (MAP) – ICP. Additionally, a decrease in MAP (i.e., systemic hypotension) due to hemorrhagic shock will also lead to a decrease in CPP. Whether due to increased ICP or decreased MAP, the fall in CPP will cause a secondary fall in cerebral blood flow (CBF) and oxygen delivery, which, if severe, can lead to an ischemic insult to the brain tissue on top of the mechanical TBI.

Intracerebral (aka Intraparenchymal) Hemorrhage Models

In regard to hemorrhagic stroke paradigms that are also relevant to the pathophysiology and neuroprotective therapeutic investigation of TBI, there are two types of insults to consider. The first is intracerebral hemorrhage (ICH), where cerebral blood vessels may be ruptured by the mechanical trauma with blood being released into brain parenchyma where it produces secondary brain damage by triggering brain edema (swelling) and mass effects that result in secondary ischemia within the tissue surrounding the intracerebral mass of blood (hematoma). For the induction of ICH, usually in rats, the approach is simply to inject a volume of the animals' own blood directly into the brain parenchyma followed by an analysis of the volume of damage to the surrounding brain tissue.

Subarachnoid Hemorrhage Models

The second type of TBI-relevant hemorrhagic stroke model involves the induction of subarachnoid hemorrhage (SAH), where blood is released from mechanically ruptured vessels running through the subarachnoid space (between the subarachnoid and pial membranes covering the brain) from a burst congenital berry aneurysm ballooning out from one of the major arteries at the base of the brain. The pathophysiology of this type of hemorrhagic stroke also involves the triggering of a secondary ischemic insult due to the induction of delayed cerebral vasospasm that peaks in incidence at 4–7 days post-SAH. While both types of hemorrhagic insults are common in TBI patients, traumatic SAH (tSAH) occurs in approximately 50% of severe TBIs. To induce SAH experimentally, most of the available models involve the injection of a volume of the autologous blood, withdrawn immediately prior to SAH induction from the systemic circulation of the animal (e.g., rat, cat, rabbit, dog), into the subarachnoid space via injection into the cisterna magna or over one of the cerebral hemispheres via a small burr hole and puncture of the dura mater covering the brain. The most common endpoints for drug evaluation involve measurement of blood–brain barrier compromise, or decreases in cerebral blood flow during the first several post-SAH hours, or the assessment of cerebral vasospasm by histological or arteriographic methods between 2 to 7 days. The most sophisticated SAH model available involves the neurosurgical placement of an autologous blood clot around the base of the MCA in monkeys followed by arteriographic and histological ischemic damage measurements at 7 days.

Subdural and Epidural Hematoma Models

The rat lacks an arachnoid membrane, and thus the rat version of the SAH model involving injection of blood through the dura mater can also be thought of, and employed, as a subdural hematoma model. When the autologous, nonheparinized blood is injected through the dorsal burr hole through the dura mater, it will typically form a clot over the dorsal surface of the brain mimicking a post-traumatic subdural hematoma. Similarly, if the autologous blood is injected outside the dura, but beneath the skull it constitutes an epidural hematoma. Just as is done in the management of human post-TBI subdural and epidural hematomas, the experimental protocol involves surgical removal of the clot at a specified time followed by histological measurement of blood–brain barrier compromise and ischemic damage caused by the compression of the underlying brain tissue by the hematoma as well as the reperfusion injury caused by its removal.[4,88]

Focal Ischemic Stroke Models

Various ischemic stroke models have been developed during the past 30 years.[89] However, the main models in use today are the unilateral middle cerebral artery occlusion (MCAO) models employed in either rats or mice. Since these models were first developed in the 1980s, the MCAO has been variably induced by surgical ligation or cauterization via a small craniotomy over the MCA, passage of a intraluminal nylon suture up into the ipsilateral cerebral circulation via the external carotid in the neck, or via injection of a small autologous thrombus into the common carotid artery. The latter two are the most commonly employed today, and the thromboembolic paradigm is the most clinically relevant, since the majority of human focal ischemic strokes involve a thromboembolic occlusion of the MCA. The MCAO models come in two varieties, temporary and permanent. The temporary MCAO involves removal of the vascular occlusion at varying times (30, 60, 90, 120, 180 min) after the onset in order to allow reperfusion of the ischemic tissue to take place. This experimental scenario, which is accomplished by surgical removal of the extraluminal or intraluminal occlusion device, mimics either the instance where spontaneous thrombus dissolution may take place during the first 3 hours after the beginning of the stroke due to the activation of endogenous thrombolytic processes (believed to be a fairly rare occurrence), or the situation where the stroke victim is treated with the thrombolytic agent tissue plasminogen activator (tPA) for the purpose of dissolving the clot and restoring recirculation.

Although removal of the vascular occlusion and re-establishment of the normal cerebral circulation is an obviously desirable therapeutic goal, it is well known to be a two-edged sword that can lead to *reperfusion injury*, which is caused by a burst of reactive oxygen species in the previously ischemic brain tissue. Thus, there is a need in this situation for a neuroprotective agent to reduce both the pathophysiological events set in motion by the ischemic insult as well as the subsequent deleterious side effects of pharmacological recirculation. Accordingly, the temporary MCAO models are most useful for evaluating neuroprotective strategies that may be used in conjunction with tPA and other thrombolytic agents. However, since pharmacological thrombolysis can lead to secondary cerebral hemorrhage if used beyond the first few hours, it is only able to be safely employed in patients who are available for emergency treatment during the first few hours after the onset of their strokes. This is only feasible in a small fraction of ischemic stroke patients. Furthermore, MCAO animal studies have shown that reperfusion beyond the first 3 hours is unable to lessen the extent of ischemic damage. Thus, the temporary MCAO models, although widely used in stroke research, actually have limited relevance to the majority of MCA territory strokes.

The second variety of focal ischemic stroke model, the permanent MCAO, where the occlusion is permanently left in place, is therefore a better model of the vast majority of strokes where recirculation has not been re-established either spontaneously or pharmacologically during the critical first few hours after stroke onset. In this instance, the therapeutic goal is simply to try to reduce the expansion of the ischemic damage from the severely ischemic “core” area, which is doomed to infarction if reperfusion does not occur during the first 3 hours, into the surrounding “penumbral” area. The *ischemic penumbra* is potentially salvageable for several hours due to its partial circulation from collateral blood vessels. In this author’s opinion, the permanent MCAO version is the best option for preclinical evaluation of potential neuroprotective agents since it is more relevant to the overall focal ischemic stroke population where early reperfusion is not all that common. However, the testing of compounds in the temporary MCAO paradigm is probably more relevant to post-TBI ischemic insults caused by subdural or epidural hematomas, since their surgical removal leading to reperfusion of the previously compressed ischemic tissue underlying the hematoma is generally attempted as soon as possible.

Most stroke research with either the temporary or permanent MCAO models is carried out in mice or rats. The primary endpoints are histological demonstration of a reduction in brain infarct size and improvement in motor function typically determined between 24 to 72 hours or sometimes as long as 1 week after stroke onset. However, MCAO models have been developed and are occasionally used in higher species including the cat, monkey, and baboon. Obviously, the use of these for neuroprotective drug evaluation carries considerable expense.

Some investigators, a minority, believe that it is important to replicate pharmacological neuroprotective actions in these gyrencephalic species prior to movement of the compound into human clinical trials. In actuality, there is no solid comparative evidence that supports the notion that neuroprotective effects seen in rodent stroke models may not be predictive of human efficacy. Furthermore, there is presently no firm data that supports the commonly held idea that the therapeutic time window for a particular neuroprotective mechanism in a rat stroke model (e.g., 1 hour) may be longer in nonhuman primates or humans (e.g., 6 hours). On the contrary, the fact that various neuroprotective compounds that demonstrated a rather limited (1–2 hours) therapeutic window for reduction of infarct size in rat MCAO models, subsequently failed to improve outcome of stroke patients in clinical trials where the treatment initiation time varied from 6 to 24 hours, and is consistent with the concept that the therapeutic window for neuroprotective effects may not be all that different between rodents and primates. At least, no difference has been firmly demonstrated.

ISSUES THAT NEED TO BE ADDRESSED IN PRECLINICAL NEUROPROTECTIVE DRUG EVALUATION

Last, since the main purpose of TBI modeling is to enable the discovery of neuroprotective approaches that are mostly pharmacological, this chapter closes with the following list of issues or questions that need to be addressed during the preclinical evaluation of drugs for acute neuroprotection.

1. A thorough demonstration of the time course of the target pathophysiological mechanism in relevant animal models. This is needed to determine when treatment needs to begin and how long it must be maintained. This should be done in both male and female animals based upon several studies showing that the time course and magnitude of post-traumatic pathophysiology and neurodegeneration may differ greatly between genders in certain models.[53,90,91]
2. A rigorous dose-response analysis in regard to effects on the target mechanism, ability to reduce post-traumatic neurodegeneration, and improve behavioral recovery.
3. A correlation of the neuroprotective action with plasma and CNS tissue pharmacokinetics; that is, a definition of the effective neuroprotective concentration and a dosing protocol that is adequate to maintain the therapeutic concentration for as long as the target secondary injury mechanism is active.
4. A comparison of single versus multiple dose regimens in order to establish an optimum treatment regimen (i.v., bolus plus infusion makes the most sense).

5. A determination of the therapeutic window in order to know how early treatment must begin. It has been argued that even if a particular agent only has an hour window in a rat stroke, TBI, or SCI model, the window in humans with the corresponding condition is likely to be much longer. However, there is little evidence to support this assumption. Consequently, clinical trial design should probably take the preclinical therapeutic window definition for a particular agent seriously in regard to how soon the compound may need to be given to patients. With this in mind, a failure to demonstrate a clinically practical therapeutic window for a particular agent in an animal model may mean that this agent and its corresponding secondary injury mechanism may be too short to be effectively addressed in real-world therapeutics.
6. The above parameters' dose-response, optimum treatment duration, and therapeutic window are most likely to vary between TBI, ischemic stroke, cardiac arrest/resuscitation, SAH, and SCI models.
7. A comparison of the neuroprotective pharmacology (e.g., dose-response, optimum treatment duration and therapeutic window) in multiple injury models (e.g., focal vs. diffuse TBI) in order to determine whether the agent in question only works in certain types of injuries.
8. A comparison of the neuroprotective pharmacology in male versus female animals.
9. A determination of pharmacodynamic and pharmacokinetic interactions with other commonly used ancillary treatments (e.g., anticonvulsants, minor and major tranquilizers).

REFERENCES

1. Cernak, I. 2005. Animal models of head trauma. *NeuroRx* 2(3):410–422.
2. Kline, A.E., Dixon, C.E. 2001. Contemporary in vivo models of brain trauma and a comparison of injury responses. In *Head Trauma: Basic, Preclinical and Clinical Directions*, eds. L.P. Miller, R.L. Hayes, 65–84. New York: Wiley-Liss.
3. LaPlaca, M.C. et al. 2007. CNS injury biomechanics and experimental models. *Prog Brain Res* 161:13–26.
4. Marklund, N., Hillered, L. 2011. Animal modelling of traumatic brain injury in preclinical drug development: Where do we go from here? *Br J Pharmacol* 164(4):1207–1229.
5. Morales, D.M. et al. 2005. Experimental models of traumatic brain injury: Do we really need to build a better mousetrap? *Neuroscience* 136(4):971–989.
6. Morganti-Kossmann, M.C., Yan, E., Bye, N. 2010. Animal models of traumatic brain injury: Is there an optimal model to reproduce human brain injury in the laboratory? *Injury* 41(Suppl 1):S10–S13.
7. O'Connor, W.T., Smyth, A., Gilchrist, M.D. 2011. Animal models of traumatic brain injury: A critical evaluation. *Pharmacol Ther* 130(2):106–113.
8. Spaethling, J.M. et al. 2007. Linking impact to cellular and molecular sequelae of CNS injury: Modeling in vivo complexity with in vitro simplicity. *Prog Brain Res* 161:27–39.
9. Povlishock, J.T. 2000. Pathophysiology of neural injury: Therapeutic opportunities and challenges. *Clin Neurosurg* 46:113–126.

10. Povlishock, J.T. et al. 1999. Initiating mechanisms involved in the pathobiology of traumatically induced axonal injury and interventions targeted at blunting their progression. *Acta Neurochir Suppl* 73:15–20.
11. Povlishock, J.T., Jenkins, L.W. 1995. Are the pathobiological changes evoked by traumatic brain injury immediate and irreversible? *Brain Pathol* 5(4):415–426.
12. Sullivan, P.G. et al. 2002. Cytochrome c release and caspase activation after traumatic brain injury. *Brain Res* 949(1–2):88–96.
13. Sullivan, P.G. et al. 1998. Traumatic brain injury alters synaptic homeostasis: Implications for impaired mitochondrial and transport function. *J Neurotrauma* 15(10):789–798.
14. Sullivan, P.G. et al. 2005. Mitochondrial permeability transition in CNS trauma: Cause or effect of neuronal cell death? *J Neurosci Res* 79(1–2):231–239.
15. Fournier, A.E., Strittmatter, S.M. 2001. Repulsive factors and axon regeneration in the CNS. *Curr Opin Neurobiol* 11(1):89–94.
16. Gallo, G., Letourneau, P.C. 2004. Regulation of growth cone actin filaments by guidance cues. *J Neurobiol* 58(1):92–102.
17. McKerracher, L. 2001. Spinal cord repair: Strategies to promote axon regeneration. *Neurobiol Dis* 8(1):11–18.
18. Ren, J.M., Finklestein, S.P. 2005. Growth factor treatment of stroke. *Curr Drug Targets CNS Neurol Disord* 4(2):121–125.
19. Murray, G.D., Teasdale, G.M., Schmitz, H. 1996. Nimodipine in traumatic subarachnoid haemorrhage: A re-analysis of the HIT I and HIT II trials. *Acta Neurochir (Wien)* 138(10):1163–1167.
20. Pickard, J.D. et al. 1989. Effect of oral nimodipine on cerebral infarction and outcome after subarachnoid haemorrhage: British aneurysm nimodipine trial. *BMJ* 298(6674):636–642.
21. Bains, M., Hall, E.D. 2012. Antioxidant therapies in traumatic brain and spinal cord injury. *Biochim Biophys Acta* 1822(5):675–684.
22. Narayan, R.K. et al. 2002. Clinical trials in head injury. *J Neurotrauma* 19(5):503–557.
23. Yurkewicz, L. et al. 2005. The effect of the selective NMDA receptor antagonist traxoprodil in the treatment of traumatic brain injury. *J Neurotrauma* 22(12):1428–1443.
24. Faden, A.I. et al. 1989. The role of excitatory amino acids and NMDA receptors in traumatic brain injury. *Science* 244(4906):798–800.
25. Hinzman, J.M. et al. 2010. Diffuse brain injury elevates tonic glutamate levels and potassium-evoked glutamate release in discrete brain regions at two days post-injury: An enzyme-based microelectrode array study. *J Neurotrauma* 27(5):889–899.
26. Hinzman, J.M. et al. 2012. Disruptions in the regulation of extracellular glutamate by neurons and glia in the rat striatum two days after diffuse brain injury. *J Neurotrauma* 29(6):1197–1208.
27. Biegon, A. et al. 2004. Dynamic changes in N-methyl-D-aspartate receptors after closed head injury in mice: Implications for treatment of neurological and cognitive deficits. *Proc Natl Acad Sci U S A* 101(14):5117–5122.
28. Adeleye, A. et al. 2010. D-cycloserine improves functional outcome after traumatic brain injury with wide therapeutic window. *Eur J Pharmacol* 629(1–3):25–30.
29. Muizelaar, J.P., Kupiec, J.W., Rapp, L.A. 1995. PEG-SOD after head injury. *J Neurosurg* 83(5):942.
30. Hall, E.D., McCall, J.M., Means, E.D. 1994. Therapeutic potential of the lazaroids (21-aminosteroids) in acute central nervous system trauma, ischemia and subarachnoid hemorrhage. *Adv Pharmacol* 28:221–268.
31. Hall, E.D. 1997. Lazaroid: Mechanisms of Action and Implications for Disorders of the CNS. *Neuroscientist* 3:42–51.

32. Marshall, L.F. et al. 1998. A multicenter trial on the efficacy of using tirilazad mesylate in cases of head injury. *J Neurosurg* 89(4):519–525.
33. Kassell, N.F. et al. 1996. Randomized, double-blind, vehicle-controlled trial of tirilazad mesylate in patients with aneurysmal subarachnoid hemorrhage: A cooperative study in Europe, Australia, and New Zealand. *J Neurosurg* 84(2):221–228.
34. Cheng, Y.D., Al-Khoury, L., Zivin, J.A. 2004. Neuroprotection for ischemic stroke: Two decades of success and failure. *NeuroRx* 1(1):36–45.
35. Labiche, L.A., Grotta, J.C. 2004. Clinical trials for cytoprotection in stroke. *NeuroRx* 1(1):46–70.
36. Bracken, M.B. et al. 1997. Administration of methylprednisolone for 24 or 48 hours or tirilazad mesylate for 48 hours in the treatment of acute spinal cord injury. Results of the Third National Acute Spinal Cord Injury Randomized Controlled Trial. National Acute Spinal Cord Injury Study. *JAMA* 277(20):1597–1604.
37. Lanzino, G., Kassell, N.F. 1999. Double-blind, randomized, vehicle-controlled study of high-dose tirilazad mesylate in women with aneurysmal subarachnoid hemorrhage. Part II. A cooperative study in North America. *J Neurosurg* 90(6):1018–1024.
38. Lanzino, G. et al. 1999. Double-blind, randomized, vehicle-controlled study of high-dose tirilazad mesylate in women with aneurysmal subarachnoid hemorrhage. Part I. A cooperative study in Europe, Australia, New Zealand, and South Africa. *J Neurosurg* 90(6):1011–1017.
39. Teasdale, G., Jennett, B. 1974. Assessment of coma and impaired consciousness. A practical scale. *Lancet* 2(7872):81–84.
40. Teasdale, G., Jennett, B. 1978. Assessment of coma and severity of brain damage. *Anesthesiology* 49(3):225–226.
41. Saatman, K.E. et al. 2008. Classification of traumatic brain injury for targeted therapies. *J Neurotrauma* 25(7):719–738.
42. Adams, J.H., Jennett, B., Murray, L.S., Teasdale, G.M., Gennarelli, T.A. Graham, D.I. 2011. Neuropathological findings in disabled survivors of a head injury. *J Neurotrauma* 28(5):701–709.
43. Namjoshi, D.R. et al. 2013. Towards clinical management of traumatic brain injury: A review of models and mechanisms from a biomechanical perspective. *Dis Model Mech* 6(6):1325–1338.
44. Kontos, H.A., Wei, E.P. 1986. Superoxide production in experimental brain injury. *J Neurosurg* 64(5):803–807.
45. Dixon, C.E. et al. 1987. A fluid percussion model of experimental brain injury in the rat. *J Neurosurg* 67(1):110–119.
46. McIntosh, T.K. et al. 1987. Traumatic brain injury in the rat: Characterization of a midline fluid-percussion model. *Cent Nerv Syst Trauma* 4(2):119–134.
47. Hall, K.D., Lifshitz, J. 2010. Diffuse traumatic brain injury initially attenuates and later expands activation of the rat somatosensory whisker circuit concomitant with neuroplastic responses. *Brain Res* 1323:161–173.
48. Foda, M.A., Marmarou, A. 1994. A new model of diffuse brain injury in rats. Part II: Morphological characterization. *J Neurosurg* 80(2):301–313.
49. Marmarou, A. et al. 1994. A new model of diffuse brain injury in rats. Part I: Pathophysiology and biomechanics. *J Neurosurg* 80(2):291–300.
50. Hall, E.D. 1995. The mouse head injury model: Utility in the discovery of acute cerebroprotective agents. In *Central Nervous System Trauma Research Techniques*, edited by S T and Onishi Onishi, T. 213–233, Boca Raton, FL: CRC Press.
51. Hall, E.D. 1985. High-dose glucocorticoid treatment improves neurological recovery in head-injured mice. *J Neurosurg* 62(6):882–887.

52. Hall, E.D. 1986. Beneficial effects of acute intravenous ibuprofen on neurological recovery of head injured mice: Comparison of cyclooxygenase inhibition of thromboxane A2 synthetase or 5-lipoxygenase. *CNS Trauma* 2:75–83.
53. Hall, E.D. 1987. Beneficial effects of the 21-aminosteroid U74006F in acute CNS trauma and hypovolemic shock. *Acta Anaesthesiol Belg* 38(4):421–425.
54. Hall, E.D. et al. 2004. Peroxynitrite-mediated protein nitration and lipid peroxidation in a mouse model of traumatic brain injury. *J Neurotrauma* 21(1):9–20.
55. Kupina, N.C. et al. 2003. Cytoskeletal protein degradation and neurodegeneration evolves differently in males and females following experimental head injury. *Exp Neurol* 180(1):55–73.
56. Mesenge, C. et al. 1998. Reduction of tyrosine nitration after N(omega)-nitro-L-arginine-methylester treatment of mice with traumatic brain injury. *Eur J Pharmacol* 353(1):53–57.
57. Mesenge, C. et al. 1998. Protective effect of melatonin in a model of traumatic brain injury in mice. *J Pineal Res* 25(1):41–46.
58. Mesenge, C. et al. 1996. Reduction of the neurological deficit in mice with traumatic brain injury by nitric oxide synthase inhibitors. *J Neurotrauma* 13(4):209–214.
59. Chen, Y. et al. 1996. An experimental model of closed head injury in mice: Pathophysiology, histopathology, and cognitive deficits. *J Neurotrauma* 13(10):557–568.
60. Flierl, M.A. et al. 2009. Mouse closed head injury model induced by a weight-drop device. *Nat Protoc* 4(9):1328–1337.
61. Laskowitz, D.T. et al. 2007. COG1410, a novel apolipoprotein E-based peptide, improves functional recovery in a murine model of traumatic brain injury. *J Neurotrauma* 24(7):1093–1107.
62. Tang, Y.P. et al. 1997. A concussive-like brain injury model in mice (I): Impairment in learning and memory. *J Neurotrauma* 14(11):851–862.
63. Tang, Y.P. et al. 1997. A concussive-like brain injury model in mice (II): Selective neuronal loss in the cortex and hippocampus. *J Neurotrauma* 14(11):863–873.
64. Smith, D.H., Meaney, D.F., Shull, W.H. 2003. Diffuse axonal injury in head trauma. *J Head Trauma Rehabil* 18(4):307–316.
65. Lighthall, J.W. 1988. Controlled cortical impact: A new experimental brain injury model. *J Neurotrauma* 5(1):1–15.
66. Dixon, C.E. et al. 1991. A controlled cortical impact model of traumatic brain injury in the rat. *J Neurosci Methods* 39(3):253–262.
67. Smith, D.H. et al. 1995. A model of parasagittal controlled cortical impact in the mouse: Cognitive and histopathologic effects. *J Neurotrauma* 12(2):169–178.
68. Hannay, H.J. et al. 1999. Validation of a controlled cortical impact model of head injury in mice. *J Neurotrauma* 16(11):1103–1114.
69. Hall, E.D. et al. 2005. Spatial and temporal characteristics of neurodegeneration after controlled cortical impact in mice: More than a focal brain injury. *J Neurotrauma* 22(2):252–265.
70. Hall, E.D. et al. 2008. Evolution of post-traumatic neurodegeneration after controlled cortical impact traumatic brain injury in mice and rats as assessed by the de Olmos silver and fluorojade staining methods. *J Neurotrauma* 25(3):235–247.
71. Stein, D.G., Hoffman, S.W. 2003. Estrogen and progesterone as neuroprotective agents in the treatment of acute brain injuries. *Pediatr Rehabil* 6(1):13–22.
72. McIntosh, T.K. et al. 1989. Traumatic brain injury in the rat: Characterization of a lateral fluid-percussion model. *Neuroscience* 28(1):233–244.
73. Maegele, M. et al. 2005. Characterization of a new rat model of experimental combined neurotrauma. *Shock* 23(5):476–481.

74. Cunningham, T.L. et al. 2014. Correlations between blood-brain barrier disruption and neuroinflammation in an experimental model of penetrating ballistic-like brain injury. *J Neurotrauma* 31(5):505–514.
75. Leung, L.Y. et al. 2013. The acute effects of hemorrhagic shock on cerebral blood flow, brain tissue oxygen tension, and spreading depolarization following penetrating ballistic-like brain injury. *J Neurotrauma* 30(14):1288–1298.
76. Shear, D.A. et al. 2011. Severity profile of penetrating ballistic-like brain injury on neurofunctional outcome, blood-brain barrier permeability, and brain edema formation. *J Neurotrauma* 28(10):2185–2195.
77. Williams, A.J., Ling, G.S., Tortella, F.C. 2006. Severity level and injury track determine outcome following a penetrating ballistic-like brain injury in the rat. *Neurosci Lett* 408(3):183–188.
78. Williams, A.J. et al. 2007. Acute and delayed neuroinflammatory response following experimental penetrating ballistic brain injury in the rat. *J Neuroinflammation* 4:17.
79. Zoltewicz, J.S. et al. 2013. Biomarkers track damage after graded injury severity in a rat model of penetrating brain injury. *J Neurotrauma* 30(13):1161–1169.
80. Wei, H.H. et al. 2009. NNZ-2566 treatment inhibits neuroinflammation and pro-inflammatory cytokine expression induced by experimental penetrating ballistic-like brain injury in rats. *J Neuroinflammation* 6:19.
81. Ahmed, F. et al. 2012. Time-dependent changes of protein biomarker levels in the cerebrospinal fluid after blast traumatic brain injury. *Electrophoresis* 33(24):3705–3711.
82. Ahmed, F.A. et al. 2013. Long-term consequences of single and multiple mild blast exposure on select physiological parameters and blood-based biomarkers. *Electrophoresis* 34(15):2229–2233.
83. Gyorgy, A. et al. 2011. Time-dependent changes in serum biomarker levels after blast traumatic brain injury. *J Neurotrauma* 28(6):1121–1126.
84. Kamnaksh, A. et al. 2011. Factors affecting blast traumatic brain injury. *J Neurotrauma* 28(10):2145–2153.
85. Kamnaksh, A. et al. 2012. Neurobehavioral, cellular, and molecular consequences of single and multiple mild blast exposure. *Electrophoresis* 33(24):3680–3692.
86. Kovacs, S.K., Leonessa, F., Ling, G.S. 2014. Blast TBI models, neuropathology, and implications for seizure risk. *Front Neurol* 5:47.
87. Turner, R.C. et al. 2013. Modeling clinically relevant blast parameters based on scaling principles produces functional and histological deficits in rats. *Exp Neurol* 248:520–529.
88. Miller, J.D. et al. 1990. Ischemic brain damage in a model of acute subdural hematoma. *Neurosurgery* 27(3):433–439.
89. Ginsberg, M.D., Busto, R. 1998. Small animal models of global and focal cerebral ischemia. In *Cerebrovascular Disease: Pathophysiology, Diagnosis and Management*, eds. M.D. Ginsberg, J. Bogousslavsky, 14–35. Malden, MA: Blackwell Science.
90. Bramlett, H.M., Dietrich, W.D. 2001. Neuropathological protection after traumatic brain injury in intact female rats versus males or ovariectomized females. *J Neurotrauma* 18(9):891–900.
91. Bramlett, H.M. et al. 1997. Chronic histopathological consequences of fluid-percussion brain injury in rats: Effects of post-traumatic hypothermia. *Acta Neuropathol* 93(2): 190–199.

12 Biomarkers of Traumatic Brain Injury and Their Relationship to Pathology

Jin Zhang, Vikram Puvenna, and Damir Janigro

CONTENTS

RADIOLOGICAL MARKERS OF TRAUMATIC BRAIN INJURY (TBI) AND THEIR RELATIONSHIP TO SERUM OR CEREBROSPINAL FLUID MARKERS

Traumatic brain injuries (TBIs) can occur in a number of ways and can be classified by the mechanism of injury, the clinical severity as graded by the Glasgow Coma Scale (GCS), or by the characterization of structural damage. The heterogeneity of the disease makes it difficult to accurately assess the level of trauma and predict the clinical outcome of the patient. Generally, injuries are classified as mild, moderate, or severe depending on the GCS score, which utilizes motor, eye, and verbal responses to evaluate the level of consciousness of the patient. In addition to the GCS, TBI may be classified according to a clinical injury severity score based on the level of injury incurred to various key body regions, according to the level of structural damage incurred, and according to patient prognosis based on various prognostic models.[1] TBI injury is heterozygous in nature, and no single classification scheme is sufficient in characterizing injury for the purposes of diagnosis and prognosis. Depending on the severity of the initial insult, different imaging modalities are used to obtain the necessary information for patient care and prognosis. Whereas evidence of moderate to severe TBI is usually reliably visible as structural abnormalities using techniques such as computed tomography (CT) or conventional magnetic resonance imaging (MRI), more subtle disturbances characteristic of mild TBI are not so easily demonstrated by these imaging modalities. Mild TBI results from the main etiologies of neural contusion and axonal injury, which subsequently results in biochemical, metabolic, and cellular changes that may be responsible for some of the long-term problems seen in patients who develop postconcussion syndrome (PCS).[2]

Since several if not all imaging modalities employed in TBI diagnosis are dependent on structural of functional rearrangement of cellular or extracellular components or assessment of blood–brain barrier (BBB) disruption, it is not surprising that surrogate peripheral or central nervous system (CNS) correlates of CT or MRI have been described.[3–10] In the following sections we will describe, side by side, the use of radiological markers and their surrogates in the diagnosis of TBI. We will first, however, illustrate the rationale for employing peripheral or CNS markers as surrogates of radiologic investigations.

HOW CAN A BLOOD TEST COMPLEMENT MAGNETIC RESONANCE IMAGING OR COMPUTED TOMOGRAPHY?

CT scans administered after suspected TBI are primarily used to assess cerebrovascular integrity (micro- and macrobleeds) or to determine gross anatomical changes induced to the skull or brain/meninges. Unlike CT, MRI is often used to visualize regions of increased brain activity (functional MRI and autoregulation studies) but can, similar to CT, be employed to measure cerebral blood flow (MRA) or to unveil and localize regions of blood–brain barrier disruption (contrast MRI; measurements of endothelial tight junction integrity in normal or diseased brain). Positron-emission

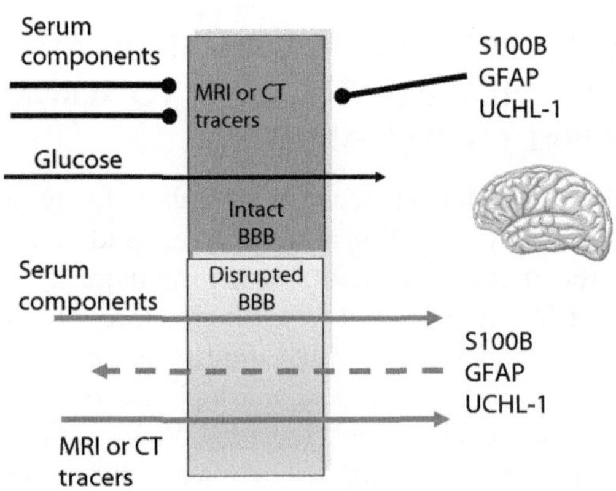

FIGURE 12.1 **(See color insert.)** Under conditions of an intact BBB, serum extravasation of brain markers and leakage of radiological tracers is prevented by the tight junction. When the BBB is disrupted, both radiologic and protein markers cross the endothelial layer.

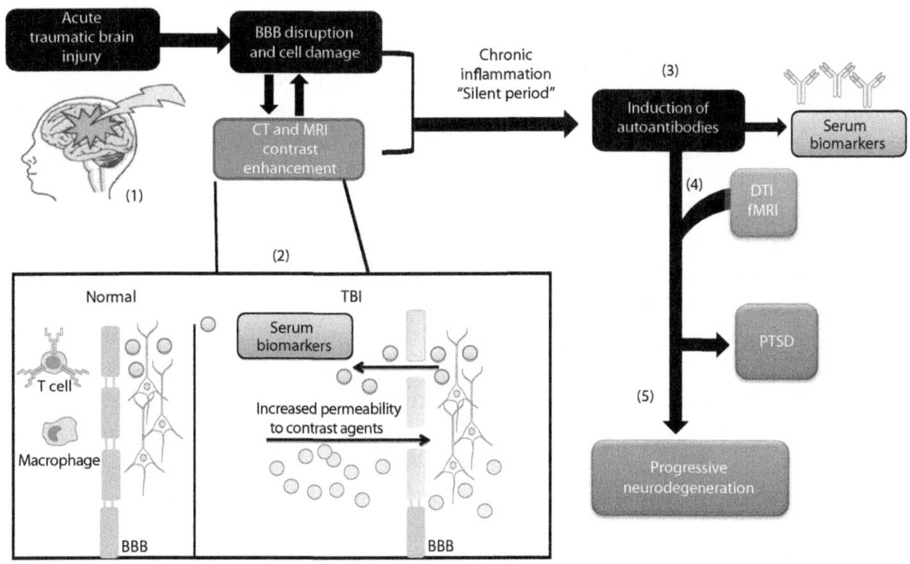

FIGURE 12.2 **(See color insert.)** Schematic of research hypothesis: (1) Single or repetitive TBI triggers CNS damage, mechanical BBB disruption (BBBD) and inflammation. (2) Release of pro-inflammatory cytokines including IL-1β, TNF-α, and IL-6 induces activation of peripheral immune cells and activation of resident microglia. Increased BBB permeability due to cytokine release, apoptosis, and mechanical BBBD facilitates extravasation of leukocytes into the parenchyma. (3) Factors such as genetic background (e.g., HLA), injury severity, history of infection, or excessive immune activation (e.g., due to poor clearance of debris) predispose an individual to chronic inflammation and autoantibody induction. These autoantibodies may serve as biomarkers for increased Alzheimer's disease risk. (4) CNS autoreactive antibodies circulate in the periphery and contribute to the development of Alzheimer's disease in conjunction with other risk factors that increase inflammation, BBB permeability, etc. (5) Identification of biomarker autoantibodies paves the way for early interventions including immunosuppressive therapy or cognitive therapy.

tomography (PET) scans can exploit brain-directed glucose transport to determine brain metabolism and to identify regions of abnormal uptake (hypo- or hypermetabolism seen in epilepsy or neoplasms). Although functional MRI has not (yet) produced viable surrogate blood markers, angiographic techniques using contrast agents have been exploited as a means to discover nonradiologic surrogates of imaging techniques. For example, since gadolinium extravasation ultimately depends on exchange of matter between the peripheral circulation and the brain, it was postulated that following the fate of extravasated (brain to blood) cerebrospinal fluid (CSF) or parenchymal protein, one may predict whether contrast extravasation would be detected. This was confirmed in several studies comparing systemic levels of S100B to enhanced MRI or CT scans (see Figures 12.1 and 12.2).[6,11–16]

Opening of the BBB provides molecules normally present in blood with open passage into the CNS. However, this "opening," unless involving specific transporters, does not only work in one direction. Proteins normally present in blood are free to diffuse into the CNS, and in turn, proteins normally present in high concentrations in the CNS are free to diffuse down concentration gradients into the blood. These peripheral markers of BBB can be detected in the blood in order to evaluate the permeability characteristics of the BBB at any given time. Marchi et al.[9] discussed the ideal properties of a peripheral marker of BBB disruption. Such proteins should have low or undetectable plasma levels in normal subjects, be normally present in CSF, and have a normally higher concentration in the CSF than in plasma. Additionally, the CSF concentration of the protein should increase in response to insult. The protein should be normally blocked by the BBB and exhibit flux across the BBB during barrier disruption. Several proteins, including S100B, neuron-specific enolase (NSE), and glial fibrillary acidic protein (GFAP) have been evaluated for this purpose, but only S100B is used so far, perhaps owing to very low plasma levels with a concentration tenfold less than that found in the CSF in normal subjects.

Figures 12.1 and 12.2 summarize the rationale used to develop surrogate markers of contrast-enhanced CT or MRI scans. Note that the comparison between tracer entry into the parenchyma versus egress of S100B or GFAP from brain to blood has been greatly simplified, since the properties of brain and blood are not easily described by static, well-defined topographic entities.

X-RAY COMPUTED TOMOGRAPHY (CT)

X-ray computed tomography (CT) is a traditionally and frequently used imaging modality for assessing damage in TBI, especially during the acute phase of injury. The technique of CT imaging relies on the emission of x-ray beams in thin axial slices along the object and an x-ray detector positioned on the opposite side that measures x-ray attenuation per voxel through the object. Using this information, a 3D image of the scanned object will be digitally reconstructed. CT is best suited for detection of structural damage and abnormalities in the acute setting and more sensitive for cases of moderate or severe TBI. CT has shown to be effective in detecting macroscopic lesions and physical changes in the skull and brain, including intracranial lesions, cerebral contusions, edema, epidural and subdural hematomas, skull fractures, pneumocephalus, and brain swelling.[17]

In the acute period of TBI, CT is sensitive in detecting skull fractures, intracranial lesions, intracranial hematomas such as subarachnoid hemorrhages and subdural hematomas, and mass effect.[17] Detection of these macroscopic lesions in the acute setting of a trauma is crucial for patient prognosis and provides essential information for determining correct patient management. Intracranial hemorrhages as a result of trauma will appear hyperdense compared to gray matter in the acute setting, but transitions become isodense during the subacute and chronic periods. However, CT is not highly sensitive in cases of mild TBI (mTBI) due to the frequently diffuse and microscopic nature of injury in mTBI. A 2004 report from the WHO mTBI task force indicated that CT imaging was able to detect abnormalities indicative of mild subarachnoid hemorrhage, subdural hemorrhage, or contusions in about 10% of mTBI patients but 30%–80% of mTBI patients will later experience symptoms congruent with PCS.[18] The discrepancy between CT abnormalities and eventual symptoms of mTBI indicates that CT does not have sufficient sensitivity in detecting damage incurred in mTBI. A study comparing sensitivity of MRI to CT found that CT detected 22% of hemorrhagic diffuse axonal injury (DAI) cases and 0% of non-hemorrhagic cases.[19] Due to the radiation exposure of this imaging modality and low sensitivity for mTBI, it is questionable whether it is recommended for patients with minor head injuries. Thus, guidelines have been formulated for clinical decision making for use of CT such as the Canadian CT Head Rule and the New Orleans Criteria. These guidelines incorporate the mechanism of injury; symptoms after injury such as headache, vomiting, and amnesia; the GCS score, and visible signs of trauma in the evaluation of head injuries; and the necessity of CT. Current guidelines recommend that all TBI patients with a GCS score of 14 and below to undergo CT examination and TBI patients with a GCS score of 15 in conjunction with predetermined risk factors to undergo CT examination. CT findings are graded according to the Marshall CT classification or Rotterdam CT score in order to classify the severity of the TBI and can be used to predict clinical outcomes of the patient.

SERUM AND CEREBROSPINAL FLUID MARKERS FOR TBI IN THE ACUTE SETTING

With consideration to the radiation exposure of CT imaging, it would be advantageous to develop and validate peripheral serum markers for the diagnosis and prognosis of TBI in cases of acute trauma. In addition, pathophysiological changes on the microscopic level induced by mild TBI are often not visible via CT or MRI imaging but prove to be clinically significant in the long-term prognosis of the patient. Extensive literature is available on surrogate serum and CSF markers of severe TBI correlating levels of measured biomarkers to both the severity of the trauma and cognitive performance and outcome. Serum biomarkers that can be obtained through a blood draw and analyzed soon after injury can provide a highly convenient method of diagnosis and prognosis. Currently, the selection of viable biomarkers for TBI is based on the hypothesized sites and mechanisms of injury induced by trauma, such as acute axonal injury, acute neuronal injury, amyloid-related processes, acute astroglial injury, and neuroinflammation. Various markers may be detectable in both the CSF and peripheral blood, depending on the integrity of the BBB. Integrity of the BBB in response to trauma and other pathological causes can be assessed using the CSF:serum albumin ratio. When the BBB has been disrupted, the CSF:serum albumin ratio will be elevated, as is often the case in severe TBI. In mTBI, the CSF:serum albumin ratio is largely normal. Biomarkers of acute astroglial injury include elevated S100B and elevated GFAP, which can be detected in both CSF and peripheral blood.[20] S100B is traditionally thought of as an astrocytic protein, but evidence now shows that this protein can also be found in oligodendrocytes and peripheral cells such as chondrocytes and adipocytes.[21] The potential for normal extracerebral production of S100B must be taken into account when measuring peripheral levels of S100B. Nonetheless, the literature supports the potential use of S100B as a peripheral serum marker of TBI and such markers have been shown to improve the predictive power of clinical outcome calculating models.[22] A study of amateur boxers who experienced minor head injury found that both S100B and GFAP were slightly elevated upon lumbar puncture.[20] As a CSF marker, S100B and

GFAP are less sensitive than other CSF markers such as NLF or NFH. However, S100B and GFAP hold promise as potential peripheral markers for TBI. The magnitude of S100B and GFAP elevation has been correlated to the severity of TBI, scoring on the Glasgow Coma Scale, as well as radiological findings at hospital admission.[23] Peripheral measurements of S100B and GFAP may potentially be used to help differentiate between severe and mild TBI. In addition, serum S100B may be measured to determine the necessity of a CT scan in acute settings and can reduce unnecessary radiation exposure and costs associated with CT. A study found that low serum levels of S100B (below 0.10 μg/L) had a negative predictive value of 90% to 100% for a normal CT scan in patients with minor head injuries.[15] This study supports the potential of S100B to act as a biomarker for clinical decision making and reducing the number of unnecessary CTs routinely ordered.

Biomarkers of acute neuronal injury in TBI include γ-enolase (NSE) and the more novel biomarkers α-II spectrin and UCH-L1. In studies of severe TBI, NSE was found to be elevated in both ventricular CSF and peripheral serum, with a higher magnitude of elevation corresponding to higher mortality and a more severe score on the GCS for both adults and children.[23] Studies of NSE in mTBI are lacking. However, a major limitation of NSE as a biomarker is its low specificity largely due to its high sensitivity for elevations due to hemolysis. Despite its name of neuron-specific enolase, NSE is also naturally expressed in erythrocytes and may be elevated due to hemolysis or contamination of CSF with peripheral blood. UCH-L1 and α-II spectrin breakdown products are novel biomarkers naturally expressed in neurons, which have been found to be elevated in CSF and peripheral blood in response to TBI. The elevation of spectrin breakdown products and UCH-L1 (as well as GFAP) has been correlated to the severity of trauma and provides more predictive power for the IMPACT outcome calculator for patients with severe TBI.[22] A pilot study of UCH-L1 in mild to moderate TBI, UCH-L1 was shown to be elevated 1 hour postinjury and was found to be correlated with injury severity, the GCS score, and positive lesion findings on CT imaging.

Fluid biomarkers of acute axonal injury include tau proteins, neurofilament light (NFL) peptide, and phosphorylated neurofilament heavy (NFH) peptide. Tau proteins are naturally expressed in thin, nonmyelinated axons of cortical interneurons, whereas NFL are expressed in large-caliber myelinated axons in the spinal cord and deeper brain layers. In severe TBI, total tau protein levels are elevated in ventricular CSF and levels correspond to the lesion size, injury severity, and clinical outcome.[24–26] In mTBI, total tau protein levels are elevated and peak at 4–10 days postinjury in lumbar puncture CSF but normalize 8–12 weeks postinjury. Similarly both NFL and NFH are elevated in CSF in severe TBI.[20,23] CSF measurements of NFL are regarded as the most sensitive marker of axonal injury.

Finally, TBI induces a neuroinflammatory process that can be detected through inflammatory protein markers such as IL-6, IL-8, and IL-10 in CSF. Magnitude of ventricular CSF elevations of inflammatory cytokines has been correlated with patient prognosis and level of BBB dysfunction as measured by CSF:serum albumin ratios. However, few studies are available for inflammatory markers in mTBI or using lumbar punctures rather than ventricular CSF. Ventricular CSF measurements are not usually obtained in the clinical setting for cases of mTBI, and so the applicability of these findings is questionable.

CONVENTIONAL MRI

Conventional MRI is typically not used in the acute evaluation of TBI but is often utilized in subacute and chronic cases when postconcussive symptoms are persistent. In acute settings, CT is usually used in favor of MRI due to cost and equipment availability constraints. MRI provides similar structural information compared to CT but offers advantages in subacute and chronic settings by providing additional information on white matter lesions and is superior in detecting small intraparenchymal abnormalities such as small contusions. In addition, MRI has been shown to be effective in the detection of structural hemorrhagic lesions, axonal injuries, hemosiderin deposits, and post-traumatic encephalomalacia.[27] Compared to CT, MRI is more sensitive in detection of lesions associated with TBI. A study on patients with a Glasgow score between 13 and 15, loss of consciousness fewer than 30 min, and post-traumatic amnesia showed that conventional MRI was able to detect parenchymal lesions in 75% of patients, nonhemorrhagic DAI in 11% of patients, hemorrhagic DAI in 47% of patients, and contusion in 57% of patients, as opposed to 50%, 0%, 22%, and 35%, respectively, for CT.[19]

MRI imaging provides fine anatomical detail that is not readily available on CT imaging and therefore has greater sensitivity compared to CT in detection of various types of lesions due to TBI. MRI imaging is conceptually based on the nuclear magnetic resonance (NMR) phenomenon and relies on the excitation and relaxation of proton spins in a powerful external magnetic field. The quantum behavior of the protons varies depending on its environment, thus allowing for the differentiation of various tissue types upon imaging. Conventional MRI can be adjusted to provide various types of images including T1-weighted, T2-weighted, T2-fast spin echo, T2-gradient echo, and fluid-attenuated inversion recovery (FLAIR) sequences, each highlighting different tissue structures and components of the brain. For example, FLAIR imaging suppresses hyperintense signals from free-flowing water such as in the CSF and therefore provides more sensitivity in detecting hemorrhagic lesions compared to T2-weighted imaging. In T2-weighted imaging, the hyperintensity of CSF under T2-weight images may mask abnormalities in the parenchyma adjacent to the ventricles or pia. Conventional MRI imaging is effective in detecting structural abnormalities of TBI in finer detail when compared to CT and has the advantage of not utilizing ionizing radiation. However, it is limited in detecting bone damage and fresh hemorrhages when compared to CT.

ADVANCED MRI IMAGING

Exciting advancements have been made in MRI imaging that provide more detailed information on cellular or microscopic damage or changes induced by TBI. Microscopic lesions such as nonhemorrhagic DAI are not readily detected on neither conventional MRI nor CT but has long-term clinical impact in terms of symptoms related to PCS, including headaches, dizziness, memory deficits, cognitive deficits, anxiety, and depression months to years after initial injury. We will focus our discussion on the modalities of diffusion-weighted imaging (DWI), susceptibility-weighted imaging (SWI), and diffusion tensor imaging

(DTI). These modalities are particularly useful in detecting damage sooner after the traumatic event or by detecting subtle damage such DAI, which would not be readily apparent by CT or conventional MRI.

Diffusion-Weighted Imaging

The concept of DWI relies on measuring the diffusion of water protons throughout brain tissue and sensing any disturbances in the natural Brownian motion of water protons caused by injury. Areas of disturbed diffusion caused by injury appear as bright white regions on a DWI. DTI has important clinical applications in detecting cellular changes caused by acute ischemic stroke within 30 minutes of stroke onset. More recently, clinical applications of DTI have expanded beyond acute stroke as a way to characterize white matter damage or abnormalities that present as diffusional abnormalities. In the context of TBI, DWI is useful for shearing injuries, hemorrhages, and edema. An advantage of DWI over conventional MRI is that abnormalities in diffusion can be telltale for microscopic damage in white matter tracts that are not easily visible through CT or conventional MRI. DAIs are common among patients of mild TBI and are believed to be responsible for some of the persistent and lingering neurological symptoms and problems long after the traumatic incident. DAI is not well suited for imaging by CT or conventional MRI due to the subtle and diffuse nature of the injury and are not characterized by localized lesions. DWI is much more effective at detection of DAI and can be utilized in the acute phase of mild TBI.

Diffusion Tensor Imaging

DTI is an extension of DWI that has received an increasing amount of attention for its potential in providing more accurate diagnosis and prognosis for mild TBI patients, in particular regarding DAI and PCS. DTI measures the diffusion of water molecules along white matter tracts and collects information on directionality and rate of diffusion to access the structural integrity of white matter tracts. In a healthy individual, water diffusion should occur along the length of the axons due to natural barriers to orthogonal diffusion such as intact myelin sheaths, axonal membrane, neurofilaments, and microtubules. This normal directional preference for diffusion to occur along the length of white matter tract rather than perpendicular is also referred to as anisotropy. DTI creates a three-dimensional image of diffusion and calculates the apparent diffusion coefficient (ADC) and the fractional anisotropy (FA). The ADC is a measurement of mean diffusivity or the average rate of diffusion across all directions. The fractional anisotropy is measurement that describes the fraction of total diffusion, which lies in the direction of the white matter tracts. An FA of 0 represents total lack of directionality in diffusion and an FA of 1 represents completely anisotropic diffusion along the white matter tracts. Analysis of ADC and FA values using DTI in TBI patients has shown that changes in ADC and FA are indicative of TBI and can be used to predict clinical outcomes of patients in terms of cognitive defects.[28] Patients with moderate to severe TBI have been found to have lower fractional anisotropy on a global level compared to healthy controls.[29] This is unsurprising as lower fractional anisotropy is indicative of white matter pathology, which disrupts the structural organization of white matter and leads to more disorganized, random diffusion of water. Furthermore, white matter load as defined as the total number of regions of interest with reduced FA was shown to be negatively correlated with long-term cognitive deficits in executive and memory domains. A prospective observational study of mild TBI patients demonstrated that mTBI patients have lower trace (mean diffusivity) values and higher FA values when compared to controls in both whole brain and region of interest analyses.[30] These abnormalities correlated with poor clinical outcomes in neurocognitive functions, postconcussive symptoms, and quality of life, and were attributed to the axonal swelling that occurs as an early step in axonal injury in mTBI. A meta-analysis of mTBI across a spectrum of neuroimaging modalities by Eierud et al.[2] supports the finding that acute mTBI is associated with elevated FA, whereas chronic mTBI is associated with depressed FA.[2] In addition, meta-analysis showed that cognitive performance is negatively correlated with FA in the acute phases of mTBI and positively correlated with FA in the chronic phases. Furthermore, it is consistent throughout literature that abnormal DTI findings are concentrated in the anterior regions of the brain, which is purposed to be more susceptible to injury in mTBI.[2]

Even more important than the sensitivity of an imaging modality for detecting changes after TBI is its value in outcome prediction and clinical implications for the patient. The outcome prediction capabilities of DTI are not completely established, but some studies support the potential usage of DTI in the acute and subacute stage for long-term clinical prognosis. The prediction capabilities of DTI appear to be highly dependent on patient characteristics such as prior substance abuse and preexisting neuropsychiatric disorders. In studies where patients with a prior history of substance abuse and neuropsychiatric disorders were excluded, white matter abnormalities were detectable on DTI in patients with negative CT and conventional MRI compared to controls and abnormalities were associated with differences in clinical outcome. Yuh et al.[31] conducted a study evaluating outcome prediction capabilities of DTI in both an all-inclusive patient population and a subset with prior history of substance abuse and psychiatric disorders excluded. In the all-inclusive population, DTI did not provide additional diagnostic information for patients negative in both CT and conventional MRI. DTI was predictive in extended GOS (GOS-E) outcome but not more so than detectable lesions in conventional MRI.[31] However, when analyzing patients without a prior history of substance abuse or neuropsychiatric disorders, the correlation between both DTI and conventional MRI findings and 3-month and 6-month GOS-E outcome measures was strengthened. The differences in prediction capabilities may be due to the fact that DTI is sensitive to white matter abnormalities result from preexisting conditions and thus diminishes the effect of traumatic lesions on DTI findings. Ultimately, the study found that for the subset of patients, DTI was a stronger predictor of 3- and 6-month outcome than CT findings and known clinical, socioeconomic, and demographic factors. DTI is not yet routinely used clinically but appears to show promise as a future modality for assessing mild TBI lesions not readily characterized by CT or conventional MRI.[31]

Correlating DTI with Peripheral Biomarkers

In conjunction to the utilization of DTI for diagnosis and outcome prediction in mTBI, the use of serum markers in mTBI has been explored as a more clinically practical approach. Peripheral elevation of S100B is recognized as a biomarker of TBI and breaches in the BBB. Marchi et al.[9] conducted a study on 67 college football players with no prior concussions and obtained blood samples pre- and post-games throughout the season to measure serum levels of S100B and the autoantibody against S100B.[9] A subset of the population received DTI imaging. It was found that elevated S100B and S100Bab were correlated with the severity and quantity of head hits received during games. In addition, S100B elevations were correlated to abnormalities in mean diffusivity (MD) in DTI imaging and deficits in cognitive abilities based on cognitive assessments. This study suggests that S100B may be a viable biomarker for the quantification of risk following mild TBI and provides support for the hypothesis that TBI initiates a breach in BBB, serum elevation of S100B, and a subsequent autoimmune process toward S100B and possibly other antigens native to the brain. Similar findings, albeit focused on GFAP and autoantibodies against the same, were recently published.[32] These results, taken together, demonstrate that long-term effects of TBI encompass a robust autoimmune response against astrocyte specific proteins GFAP and S100B. Interestingly, in early stages of Alzheimer's disease, serum IgGs against S100B have been described.[33] The pathogenic significance of brain autoantibodies has been discussed in recent reviews.[34,35] The mechanisms by which these glial or neuronal proteins may induce an autoimmune response are discussed.[3,36–38]

FUNCTIONAL MRI

Whereas the previously mentioned imaging modalities reveal abnormalities in structure, functional MRI (fMRI) focuses on the detection of abnormal brain function through the measurement of blood flow throughout the brain. Measurements for fMRI are taken while a patient is asked to perform a specific task are based on blood oxygen level–dependent (BOLD) signals. Areas of the brain that are activated during the task are highlighted due to a change in the ratio of oxyhemoglobin to deoxyhemoglobin. FMRI is able to detect abnormalities in brain activity in TBI patients who have appeared to make a full recovery when compared to controls. In a study of high school football players, persistent fMRI abnormalities were found in players who were outwardly normal and performed well under neuropsychological examinations.[39] In these patients, the brain's ability to compensate and reallocate resources leads to the overactivation and underactivation of regions in patterns not typical of healthy control.

Of more clinical importance, fMRI can be used to study differences in functional activity between TBI patients who still experience symptoms such as memory and attention deficits. Functional MRI may provide insights into the clinical presentation of TBI and help explain and localize the deficits commonly observed in patients. A meta-analysis showed a general trend for frontal vulnerability in mTBI with a decreased signal in the prefrontal cortex compared to controls, more specifically in the dorsolateral prefrontal cortex (DLPFC).[2] Decreases in fMRI signal in the DLPFC are believed to be correlated with deficits in working memory. Along with the prefrontal cortex, the right cerebellar tonsil and left culmen are commonly affected areas in mTBI as detected by fMRI with implications that the cerebellum is involved in behavioral regulation, working memory, and executive control. In study of subjects with memory deficits and trouble concentrating, the mild TBI patients asked to perform increasingly difficult memory tasks showed variable activation on fMRI rather than increasing activation corresponding to task difficulty. This suggests that the cause of working memory deficits in mTBI patients is due to abnormalities in allocating and modulating resources in response to load rather than a general loss of neurons.

Interestingly, recent evidence suggests that fMRI may also be used as a prognostic measure in predicting sustained attention deficits after TBI.[40] TBI patients who go on to develop sustained attention deficits were found to have an increase in default mode network (DMN) activation, especially in the precuneus and posterior cingulate cortex, compared to controls. This finding is significant in that the increased DMN activation can be detected prior to exhibition of any symptoms related to attention deficits enabling fMRI to serve as a predictive tool.

Another variation of fMRI that shows promise in its diagnostic and predictive value in TBI is resting state function MRI (rs-fMRI), which provides connectivity information of the brain at rest. A study by Mayer et al.[41] demonstrated that rs-fMRI can serve as a radiological marker for mTBI in the absence of T1- or T2-weighted MRI findings as well as predict clinical cognitive, emotional, and somatic symptoms during the semiacute (less than 3 weeks after injury) and late recovery (3–5 months) periods. Mild TBI patients in this study showed decreased functional connectivity in the DMN and increased functional connectivity between the rostral anterior cingulate gyrus (rACC) and lateral prefrontal cortical areas and between the right prefrontal cortex and posterior parietal cortex.[41] Rs-fMRI imaging highlights the importance of baseline functional connectivity in cognitive deficits experienced by TBI patients.

MR SPECTROSCOPY

MR spectroscopy (MRS) is a variation of MR imaging that allows for the detecting and profiling of chemicals and metabolites indicative of neuronal injury or the normal energy metabolism of the brain. The cellular basis for injury in mTBI is primarily cellular degradation, demyelination due to reactive changes in astrocytes, and neuroinflammation. MRS is highly sensitive to damage from TBI and may be useful in prognosis due to its ability to differentiate abnormalities not detectable over CT or MRI.[42] Mild TBI may result in the apparent quick "recovery" of motor and sensory abilities, but patients may have persistent underlying pathophysiological effects of injury long after, involving higher order functions of the brain. In a study of sports concussions, 11 athletes who passed physical and neurocognitive screening for return to play were found to have MRS abnormalities 11 days postinjury.[43] Although these athletes had apparently "recovered" from mTBI based on medical, cognitive, and behavioral indices, there were metabolic alterations in the corpus callosum revealed by MRS indicative of disrupted brain functional connectivity. Likewise, MRS has been shown to be highly sensitive in detecting axonal injury from TBI based on experiments conducted in the corpus callosum.[44–46] MRS can also be potentially used as a prognostic tool through the measurement of metabolite ratios such as the N-acetyl aspartate/creatine (NAA/Cr) ratio measured in one study on severe TBI.[46] Finally, beyond its diagnostic capacity, MRS can be used to track recovery from TBI over time on a molecular pathophysiological level. Ensuring full recovery before clearing athletes for return to play may help prevent the *second-impact syndrome*, in which individuals who have not fully recovered from one mTBI are more susceptible to getting reinjured.[2,42]

SUMMARY

Recent advancements in neuroimaging techniques have allowed us to gain insight into the pathogenesis and prognosis of traumatic brain injury. Techniques such as DTI and rs-fMRI have revealed previously undetectable microscopic structural and connectomic changes in the brain induced by traumatic brain injury, which may explain some of the long-term sequelae observed in TBI patients. Changes detected via advanced neuroimaging techniques have been correlated with clinical outcomes and improve the predictive power of outcome models. Currently, these advanced imaging techniques are used primarily in a research setting, whereas traditional CT and MRI techniques are dominant in clinical situations. MRI techniques, while informative and radiation free, are time consuming and currently high cost. CT, while lower cost compared to MRI, has the disadvantage of radiation exposure. A promising direction for clinical diagnosis and prognosis is the use of fluid biomarkers, with preference for serum biomarkers. Serum biomarkers would provide a potentially very low-cost and time-efficient method for detection, evaluation, and prognosis of TBI. Many serum biomarkers such as S100B and GFAP, among others, have proven to be indicative of molecular and biochemical changes induced by TBI and breaches in the BBB. Furthermore, quantitative levels of CSF and serum markers have been correlated with severity of injury and long-term clinical outcomes. With further validation of serum and CSF biomarkers of TBI, a reliable and cost-effective system of diagnosis and prognosis may be developed. Currently, a need exists for improved point-of-care devices in order to complement fluid biomarkers of TBI. With the refinement and validation of serum biomarkers, we have the potential to use serum biomarkers as a convenient and cost-effective surrogate for currently used imaging modalities for the evaluation of TBI in the clinical setting.

REFERENCES

1. Maas, A.I., Stocchetti, N., Bullock, R. 2008. Moderate and severe traumatic brain injury in adults. *Lancet Neurol* 7(8):728–741.
2. Eierud, C. et al. 2014. Neuroimaging after mild traumatic brain injury: Review and meta-analysis. *Neuroimage Clin* 4:283–294.
3. Bazarian, J.J. et al. 2014. Persistent, long-term cerebral white matter changes after sports-related repetitive head impacts. *PLoS One* 9(4):e94734.

4. Blyth, B. et al. 2011. Elevated serum ubiquitin carboxy-terminal hydrolase L1 is associated with abnormal blood brain barrier function after traumatic brain injury. *J Neurotrauma* 28(12):2453–2462.
5. Janigro, D., Barnett, G., Mayberg, M. 2005. Peripheral marker of blood brain barrier permeability. US patent 6,884,591, filed June 25, 2001, and issued April 26, 2005.
6. Marchi, N. et al. 2003. Peripheral markers of brain damage and blood-brain barrier dysfunction. *Restor Neurol Neurosci* 21(3–4):109–121.
7. Marchi, N. et al. 2004. Peripheral markers of blood-brain barrier damage. *Clin Chim Acta* 342(1–2):1–12.
8. Marchi, N. et al. 2012. Blood-brain barrier dysfunction and epilepsy: Pathophysiologic role and therapeutic approaches. *Epilepsia* 53(11):1877–1886.
9. Marchi, N. et al. 2013. Consequences of repeated blood-brain barrier disruption in football players. *PLoS One* 8(3):e56805.
10. Puvenna, V. et al. 2014. Significance of ubiquitin carboxy-terminal hydrolase l1 elevations in athletes after sub-concussive head hits. *PLoS One* 9(5):e96296.
11. Kanner, A.A. et al. 2003. Serum S100beta: A noninvasive marker of blood-brain barrier function and brain lesions. *Cancer* 97(11):2806–2813.
12. Korfias, S. et al. 2006. Serum S-100B protein as a biochemical marker of brain injury: A review of current concepts. *Curr Med Chem* 13(30):3719–3731.
13. Nylen, K. et al. 2008. Serum levels of S100B, S100A1B and S100BB are all related to outcome after severe traumatic brain injury. *Acta Neurochir (Wien)* 150(3):221–227.
14. Ruan, S., Noyes, K., Bazarian, J.J. 2008. S100 beta screening: A cost-minimization study for managing adult patients with mild traumatic brain injury in the emergency department. *Ann Emerg Med* 51(4):474.
15. Unden, J., Romner, B. 2010. Can low serum levels of S100B predict normal CT findings after minor head injury in adults?: An evidence-based review and meta-analysis. *J Head Trauma Rehabil* 25(4):228–240.
16. Vogelbaum, M.A. et al. 2005. S100beta as a predictor of brain metastases: Brain versus cerebrovascular damage. *Cancer* 104(4):817–824.
17. Edlow, B.L., Wu, O. 2012. Advanced neuroimaging in traumatic brain injury. *Semin Neurol* 32(4):374–400.
18. Borg, J. et al. 2004. Diagnostic procedures in mild traumatic brain injury: Results of the WHO Collaborating Centre Task Force on Mild Traumatic Brain Injury. *J Rehabil Med* 36(43 Suppl):61–75.
19. Lee, H. et al. 2008. Focal lesions in acute mild traumatic brain injury and neurocognitive outcome: CT versus 3T MRI. *J Neurotrauma* 25(9):1049–1056.
20. Neselius, S. et al. 2012. CSF-biomarkers in Olympic boxing: Diagnosis and effects of repetitive head trauma. *PLoS One* 7(4):e33606.
21. Olsson, B. et al. 2011. Biomarker-based dissection of neurodegenerative diseases. *Prog Neurobiol* 95(4):520–534.
22. Czeiter, E. et al. 2012. Brain injury biomarkers may improve the predictive power of the IMPACT outcome calculator. *J Neurotrauma* 29(9):1770–1778.
23. Zetterberg, H., Smith, D.H., Blennow, K. 2013. Biomarkers of mild traumatic brain injury in cerebrospinal fluid and blood. *Nat Rev Neurol* 9(4):201–210.
24. Franz, G. et al. 2003. Amyloid beta 1-42 and tau in cerebrospinal fluid after severe traumatic brain injury. *Neurology* 60(9):1457–1461.
25. Ost, M. et al. 2006. Initial CSF total tau correlates with 1-year outcome in patients with traumatic brain injury. *Neurology* 67(9):1600–1604.
26. Zemlan, F.P. et al. 2002. C-tau biomarker of neuronal damage in severe brain injured patients: Association with elevated intracranial pressure and clinical outcome. *Brain Res* 947(1):131–139.

27. Gonzalez, P.G., Walker, M.T. 2011. Imaging modalities in mild traumatic brain injury and sports concussion. *PM R* 3(10 Suppl 2):S413–S424.
28. Irimia, A. et al. 2012. Neuroimaging of structural pathology and connectomics in traumatic brain injury: Toward personalized outcome prediction. *Neuroimage Clin* 1(1):1–17.
29. Kraus, M.F. et al. 2007. White matter integrity and cognition in chronic traumatic brain injury: A diffusion tensor imaging study. *Brain* 130(Pt 10):2508–2519.
30. Bazarian, J.J. et al. 2007. Diffusion tensor imaging detects clinically important axonal damage after mild traumatic brain injury: A pilot study. *J Neurotrauma* 24(9):1447–1459.
31. Yuh, E.L. et al. 2014. Diffusion tensor imaging for outcome prediction in mild traumatic brain injury: A TRACK-TBI study. *J Neurotrauma* 31(17):1457–1477.
32. Zhang, Z. et al. 2014. Human traumatic brain injury induces autoantibody response against glial fibrillary acidic protein and its breakdown products. *PLoS One* 9(3):e92698.
33. Gruden, M.A. et al. 2007. Differential neuroimmune markers to the onset of Alzheimer's disease neurodegeneration and dementia: Autoantibodies to A beta(25-35) oligomers, S100b and neurotransmitters. *J Neuroimmunol* 186(1–2):181–192.
34. Lee, R. et al. 2012. Autoantibody testing in encephalopathies. *Pract Neurol* 12(1):4–13.
35. Vincent, A. et al. 2011. Autoantibodies associated with diseases of the CNS: New developments and future challenges. *Lancet Neurol* 10(8):759–772.
36. Bargerstock, E. et al. 2014. Is peripheral immunity regulated by blood-brain barrier permeability changes? *PLoS One* 9(7):e101477.
37. Iffland, P.H. et al. 2013. Intracellular and circulating neuronal antinuclear antibodies in human epilepsy. *Neurobiol Dis* 59:206–219.
38. Michalak, Z. et al. 2012. IgG leakage may contribute to neuronal dysfunction in drug-refractory epilepsies with blood-brain barrier disruption. *J Neuropathol Exp Neurol* 71(9):826–838.
39. Talavage, T.M. et al. 2014. Functionally-detected cognitive impairment in high school football players without clinically-diagnosed concussion. *J Neurotrauma* 31(4):327–338.
40. Bonnelle, V. et al. 2011. Default mode network connectivity predicts sustained attention deficits after traumatic brain injury. *J Neurosci* 31(38):13442–13451.
41. Mayer A.R. et al. 2011. Functional connectivity in mild traumatic brain injury. *Hum Brain Mapp* 32:1825–1835.
42. Bigler, E.D. 2012. Mild traumatic brain injury: The elusive timing of "recovery." *Neurosci Lett* 509(1):1–4.
43. Johnson, B. et al. 2012. Metabolic alterations in corpus callosum may compromise brain functional connectivity in MTBI patients: An 1H-MRS study. *Neurosci Lett* 509(1):5–8.
44. Babikian, T. et al. 2010. Metabolic levels in the corpus callosum and their structural and behavioral correlates after moderate to severe pediatric TBI. *J Neurotrauma* 27(3):473–481.
45. Cecil, K.M. et al. 1998. Proton magnetic resonance spectroscopy for detection of axonal injury in the splenium of the corpus callosum of brain-injured patients. *J Neurosurg* 88(5):795–801.
46. Tollard, E. et al. 2009. Experience of diffusion tensor imaging and 1H spectroscopy for outcome prediction in severe traumatic brain injury: Preliminary results. *Crit Care Med* 37(4):1448–1455.

13 Advanced Neuroimaging of Mild Traumatic Brain Injury

Inga K. Koerte, Jakob Hufschmidt, Marc Muehlmann, Alexander P. Lin, and Martha E. Shenton

CONTENTS

INTRODUCTION TO NEUROIMAGING OF MILD TRAUMATIC BRAIN INJURY

In the acute clinical setting both computed tomography (CT) and conventional magnetic resonance imaging (MRI) are established imaging modalities to rule out severe complications resulting from traumatic brain injury (TBI) such as skull fracture, intracranial hemorrhage, and brain edema. However, when applied to mild traumatic brain injury (mTBI) only about 10% of CT scans and 30% of MR scans reveal abnormalities such as subdural or subarachnoidal hemorrhage.[1–3] Moreover, conventional CT and MRI have proven insensitive to more subtle changes in the brain such as diffuse axonal injury, the most common injury present in mTBI.[4] Furthermore, these conventional imaging modalities do not provide accurate information relevant to long-term prognosis.[3,5,6]

In search of radiological evidence for mTBI, what are needed are highly sensitive, as well as objective measures of early detection (diagnosis) and accurate prognosis. Several promising advances in neuroimaging techniques are currently available and will be evaluated here. We review promising imaging techniques and their application to understanding brain abnormalities following mTBI and repetitive brain trauma. These imaging techniques include techniques to understand brain structure (high-resolution structural MR imaging) and tissue architecture (diffusion tensor imaging), techniques to understand regional blood flow (dynamic susceptibility contrast MR imaging, arterial spin labeling, single photon emission tomography), techniques to evaluate microhemorrhages (susceptibility-weighted imaging), and, finally, techniques used as indirect measures of neuronal activation (functional MR imaging) and brain metabolism (positron-emission tomography, MR spectroscopy).

HIGH-RESOLUTION STRUCTURAL MAGNETIC RESONANCE IMAGING

Structural information about the brain can be quantified using brain volume based on T1-weighted MRI. To date, automated software tools such as SPM and FreeSurfer[7] are available to investigate specific regions of interest (ROIs) of gray and white matter or the entire brain using a whole-brain voxel-based approach. Volumetric measures can then be compared across subjects and groups or longitudinally within the individual. Subcortical structures and cortical thickness are, for example, a common focus of volumetric analyses. Volumetric measures have been shown to be related to cognitive performance[8–11] and have been demonstrated to be useful in the diagnosis and prognosis of neurodegenerative diseases such as Alzheimer's disease[10,12,13] and Parkinson's disease,[14,15] to name just two. Specifically, pronounced cortical thinning has been reported to be a risk factor for the development of neurodegenerative diseases.[12–17]

To investigate the effects of mTBI on cortical thickness, a study by Lewen et al.[18] induced brain trauma in a murine model in mice.[18] At day 1 a significant increase was reported in cortical thickness at the region of impact (frontoparietal cortex), followed by a decrease in cortical thickness of 15%–20% after 21 days.[18] In another study using a murine model, animals were treated with fluid percussion to induce brain trauma.[19] Seventeen days postinjury, cortical thinning was found in the frontal and occipital regions of the ipsilateral hemisphere when compared to a control group. The results of these studies suggest an initial increase in cortical thickness due to edema of the cortex in the acute phase, followed by Wallerian degeneration and reactive astrocytosis, with a decrease in cortical thickness in the chronic stage.

A study investigating the effects of TBI on cortical thickness in children (age range 9–16 years) who sustained a head injury from motor vehicle accidents reported widespread significant cortical thinning in the frontal, parietal, temporal, and occipital lobes 3 years postinjury.[20] In addition, a correlation between memory performance and cortical thickness was also reported.[20] Wilde et al.[21] reported cortical thinning in the frontal lobes and in the right temporal lobe in children with moderate to severe TBI compared to a control group with orthopedic injury. The recent literature suggests a link between a pronounced decline in cortical thickness with age, and exposure to concussive or repetitive subconcussive brain trauma in athletes participating in contact sports.[22] The underlying mechanism is not completely understood, although shear injury of the axons with consecutive Wallerian degeneration and astrocytosis may play a role. Another recent study reports a significant inverse relationship between exposure to concussion and hippocampal volume, as well as slower reaction time in college-level football players.[11]

High-resolution structural MR imaging may thus be a useful technique in the assessment of more chronic stages following mTBI. However, further research is needed to detect more acute structural brain changes following mTBI so as to provide an early diagnosis in order to have a window of opportunity for possible preventative treatment, as well as better prognosis for long-term outcome.

SUSCEPTIBILITY-WEIGHTED IMAGING

Susceptibility-weighted imaging (SWI) is a special gradient-recalled echo MRI that SWI uses a special gradient-recalled echo (GRE) pulse sequence that takes advantage of the susceptibility differences between tissues. The phase data is then combined with the magnitude data resulting in images with accentuated local susceptibility. SWI is sensitive to venous blood, hemorrhage, and iron in the brain.[23]

SWI has proven to be an effective technique to identify microhemorrhages resulting from mTBI.[24–26] A greater number and volume of microhemorrhages, for example, has been shown to correlate with neurological outcome following mTBI and TBI in children.[27,28] Evidence also supports the use of SWI as an effective tool in predicting long-term neurological outcomes following TBI.[29]

A few studies have evaluated the use of SWI in the detection of microstructural changes in adult patients with acute and subacute mTBI. More specifically, Toth et al.[30] evaluated brain microhemorrhages in 14 patients with mTBI using diffusion tensor imaging (DTI) and SWI 72 hours and 1 month postinjury. Although, DTI revealed changes in mean diffusivity (MD) and fractional anisotropy (FA), SWI failed to detect microhemorrhages in any of the subjects.[30] Two separate studies applied SWI to search for microhemorrhages in boxers and found microhemorrhages in 3 out of 42 boxers[31] and in 2 out of 21 boxers.[32] Although, there were no observable lesions in the control group, the prevalence was not significantly different. However, in a recent prospective study on incidence of mTBIs in male and female ice hockey players, a new analysis technique was used that measures hypointensity burden (HIB) to detect microhemorrhages in the acute and subacute phases of mTBI (Figure 13.1).[33]

This method makes it possible to detect more subtle intensity changes in the sensitized images than conventional analysis techniques. In this study, 11 college level ice hockey players were evaluated 72 hours, 2 weeks, and 2 months postconcussion

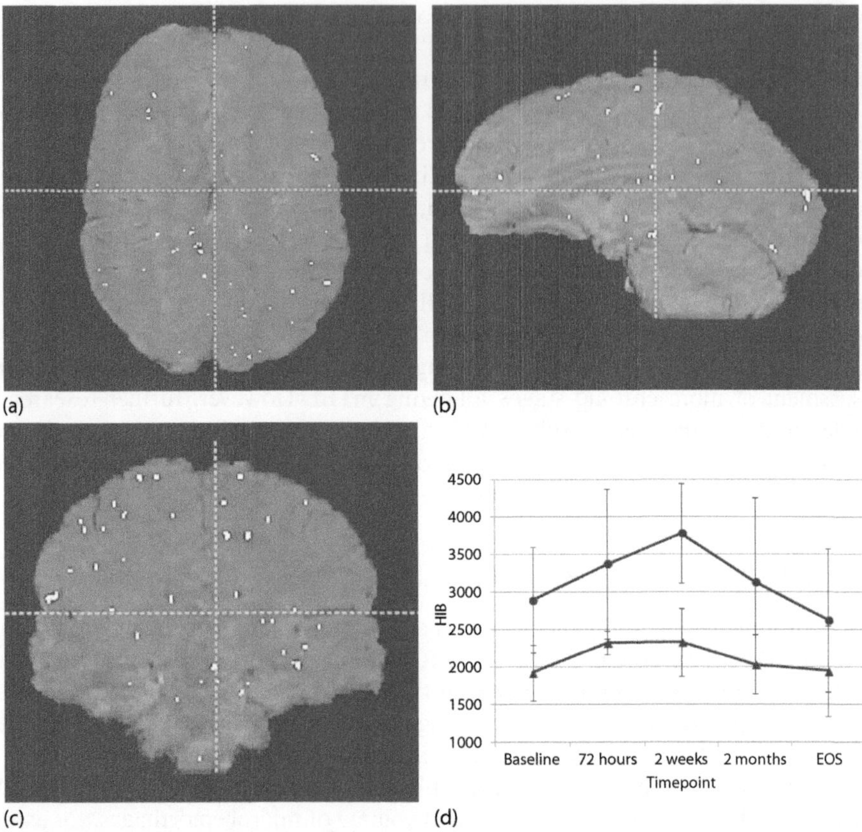

FIGURE 13.1 Susceptibility-weighted imaging (SWI) is sensitive to intracranial hemorrhage and iron deposition. Pictures (a) through (c) show the hypointensity burden (HIB) in one contact-sport athlete 2 weeks following a concussion. In these images the hypointensities are masked to appear bright, in order to allow for easy identification, especially by introducing a contrast to concomitantly hypointense appearing venous vessels. Graph (d) shows the longitudinal course of the HIB starting at the beginning of the season, 72 hours, 2 weeks, and 2 months after concussion, as well as at the end of the season. While no difference in HIB between baseline and end-of-season scans could be observed, a significant increase of hypointensity lesions with a peak around 2 weeks after concussion was found. Male participants had a higher HIB compared to female participants. HIB = hypointensity burden; circle = mean HIB of male participants; triangle = mean HIB of female participants.

using SWI. All concussed athletes showed microhemorrhages. In addition, a significantly higher HIB was found in male ice hockey players compared to female athletes.[33]

Thus, SWI may be a useful technique for detecting microhemorrhages following mTBI. Nonetheless, further research is needed, as early identification of these changes may help in determining prognosis and the likelihood of developing long-term symptoms.

FUNCTIONAL MAGNETIC RESONANCE IMAGING (fMRI) AND RESTING STATE FUNCTIONAL MAGNETIC RESONANCE IMAGING (rsfMRI)

Functional magnetic resonance imaging measures cerebral blood flow using the blood oxygen level-dependent contrast that results from the paramagnetic deoxyhemoglobin in the blood. The assumption is that increased flow is representative of neuronal activity. Neuronal activation can be assessed either while subjects are performing a task or while they are in a resting state. Both fMRI and rsfMRI have been applied in research and, to lesser extent, in clinical settings.

Altered activation patterns in a variety of brain regions have been reported in patients with mTBI and repetitive traumatic brain trauma.[4,34] McAllister et al.[35] evaluated patients who suffered from mTBI using auditory–verbal and visual–verbal N-back tasks. When compared to a control group, the task performance revealed no significant differences. However, one month postinjury, patients with mTBI showed increased activation while performing tasks of moderate difficulty and decreased activation when the tasks where more complex compared to controls. Another study by McAllister et al.[36] investigated patients at around one year postinjury.[36] Although, the postconcussive symptoms had resolved, increased activation patterns in the right frontal lobe were observed. These results suggest that persistent alterations in brain activation may be evident even when postconcussive symptoms are no longer present.

In another study by Smits et al.,[37] significantly lower N-back task performance was found in patients with postconcussive syndrome following mTBI. In addition, the severity of postconcussive symptoms was associated with atypical higher activation patterns outside the normal activation circuit. These findings suggest that damaged brain regions require backup from brain regions with lesser or no injury. Another study reported both increased and decreased brain activation in several brain regions in patients with more postconcussive symptoms following mTBI.[38] In another study, Matthews et al.[39] compared mTBI patients who had suffered from loss of consciousness with patients who had only altered consciousness 3 years postinjury. Patients with loss of consciousness showed decreased left frontal activations. In addition, the activation in this area correlated with the reported symptoms, which indicates that this area is a neural correlate of impaired self-awareness after loss of consciousness. In a study investigating the long-term effects of mTBI, reduced activity of the posterior parietal cortex during a hippocampal memory task was found, indicating the functional long-term significance of a previous mTBI.[40]

Over the last decade rsfMRI has gained increasing attention.[41–44] Resting state measures the activation in a state of relaxed consciousness[45,46] for which it is important to note that the brain still uses about 16% of the body's total energy. The observed patterns of correlated activation across regions can be interpreted as networks,[47] and there is evidence that their integrity and strength is relevant for behavior and cognition.[48–52]

Consequently fMRI has been used in a few recent studies investigating the effects of TBI. A disruption in activity was found in the default mode network in patients with blast-related TBI[53,54] and also in patients with mTBI.[55,56] Increased amplitudes of low fluctuation were related to better neurocognitive performance following TBI with chronic axonal injury. In addition, increases in nodes of the default mode network (DMN) may be interpreted as compensatory mechanisms after the loss of functional integrity in other brain areas.[54] Patients with mTBI have also shown a dysfunction of the DMN compared to healthy controls.[56] The disruptions between the DMN and the Task Positive Network (TPN) may be responsible for the memory dysfunctions observed in patients with mTBI.[55] Stevens et al.[57] reported abnormal functional connectivity in visual processing, motor, limbic, and numerous circuits purported to be involved in executive cognition.

The role of functional MRI and resting state fMRI in diagnosis, prognosis, and treatment monitoring of long-term functional deficits following mTBI is thus promising, and interest in this area of research is expected to increase over the coming years.[58]

DYNAMIC SUSCEPTIBILITY CONTRAST MRI

Dynamic susceptibility contrast MRI (DSC-MRI), also known as dynamic contrast-enhanced MRI (DCE-MRI), dynamically quantifies cerebral hemodynamics and regional cerebral blood flow (rCBF) by measuring the tissue concentration time curve of a contrast agent applied through intravenous bolus injection. However, the critical part of this technique is the exact definition of the arterial input function (AIF) over time to calculate the concentration of contrast agent in the blood vessel feeding the region of interest.[59] Although most common contrast agents are based on gadolinium, new pharmaceuticals (for example, gadobenate dimeglumine, Gd-BOPTA) have been developed with higher T1 and T2 relaxivity to improve signal intensity enhancement and thereby improving lesion visualization.[60]

In a study of 27 soldiers returning from deployment with a diagnosis of mTBI, multiple clusters of decreased brain perfusion were found compared to a control group. These clusters included the right anterior and middle cingulate gyrus, the left hemisphere of the cerebellum, and the left cuneus. Voxelwise statistics showed altered perfusion in the cerebellum and the right anterior cingulate gyrus were correlated with neurocognitive measures and symptoms.[61] Data from a TBI animal model suggests that DCE-MRI has the potential to detect accurately breaks in the blood–brain barrier (BBB).[62] However, other investigators report that a break in the BBB was not involved in the development of a cerebral edema following diffuse traumatic brain injury.[63]

Further cross-sectional and longitudinal studies on mTBI are needed to estimate the role of DCE-MRI in the diagnosis and prognosis of mTBI.

ARTERIAL SPIN LABELING

Arterial spin labeling (ASL) is a noninvasive variant of MR perfusion imaging. It uses water in the blood as an internal contrast agent by electromagnetically labeling it proximal to a region of interest. The results are obtained by measuring the difference between the images acquired with labeling and those without. Compared to DCE-MRI, the advantage of not needing a contrast agent is a tradeoff for a lower signal-to-noise ratio and lower spatial resolution. However, progress is being made with new ASL acquisition protocols to address these issues.[64]

To date, there are several studies using ASL in rodent models, yet it has also been applied to patients in the chronic stage of TBI.[64] A study on 21 patients with a history of mTBI (mean 24.6 months postinjury) reported reduced thalamic perfusion, which correlated with neurocognitive measures.[65] Similarly, in a study on 27 patients with a history of moderate to severe TBI, a global reduction of cerebral blood flow was observed, with more pronounced changes in the posterior cingulate cortices and in the thalami.[66] Multiple affected clusters in the superior and middle frontal cortices were also associated with focal lesions. The authors of this study hypothesized that the changes in cerebral blood flow seen in the posterior cingulate gyrus may also reflect changes in the default mode network and in attentional deficits observed in the same cohort.[66]

ASL has the advantage of being noninvasive, and efforts that are currently being made toward technical improvements will likely make it a promising tool for the evaluation of patients with mTBI.

DIFFUSION TENSOR IMAGING

DTI is an advanced MRI technique measuring diffusion properties of water molecules in the brain.[67] Diffusion measures give indirect information about how the tissue's microstructure influences the motion of water molecules. In the cerebrospinal fluid (CSF), water molecules diffuse with the same probability and speed in all possible directions. This is called isotropic diffusion. In contrast, restriction of movement due to axonal membranes, filaments, and myelin sheaths leads to directionality in the motion of water molecules. This is called anisotropic diffusion.

A DTI scan thus provides information about the diffusion of water molecules. Based on this information, the diffusion tensor can be quantified and characterized in each voxel (Figure 13.2). Common parameters are mean diffusivity (MD), which

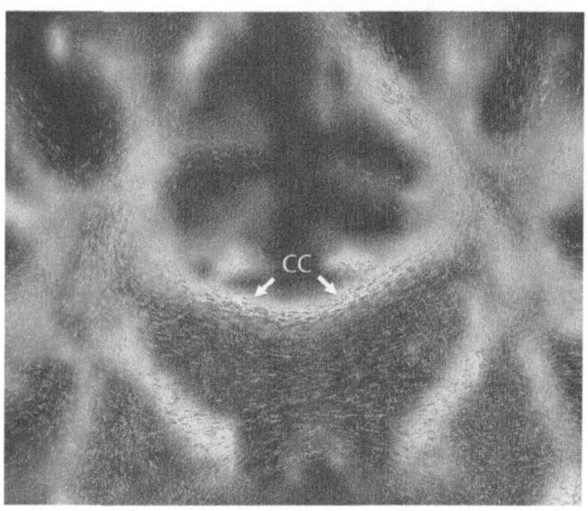

FIGURE 13.2 (See color insert.) Diffusion tensor imaging makes it possible to trace the preferred diffusion direction of water molecules. On a coronal view, the higher directionality of diffusion can be seen within highly organized brain structures such as the corpus callosum (CC). The blue color corresponds to high FA values, which translates to high directionality of diffusion.

is the average of the three principal diffusion directions, and fractional anisotropy (FA), which takes into account the shape of the diffusion ellipsoid. FA and MD are often found to be inversely related. Reduced FA is purported to reflect microscopic damage to myelin sheaths or axon membranes and/or axonal packing density.[68,69] In addition, radial diffusivity (RD) and axial diffusivity (AD) can be calculated relative to the main vector as two additional measures of diffusivity within each voxel.[67] RD is purported to measure myelin and AD is purported to measure axonal diffusion.[70]

DTI was first used to investigate mTBI in 2002.[71] Since then several studies have been conducted and many different regions have been found to be affected, including the corpus callosum and the corona radiata (for a review, see Ref. 4). Diffusion measures have been found to correlate with postconcussive symptoms and cognitive performance as in one study where reduced FA in the uncinate fasciculus correlated with memory performance and attentional control correlated with reduced FA in the left anterior corona radiata.[72] Of note, a study of professional soccer players, compared to noncontact sport athletes, demonstrated, for the first time, white matter alterations that showed increased radial diffusivity in the absence of mTBI (Figure 13.3).[73]

These results suggest that repetitive *subconcussive* blows to the head, such as those experienced by professional soccer players while heading the ball, can have a cumulative effect on neural structures.[73] Similar effects with increase in MD, RD, and AD were observed in ice hockey players evaluated before and after one play season. Interestingly, the three athletes who had sustained a concussion during the season presented with the most pronounced white matter alteration.[74]

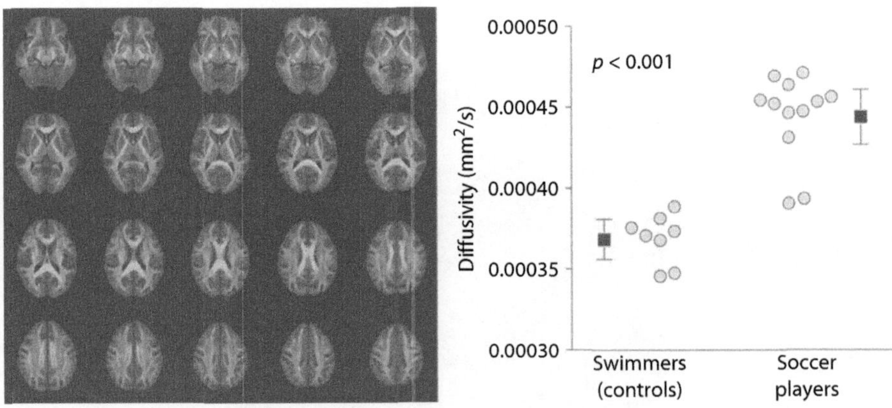

FIGURE 13.3 **(See color insert.)** Changes in radial diffusivity in asymptomatic soccer players frequently exposed to subconcussive impacts to their heads. Soccer players showed increased radial diffusivity when compared to age- and gender-matched swimmers using tract-based spatial statistics (TBSS, FSL 4.03, FMRIB, UK). (From Koerte I.K. et al., *JAMA* 308:1859–1861, 2012. With permission.)

Future Directions of DTI

DTI has been an important imaging tool to study mTBI because it is the main imaging tool to reveal diffuse axonal injury, the main injury observed in mTBI. Future directions for research include increasing the sensitivity and specificity of DTI findings in mTBI. Here, there are three important foci: (1) technical improvements in DTI, (2) novel diffusion measures, and (3) innovative analysis techniques that reflect a move toward more individualized medicine.

Technical Improvements of DTI

One of the new DTI sequences is high angular resolution diffusion imaging (HARDI), which is performed by measuring multiple diffusion directions. A way to shorten image acquisition time is to use compressed sensing, which makes it possible to use HARDI in clinical practice.[75] Multishell imaging is another way to modify the MRI sequence by using multiple b-values in one session to measure diffusion properties that vary at different gradient strengths.[76] With multiple b-values one can also perform diffusional kurtosis imaging (DKI), which is used to quantify the deviation of diffusion properties from Gaussian behavior.[77] This is an extension of DTI that provides measures that may be more specific than FA and MD, and may provide specific information about myelin alterations in the brain.

Novel Tools

In addition to improvements in acquisition techniques, novel analysis tools are also currently being developed. For example, new tractography algorithms increase sensitivity to crossing fibers by fitting multitensor models,[78] which enable a tracing of smaller fiber bundles in the periphery of the brain (Figure 13.4). This is especially important in mTBI as white matter damage is commonly seen at the junction between white matter and gray matter.

Several new diffusion measures are also being developed to further enhance the sensitivity and specificity of DTI findings regarding axonal and myelin pathologies. These new measures include mode, which further defines the shape of the diffusion

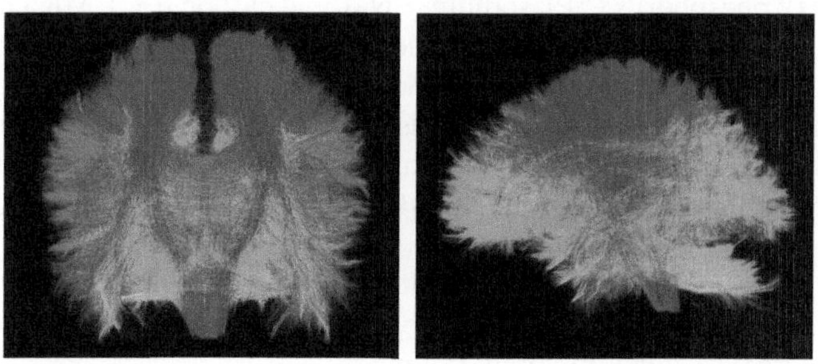

FIGURE 13.4 **(See color insert.)** Multitensor algorithms allow for a more precise differentiation between the end of a fiber tract (FA <0.15) and crossing fiber tracts by calculating multiple tensors within each voxel. A full brain tractography, color coded for the main orientation along the xyz-axis (z-axis, blue; x-axis, red; y-axis, green) is shown.

ellipsoid[79]; return to origin probability, which is reported to be more sensitive to demyelinization than conventional DTI measures[80]; mean square displacement, which analyzes the main distance traveled by a water molecule within a given period[81]; and free water imaging, which differentiates between intra- and extracellular water thereby enabling the differentiation between cytotoxic and vasogenic edema.[82,83]

A Focus on Individual Profiles of Injury

To address the heterogeneity of patterns of traumatic brain injury across subjects, it is important to characterize and to quantify individual profiles of injury for each patient, which represents a step toward personalized medicine that fits the patient's needs. The first step in this process involves the creation of a normative atlas, based on data from healthy control subjects. An example of this approach is evident in the work by Bouix and colleagues[84] who compared FA measures of individual mTBI patients experiencing chronic symptoms, with a normative atlas in order to detect the profile of injury for individual patients. This approach, which can assess different dependent measures, and not just FA, may thus be used to quantify and map injury patterns, thereby providing important new information to the treating physician in order to achieve the best outcome possible. This approach is much more sensitive than a group comparison, especially given the heterogeneous nature of mTBI.[84]

Overall, DTI has proven to be highly sensitive to white and gray matter changes following mTBI. Given the heterogeneity of methodologies and study designs, it is quite remarkable that DTI is a very, if not even the most, sensitive, noninvasive, in vivo method for detecting subtle white matter changes in mTBI.[4] Future research must include longitudinal studies to understand the dynamic nature of microstructural brain changes following mTBI, and to identify markers for diagnosis and prognosis, as well as possibilities for early interventional treatment.

MR SPECTROSCOPY

MR spectroscopy (MRS) is an in vivo means of quantifying the concentration of metabolites in human tissue based on the resonance frequencies of isotopes such as hydrogen (^{1}H), phosphorus (^{31}P), sodium (^{23}Na), and carbon (^{13}C). MRS can either be applied to a single voxel or to a larger region of interest. Spatial variation in concentration can be characterized using chemical shift imaging. To increase specificity without extracting spatial information it is also possible to use different echo times and create spectral information in two dimensions using Fourier transformation.[85]

The major metabolites measured by MRS include the following (Figure 13.5):

- *Lipid*—Although lipids are present throughout the brain in the form of membranes, they are not "MR visible" unless liberated by a severe pathological process including brain trauma.[86]
- *Lactate*—Lactate is the end product of anaerobic glycolysis and therefore a direct indicator of hypoxic events in the brain. The presence of lactate in spectra indicates impairment of perfusion and is indicative of poor outcome in severe TBI.[87]

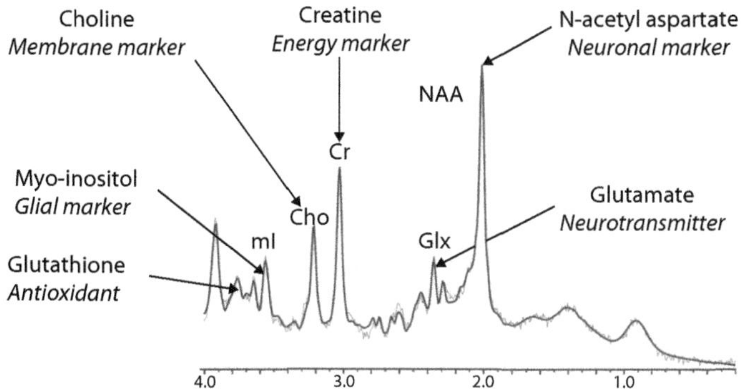

FIGURE 13.5 Spectral information in a single voxel obtained by MR spectroscopy. Arrows identify the peaks for several brain metabolites, each of which has an important biological role that is summarized in italics below the name of the metabolite. The x-axis is the chemical shift in parts per million, which provides the location of each peak. The abbreviation for each metabolite is given next to the peak.

- *N-acetyl aspartate (NAA)*—An amino-acid derivative synthesized in neurons and transported down axons, NAA is a marker of viable neurons, axons, and dendrites. Brain injury is associated with decreased NAA.[88–90]
- *Glutamate and glutamine (Glx)*—Glutamate and glutamine are difficult to distinguished at field strengths of 3T or less, hence the use of the abbreviation, Glx. MRS measures static pools of these amino acids that are involved in excitatory neurotransmission and metabolism between the astrocyte and neuron. Studies have shown that Glx is predictive of outcome after severe TBI.[91] Glx is not visible at long echo.
- *Cr (creatine)*—Creatine is part of ATP synthesis and therefore serves as a marker of mitochondrial energetics. As it is relatively stable, it is often used as an internal reference. Earlier MRS TBI studies utilize ratios of metabolite to Cr. However, recent studies have shown changes in Cr as a result of head injury and therefore these ratios must be carefully considered.[92,93]
- *Choline (Cho)*—Choline is a constituent of the phopholipids that make up the membranes of brain tissue. Since the majority of choline-containing brain constituents are not normally soluble, pathological alterations in membrane turnover result in an increase in MRS visible Cho. In the context of head injury, Cho is often described as a marker of diffuse axonal injury.[94]
- *mI (myo-inositol)*—mI is an astrocyte marker and osmolyte. It is also involved in the metabolism of phosphatidyl inositol, a membrane phospholipid, and, similar to choline, is expected to increase after TBI due to membrane damage. Another explanation for increased mI after TBI is that mI is a purported glial marker and increases as a result of reactive astrocytosis or glial scarring.[95]

Applied to TBI, MRS has shown significant changes in brain metabolism in the acute phase following various severity levels of brain trauma.[85] Several studies in adult patients with a history of mTBI report a decrease in NAA in different brain regions including the centrum semiovale, the splenium of the corpus callosum, and the bilateral frontal white matter.[96] A decrease in NAA in the white matter suggests a decrease in axonal viability. Another common MRS finding following TBI is an increase in choline, suggesting tissue damage or glial proliferation.[97–101] It has been suggested that repetitive brain trauma may lead to a vulnerable period that is prolonged and perhaps even exacerbated by repetitive injuries, which results in a longer time for NAA to return to baseline level.[101] Of note, a pilot study on athletes with a history of repetitive brain trauma showed significantly increased glutamate and choline compared to age-matched controls.[102] Studies in the acute and subacute stage of sports-related mTBI in young adult athletes report significant decreases of Glx, NAA, and NAA/Cr in areas including the dorsolateral prefrontal cortex and the primary motor cortex.[103–105] A study in children with sports-related concussion did not find any abnormalities for NAA or Cho.[106] The authors suggested that children may either sustain different patterns of injuries compared to adults or may be protected from some alterations seen in older populations.

In summary, MRS has the potential to be a powerful tool in the diagnosis and follow-up evaluation of brain changes following mTBI. Future studies will have to define standardized protocols for MRS analyses, as there are many ways to measure metabolite concentrations.

POSITRON-EMISSION TOMOGRAPHY

Positron-emission tomography (PET) uses markers labeled with special isotopes that emit measurable radiation signals. A signal increase hence suggests an area of increased binding or uptake of the marker. The most popular PET ligand is 18-fluorodeoxyglucose (FDG) which is involved in glucose metabolism. Other possible isotopes include carbon, oxygen, and nitrogen. As these are naturally found in the body they can be incorporated in new radiopharmaceuticals. PET has high sensitivity and therefore can be acquired relatively quickly. FDG-PET can be used, much like fMRI, in resting state or as a response to stimuli.[107–110] Possible limiting factors for longitudinal studies using PET are, however, high costs and the exposure to ionizing radiation.

In veterans who sustained blast injuries, FDG-PET showed impaired glucose metabolism in the cerebellum, the vermis, the pons, and in the medial temporal lobe.[111] Another study found boxers to have hypometabolism in the cerebellum, the posterior cingulate gyrus, and in the frontal lobe.[112] After severe TBI, FDG-PET has been reported to correlate with TBI severity and with the extant of cognitive and functional impairment at follow-up.[113] In mTBI, FDG-PET revealed altered glucose metabolism or uptake in brain regions similar to those found in more severe cases of TBI.[114] In some studies hypometabolism in frontal areas was correlated with neuropsychological testing.[108,115] These studies were, however, criticized for lacking in details about metabolism quantification and statistical power.[84] It has also been noted that regardless of the inconsistencies of the spatial patterns and a variety of outcomes, task-based FDG-PET studies consistently report hypometabolism in frontal and temporal regions.[96] Furthermore, a possible link to post-traumatic stress disorder (PTSD) as a confounding variable, especially in military personnel, is being investigated.[111,116]

Another target for PET ligands is the peripheral benzodiazepine receptor complex that is believed to be a marker for activated microglia. Targeting this receptor may be important for evaluating neuroinflammation, for which specific probes are also being developed.[117–119]

Chronic traumatic encephalopathy (CTE) is a neurodegenerative disease marked by widespread accumulation of hyperphosphorylated tau (p-tau), predominantly as neurofibrillary and astrocytic tangles.[120] CTE has been found most often in professional athletes involved in contact sports (e.g., boxing, American football) who have been subjected to repetitive brain trauma (RBT), including mTBI or even asymptomatic, subconcussive trauma. CTE, like most other neurodegenerative diseases, can only be definitively diagnosed postmortem. The development of a method to detect and measure tau in the brains of living individuals is therefore critical for the early diagnosis of CTE, even prior to the onset of symptoms, as well as for quantifying the degree and progression of tau accumulation. PET tau ligands have only reached initial human study in recent months.[121] In fact, only a single case study has been published thus far in sports-related head injury.[122] Nonetheless, the findings are encouraging in that the patterns differ from other neurodegenerative disorders such as Alzheimer's disease. For cross-sectional studies, PET provides valuable information and could be of great value for future evaluation of tauopathies in CTE.

SINGLE-PHOTON EMISSION COMPUTER TOMOGRAPHY

Single-photon emission computer tomography (SPECT) uses the distribution of an injected radionuclide to quantify regional cerebral blood flow (rCBF) by detecting emitted photons with gamma cameras that rotate around the patient. It is available at most major hospitals and relatively low cost as the radiopharmaceuticals used are inexpensive to generate, have a longer half-life, and are readily available for the use in a clinical setting. A commonly used radionuclid in the evaluation of the brain's rCBF is technetium-99m-hexamethylpropyleneamine oxime (^{99}mTC-HMPAO). Once injected, it accumulates in areas of blood flow where it is detectable for 24 hours. With its long half-life time this probe has been thought to be especially useful for performing resting state scans as there is enough time to ensure a quiet environment and a relaxed state of mind to avoid potential stress or activity-related brain activity. Besides ^{99}mTC-HMPAO, several other radionuclides have been used in mTBI research. For example, iodine-123 N-isopropyl-p-iodoamphetamine (^{123}IMP) was approved by the FDA but was less readily available. Another radionuclide that has been approved by the FDA is technetium-99 m-ethyl cysteinate dimer (^{99}mTc-ECD). It has been reported to differently distribute across the brain compared to ^{99}mTC-HMPAO, which makes it necessary to choose the best fitting tracer for the region of interest.[123] In one study 57-cobalt chloride (^{57}CoCl^{2}) was used, as it is believed to target calcium in the brain.[124] To date, there are a few studies on SPECT in the more chronic stages of mTBI.[125–129] However, there are only a few publications on patients in the acute and subacute stages of mTBI. For the latter, two studies showed reduced perfusion of SPECT tracers in the frontal lobe following an mTBI.[130,131] Scans were performed within 72 hours of injury and in a single-case study within a week. Studies focusing on the subacute stage report similar patterns in the frontal lobe 2 days after injury,[124] as well as the parietal lobe 5 days post-mTBI.[126] However, in a study performing SPECT, MRI, and neuropsychological testing in mTBI patients, the additional tests failed to correlate well with findings in ^{99}mTC-HMPAO SPECT.[126] In two studies focusing on repetitive brain trauma, 100 retired NFL players were scanned with SPECT[132] and a subset of 30 players was evaluated for possible treatment.[133] The results showed hypoperfusion of the hippocampus, the anterior and posterior cingulate gyrus, the occipital lobe, the temporal poles, and the prefrontal poles. Other studies have investigated the link between SPECT findings and neuropsychological measures with mixed results.[127,129,134–136] It has been noted that SPECT shows nonspecific changes that are also seen in chronic pain, drug or alcohol abuse, and headache—many of which are comorbidities especially in professional athletes.[137] Another rather fundamental problem of SPECT in mTBI or repetitive brain trauma might be that the SPECT analyzes a region of interest compared to a control region within the same brain. This approach may introduce a systematical error given the often diffuse and heterogeneous nature of mTBI.[138] In a clinical setting the use of SPECT alone is not believed to be sufficient in the evaluation of a patient with mTBI, because of a lack of sensitivity, although the absence of SPECT findings is considered to be a prognostic factor for a better outcome.

SUMMARY

CT and conventional MRI have proven to be effective in ruling out complications of severe TBI. They lack, however, sensitivity when applied to mTBI. To date there exists a variety of advanced neuroimaging modalities that have provided helpful insights into many aspects of the pathophysiology following mTBI (for review see Ref. 4), as well as repetitive brain trauma.[34] The recent progress in neuroimaging was made possible by the development of more sensitive methods and measures that make it possible to quantify the brain's metabolism, perfusion, function, and microstructure. However, there is still a need for further longitudinal studies with carefully selected cohorts in order to understand the dynamic nature of mTBI and the influence of possibly modulating factors such as the biomechanics of the injury, lesion location, genetics, and other factors that are currently not known. Further, the heterogeneity of mTBI calls for new postprocessing approaches to identify individual injury patterns, where the use of a normative atlas is most promising. This approach, regardless of the imaging modality applied, presents a new path toward a personalized medicine approach that fits the patient's needs and provides the treating physician with new information that will likely achieve the best outcome. Moreover, there is a need for studies using a multimodal approach that combines the advantages of different imaging modalities, clinical evaluation, neuropsychological evaluation, as well as biomarkers from blood and CSF, all acquired in the same patients, in order to identify sensitive biomarkers for diagnosis, therapeutic decision making, and prognosis.

ACKNOWLEDGMENTS

Inga Koerte and Marc Muehlmann are supported by the Else Kröner-Fresenius Stiftung, Germany. This work was also partially funded by grants from the Department of Defense (W81XWH-10-1-0835: APL; W81XWH-07-CC-CSDoD: MES), the National Institutes of Health (R01-NS078337: APL, MES), and a VA Merit Award (MES).

REFERENCES

1. Borg, J. et al. 2004. Non-surgical intervention and cost for mild traumatic brain injury: Results of the WHO Collaborating Centre Task Force on Mild Traumatic Brain Injury. *J Rehabil Med* 76–83.
2. Hesselink, J.R. et al. 1988. MR imaging of brain contusions: A comparative study with CT. *AJR Am J Roentgenol* 150:1133–1142.
3. Mittl, R.L. et al. 1994. Prevalence of MR evidence of diffuse axonal injury in patients with mild head injury and normal head CT findings. *AJNR Am J Neuroradiol* 15:1583–1589.
4. Shenton, M.E. et al. 2012. A review of magnetic resonance imaging and diffusion tensor imaging findings in mild traumatic brain injury. *Brain Imaging Behav* 6:137–192.
5. Iverson, G.L. 2005. Outcome from mild traumatic brain injury. *Curr Opin Psychiatry* 18:301–317.
6. Le, T.H., Gean, A.D. 2009. Neuroimaging of traumatic brain injury. *Mt Sinai J Med* 76:145–162.
7. Khan, A.R., Wang, L., Beg, M.F. 2008. FreeSurfer-initiated fully-automated subcortical brain segmentation in MRI using Large Deformation Diffeomorphic Metric Mapping. *Neuroimage* 41:735–746.
8. Chang, Y.L. et al. 2010. Level of executive function influences verbal memory in amnestic mild cognitive impairment and predicts prefrontal and posterior cingulate thickness. *Cereb Cortex* 20:1305–1313.
9. Seo, S.W. et al. 2010. Cortical thinning in vascular mild cognitive impairment and vascular dementia of subcortical type. *J Neuroimaging* 20:37–45.
10. Singh, V. et al. 2006. Spatial patterns of cortical thinning in mild cognitive impairment and Alzheimer's disease. *Brain* 129:2885–2893.
11. Singh, R. et al. 2014. Relationship of collegiate football experience and concussion with hippocampal volume and cognitive outcomes. *JAMA* 311:1883–1888.
12. Dickerson, B.C., Wolk, D.A., Alzheimer's Disease Neuroimaging Initiative. 2012. MRI cortical thickness biomarker predicts AD-like CSF and cognitive decline in normal adults. *Neurology* 78:84–90.
13. Park, H. et al. 2013. Dimensionality reduced cortical features and their use in predicting longitudinal changes in Alzheimer's disease. *Neurosci Lett* 550:17–22.
14. Ibarretxe-Bilbao, N. et al. 2012. Progression of cortical thinning in early Parkinson's disease. *Mov Disord* 27:1746–1753.
15. Jubault, T. et al. 2011. Patterns of cortical thickness and surface area in early Parkinson's disease. *Neuroimage* 55:462–467.
16. Agosta, F. et al. 2012. The cortical signature of amyotrophic lateral sclerosis. *PLoS One* 7:e42816.
17. Verstraete, E. et al. 2012. Structural MRI reveals cortical thinning in amyotrophic lateral sclerosis. *J Neurol Neurosurg Psychiatry* 83:383–388.
18. Lewen, A. et al. 1995. Traumatic brain injury in rat produces changes of beta-amyloid precursor protein immunoreactivity. *Neuroreport* 6:357–360.

19. Fineman, I. et al. 2000. Inhibition of neocortical plasticity during development by a moderate concussive brain injury. *J Neurotrauma* 17:739–749.
20. Merkley, T.L. et al. 2008. Diffuse changes in cortical thickness in pediatric moderate-to-severe traumatic brain injury. *J Neurotrauma* 25:1343–1345.
21. Wilde et al. 2012. Longitudinal changes in cortical thickness in children after traumatic brain injury and their relation to behavioral regulation and emotional control. *Int J Dev Neurosci.*
22. Tremblay, S. et al. 2013. Sports concussions and aging: A neuroimaging investigation. *Cereb Cortex* 23:1159–1166.
23. Haacke, E.M. et al. 2004. Susceptibility weighted imaging (SWI). *Magn Reson Med* 52:612–618.
24. Barnes, S.R., Haacke, E.M. 2009. Susceptibility-weighted imaging: Clinical angiographic applications. *Magn Reson Imaging Clin N Am* 17:47–61.
25. Benson, R.R. et al. 2012. Detection of hemorrhagic and axonal pathology in mild traumatic brain injury using advanced MRI: Implications for neurorehabilitation. *NeuroRehabilitation* 31:261–279.
26. McAllister, T.W. et al. 2001. Neuroimaging findings in mild traumatic brain injury. *J Clin Exp Neuropsychol* 23:775–791.
27. Ashwal, S. et al. 2006. Susceptibility-weighted imaging and proton magnetic resonance spectroscopy in assessment of outcome after pediatric traumatic brain injury. *Arch Phys Med Rehabil* 87:S50–S58.
28. Beauchamp, M.H. et al. 2013. Susceptibility weighted imaging and its relationship to outcome after pediatric traumatic brain injury. *Cortex* 49:591–598.
29. Colbert, C.A. et al. 2010. Value of cerebral microhemorrhages detected with susceptibility-weighted MR Imaging for prediction of long-term outcome in children with nonaccidental trauma. *Radiology* 256:898–905.
30. Toth, A. et al. 2013. Multi-modal magnetic resonance imaging in the acute and sub-acute phase of mild traumatic brain injury: Can we see the difference? *J Neurotrauma* 30:2–10.
31. Hahnel, S. et al. 2008. Prevalence of cerebral microhemorrhages in amateur boxers as detected by 3T MR imaging. *AJNR Am J Neuroradiol* 29:388–391.
32. Hasiloglu, Z.I. et al. 2011. Cerebral microhemorrhages detected by susceptibility-weighted imaging in amateur boxers. *Am J Neuroradiol* 32:99–102.
33. Helmer, K.G. et al. 2014. Hockey Concussion Education Project, Part 1. Susceptibility-weighted imaging study in male and female ice hockey players over a single season. *J Neurosurg* 120:864–872.
34. Koerte, I.K. et al. 2015. *Brain Pathol* 25(3):318–349.
35. McAllister, T.W. et al. 2001. Differential working memory load effects after mild traumatic brain injury. *Neuroimage* 14:1004–1012.
36. McAllister, T.W. et al. 2002. Working memory activation patterns 1 month and 1 year after mild traumatic brain injury: A longitudinal fMRI study. *J Neuropsychiatry Clin Neurosci* 14:116.
37. Smits, M. et al. 2009. Postconcussion syndrome after minor head injury: Brain activation of working memory and attention. *Hum Brain Mapp* 30(9):2789-803.
38. Gosselin, N. et al. 2011. Electrophysiology and functional MRI in post-acute mild traumatic brain injury. *J Neurotrauma* 28:329–341.
39. Matthews, S., Simmons, A., Strigo, I. 2011. The effects of loss versus alteration of consciousness on inhibition-related brain activity among individuals with a history of blast-related concussion. *Psychiatry Res* 30;191(1):76–79.
40. Monti, J.M. et al. 2013. History of mild traumatic brain injury is associated with deficits in relational memory, reduced hippocampal volume, and less neural activity later in life. *Front Aging Neurosci* 5:41.

41. Biswal, B.B. et al. 2010. Toward discovery science of human brain function. *Proc Natl Acad Sci U S A* 107:4734–4739.
42. Buckner, R.L., Andrews-Hanna, J.R., Schacter, D.L. 2008. The brain's default network: Anatomy, function, and relevance to disease. *Ann N Y Acad Sci* 1124:1–38.
43. Buckner, R.L., Vincent, J.L. 2007. Unrest at rest: Default activity and spontaneous network correlations. *Neuroimage* 37:1091–1096; discussion 1097–1099.
44. Damoiseaux, J.S. et al. 2006. Consistent resting-state networks across healthy subjects. *Proc Natl Acad Sci U S A* 103:13848–13853.
45. Fox, M.D., Raichle, M.E. 2007. Spontaneous fluctuations in brain activity observed with functional magnetic resonance imaging. *Nat Rev Neurosci* 8:700–711.
46. Raichle, M.E. et al. 2001. A default mode of brain function. *Proc Natl Acad Sci U S A* 98:676–682.
47. Shulman, R.G. et al. 2004. Energetic basis of brain activity: Implications for neuroimaging. *Trends Neurosci* 27:489–495.
48. Ferrarelli, F. et al. 2010. Breakdown in cortical effective connectivity during midazolam-induced loss of consciousness. *Proc Natl Acad Sci U S A* 107:2681–2686.
49. Hampson, M. et al. 2006. Brain connectivity related to working memory performance. *J Neurosci* 26:13338–13343.
50. He, B.J. et al. 2007. Breakdown of functional connectivity in frontoparietal networks underlies behavioral deficits in spatial neglect. *Neuron* 53:905–918.
51. Kelly, A.M. et al. 2008. Competition between functional brain networks mediates behavioral variability. *Neuroimage* 39:527–537.
52. Wirth, M. et al. 2011. Semantic memory involvement in the default mode network: A functional neuroimaging study using independent component analysis. *Neuroimage* 54:3057–3066.
53. Han, K. et al. 2014. Disrupted modular organization of resting-state cortical functional connectivity in U.S. military personnel following concussive "mild" blast-related traumatic brain injury. *Neuroimage* 84:76–96.
54. Palacios, E.M. et al. 2013. Resting-state functional magnetic resonance imaging activity and connectivity and cognitive outcome in traumatic brain injury. *JAMA Neurol* 70:845–851.
55. Sours, C. et al. 2013. Default mode network interference in mild traumatic brain injury—A pilot resting state study. *Brain Res* 1537:201–215.
56. Zhou, Y. et al. 2012. Default-mode network disruption in mild traumatic brain injury. *Radiology* 265:882–892.
57. Stevens M.C. et al. 2012. Multiple resting state network functional connectivity abnormalities in mild traumatic brain injury. *Brain Imaging Behav* 6(2):293–318.
58. McDonald, B.C., Saykin, A.J., McAllister, T.W. 2012. Functional MRI of mild traumatic brain injury (mTBI): Progress and perspectives from the first decade of studies. *Brain Imaging Behav* 6:193–207.
59. Conturo, T.E. et al. 2005. Arterial input functions for dynamic susceptibility contrast MRI: Requirements and signal options. *J Magn Reson Imaging* 22:697–703.
60. Maravilla, K.R. 2006. Gadobenate dimeglumine-enhanced MR imaging of patients with CNS diseases. *Eur Radiol* 16(Suppl 7):M8–M15.
61. Liu, W. et al. 2013. Perfusion deficits in patients with mild traumatic brain injury characterized by dynamic susceptibility contrast MRI. *NMR Biomed* 26:651–663.
62. Wei, X.-E. et al. 2011. A useful tool for the initial assessment of blood-brain barrier permeability after traumatic brain injury in rabbits: Dynamic contrast-enhanced magnetic resonance imaging. *J Trauma* 71:1645–1650; discussion 1650–1651.
63. Beaumont, A. et al. 2006. Bolus tracer delivery measured by MRI confirms edema without blood-brain barrier permeability in diffuse traumatic brain injury. *Acta Neurochir* 96:171–174.

64. Hunter, J.V. et al. 2012. Emerging imaging tools for use with traumatic brain injury research. *J Neurotrauma* 29:654–671.
65. Ge, Y. et al. 2009. Assessment of thalamic perfusion in patients with mild traumatic brain injury by true FISP arterial spin labelling MR imaging at 3T. *Brain Inj* 23:666–674.
66. Kim, J. et al. 2010. Resting cerebral blood flow alterations in chronic traumatic brain injury: An arterial spin labeling perfusion FMRI study. *J Neurotrauma* 27:1399–1411.
67. Basser, P.J., Jones, D.K. 2002. Diffusion-tensor MRI: Theory, experimental design and data analysis—A technical review. *NMR Biomed* 15:456–467.
68. Basser, P.J., Pierpaoli, C. 1996. Microstructural and physiological features of tissues elucidated by quantitative-diffusion-tensor MRI. *J Magn Reson B* 111:209–219.
69. Pierpaoli, C. et al. 2001. Water diffusion changes in Wallerian degeneration and their dependence on white matter architecture. *Neuroimage* 13:1174–1185.
70. Song, S.K. et al. 2003. Diffusion tensor imaging detects and differentiates axon and myelin degeneration in mouse optic nerve after retinal ischemia. *Neuroimage* 20:1714–1722.
71. Arfanakis, K. et al. 2002. Diffusion tensor MR imaging in diffuse axonal injury. *AJNR Am J Neuroradiol* 23:794–802.
72. Niogi, S.N. et al. 2008. Structural dissociation of attentional control and memory in adults with and without mild traumatic brain injury. *Brain* 131:3209–3221.
73. Koerte, I.K. et al. 2012. White matter integrity in the brains of professional soccer players without a symptomatic concussion. *JAMA* 308:1859–1861.
74. Koerte, I.K. et al. 2012. A prospective study of physician-observed concussion during a varsity university hockey season: White matter integrity in ice hockey players. Part 3 of 4. *Neurosurg Focus* 33(E3)1–7.
75. Michailovich, O., Rathi, Y., Dolui, S. 2011. Spatially regularized compressed sensing for high angular resolution diffusion imaging. *IEEE Trans Med Imaging* 30:1100–1115.
76. Rathi, Y. et al. 2011. Sparse multi-shell diffusion imaging. *Med Image Comput Comput Assist Interv* 14:58–65.
77. Jensen, J.H., Helpern, J.A. 2010. MRI quantification of non-Gaussian water diffusion by kurtosis analysis. *NMR Biom* 23:698–710.
78. Malcolm, J.G., Shenton, M.E., Rathi, Y. 2010. Filtered multitensor tractography. *IEEE Trans Med Imaging* 29:1664–1675.
79. Ennis, D.B., Kindlmann, G. 2006. Orthogonal tensor invariants and the analysis of diffusion tensor magnetic resonance images. *Magn Reson Med* 55:136–146.
80. Assaf, Y. et al. 2005. White matter changes in multiple sclerosis: Correlation of q-space diffusion MRI and 1H MRS. *Magn Reson Imaging* 23:703–710.
81. Assaf, Y. et al. 2002. High b-value q-space analyzed diffusion-weighted MRI: Application to multiple sclerosis. *Magn Reson Med* 47:115–126.
82. Pasternak, O. et al. 2009. Free water elimination and mapping from diffusion MRI. *Magn Reson Med* 62:717–730.
83. Pasternak, O. et al. 2014. Hockey Concussion Education Project, Part 2. Microstructural white matter alterations in acutely concussed ice hockey players: A longitudinal free-water MRI study. *J Neurosurg* 120:873–881.
84. Bouix, S. et al. 2013. Increased gray matter diffusion anisotropy in patients with persistent post-concussive symptoms following mild traumatic brain injury. *PLoS One* 8:e66205.
85. Ng, T.S. et al. 2014. Neuroimaging in repetitive brain trauma. *Alzheimers Res Ther* 6:10.
86. Haseler, L.J. et al. 1997. Evidence from proton magnetic resonance spectroscopy for a metabolic cascade of neuronal damage in shaken baby syndrome. *Pediatrics* 99:4–14.

87. Makoroff, K.L. et al. 2005. Elevated lactate as an early marker of brain injury in inflicted traumatic brain injury. *Pediatr Radiol* 35:668–676.
88. Moffett, J.R. et al. 2013. N-Acetylaspartate reductions in brain injury: Impact on post-injury neuroenergetics, lipid synthesis, and protein acetylation. *Front Neuroenergetics* 5:11.
89. Ross, B.D. et al. 1998. 1H MRS in acute traumatic brain injury. *J Magn Reson Imaging* 8:829–840.
90. Signoretti, S. et al. 2001. N-Acetylaspartate reduction as a measure of injury severity and mitochondrial dysfunction following diffuse traumatic brain injury. *J Neurotrauma* 18:977–991.
91. Shutter, L., Tong, K.A., Holshouser, B.A. 2004. Proton MRS in acute traumatic brain injury: Role for glutamate/glutamine and choline for outcome prediction. *J Neurotrauma* 21:1693–1705.
92. Gasparovic, C. et al. 2009. Neurometabolite concentrations in gray and white matter in mild traumatic brain injury: An 1H-magnetic resonance spectroscopy study. *J Neurotrauma* 26:1635–1643.
93. Yeo, R.A. et al. 2011. A longitudinal proton magnetic resonance spectroscopy study of mild traumatic brain injury. *J Neurotrauma* 28:1–11.
94. Holshouser, B.A., Tong, K.A., Ashwal, S. 2005. Proton MR spectroscopic imaging depicts diffuse axonal injury in children with traumatic brain injury. *AJNR Am J Neuroradiol* 26:1276–1285.
95. Ashwal, S. et al. 2004. Proton spectroscopy detected myoinositol in children with traumatic brain injury. *Pediatr Res* 56:630–638.
96. Lin, A.P. et al. 2012. Metabolic imaging of mild traumatic brain injury. *Brain Imaging Behav* 6:208–223.
97. Cecil, K.M. et al. 1998. Proton magnetic resonance spectroscopy for detection of axonal injury in the splenium of the corpus callosum of brain-injured patients. *J Neurosurg* 88:795–801.
98. Cimatti, M. 2006. Assessment of metabolic cerebral damage using proton magnetic resonance spectroscopy in mild traumatic brain injury. *J Neurosurg Sci* 50:83–88.
99. Cohen, B.A. et al. 2007. Proton MR spectroscopy and MRI-volumetry in mild traumatic brain injury. *AJNR Am J Neuroradiol* 28:907–913.
100. Garnett, M.R. et al. 2000. Early proton magnetic resonance spectroscopy in normal-appearing brain correlates with outcome in patients following traumatic brain injury. *Brain* 123(Pt 1):2046–2054.
101. Vagnozzi, R. et al. 2008. Temporal window of metabolic brain vulnerability to concussion: A pilot 1H-magnetic resonance spectroscopic study in concussed athletes—Part III. *Neurosurgery* 62:1286–1295; discussion 1295–1296.
102. Lin, A.P. et al. 2015. Changes in the neurochemistry of athletes with repetitive brain trauma: Preliminary results using localized correlated spectroscopy. *Alzheimers Res Ther* 7:13.
103. Henry, L.C. et al. 2010. Neurometabolic changes in the acute phase after sports concussions correlate with symptom severity. *J Neurotrauma* 27:65–76.
104. Henry, L.C. et al. 2011. Metabolic changes in concussed American football players during the acute and chronic post-injury phases. *BMC Neurol* 11:105.
105. Henry, L.C. et al. 2011. Acute and chronic changes in diffusivity measures after sports concussion. *J Neurotrauma* 28:2049–2059.
106. Maugans, T.A. et al. 2012. Pediatric sports-related concussion produces cerebral blood flow alterations. *Pediatrics* 129:28–37.
107. Garcia-Panach, J. et al. 2011. A voxel-based analysis of FDG-PET in traumatic brain injury: Regional metabolism and relationship between the thalamus and cortical areas. *J Neurotrauma* 28:1707–1717.

108. Gross, H. et al. 1996. Local cerebral glucose metabolism in patients with long-term behavioral and cognitive deficits following mild traumatic brain injury. *J Neuropsychiatry Clin Neurosci* 8:324–334.
109. Humayun, M.S. et al. 1989. Local cerebral glucose abnormalities in mild closed head injured patients with cognitive impairments. *Nucl Med Commun* 10:335–344.
110. Kato, T. et al. 2007. Statistical image analysis of cerebral glucose metabolism in patients with cognitive impairment following diffuse traumatic brain injury. *J Neurotrauma* 24:919–926.
111. Peskind, E.R. et al. 2011. Cerebrocerebellar hypometabolism associated with repetitive blast exposure mild traumatic brain injury in 12 Iraq war veterans with persistent post-concussive symptoms. *Neuroimage* 54(Suppl 1):S76–S82.
112. Provenzano, F.A. et al. 2010. F-18 FDG PET imaging of chronic traumatic brain injury in boxers: A statistical parametric analysis. *Nucl Med Commun* 31:952–957.
113. de la Cueva-Barrao, L. et al. 2009. [The clinical relevance of FDG-PET imaging in severe traumatic brain injuries]. *Rev Neurol* 49:58–63.
114. Byrnes, K.R. et al. 2014. FDG-PET imaging in mild traumatic brain injury: A critical review. *Front Neuroenergetics* 5:13.
115. Ruff, R.M. et al. 1994. Selected cases of poor outcome following a minor brain trauma: Comparing neuropsychological and positron emission tomography assessment. *Brain Inj* 8:297–308.
116. Petrie, E.C. et al. 2014. Neuroimaging, behavioral, and psychological sequelae of repetitive combined blast/impact mild traumatic brain injury in Iraq and Afghanistan war veterans. *J Neurotrauma* 31:425–436.
117. Cagnin, A. et al. 2007. Positron emission tomography imaging of neuroinflammation. *Neurotherapeutics* 4:443–452.
118. Médran-Navarrete, V. et al. 2014. [18F]DPA-C5yne, a novel fluorine-18-labelled analogue of DPA-714: Radiosynthesis and preliminary evaluation as a radiotracer for imaging neuroinflammation with PET. *J Labelled Comp Radiopharm* 57:410–418.
119. Van Camp, N. et al. 2010. In vivo imaging of neuroinflammation: A comparative study between [(18)F]PBR111, [(11)C]CLINME and [(11)C]PK11195 in an acute rodent model. *Eur J Nucl Med Mol Imaging* 37:962–972.
120. McKee, A.C. et al. 2013. The spectrum of disease in chronic traumatic encephalopathy. *Brain* 136:43–64.
121. Xia, C.F. et al. 2013. [(18)F]T807, a novel tau positron emission tomography imaging agent for Alzheimer's disease. *Alzheimers Dement* 9:666–676.
122. Mitsis, E.M. et al. 2014. Tauopathy PET and amyloid PET in the diagnosis of chronic traumatic encephalopathies: Studies of a retired NFL player and of a man with FTD and a severe head injury. *Transl Psychiatry* 4:e441.
123. Asenbaum, S. et al. 1998. Imaging of cerebral blood flow with technetium-99m-HM-PAO and technetium-99m-ECD: A comparison. *J Nucl Med* 39:613–618.
124. Audenaert, K. et al. 2003. Imaging of mild traumatic brain injury using 57Co and 99mTc HMPAO SPECT as compared to other diagnostic procedures. *Med Sci Monit* 9:MT112–M1177.
125. Gray, B.G. et al. 1992. Technetium-99m-HMPAO SPECT in the evaluation of patients with a remote history of traumatic brain injury: A comparison with x-ray computed tomography. *J Nucl Med* 33:52–58.
126. Hofman, P.A. et al. 2001. MR imaging, single-photon emission CT, and neurocognitive performance after mild traumatic brain injury. *AJNR Am J Neuroradiol* 22:441–449.
127. Ichise, M. et al. 1994. Technetium-99m-HMPAO SPECT, CT and MRI in the evaluation of patients with chronic traumatic brain injury: A correlation with neuropsychological performance. *J Nucl Med* 35:217–226.

128. Kant, R. et al. 1997. Tc-HMPAO SPECT in persistent post-concussion syndrome after mild head injury: Comparison with MRI/CT. *Brain Inj* 11:115–124.
129. Lewine, J.D. et al. 2007. Objective documentation of traumatic brain injury subsequent to mild head trauma: Multimodal brain imaging with MEG, SPECT, and MRI. *J Head Trauma Rehabil* 22:141–155.
130. Abu-Judeh, H.H. et al. 1998. Discordance between FDG uptake and technetium-99m-HMPAO brain perfusion in acute traumatic brain injury. *J Nucl Med* 39:1357–1359.
131. Gowda, N.K. et al. 2006. Technetium Tc-99m ethyl cysteinate dimer brain single-photon emission CT in mild traumatic brain injury: A prospective study. *AJNR Am J Neuroradiol* 27:447–451.
132. Amen, D.G. et al. 2011. Impact of playing American professional football on long-term brain function. *J Neuropsychiatry Clin Neurosci* 23:98–106.
133. Amen, D.G. et al. 2011. Reversing brain damage in former NFL players: Implications for traumatic brain injury and substance abuse rehabilitation. *J Psychoactive Drugs* 43:1–5.
134. Bonne, O. et al. 2003. Cerebral blood flow in chronic symptomatic mild traumatic brain injury. *Psychiatry Res* 124:141–152.
135. Umile, E.M., Plotkin, R.C., Sandel, M.E. 1998. Functional assessment of mild traumatic brain injury using SPECT and neuropsychological testing. *Brain Inj* 12:577–594.
136. Umile, E.M. et al. 2002. Dynamic imaging in mild traumatic brain injury: Support for the theory of medial temporal vulnerability. *Arch Phys Med Rehabil* 83:1506–1513.
137. Wortzel, H.S. et al. 2008. Forensic applications of cerebral single photon emission computed tomography in mild traumatic brain injury. *J Am Acad Psychiatry Law* 36:310–322.
138. Belanger, H.G. et al. 2007. Recent neuroimaging techniques in mild traumatic brain injury. *J Neuropsychiatry Clin Neurosci* 19:5–20.

14 Epilepsy after Traumatic Brain Injury

Kan Ding, Puneet K. Gupta, and Ramon Diaz-Arrastia

CONTENTS

DEFINITIONS

An epileptic seizure is a "transient occurrence of signs and/or symptoms due to abnormal excessive or synchronous neuronal activity in the brain."[1] Epilepsy is a disorder of the brain characterized by an enduring predisposition to generate epileptic seizures and by the neurobiological, cognitive, psychological, and social consequences of this condition.[1] The widely accepted operational definition of epilepsy requires that an individual have at least two unprovoked seizures on separate days, generally 24 hours apart. An unprovoked seizure refers to a seizure that occurs in the absence of an acute brain insult or systemic disorder.[2] This nosology is based on the observation that while a single unprovoked seizure is often an isolated event that does not recur during an individual's lifetime, two unprovoked seizures rarely occur in isolation and are associated with high risk of the individual experiencing more seizures. In some cases, such as when the seizure occurs in the setting of a potentially epileptogenic brain insult such as an episode of encephalitis or a traumatic brain injury (TBI), it is possible to recognize the specific form of epilepsy at its earliest presentation. In such cases, the diagnosis of epilepsy can be made after the very first unprovoked seizure.[2] Based on its definition, epilepsy is not a single disorder but rather a syndrome or syndromes. Different forms of epilepsy syndrome(s) have different causes, different manifestations, different implications for short- and long-term management and treatment, and different outcomes.

Post-traumatic epilepsy (PTE) refers to epilepsy that develops after TBI. Most investigators agree that PTE is to be distinguished from repeated seizures in the early stage following TBI, while the brain is acutely traumatized, inflamed, and metabolically disrupted. Therefore, a common set of definitions adopted by many researchers is the following: (1) immediate seizures, usually defined as those occurring within 24 h after injury; (2) early seizures, which occur less than 1 week after injury; and (3) late seizures, which occur more than a week after injury. Since the risk of recurrence after a single late post-traumatic seizure is over 70%, most investigators consider a single late post-traumatic seizure as being sufficient for the diagnosis of PTE.[3] Although these are the most widely accepted definitions, there is controversy. Some narrow the definition of immediate seizures to those occurring at impact or within minutes of the injury, classifying seizures occurring hours after trauma as early seizures. Others extend the time of early seizures to as long as 30 days after the injury.

Another set of considerations related to PTE is the severity of head trauma. A variety of classifications of TBI severity exist, but most investigators currently use the following schema: (1) mild TBI, characterized by loss of consciousness less than 30 min, post-traumatic amnesia less than one hour, and normal neuroimaging; (2) moderate TBI, characterized by loss of consciousness more than 30 minutes and less than 24 hours, post-traumatic amnesia between 1 and 7 days, with or without abnormal neuroimaging; and (3) severe TBI, characterized by loss of consciousness greater than 24 hours, post-traumatic amnesia more than 7 days, usually with abnormal neuroimaging such as contusion, cerebral hematoma, or extra-axial hematoma.[3] This nosology, while robust and usually easily applied, has many limitations, foremost among them the fact that it does not take into account findings from modern neuroimaging studies.

EPIDEMIOLOGY

The prevalence of epilepsy in the general United States population is estimated at 0.5%–2%, and the cumulative incidence of PTE ranges widely, from 2% to over 50% depending on injury severity (Tables 14.1 and 14.2).[4] Studies on hospitalized TBI patients report a cumulative incidence of 5%–7%,[5] and the risk of PTE is as high as 53% in a military series that included penetrating head injuries.[6,7] In a large population-based study done out of Rochester, Minnesota, head trauma was identified as the cause of epilepsy in 6% of the population (see Figure 14.1).[8] There are important differences among age groups. Of cases where the cause of epilepsy could be identified, trauma is the cause of epilepsy in almost 30% of individuals who develop seizures between ages 15 and 34 years, whereas it is a cause in approximately 14% in children

TABLE 14.1
Frequency of Post-Traumatic Epilepsy (PTE) Stratified by Injury Severity in Civilian Populations

Series	Injury Severity					
	Adults or Mixed Series					
	PTA <24 h				PTA >24 h	
References	**Early**		**Late**		**Early**	**Late**
Mares et al.[28]	2.7%		6.7%		12%	14.2%
	Mild		**Moderate**		**Severe**	
Salazar et al.[7]	1.5 (SIR)		2.9 (SIR)		17.0 (SIR)	
Young[19]	2.8% (Early)		1.7% (Early)		21.6% (Early)	
	Pediatric Series					
	Mild		**Moderate**		**Severe**	
	Early	**Late**	**Early**	**Late**	**Early**	**Late**
Caveness et al.[6]	1.0%	0.2%	1.1%	1.6%	30.5%	7.4%
Hahn et al.[10]	5.9%		26.9%		35.3%	

Source: Frey L.C., *Epilepsia*, 2003, 44(Suppl 10):11–17.
Note: PTA, post-traumatic amnesia; SIR, standardized incidence ratio.

younger than 14 years and 8% in adults older than 65 years.[8] About 4% patients who were seen at a specialized epilepsy center reported a history of moderate to severe head trauma.[9]

Early seizures occur in 2%–17% of all patients with head injuries, being more common in children than adults, and correlated with the distribution of head-injury severity.[4,10] The frequency of early seizures after severe head injuries is typically higher (10%–15% for adults; 30%–35% for children).[4,10] Of the seizures that occur within the first week of trauma, about 50% occur within the first 24 hours.[11] Of the seizures that occur within the first 4 weeks of injury, about 10% occur within the first week.[4] Clinically apparent early seizures with typical convulsive features are likely the tip of the iceberg, and the use of continuous electroencephalography (EEG) monitoring reveals that nonconvulsive seizures are much more common than convulsive seizures in the first week after severe TBI.[12] The clinical significance of nonconvulsive seizures identified through continuous EEG monitoring remains a topic of investigation. While they are associated with poor outcome, it is unclear whether aggressively treating them improves recovery.

The cumulative incidence of late seizure over 30 years after TBI is 2% for mild injuries, 4% for moderate injuries, and over 15% for severe injuries.[13] The incidence was about 5% in people hospitalized with TBI.[14] In the setting of severe head trauma with cortical injury and neurologic sequelae, the incidence of PTE increases reaches

TABLE 14.2
Frequency of Post-Traumatic Epilepsy (PTE) Stratified by Injury Severity from Select Military Series

References	Length of Study	Overall Frequency of PTSs	Frequency of Early PTS	Frequency of Late PTS	Population Characteristics
		Adults or Mixed Series			
Salazar et al.[7] (n = 4541)	10+ yr	4.4%	2.6%	2.1%	Population based
Young[19] (n = 702)	7+ days	–	4.1%	–	Hospital admissions
Pagni[11] (n = 1000)	1–4 yr	–	4.6%	5%	Hospital admissions
Frey[4] (n = 490)	5+ yr	–	16.9%	24.5%	Required intensive rehabilitation
		Pediatric Series			
Jennett[20] (n = 4465)	4–12 yr	7.03%	6.5%	1.3%	Pediatric hospital admissions
Hahn et al.[10] (n = 937)	7+ mo	9.8%	9.6%	–	Pediatric hospital admissions

Source: Frey L.C., *Epilepsia*, 2003, 44(Suppl 10):11–17.

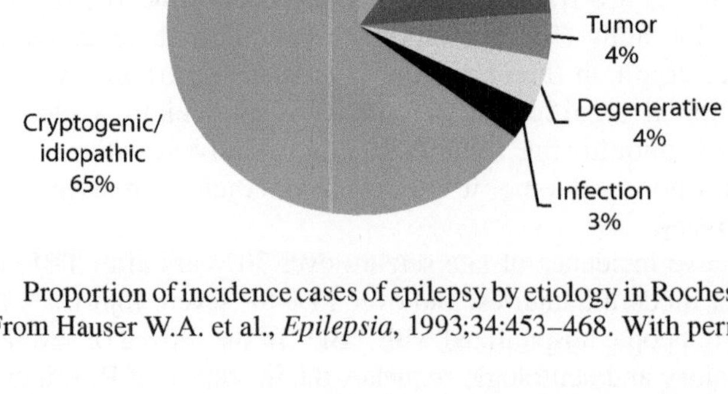

FIGURE 14.1 Proportion of incidence cases of epilepsy by etiology in Rochester, Minnesota, 1935–1983. (From Hauser W.A. et al., *Epilepsia*, 1993;34:453–468. With permission.)

over 50%, particularly if the dura is penetrated.[15] Late seizures can have a very long latency period but usually start in the first few years after injury.[16,17] About 40% start within the first 6 months of injury, 50%–60% within the first 12 months, and approximately 80% of first late seizures occur within 2 years of the injury. The risk of incident late seizure decreases with time and reaches background risk for the population at 10–15 years after the head injury.[13,18] Up to 86% of TBI survivors with a first post-traumatic seizure will have a second within 2 years, providing the rationale for making the diagnosis of post-traumatic epilepsy after a single late post-traumatic seizure. A certain percentage of PTE cases remain refractory to antiseizure medications.[4]

RISK FACTORS

Risk Factors for Early Seizures

Risk factors for early seizures include younger age (especially <5 years), acute intracerebral hematoma, acute subdural hematoma, diffuse cerebral edema, intracranial metal fragment retention, focal neurologic deficits, depressed/linear skull fractures, and loss of consciousness (LOC)/amnesia for >30 minutes.[4] Early seizures may be a weaker predictor for PTE in children; the portion of children with early seizure who have late seizures is less than one-fifth in children and may be as low as one-tenth.[19] On the other hand, about 3% of patients with no early seizures develop late PTE; this number is 25% in those who do have early PTS, and the distinction is greater if other risk factors for developing PTE are excluded.[20]

Risk Factors for Late Seizures

Factors that increase the risk of late post-traumatic seizures are age at injury >65, the presence of early post-traumatic seizures (in adults), subdural hematoma, brain contusion, premorbid chronic alcoholism, penetrating head injury, retained metal fragments in brain, depressed skull fracture, focal neurological deficits, degree of loss of brain tissue, and severity of injury (Table 14.3).[4,7,13,21]

A population-based study in Olmsted County, Minnesota, found that the relative risk of late seizures was 12.7% in the first year, 4.4% during the next 5 years, and 1.4% thereafter.[22] The head-injured victims were divided into groups with respect to severity of injury.[13,22] Those with a severe head injury (brain contusion, intracerebral or intracranial hematoma, or 24 hours of unconsciousness or amnesia) had a risk factor of 7.1% within the first year after injury and 11.5% in 5 years with an overall incidence ratio of 17.0. Those with a moderate head injury (skull fracture or 30 minutes to 24 hours of unconsciousness or amnesia) had a risk factor of 0.7% within the first year and 1.6% within 5 years with an overall incidence ratio of 2.9. Those with a mild head injury (shorter periods of unconsciousness or amnesia) had a risk factor of 0.1% within the first year and 0.6% within 5 years with an overall incidence ratio of 1.5. Recent data suggest that genomic information (e.g., haptoglobin genotypes, apolipoprotein E levels) may be helpful in predicting an individual's risk for PTE as a possible genetic predisposition has been observed.[23,24]

TABLE 14.3
Risk Factors for Epileptic Seizures and Epilepsy after Nonmissile Head Trauma

Early Seizure	Later Seizure
Intracranial hematoma	Intracranial hematoma
Focal neurological signs	Early seizures, especially when delayed
Post-traumatic amnesia >24 hours	Post-traumatic amnesia >24 hours
Any neurological signs	At least one nonreactive pupil
Depressed skull fracture	Depressed skull fracture
Subarachnoid hemorrhage	Dural penetration
Injury before 5 years of age	Injury after 16 years of age
Linear skull fracture	Glasgow Coma Scale score of 3–8
	Time to following commands of a week or more
	Parietal lesion on CT

Source: Jennett B., *Epilepsy after Non-Missile Head Injuries*, 2nd ed., London: William Heinemann Medical Books, 1975; Engel J., Jr., *Seizure and Epilepsy*, New York: Oxford University Press, 2013; Temkin N., *Epilepsia* 44(Suppl 10):18–20, 2003.

Genetic Factors

Of note, not every patient with clinical risk factors for PTE goes on to develop PTE, and this variability in occurrence may be attributable to genetic variation in genes involved in the epileptogenic process. According to Jennett,[20] the incidence of a family history of epilepsy in patients with closed head trauma who develop late seizures is 6%–17%, compared with only 3%–4% in those who do not. Furthermore, the risk for late epilepsy is greater and the latency between head injury and the first late spontaneous seizure is shorter for patients who experience early reactive seizures. These findings strongly suggest that innate, presumably genetically determined, predisposing factors contribute to the development of PTE. That a family history of epilepsy may not be a significant risk factor for late epilepsy following penetrating head injuries could indicate that genetic influence is relatively weak compared with the effects of very extensive intracranial trauma.[25] Although advances in genetics have now revealed numerous susceptibility genes, predominantly involving ion channels and GABA receptors, which appear to play an important role in the manifestation of acquired epilepsy in humans,[25] relatively little has been conclusively established about genetic susceptibility to PTE.

Although there have not been any genome-wide allelic association studies designed to identify genetic variants associated with PTE, several candidate genes have been studied using allelic association. Inheritance of the apolipoprotein E (ApoE) epsilon 4 allele is associated with increased risk of Alzheimer's disease, progression to disability in multiple sclerosis, and poor outcome after traumatic brain injury. Diaz-Arrastia et al.[26] prospectively collected DNA samples from 106 patients in a neurocritical care unit who had suffered severe traumatic brain injury. Six months after injury, 22 patients (20%) had at least one later post-traumatic seizure. The relative risk of late post-traumatic seizures for patients with the epsilon 4 allele was 2.41 (95% confidence interval). In the other recent retrospective study, Miller et al. invested ApoE genotype in 322 adult Caucasians with a severe TBI.[27] Although no statistical associations were found, two out of the four individuals (50%) with the E4/E4 had late post-traumatic seizures. Sample size calculations indicate that over 2000 individuals would be needed to have a cohort large enough to definitively evaluate the E4/E4 genotype in relation to PTE.[27]

C677T variant in the methylenetetrahydrofolate reductase (MTHFR) enzyme is a biologically plausible genetic risk factor for seizure or epilepsy. This variant is associated with elevated homocysteine levels, and homocysteine is recognized to lower seizure thresholds and has been used as a proconvulsant in animal epilepsy models.[28–30] Homozygote TT genotype may be overrepresented in epilepsy patients.[31,32] Scher et al. genotyped C677T in a military cohort and found that the odds of epilepsy were increased in subjects with the TT versus CC genotype (adjusted OR = 1.57).[33] Adenosine A1 receptors (A1AR) are located on neurons in regions susceptible to both seizures and TBI (e.g., hippocampus, cortex) and are spatially associated with NMDA receptors.[34] A1AR activation influences the extent of status epilepticus in both ex vivo and in vivo seizure models.[35] Importantly, profound PTE, leading to lethal status epilepticus, occurs in A1AR knockout (KO) mice when subjected to the controlled cortical impact (CCI) model of experimental TBI.[36] Tagging single nucleotide polymorphisms (tSNPs) analysis in the A1AR gene in 206 subjects with severe TBI demonstrated that genetic variability within the rs3766553 and the rs10920573 regions are associated with increased susceptibility for post-traumatic seizures. Subjects with both risk genotypes had a 47% chance of late post-traumatic seizures.

MECHANISM

Epileptogenesis is defined as the development and extension of tissue capable of generating spontaneous seizures, including development of an epilepsy condition and progression after the condition is established.[25,37,38] TBI can produce both focal and diffuse insults to the brain, and both mechanisms often coexist in given patients. Focal insults such as contusions or intracranial hematomas result in cicatrix in the cortex or subcortical regions, and the neighboring neural tissue experiences inflammation, gliosis, and ultimately neuronal sprouting and neurogenesis. These secondary tissue responses are believed to be the initiating events in a cascade of processes that result in epileptogenesis. However, the exact underlying pathophysiologic epileptogenic process is not clearly understood after brain trauma.

Immediate post-traumatic seizures likely occur because the impact from the injury stimulates brain tissue that has a low threshold for seizures when stimulated.[39] Early post-traumatic seizures can be the result of the secondary effects of the head trauma such as cerebral edema, intracranial hemorrhage, cerebral contusion or laceration, alterations in the blood–brain barrier, changes in extracellular ions, excessive release of excitatory neurotransmitters such as glutamate, damage to tissues caused by free radicals, and changes in the way cells produce energy.[39,40] Late seizures are thought to indicate permanent changes in the brain's structure that are thought to result from neuronal and synaptic loss, aberrant sprouting, and rewiring.

Enhanced excitatory connectivity and decreases in GABAergic inhibition are important mechanisms underlying injury-induced epileptogenesis in many animal models and in humans. Sprouting of excitatory axons and establishment of new synapses is a ubiquitous epileptogenic response to cortical injury.[41] The partially isolated neocortical island with intact pial circulation ("undercut") is an established in vivo and in vitro model for development of chronic post-traumatic hyperexcitability and epileptogenesis.[42] This model provides a high yield of animals with epileptogenic cortex that may be studied in vitro.[43,44] In this model, the treatment with tetrodotoxin (TTX) during a critical period early after lesion placement can prevent epileptogenesis.[41] The immunocytochemical experiments show that such treatment markedly attenuates histologic indices of axonal and terminal sprouting and presumably associated aberrant excitatory connectivity.[42] Another finding in the undercut model is a decrease in spontaneous inhibitory events. This is accompanied by regressive alterations in fast-spiking γ-aminobutyric acid (GABA) ergic interneurons, including shrinkage of dendrites, marked decreases in axonal length, structural changes in inhibitory boutons, and loss of inhibitory synapse on pyramidal cells.[42] A potential underlying mechanism is loss of trophic support from brain-derived neurotrophic factor (BDNF) released by pyramidal neurons acting on interneuronal TrkB receptors.[41]

In summary, there are many processes initiated by cortical injury that are ongoing in parallel. Each of them, or combinations of several, might well be important contributors to epileptogenesis and targets for prophylaxis.

CLINICAL PRESENTATION

Post-traumatic seizures may present anywhere in the spectrum, including partial seizure without alteration of consciousness, partial seizure with alteration of consciousness, and symptomatic and secondarily generalized seizures. However, primary generalized seizures, such as absence seizures, are not thought to be caused by head injury.[45,46]

The temporal lobes are the most common site for late post-traumatic seizures.[9,47] Temporal lobe seizures are associated with an aura approximately two-thirds of the time. The auras can be autonomic (abdominal discomfort, nausea, abdominal rising feeling), psychic (fear or sense of impending doom, anxiety, feelings of déjà vu or jamais vu), or olfactory and gustatory hallucinations (usually of an obnoxious smell or taste). As the seizure evolves, alteration of consciousness may occur and patients may stare and be unresponsive, and may also have stereotyped behaviors such as chewing, lip smacking, or grunting, as well as automatisms such as self-polishing or fumbling with their clothes. Secondarily generalizations are relatively rare with seizures of temporal lobe onset.

The frontal lobe is the second common site for post-traumatic epilepsy.[9,47] Auras are rare in frontal lobe epilepsy. Typical behaviors during frontal seizures include hyperkinetic motor movements, bicycling, hip thrusting, thrashing about, and asymmetric tonic posturing. Secondarily generalizations are common with frontal lobe onset seizures.[9]

Seizures arising from other lobes are less common.[47] Seizures starting in the occipital lobes are often associated with elementary visual hallucinations, such as seeing bright lights, zigzagging colored lines, or kaleidoscopic shapes. Seizures starting in the more rostral parts of the occipital lobe and occipitotemporal junction are often associated with formed visual hallucinations. Seizures starting in the parietal lobes are often associated with a vertiginous aura, and if they arise near the postcentral gyrus, then elementary sensory symptoms are experienced, which can be painful.

Any partial onset seizure can secondarily generalize. Secondarily generalized tonic–clonic seizures are characterized by tonic extensor posturing of the arms and legs, followed by rhythmic clonic movements of the arms, legs, and trunk. Generalized tonic–clonic seizures (GTCs) are often associated with transient apnea, vomiting, tongue biting, and sphincter incontinence. Partial seizures with alteration of consciousness and secondarily generalized seizures are characterized by a postictal period, during which the patient is obtunded and difficult to arouse. This stage usually lasts a few minutes but even after regaining consciousness patients are often confused and amnestic for up to several more hours. Patients often report headaches, dizziness, and sleepiness after a seizure, particularly GTCs.

PHYSICAL EXAMINATION

No specific findings are noted on physical examination and the neurological examination of patients with PTE may be nonfocal. However, if present, focal neurological findings may correlate neuroanatomically with the traumatic brain lesion, which may be in the epileptogenic zone. If the patient can be examined immediately after a seizure, finding a focal neurologic deficit that later resolves (Todd's paralysis) is often very helpful for localizing the site of seizure onset.

DIAGNOSTIC EVALUATION

The differential diagnosis for post-traumatic spells needs to include PTS/PTE, concussive convulsions, psychogenic nonepileptic seizures (PNSs) (e.g., pseudoseizures), syncope (e.g., concussive syncope), confusional states (i.e., delirium), acute memory disorders (i.e., fugue state), dizziness, and imbalance. Concussive convulsions are not actual PTS and are not predictive of post-traumatic epilepsy.[48] They are thought to result from temporary loss of brain function rather than from structural damage and are usually associated with a good outcome.[49] PNS are seizure-like episodes that may occur after head injury, but video EEG (VEEG) monitoring, which is often required to definitively make the diagnosis, shows that the nature of the seizures is psychogenic rather than epileptic. Features that distinguish PNS from epileptic seizures on VEEG monitoring include lack of stereotypy in semiology and duration of the seizures, lack of scalp EEG correlate to the events, and atypical evolution of the behaviors, such as start-and-stop phenomena. In patients with moderate to severe traumatic brain injury referred to a Comprehensive Epilepsy Center for evaluation of refractory PTE, about 30% were found to have been misdiagnosed and have psychogenic attacks.[46] This percentage is similar to patients with epilepsy after nontraumatic etiologies. Therefore, if atypical features and seizures continue despite treatment, the diagnosis should be verified by VEEG rather than assuming the patient has PTE.

LABORATORY STUDIES

In patients hospitalized after a recent head injury, investigation of a seizure should focus on determining whether an intracranial bleed or a change in clinical condition (e.g., hyponatremia) caused the seizure. Acutely injured patients whose sensorium fails to return to normal after a seizure should be evaluated with EEG. For patients in otherwise stable condition whose serum electrolytes are within the normal range and whose neurologic findings are the same as those before the seizure, further laboratory studies are not needed. Some academic centers perform continuous VEEG monitoring routinely in patients who are comatose or stuperous after TBI, and routinely find that up to 20% have unrecognized nonconvulsive seizures.[50,51]

In a patient presenting months or years after the injury, the usual investigations that are applicable for evaluation of first epileptic seizure should be performed (i.e., electrolytes, liver function test, urine drug screen). EEG is very helpful for differential diagnosis. An interictal EEG showing epileptiform discharges (spikes or sharp waves) has a specificity of >97% for epilepsy. However, the sensitivity of interictal EEG is low, in the range of 30%–50%. VEEG monitoring may help in narrowing the differential diagnosis, in localizing seizure foci and in prognosticating their severity, and in predicting relapse before anticonvulsant medication is withdrawn. VEEG evaluation should be considered in every patient with frequent, disabling spells who has not responded to appropriate antiepileptic medications.

RADIOGRAPHIC ASSESSMENT

Patients with a TBI and a seizure should be imaged with a CT scan immediately, and the study should be repeated if the condition of the patient does not improve or worsens (e.g., acute change in mental status). CT is less sensitive but more accessible and less expensive than MRI and should be able to depict all pathology (intracranial bleed, midline shift) that needs urgent intervention. Brain MRI is the study of choice for nonacute evaluations of PTS/PTE. Findings on certain sequences may convey important information relating to recent seizure or epileptogenic focus. Fluid attenuated inversion recovery (FLAIR) may show diffuse axonal injury (DAI), contusion, blood, edema, or encephalomalacia. Gradient echo (GRE) and susceptibility-weighted imaging (SWI) is more sensitive than CT and conventional MR techniques for demonstrating microhemorrhages resulting from shear injury. Diffusion tenor imaging (DTI) appears to be very sensitive for identifying disruption in white matter tracts, but its use is still primarily investigational. Of note, diffusion-weighted imaging (DWI) and FLAIR can demonstrate transient abnormalities after a seizure that later disappears, and does not necessarily reflect structural injury.

TREATMENT

The occurrence of PTS significantly worsens functional outcome, and prevention of PTS is an important goal.[52] Prophylaxis treatment with antiseizure drugs is often initiated as soon as possible after moderate to severe TBI.[16] A landmark study indicates that antiseizure medication (e.g., phenytoin) given within a day of injury prevents early seizures (within the first week of injury) but not late post-traumatic seizures or post-traumatic epilepsy.[53] Prophylactic treatment with antiepileptic drugs (AEDs) also does not improve functional outcome from TBI. There is also the suspicion that prophylactic use of AEDs over a long period is associated with an increased risk for seizures. For these reasons, AEDs are commonly used for a short time after head trauma to prevent immediate and early, but not late, seizures.[54] Thus, the use antiseizure drugs for prophylaxis during the first week, while a widespread practice, is an option and not a guideline. No treatment is established to prevent the development of epilepsy.[55] However, medications may be given to repress more seizures if late seizures do occur. In children, AEDs may be ineffective for both early and late seizures.[5] Treatment of PTE does not require hospitalization, but admission may be needed for the treatment of status epilepticus or for VEEG to assist in the diagnosis.

INITIAL MANAGEMENT

In those with a single unprovoked seizure, the decision whether to begin antiepileptic drug treatment depends on the risk of the individual for developing further seizures. For those with early PTS, the recommendation is that antiepileptic drugs should be started promptly and continued for a few weeks. Acutely, lorazepam, diazepam, fosphenytoin, sodium valproate, and levetiracetam are the drugs of choice and usually effective in stopping an ongoing seizure. In most cases, administering the medication via the intravenous (IV) route is desirable, as the patient is still in the recovery stage from the head injury. Chronically, one first or second generation antiepileptic medication can be used. Typically, the second generation medications are just as efficacious and better tolerated (fewer drug–drug interactions and side effects). No known randomized controlled studies have been performed to prove that one is better than the other. Phenytoin is widely used, but it seems to increase the risk of impairing cognitive function.[56] Levetiracetam is better tolerated and there is evidence that it is equivalently effective as phenytoin after craniotomy and in severe TBI.[57,58] The dosage prescribed should lead to a therapeutic blood level. If control is not achieved, the dose is increased until the patient has toxic side effects. The dosage is then decreased just enough to prevent toxicity. If seizure control is not achieved with one drug, a second drug is added. For those with late PTS, because of the high risk of seizure recurrence after a first late PTS, chronic use of AEDs is recommended after even a single late seizure. The choice of antiepileptic medications is the same as noted earlier.

ONGOING CARE

Regular medical follow-up should be performed to review seizure control, medication side effects, and neurologic/neuropsychological assessment. Once a therapeutic medication regime is achieved, the individual is typically maintained on the same dosage for a period of 2 years. After 2 years, the individual should be evaluated for the possibility of withdrawal from the antiepileptic therapy, which if done should occur gradually over a period of several months. Factors such as the presence of focal neurologic deficits, CT evidence of structural brain disease, and persistent EEG abnormalities increase the risk of recurrence after tapering antiepileptic drugs, even after a prolonged period of seizure control while on medications. These factors should be assessed and used to inform the decision of whether to stop medication. If seizures remain intractable, then consultation with a neurologist or referral to an epilepsy center may be needed to confirm the diagnosis of PTS/PTE or for consideration further management, such as possible surgical therapy.

Surgical treatment may be an option for PTE refractory to medication. The aim of resective epilepsy surgery is to identify and remove the epileptogenic focus. This is sometimes more difficult in cases of PTE than in other types of epilepsy and may require intracranial EEG evaluation. Other surgical options may include implantation of a vagus nerve stimulator (Cyberonics), a responsive neurostimulator (NeuroPace), or a deep brain stimulator (Medtronic), some of which are experimental at the moment.

TREATMENT CONTROVERSIES

Prophylactic treatment of PTE has been common practice for many years and remains an issue of controversy. The theory that an antiepileptic medication prevents the epileptic foci from developing has been disproven for drugs that have been rigorously studied. As there remains a significant risk to developing epilepsy after severe head trauma, it remains common practice to initiate prophylactic antiepileptic therapy. However, since prevention of early seizures does not translate into improved neurologic recovery, the use of prophylactic medications during the first is optional.

Another area of controversy is whether treatment is useful in patients who experienced early PTS. There is little high-quality evidence to provide therapeutic guidance in this situation. Although an early PTS is not associated with such a high risk of recurrence that it is equivalent to PTE, most neurologists treat with antiepileptic drugs for several weeks to several months, and taper the medications off slowly if no late PTS is noted.

ADDITIONAL CONSIDERATIONS

Patients must be warned to exercise caution during bathing, swimming, and climbing heights. They should never be alone during these activities. In all situations, appropriate steps should be taken to ensure the safety of the person if a seizure occurs. Patients must also be counseled about the limitations in obtaining a driver's license, based on the laws in their state of residence. Psychological problems related to social isolation and the stigma of epilepsy are common and must be addressed. Consultation with psychiatrists, counselors, and social workers is very important in the management of post-traumatic epilepsy. The medical/legal aspect is an important issue in cases of PTE, as some patients pursue legal actions against various authorities and individuals responsible for the circumstances of the accident. Clinicians are often asked to estimate the risk of a patient developing PTE in the future as a result of sustained brain injury. This is a difficult task and should be left to an experienced senior specialist.

REFERENCES

1. Fisher, R.S. et al. 2005. Epileptic seizures and epilepsy: Definitions proposed by the International League Against Epilepsy (ILAE) and the International Bureau for Epilepsy (IBE). *Epilepsia* 46:470–472.
2. Elaine, W. 2011. *Wyllie's Treatment of Epilepsy*. Philadephia, PA: Lippincott Williams & Wilkins.
3. Lowenstein, D.H. 2009. Epilepsy after head injury: An overview. *Epilepsia* 50(Suppl 2):4–9.
4. Frey, L.C. 2003. Epidemiology of posttraumatic epilepsy: A critical review. *Epilepsia* 44(Suppl 10):11–17.
5. Teasell, R. et al. 2007. Post-traumatic seizure disorder following acquired brain injury. *Brain Inj* 21:201–214.
6. Caveness, W.F., Walker, A.E., Ascroft, P.B. 1962. Incidence of posttraumatic epilepsy in Korean veterans as compared with those from World War I and World War II. *J Neurosurg* 19:122–129.
7. Salazar, A.M. et al. 1985. Epilepsy after penetrating head injury. I. Clinical correlates: A report of the Vietnam head injury study. *Neurology* 35:1406–1414.
8. Hauser, W.A., Annegers, J.F., Kurland, L.T. 1993. Incidence of epilepsy and unprovoked seizures in Rochester, Minnesota: 1935–1984. *Epilepsia* 34:453–468.
9. Gupta, P. et al. 2014. Subtypes of post-traumatic epilepsy: Clinical, electrophysiologic, and imaging features. *J Neurotrauma* 31:1439–1443.
10. Hahn, Y.S. et al. 1988. Factors influencing posttraumatic seizures in children. *Neurosurgery* 22:864–867.
11. Pagni, C.A. 1990. Posttraumatic epilepsy. Incidence and prophylaxis. *Acta Neurochir* 50:38–47.
12. Vespa, P.M. et al. 1999. Increased incidence and impact of nonconvulsive and convulsive seizures after traumatic brain injury as detected by continuous electroencephalographic monitoring. *J Neurosurg* 91:750–760.
13. Annegers, J.F. et al. 1998. A population-based study of seizures after traumatic brain injuries. *N Engl J Med* 338:20–24.
14. Cuccurullo, S. 2004. *Physical Medicine and Rehabilitation Board Review*. Demos Medical, New York.
15. Caveness, W.F. 1976. Epilepsy, a product of trauma in our time. *Epilepsia* 17: 207–215.
16. Chang, B.S., Lowenstein, D.H., Quality Standards Subcommittee of the American Academy of Neurology. 2003. Practice parameter: Antiepileptic drug prophylaxis in severe traumatic brain injury: Report of the Quality Standards Subcommittee of the American Academy of Neurology. *Neurology* 60:10–16.
17. De Santis, A. et al. 1992. Risk factors for late posttraumatic epilepsy. *Acta Neurochir* 55:64–67.
18. Weiss, G.H. et al. 1986. Predicting posttraumatic epilepsy in penetrating head injury. *Arch Neurol* 43:771–773.
19. Young, B. 1992. Post-traumatic epilepsy. In *Complications and Sequelae of Head Injury*, ed. D.L. Barrow, 127–132. New York: Thieme, American Association of Neurological Surgeons.

20. Jennett, B. 1975. *Epilepsy after Non-Missile Head Injuries*, 2nd ed. London: William Heinemann Medical Books.
21. Temkin, N. 2003. Risk factors for post-traumatic seizures in adults. *Epilepsia* 44(Suppl 10):18–20.
22. Annegers, J.F. et al. 1980. Seizures after head trauma: A population study. *Neurology* 30:683–689.
23. Schaumann, B.A. et al. 1994. Family history of seizures in posttraumatic and alcohol-associated seizure disorders. *Epilepsia* 35:48–52.
24. Panter, S.S. et al. 1985. Hypohaptoglobinemia associated with familial epilepsy. *J Exp Med* 161:748–754.
25. Engel, J., Jr. 2013. *Seizure and Epilepsy*. New York: Oxford University Press.
26. Diaz-Arrastia, R. et al. 2003. Increased risk of late post-traumatic seizures with inheritance of APOE epsilon4 allele. *Arch Neurol* 60:818–822.
27. Miller, M.A. et al. 2010. Apoe genetic associations with seizure development after severe traumatic brain injury. *Brain Inj* 24:1468–1477.
28. Diaz-Arrastia, R. 2000. Homocysteine and neurologic disease. *Arch Neurol* 57:1422–1427.
29. Kubova, H., Folbergrova, J., Mares, P. 1995. Seizures induced by homocysteine in rats during ontogenesis. *Epilepsia* 36:750–756.
30. Mares, P., Haugvicova, R., Kubova, H. 2002. Interaction of excitatory amino acid agonists with cortical afterdischarges in developing rats. *Epilepsia* 43(Suppl 5):61–67.
31. Dean, J.C. et al. 2008. A high frequency of the mthfr 677c>t polymorphism in Scottish women with epilepsy: Possible role in pathogenesis. *Seizure* 17:269–275.
32. Ono, H. et al. 2000. Methylenetetrahydrofolate reductase 677c > t mutation and epilepsy. *J Inherit Metab Dis* 23:525–526.
33. Scher, A.I. et al. 2011. Mthfr c677t genotype as a risk factor for epilepsy including post-traumatic epilepsy in a representative military cohort. *J Neurotrauma* 28:1739–1745.
34. Deckert, J., Jorgensen, M.B. 1988. Evidence for pre- and postsynaptic localization of adenosine a1 receptors in the ca1 region of rat hippocampus: A quantitative autoradiographic study. *Brain Res* 446:161–164.
35. Fedele, D.E. et al. 2006. Adenosine a1 receptors are crucial in keeping an epileptic focus localized. *Exp Neurol* 200:184–190.
36. Kochanek, P.M. et al. 2006. Adenosine a1 receptor knockout mice develop lethal status epilepticus after experimental traumatic brain injury. *J Cereb Blood Flow Metab* 26:565–575.
37. Lowenstein, D.H. 1996. Recent advances related to basic mechanisms of epileptogenesis. *Epilepsy Res* 11:45–60.
38. Ronne-Engstrom, E. et al. 1992. Intracerebral microdialysis of extracellular amino acids in the human epileptic focus. *J Cereb Blood Flow Metab* 12:873–876.
39. Payan, H., Toga, M., Berard-Badier, M. 1970. The pathology of post-traumatic epilepsies. *Epilepsia* 11:81–94.
40. Yablon, S.A. 1993. Posttraumatic seizures. *Arch Phys Med Rehabil* 74:983–1001.
41. Prince, D.A., Parada, I., Graber, K. 2012. Traumatic brain injury and posttraumatic epilepsy. In *Jasper's Basic Mechanisms of the Epilepsies*, eds. J.L. Noebels, M. Avoli, M.A. Rogawski, R.W. Olsen, A.V. Delgado-Escueta. Bethesda, MD: National Center for Biotechnology Information.
42. Prince, D.A. et al. 2009. Epilepsy following cortical injury: Cellular and molecular mechanisms as targets for potential prophylaxis. *Epilepsia* 50(Suppl 2):30–40.
43. Graber, K.D., Prince, D.A. 2006. Chronic partial cortical isolation. In *Models of Seizures and Epilepsy*, eds. A. Pitkanen, P. Schwartzkroin, 477–493. San Diego, CA: Elsevier Academic Press.

44. Hoffman, S.N., Salin, P.A., Prince, D.A. 1994. Chronic neocortical epileptogenesis in vitro. *J Neurophysiol* 71:1762–1773.
45. Diaz-Arrastia, R. et al. 2000. Neurophysiologic and neuroradiologic features of intractable epilepsy after traumatic brain injury in adults. *Arch Neurol* 57:1611–1616.
46. Hudak, A.M. et al. 2004. Evaluation of seizure-like episodes in survivors of moderate and severe traumatic brain injury. *J Head Trauma Rehabil* 19:290–295.
47. Kellaway, P. 1980. The incidence, significance and natural history of spike foci in children. In *Current Clinical Neurophysiology: Update on EEG and Evoked Potentials*, ed. C. Henry, 151–175. Amsterdam: Elsevier/North Holland.
48. McCrory, P.R., Berkovic, S.F. 1998. Concussive convulsions. Incidence in sport and treatment recommendations. *Sports Med* 25:131–136.
49. Perron, A.D., Brady, W.J., Huff, J.S. 2001. Concussive convulsions: Emergency department assessment and management of a frequently misunderstood entity. *Acad Emerg Med* 8:296–298.
50. Vespa, P.M., Nenov, V., Nuwer, M.R. 1999. Continuous EEG monitoring in the intensive care unit: Early findings and clinical efficacy. *J Clin Neurophysiol* 16:1–13.
51. Claassen, J. et al. 2004. Detection of electrographic seizures with continuous EEG monitoring in critically ill patients. *Neurology* 62:1743–1748.
52. Asikainen, I., Kaste, M., Sarna, S. 1999. Early and late posttraumatic seizures in traumatic brain injury rehabilitation patients: Brain injury factors causing late seizures and influence of seizures on long-term outcome. *Epilepsia* 40:584–589.
53. Temkin, N.R. et al. 1990. A randomized, double-blind study of phenytoin for the prevention of post-traumatic seizures. *N Engl J Med* 323:497–502.
54. Beghi, E. 2003. Overview of studies to prevent posttraumatic epilepsy. *Epilepsia* 44(Suppl 10):21–26.
55. Iudice, A., Murri, L. 2000. Pharmacological prophylaxis of post-traumatic epilepsy. *Drugs* 59:1091–1099.
56. Massagli, T.L. 1991. Neurobehavioral effects of phenytoin, carbamazepine, and valproic acid: Implications for use in traumatic brain injury. *Arch Phys Med Rehabil* 72:219–226.
57. Jones, K.E. et al. 2008. Levetiracetam versus phenytoin for seizure prophylaxis in severe traumatic brain injury. *Neurosurg Focus* 25:E3.
58. Milligan, T.A., Hurwitz, S., Bromfield, E.B. 2008. Efficacy and tolerability of levetiracetam versus phenytoin after supratentorial neurosurgery. *Neurology* 71:665–669.

15 Intracranial Pressure Monitoring and Management

Peter Le Roux

CONTENTS

INTRODUCTION

Every 15 seconds someone suffers a traumatic brain injury (TBI) in the United States. TBI causes more deaths in males <35 years old than all other diseases combined, and it is estimated that 2% of the U.S. population lives with TBI-associated disability. Despite extensive research and success in animal studies, successful drug therapies have proved elusive in clinical trials.[1] Instead, TBI care focuses on the early identification and removal of mass lesions and on the detection, prevention, and management of secondary brain insults that adversely affect outcome (e.g., hypotension, hypoxia, seizures, elevated intracranial pressure). TBI is a heterogeneous disease in cause, pathology, severity, and prognosis. Consequently, TBI care depends in large part on careful and repeated assessment of clinical and laboratory findings, imaging studies, and bedside physiological data.

A variety of physiological processes can be monitored at the bedside.[2] Traditionally, monitoring and treating intracranial pressure (ICP) that is also used to quantify cerebral perfusion pressure (CPP) has been the cornerstone of severe TBI (sTBI) management, in large part because ICP is considered an indicator of injury severity. In this chapter, the indications, technique, and safety for ICP monitoring will be discussed. In addition, we will examine the relationship between ICP and outcome, treatment thresholds, ICP management, and how ICP management influences outcome. Finally, though ICP and CPP are important, emerging evidence suggests ICP is better viewed as an indicator of an underlying pathophysiological process that needs treatment rather than an independent target. The chapter will finish with a brief discussion on how a multimodal approach may supplement the information obtained from ICP monitoring to better target and individualize care. The focus of this chapter is on adults with TBI and may not apply to pediatric TBI.

INDICATIONS FOR INTRACRANIAL PRESSURE (ICP) MONITORING

The Guidelines for the Management of Severe TBI recommend an ICP monitor in TBI patients at risk for intracranial hypertension, i.e., patients in coma (Glasgow Coma Scale [GCS] <8) and an abnormal admission head CT scan.[3] This includes a mass lesion(s), e.g., hematoma or contusion, swelling (edema), midline shift, and compressed basal cisterns, particularly the perimesencephalic cisterns.[4–6] However, up to 50% of patients who subsequently develop increased ICP may have a normal admission head CT scan.[7] In these patients, an ICP monitor is recommended if two or more of the following are present at admission: age >40 years, unilateral or bilateral motor posturing, or episodes of systolic blood pressure (BP) <90 mmHg.[3] These recommendations however are based on studies that included small numbers of patients and were performed 20–30 years ago using older generation CT scanners that may have been insensitive to pathology that can be identified now.[8,9]

The indications for an ICP monitor remain debated in several circumstances, including (a) comatose patients with an initial normal CT scan or only minimal findings, e.g., traumatic subarachnoid hemorrhage (SAH); (b) diffuse axonal injury (DAI); (c) bifrontal contusions in the noncomatose patient; and (d) following surgery such as a decompressive craniectomy or evacuation of a mass lesion. These topics were addressed in a recent consensus conference on ICP with the following conclusions.[10] First, invasive ICP monitoring is not recommended in comatose patients with an initial normal CT scan or only minor changes, e.g., small petechiae, particularly if the scan is obtained early after injury. These patients should undergo a follow-up CT scan, particularly if there is neurologic worsening, and receive an ICP monitor if there is disease progression on the CT scan. Second, an ICP monitor is indicated when the CT shows evidence of brain swelling, e.g., compressed or absent basal cisterns.[5] Third, an ICP monitor may be considered in patients with large bifrontal contusions independent of the neurological condition.[11] Fourth, an ICP monitor is recommended when sedation interruption to check neurological function may be dangerous, e.g., respiratory failure. Similarly an ICP monitor is useful when the neurological examination is not reliable, e.g., maxillofacial trauma or spinal cord injury. Fifth, elevated ICP may occur after a decompressive craniectomy performed in a delayed fashion for intracranial hypertension refractory to medical management.[12] ICP monitoring is recommended in these patients. Finally, intracranial hypertension is common in patients who undergo a craniotomy for a mass lesion, particularly an acute subdural hematoma, and when present, intracranial hypertension aggravates outcome.[13,14] Consequently an ICP monitor is recommended after a craniotomy particularly when there are other associated factors, e.g., hypoxia, hypotension, pupil abnormalities, midline shift >5 mm, brain swelling at surgery, and when patients may require other surgeries for extracranial injuries. However, the role of an ICP monitor in patients whose bone flap is left out at this initial surgery is uncertain but may be clarified by the proposed RESCUE-acute subdural hematoma trial.[15]

In the intensive care unit, ICP is best treated guided by an ICP monitor since it is difficult to diagnose elevated ICP by clinical means alone. The introduction of clinical guidelines for sTBI has increased the use of ICP monitoring,[16] however, there still is great variability in care and as many as half the patients who fulfill the criteria for an ICP monitor never receive one.[17,18] Meta-analysis of clinical series or analysis of administrative data suggests that "aggressive care," including an increase use of ICP monitors is associated with improved outcome.[18–22] Furthermore, use of an ICP monitor improves the efficiency of care[23] and decision analysis indicates that when all the costs of sTBI are considered, aggressive care including multimodal monitoring is cost-effective, even for older patients.[24]

ICP MONITORING TECHNIQUES

A variety of techniques can be used to monitor ICP including clinical examination, brain imaging, and an ICP monitor, both invasive and noninvasive monitors.[2] The best results are obtained when all three techniques are used. Technology evaluation is not ideally suited to evidence-based medicine, but the literature suggests that a ventricular catheter connected to an external strain gauge, catheter tip strain gauge devices, or catheter tip fiber optic technology inserted into the ventricles or brain parenchyma are accurate, whereas fluid-coupled or pneumatic devices placed in the subarachnoid, subdural, and epidural compartments are less accurate. Noninvasive methods, e.g., optic nerve sheath diameter or transcranial Doppler ultrasound pulsatility index among others are promising but as yet not robust enough or validated for regular clinical use.[25] These noninvasive techniques however may be used in select patients, e.g., those with coagulopathy.

Today, two techniques, an intraparenchymal strain gauge or fiber optic monitor and an intraventricular monitor using a ventriculostomy (EVD), are in common use and the preferred and recommended techniques.[26] The EVD is considered the gold standard, but this is more a matter of precedence and tradition since it was the first monitor available and for many years the only technology available rather than the result of scientific comparison or physiology. Proponents for ventricular catheters argue that cerebrospinal fluid (CSF) can be drained and hence ICP reduced. This may be true when there is hydrocephalus, but in other patients CSF drainage may alter compliance with adverse effects. Ventricular catheters are also thought to represent a "global" ICP and not be subject to local variation. While there are gradients between parenchymal and ventricular systems, none of these are of clinical significance.[27,28] Parenchymal monitors are easier to insert, e.g., insertion is not influenced by midline shift or brain swelling, and easier to use, e.g., not subject to blockage by hemorrhage or debris.

The choice of whether to use an EVD or parenchymal monitor is best decided on a case-by-case basis and factors such as pathology, presence of coagulopathy, device accuracy, reliability, complication rates, ease of insertion, and cost among others need to be considered. In the presence of hydrocephalus, however, an EVD is preferred because it can also be used for treatment.[26,29]

SAFETY OF ICP MONITORING

There are several potential complications associated with ICP monitors and, in particular, brain hemorrhage, infection, and technical failure. The overall risk of clinically significant complications is low. However, almost all studies describe a greater complication rate when using EVDs rather than parenchymal monitors.[30–33]

Hemorrhage

The risk of intracranial hemorrhage associated with EVDs is between 2% and 10% but may be higher in children.[8,32–35] The risk associated with parenchymal monitors is less than 2.5%.[36] The vast majority of EVD or parenchymal hemorrhages are associated with insertion technique and very few (0.5%–1% of insertions associated hemorrhages) are of clinical consequence.[34] Patients who require an ICP monitor may have abnormal coagulation parameters. ICP monitor insertion is feasible in patients with mild coagulopathy if an intraparenchymal monitor is used and it appears safe to place an EVD or parenchymal monitor when the international normalized ratio (INR) ≤ 1.6.[37–40] A platelet count >100,000 is preferable but this topic has had little study.

Infection

The risk of infection also is greater with EVDs (5%–20%) than parenchymal monitors (0%–1%).[41–44] In part, this difference may have to do with ease of detection since CSF is periodically sampled in clinical practice from EVDs, but routine surveillance of parenchymal monitors is not performed. The incidence may depend on definitions, i.e., contamination, colonization or infection. The consequences of an infection, however, are far greater with an EVD than a parenchymal monitor. Indeed, the incidence of EVD-related infections may be a useful and a specific indicator of quality of care in a neurocritical care unit.[26] The infection risk associated with EVDs increases with the presence of other systemic infections, longer duration of monitoring, presence of intraventricular or subarachnoid hemorrhage, open skull fracture (with or without CSF leak), catheter flushing (including frequent CSF sampling), and CSF leakage at the insertion site.[41–45] The incidence of infection can be reduced by use of long tunnels, a closed drainage system, prompt removal when monitoring is no longer required, and avoidance of flushing the system or routine CSF sampling.[44] Instead, CSF sampling should only be performed based on clinical criteria. There does not appear to be a role for routine replacement of EVDs or use of long-term prophylactic antibiotics.[45,46] The role of antibiotic- or silver-impregnated catheters is still being elucidated.

Technical Failure

EVD catheter displacement, accidental removal, or blockage (with blood or debris) may occur. When blockage occurs, the CSF waveform is flattened and the ICP is underestimated when it is monitored at the same time as CSF is drained. The system can be gently flushed with saline to reestablish flow when blocked. The role of thrombolytic agents in intraventricular hemorrhage (IVH) is still being elucidated. Once inserted, parenchymal monitors, except the pneumatic Spiegelberg ICP monitor, cannot be recalibrated. However, drift is very rare in fiber optic catheters but may occur in strain gauge catheters.[39,47] Overall technical complications, e.g., breakage or dislodgement, are observed in about 5% of intraparenchymal devices. Most occur during patient transport or when a patient is moved, but few if any appear to influence patient outcome since they are easily recognized.[36]

The safety and success of using an ICP monitor can be enhanced when specific and standardized protocols or bundles are used.[48,49] Basic core competencies and technical specifications for ICP insertion, care, and use are described elsewhere.[29]

ICP AND OUTCOME

ICP is a valuable indicator of injury severity after TBI and there is a well-described relationship between intracranial hypertension (although the definition of this varies) and mortality after sTBI (Table 15.1).[8,50–54] This relationship becomes proportionally

TABLE 15.1
Relationship (Odds Ratio and 95% Confidence Interval) between Intracranial Pressure Values and Patterns and Outcome (1-Year Glasgow Outcome Score [GOS]) after Severe TBI

ICP Value	SD/V vs. G/MD	Dead vs. G/MD	Dead vs. Alive
ICP <20 mmHg	1.0	1.0	1.0
ICP 20–40 mmHg	1.6 (0.83–3.2)	1.9 (1.05–3.38)	1.8 (1.01–3.17)
ICP >40 mmHg	2.3 (0.99–5.32)	1.9 (1.05–3.38)	6.7 (3.65–12.47)
Increased ICP but treatable	4.0 (2.3–7.0)	2.8 (1.8–4.3)	2.2 (1.4–3.3)
Refractory increased ICP	6.9 (1.13–42.8)	100.3 (34.3–293)	88 (33.5–231)

Note: The table is modified from Treggiari M.M. et al., *Neurocritical Care* 2007; 6(2): 104–112, in which the authors performed a systematic review of four studies including 409 patients and ICP values and five studies including 677 patients and ICP patterns. The data are not true natural history data since ICP was treated at thresholds. G = good outcome; MD = moderate disability; SD = severe disability; V = vegetative.

TABLE 15.2
Logistic Regression for ICP Doses and Receiver Operating Characteristic (ROC) Analysis of Predictive Power for Mortality in Severe TBI Patients

ICP Dose	Univariate p Value	AUC	95% CI	Covariate-Adjusted p Value[a]	AUC	95% CI	Covariate-Adjusted Cross-Validated AUC[a]	95% CI
D20	<0.0011	0.63	0.55–0.70	<0.0001	0.75	0.68–0.81	0.72	0.66–0.81
D25	0.0001	0.65	0.57–0.73	<0.0001	0.77	0.70–0.83	0.65	0.56–0.73
DPRx	<0.0001	0.72	0.63–0.80	<0.0001	0.81	0.74–0.87	0.77	0.68–0.89

Source: Lazaridis C. et al., *J Neurosurg*, 2014 Apr;120(4):893–900. With permission.

Note: AUC, area under the curve; CI, confidence interval; D20, ICP threshold of 20 mmHg; D25, ICP threshold of 25 mmHg; DPRx, individualized ICP thresholds calculated from the relationship between ICP and pressure reactivity index (PRx).

[a] Adjusted for baseline GCS score, age, and sex.

greater as the value of ICP increases with a sixfold increase risk of death when ICP is greater than 40 mmHg.[50] While brief episodes of elevated ICP can adversely affect outcome,[55] the relationship between ICP and mortality is greater when the pattern of elevation, e.g., early vs. late intracranial hypertension,[56,57] the "dose" of intracranial hypertension, and treatment response is included in the analysis.[50,58–60] In particular elevated ICP that is refractory to treatment significantly increases the risk of death.[8,50,61,62] In addition, using continuous dynamic ICP information improves the accuracy of outcome prediction.[63] The relationship between ICP and morbidity and overall outcome, however, is less robust and most studies find no association between ICP and morbidity.[50,52,53,64,65]

The implication of these various data is that ICP, as a simple numeric value or a single threshold, is not an independent outcome predictor, i.e., when used to predict outcome the ICP data should be interpreted with clinical and demographic characteristics, CT findings, and other physiologic data. It should be remembered, however, that these studies are confounded by treatment and decisions on futility and are not true natural history studies. Furthermore, the analysis frequently includes manually entered end hour values and uses linear statistics that may be insensitive to any potential relationship. Indeed when using more sophisticated analysis techniques, e.g., automated data and area under the curve (AUC) analysis, an independent relationship between ICP and outcome is observed.[58,66] Finally, the relationship between ICP and outcome is based on population-derived targets. Recent studies suggest that individualized or patient-specific targets may provide a more robust relationship. For example, Lazardis et al.[67] retrospectively analyzed data from 327 sTBI patients and defined individualized ICP thresholds by graphing the relationship between ICP and pressure reactivity index (PRx). They observed that individualized doses of intracranial hypertension were stronger predictors of death than doses derived from universal thresholds of 20 and 25 mm Hg (Table 15.2).

WHAT IS THE OPTIMAL ICP TREATMENT THRESHOLD?

Several different ICP thresholds, including 15,[54,68] 20,[62,69] 25,[70] and 35 mmHg[71] are described in the literature and used in clinical practice, i.e., what defines intracranial hypertension is uncertain. There are many reasons for this. First, there are no natural history studies that relate outcome to nontreated ICP in TBI. Second, described thresholds are population based rather than individualized. Third, general protocol effects, ongoing threshold-driven treatment, and futility decisions often confound the data. Fourth, sTBI has yet to be categorized on the basis of underlying pathophysiology and phenotype and so ICP treatments are empiric and phenomenological rather than mechanistic. Fifth, much of our current practice stems from research performed in the 1960s to the late 1990s under conditions that are not relevant today. During that time there was concern that systolic blood pressure (SBP) drove intracranial hypertension. Consequently patients were kept "dry," i.e., with mean arterial blood pressures (MABP) <70 mmHg. The normal autoregulatory breakpoint is at 50 mmHg and hence patients were considered at risk for cerebral ischemia when ICP was >20 mmHg. This then was adopted as a "threshold." Today however, we usually maintain a higher CPP (>60–70 mmHg) and use sedatives that reduce the risk of cerebral ischemia when CPP is <50 mmHg. Sixth, physiologic monitors in use today, e.g., brain oxygen, microdialysis, and PET among others, demonstrate that brain energy dysfunction may occur when ICP is normal or that the brain is "healthy" despite an elevated ICP.[72–74] This has led to the concept of permissive intracranial hypertension, i.e., does it matter what the ICP is (if mildly elevated, e.g., 25 mmHg) if other measures of brain health or CPP are normal? Consequently, rather than accept a generic or absolute ICP threshold, e.g., 20 mmHg, it may be preferable to individualize thresholds based on patient characteristics, pathology, other physiologic parameters, and on a risk-benefit analysis of treating the specific ICP. Conceptually, ICP values then become an epiphenomenon of cerebral compliance, ischemia, hypoxia, cellular dysfunction, or the likelihood of herniation among other pathophysiological processes. This requires multimodal monitoring that in its simplest form means interpreting the ICP value according to the clinical and CT findings or in a more sophisticated approach using other physiologic parameters as well.[75,76] This can be challenging and requires further research validation, and so for the present, it is reasonable to set the treatment threshold at 20–25 mm Hg at the onset of management but consider altering the threshold when other clinical data support this.[77]

MANAGEMENT OF INCREASED ICP

Severe TBI management is centered on avoiding increased ICP or providing an adequate CPP. An alternative to this treatment is the Lund concept, which emphasizes a reduction in microvascular pressures. However, there is no evidence that the Lund concept is a "better" treatment option.[78] Central to these approaches is use of an ICP monitor that can help guide medical and surgical therapy for sTBI; the use of which appears to improve the efficiency of care. Traditionally ICP management has followed serial box-and-arrow diagrams or stair step-type protocols.[79,80] This linear approach is based on how a specific therapy alters ICP rather than a specific pathophysiological mechanism, i.e., the treatment is largely empiric and phenomenological. This then assumes all patients with increased ICP are the same and the only difference is the resistance or response to treatment. Deleterious side effects may be observed with all ICP treatments and this linear approach generally starts with what is assumed to be the least deleterious approach and moves along the pathway to potentially more deleterious treatments. However, the propensity to harm through treatment based on the ICP is not well substantiated in the literature.[22,81] More recently, therapy for increased ICP has taken a tiered approach with several therapies grouped together.[82,83]

Potential reasons why ICP may be increased after TBI are listed in Table 15.3. The therapies for increased ICP using a tiered approach are empiric and include strategies such as sedation and analgesia, hyperosmolar agents (mannitol or hypertonic saline [HTS]), CSF drainage, induced hypocarbia, barbiturates, temperature modulation, and surgery including evacuation of mass lesion or decompressive craniectomy. These various therapies and guideline adherence, their efficacy, comparisons, and cost effectiveness in lowering ICP have all been well studied, although rigorous Level I evidence is lacking on an outcome effect. The reader is referred elsewhere for recent details and reviews.[20,79,82–93] Methodological issues and how data

TABLE 15.3
Possible Causes of Increased ICP after TBI

Intracranial
Extra-axial mass lesion—Epidural or subdural hematoma
Intraparenchymal mass lesion—Intracerebral hemorrhage, contusion(s)
Depressed skull fracture
Brain edema—Cytotoxic (intracellular) or vasogenic (extracellular)
Disturbed in CSF dynamics with or without ventricular enlargement
Hyperemia—Vasomotor paralysis or loss of autoregulation
Venous sinus thrombosis
Seizures
Extracranial
Airway obstruction
Hypoventilation—Hypoxia, hypercarbia
Agitation
Pain
Hypertension
Cranial venous outflow obstruction
Head position or posture
Fever
Increased intrathoracic pressure including Valsalva maneuver
Increased intra-abdominal pressure (including compartment syndrome)
Liver failure
Altered sodium balance
Hypoosmolarity

is collected need to be considered. For example, manually recorded ICP data can overestimate treatment effectiveness. Instead automated data collection may provide a more accurate assessment of treatment response.[84] With the advent of multimodality monitoring and better understanding of sTBI pathophysiology, newer therapies such as repeated boluses of HTS, hyperbaric oxygen (HBO_2), normobaric hyperoxia (NBH), and lactate infusions have also been used to treat increased ICP in the clinical environment.[94–96]

DOES ICP MANAGEMENT INFLUENCE PATIENT OUTCOME?

Clinical studies of whether treating ICP enhances outcome are confounded by many variables, some of which are methodological, e.g., how was ICP recorded, how was ICP data analyzed, what defines intracranial hypertension, how are ICP elevations treated, and how were decisions about futility made among others. TBI also is heterogeneous, and select rather than all patients, e.g., intracranial hematoma vs. diffuse axonal injury, benefit from ICP treatment. Furthermore, since there is good evidence that elevated ICP and mortality is associated, no randomized clinical trial (RCT) in which participants undergo ICP monitoring and some receive targeted treatment and others do not has been performed, and for ethical reasons, will likely never be performed. The BEST TRIP trial was a multicenter RCT conducted in general ICUs in Bolivia and Ecuador that included 324 patients >13 years old who had severe TBI that attempted to examine ICP monitor use.[23] Two strategies of care for severe TBI (and not ICP) were examined. An ICP monitor triggered one strategy, although care varied from established guidelines and the other was a novel strategy developed specifically for the trial that was triggered by CT and clinical examination. A similar outcome was observed in the two groups, which is not unexpected since there is no control group. In addition, many patients never developed increased ICP, i.e., BEST TRIP was not a trial of ICP monitoring or a trial of ICP management. The trial also lacks external validity and so should not alter clinical practice. It has, however, prompted research questions about TBI and ICP management.[76]

Attempts to answer how ICP management influences outcome have been examined using several methodologies, e.g., within-institution protocol studies, center-based studies, analyses of trauma or TBI registries, meta-analysis of clinical series, quality-assurance studies, or comparative effectiveness. In large part, these studies are not studies of ICP treatment per se but rather studies of "aggressiveness of care" that includes ICP treatment or guideline adherence that usually is defined by ICP monitor use. The data then are confounded by variable care and management decisions, e.g., whom to monitor. Despite this and a variable case mix, the vast majority of studies demonstrate a significant benefit to outcome, particularly mortality when TBI management protocols based in part on ICP monitoring are initiated or used.[18,19–22,93,97–103] For example, Alali et al.[18] analyzed 10,628 adults with severe TBI derived from 155 centers in the American College of Surgeons Trauma Quality Improvement Program (TQIP). There was great variability in ICP monitoring rates. Hospitals with higher rates of ICP monitoring use were associated with lower mortality. The adjusted OR of death was 0.52 (95% CI, 0.35–0.78) in the quartile of hospitals with highest use, compared to the lowest.

Rather than examine ICP monitor use or guideline adherence, several studies have examined outcome based on the response of increased ICP to treatment using either first- or second-tier therapies. This may be a better method to examine whether ICP treatment makes a difference. Treggiari et al.[50] combined data from several observational studies and observed an association between better outcome and treatment response. Similarly, Farahvar et al.[22] in a prospective TBI database of 388 patients observed that a response to treatment was independently associated with less mortality at 14 days. Data from RCTs that examine second-tier therapies, e.g., barbiturates, hypothermia also suggest that a treatment response is associated with reduced mortality.[104,105] However DECRA, an RCT of bifrontal decompressive craniectomy versus maximal medical therapy for early intractable hypertension, observed similar morality despite better control of ICP in the surgical group.[106] However, outcome was not analyzed by response to treatment in DECRA and the definition of intractable hypertension was very different from what is usually used in clinical practice.

A handful of studies have observed similar 12-month outcomes despite more ICP treatment,[107] or that ICP monitoring is associated with worse risk-adjusted hospital mortality or increased complications, e.g., pneumonia, renal failure, and infections.[108,109] These reports and the BEST TRIP trial,[23] although it lacks external validity and is not a trial of ICP per se, have made some clinicians question the usefulness of ICP monitoring. However, when all the data are taken together it appears that use of an ICP monitor is associated with reduced mortality and at least associated with more efficient care. Furthermore, successful treatment of intracranial hypertension is associated with better outcome than observed in patients who are refractory to treatment. The clinical implication of this is that ICP monitoring is necessary (who would treat blood pressure without a blood pressure cuff) and that treatment of increased ICP is beneficial. The few dissenting publications should not lead to questions of ICP monitor use in toto but rather indicate that more research is needed to better define: (a) which patients should be monitored; (b) what is intracranial hypertension and does it vary depending on patient or patient pathology; (c) is there a specific ICP level that is deleterious and is it always harmful; (d) is increased ICP as a numeric threshold simply a marker of an underlying pathophysiological process that requires specific (and mechanistic) treatment rather than an independent target; and (e) how should we best treat increased ICP among others? Perhaps the most important implication of these "negative" studies, given the preponderance of data to the contrary is that ICP treatment when a numerical threshold is breached may be an oversimplification of a complex and heterogeneous pathophysiology and that how we treat ICP requires a conceptual shift.

TOWARD A NEW CONCEPT OF ICP TREATMENT

ICP Is More than a Number

At present ICP is treated primarily for two main reasons: to prevent herniation and to preserve cerebral blood flow (CBF), i.e., prevent ischemia. Simply treating a numeric threshold, which is population derived, does not allow therapy to be targeted to an individual and generally is reactive rather than proactive. The Monro–Kellie doctrine (intracranial volume [v; constant] = v.brain + v.CSF + v.blood + v.mass lesion) provides the conceptual framework on which current ICP treatment is based. However, this doctrine describes more than a threshold since it indicates that the brain is also able to compensate for a volume increase in one of the compartments with volume changes in the others. When these compensatory mechanisms fail, a small volume change can cause a significant ICP increase or even herniation. This compensatory reserve or compliance is important since a patient with poor compliance is at much greater risk than one with good compliance even if their ICP was normal and hence management is potentially different. Compliance can be quantified by a volume challenge through an EVD, i.e., the volume–pressure response (VPR = ICP change/volume change). Logarithmic conversion of the VPR defines the volume that would produce a tenfold ICP increase, i.e., pressure–volume index (PVI). The normal PVI is between 25 and 30 mL.[110] Rather than inject fluid through an EVD with its attendant risks, compliance is better estimated from ICP waveform analysis or with derived indices such as the index of pressure–volume compensatory reserve (RAP), defined as the correlation coefficient (R) between the pulse pulsation amplitude (A) and mean ICP (P).[111] The RAP index varies between –1 and +1; higher values indicate less compliance. There are three prominent wavelets in the normal

ICP waveform, labeled P1, P2, and P3. P1 is the largest wavelet but as compliance decreases, the amplitude of P2 increases and can become greater than P1. The ICP waveform then resembles the arterial waveform (Figure 15.1).

Qualitative estimates of compliance may also be derived from observation of a patient's response to stimulation. In the ICU, patients often are subjected to noxious stimuli, e.g., suctioning or pain, to determine the level of consciousness; when intracranial compensatory reserve is compromised, the ICP increases rapidly and often remains elevated. By contrast, when compliance is adequate, there may not be an ICP increase or if it does increase it rapidly returns to normal. In addition, in the absence of stimulation, the ICP should be stable and without much variation in pulse amplitude during the respiratory cycle. Calculation of a therapeutic intensity level (TIL), i.e., what and how much treatment is needed to control ICP, also can provide insight into compensatory reserve or treatment response; a greater TIL implies worse compliance and perhaps a greater "ischemic burden."[112] This knowledge of compliance when coupled with the clinical examination and CT findings, in particular the amount of midline shift or compression of the perimesencephalic cisterns, allows herniation risk to be determined and so set an "individualized ICP threshold." However, how these measures (waveform analysis, TIL, RAP, or PVI) can best be used in TBI management is still being elucidated.

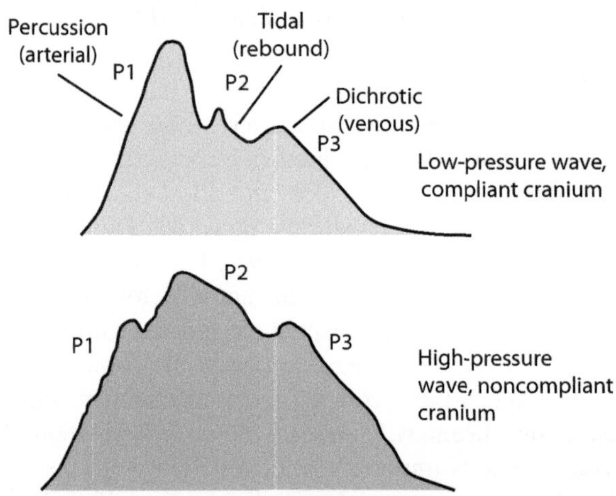

- P1: Reflects the ejection of blood from the heart transmitted through the choroid plexus in ventricles
- P2: Reflects venous compartments, brain bulk or compliance, vasomotor paralysis, brain swelling or edema
- P3: Reflects aortic valve closure

FIGURE 15.1 Intracranial pressure waveform. (Reprinted from *The Evidenced Based Practice of Critical Care*, Neligan P. and Deutschman C.S. (eds.), Oddo M. and Le Roux P., What is the etiology, pathogenesis and pathophysiology of elevated intracranial pressure? 399–405, Copyright 2009 with permission from Elsevier.)

The healthy brain maintains a constant CBF when MABP is between 40 and 160 mmHg through autoregulation (AR). However, AR frequently is compromised and there is great heterogeneity between individuals and over time after TBI. Furthermore, observational studies show an association between poor outcome and compromised AR.[113,114] TCD ultrasonography can be used to calculate AR by analyzing the blood flow velocity (BFV) changes resulting from changes in MABP. Other techniques including brain tissue oxygen ($PbtO_2$), near-infrared spectroscopy (NIRS), laser Doppler flowmetry (LDF), and thermal dilution regional CBF (td-rCBF) can all be used to calculate AR. These techniques provide intermittent assessments only and given the variability of AR indices over time may miss important information. With recent advances in data processing and computerized bedside monitoring it is now possible to perform online, real-time analysis of AR and in particular the cerebrovascular pressure reactivity index (PRx), which is the linear correlation coefficient between average arterial blood pressure and ICP over 3–4 min.[115,116] PRx is negative with intact AR and positive when AR is compromised or lost, since ICP follows MABP.

Knowledge of the PRx can help identify a CPP that builds on cerebrovascular reactivity (or cerebral autoregulation) as an intrinsic autoprotective mechanism in the brain through calculation of "optimal CPP,"[116] i.e., an individualized CPP. This is important since clinical trials that have evaluated CPP-based therapy using a single target that is population-derived demonstrate no outcome benefit in large part because of deleterious effects on the lungs by therapies used to increase CPP.[117] Instead optimal CPP values are patient and time dependent, and levels both above and below the patient's optimal level are associated with worse outcome.[118] As of yet there are no prospective studies of ICP or CPP management using "autoregulation-optimal" targets in TBI. However, clinical series suggest that the incidence of favorable outcome after TBI is increased when median CPP is close to optimal CPP. The PRx also may be used to define individualized ICP thresholds.[67] Further study is required but measurement of ICP and PRx in the future may define a patient specific threshold, allow assessment of ICP associated ischemia risk, and so facilitate targeted care.

A Role for Multimodal Monitoring

Poor outcome after TBI is associated with secondary brain injury (SBI) that includes cerebral ischemia or macrovascular dysfunction. In its current form, ICP management can help maintain CBF and prevent secondary brain injury associated with ischemia, although ICP (and CPP) monitoring does not always detect inadequate CBF.[119,120] Furthermore, several lines of evidence indicate that macrovascular disturbances and ischemia are not the only determinants of SBI. Instead, microvascular dysfunction that may be associated with diffusion rather than perfusion abnormalities or alterations in glucose utilization, e.g., mitochondrial dysfunction can contribute to SBI.[72,73,121–124] These mechanisms of SBI cannot be detected by measuring ICP or CPP alone. Furthermore, evidence for cellular injury or energy dysfunction can be observed even when ICP and CPP are normal,[74,125,126] and cellular metabolic failure often precedes an increase in ICP.[127,128]

The implication of these various data is that severe TBI management, including ICP management, requires high-tech monitoring, and use of several monitors to better understand what is happening in the brain. This should then help develop therapies that are mechanistic, rather than our current empiric approach, and targeted to an individual. A variety of techniques, including the clinical examination, imaging modalities and bedside physiologic tools, both invasive and noninvasive, are available to monitor the injured brain.[2,129,130] However, no single method can provide complete information about the brain's health given its complexity. Consequently, a combination of monitors is necessary. This platform, often called multimodality monitoring (MMM), has evolved in recent years along with the growth of bioinformatics. How best to use MMM is still being elucidated but accumulating clinical experience suggests it can help define and guide optimal use of therapies such as hyperventilation, osmotherapy, glycemic control, transfusion, CPP augmentation, therapeutic temperature modulation, and oxygen-based therapies among others, in part through identification of patient-specific thresholds and avoidance of potential deleterious side effects associated with a particular therapy.[131–133] In addition, MMM can help differentiate between pathological conditions that appear similar but require different treatments, e.g., differentiate ischemic from nonischemic mechanisms of cellular injury,[134,135] and when used provides a better and more accurate assessment of cerebral perfusion than ICP/CPP monitoring alone, including patients with diffuse injury.[136] Finally, monitors such as $PbtO_2$ and cerebral microdialysis are independent markers of mortality (and not just disease severity) and unlike ICP are also associated with morbidity in adults and children. This can augment outcome prediction and hence decisions on futility.[120,137–139]

Patient and Pathology Targeted ICP Care

An important goal in sTBI management is prevention, identification, and treatment of secondary brain injury, including cerebral ischemia, intracranial hypertension, and energy dysfunction that can occur after the primary injury and are known to aggravate patient outcome. Traditionally, ICP monitoring has been at the center of this approach, but despite much research there is still no level I evidence to support targeting a specific ICP threshold in clinical practice. However, ICP monitoring and management remain critical to the care of sTBI and a prerequisite of other forms of monitoring including MMM.[26] More important, converging evidence from diverse areas of research indicate that we should no longer simply treat a generic threshold value (i.e., more or less than 20 mmHg). To do this requires a conceptual shift that already is occurring in many ICUs around the world. The important question to ask is what is the purpose of an ICP monitor; this can then lead to targeted therapy with the realization that an ICP number per se is not a target but rather an indicator of an underlying pathophysiological process that needs treatment.

Looking ahead, successful ICP (and perhaps sTBI) management will require extraction of more information from the ICP signal, rather than just a number, and to use this information to provide patient- and pathology-specific targets and to forecast ICP increases.[63,67,116,140] In addition, ICP should be regarded as an epiphenomenon of cerebral compliance, herniation tendency, cerebral ischemia, cerebral hypoxia, and cellular dysfunction among others, i.e., ICP care needs to be directed by MMM that includes serial neurological examinations, imaging studies, ICP analysis, and measurements from other bedside physiologic monitors. As the field of bioinformatics grows, further advances in integrating this information can be expected,[141,142] but in its simplest form this currently means dynamic individualization of an ICP treatment threshold. The impact of this is not trivial. For example, threshold adjustment may facilitate extubation if it appears that the ICP increases are associated with the extubation trial and are in themselves not harmful. Similarly, threshold adjustment can avoid therapy escalation and potential deleterious side effects, e.g., optimization of volume and ventilator status in a patient with TBI and acute lung injury. Such an approach may have resulted in a different outcome in an RCT performed many years ago that compared ICP- or CPP-based care, using generic population-based thresholds, since the outcomes in the CPP group were worsened by cardiorespiratory complications even though they had a lower incidence of cerebral ischemia.[117]

Monitoring by itself does not alter outcome. Instead, it is how the information is used that contributes to patient well-being. In its more advanced form, use of MMM for ICP and sTBI care facilitates therapy that is targeted to patient-specific pathophysiology and as physiologically (and perhaps mechanistically) oriented as possible. Specific therapeutic protocols are still being elucidated,[143] but early clinical experience suggests that this approach may benefit select patients and enhance outcome.[118,144–146] For example, several but not all observational series suggest outcome is better when both $PbtO_2$ and ICP/CPP therapy rather than ICP/CPP-based care alone is used in sTBI, i.e., there still is clinical equipoise. In part, the results may depend on probe location.[147] In the past the driving question has been where to place a monitor. A more important question is where is the monitor, i.e., any data should be interpreted based on postinsertion imaging; simply doing this is MMM. Summary analysis of these various studies suggests that combined ICP/CPP- and $PbtO_2$-based therapy is associated with better outcome.[145] In addition, initial analysis from BOOST-2, a phase II RCT that compared $PbtO_2$ and ICP/CPP therapy to ICP/CPP alone in sTBI, demonstrates physiological efficacy, and although not powered for outcome also showed an outcome benefit.[146] Support for use of individualized ICP or CPP thresholds have been demonstrated in several studies using PRx to calculate an "optimal" CPP.[118,144] For example Howells et al.[144] examined the clinical utility of PRx in two ICUs, one that used a CPP-based approach and the other an ICP-based approach to sTBI, and found that a CPP-targeted approach was more successful when pressure reactivity was intact, while an ICP-based strategy was better when pressure reactivity was impaired (Figure 15.2).

In summary, even though there is no level I evidence to support the use of a generic ICP threshold, it is clear that ICP remains important to and a foundation of sTBI care. In particular, research using ICP as part of MMM and investigation of the relationship between pressure, volume, blood flow, and metabolism has improved our understanding of sTBI pathophysiology. This has opened the door on new therapies for ICP.[94] Today numeric ICP values should no longer be considered a target but rather be regarded as an indicator of targetable pathophysiology. As such, optimal ICP management requires integration of ICP data with MMM, including the clinical examination, clinical imaging, and other physiologic parameters to provide individualized and pathophysiology specific care.

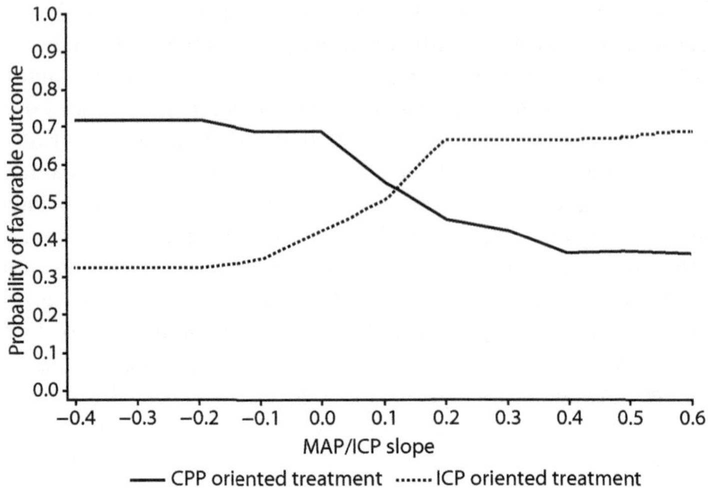

FIGURE 15.2 Clinical utility of PRx. Bayesian neural network (BANN) model to predict the likelihood of good or bad clinical outcome using the PRx index among sTBI patients managed using an ICP- or CPP-based treatment strategy. According to this data, the optimal point at which to switch from management strategy to another is when the MABP/ICP slope is 0.13. (Modified from Howells T. et al., *J Neurosurg*, 2005;102(2):311–317.)

ACKNOWLEDGMENTS AND CONFLICTS OF INTEREST

Dr. Le Roux receives research funding from Integra Lifesciences, Neurologica, the Dana Foundation, and the NIH. He is a consultant for Integra Lifesciences, Codman, Synthes, Neurologica, and a member of the scientific advisory board of Cerebrotech and Edge Therapeutics.

REFERENCES

1. Maas, A.I. et al. 2012. Re-orientation of clinical research in traumatic brain injury: Report of an international workshop on comparative effectiveness research. *J Neurotrauma* 29(1):32–46.
2. Le Roux, P. 2013. Physiological monitoring of the severe traumatic brain injury patient in the intensive care unit. *Curr Neurol Neurosci Rep* 13(3):331.
3. Brain Trauma Foundation, e. 2007. Guidelines for the management of severe traumatic brain injury, 3rd edition. VI. Indications for intracranial pressure monitoring. *J Neurotrauma* 24(Suppl 1):S37–S44.
4. Mizutani, T., Manaka, S., Tsutsumi, H. 1990. Estimation of intracranial pressure using computed tomography scan findings in patients with severe head injury. *Surg Neurol* 33(3):178–184.
5. Toutant, S.M. et al. 1984. Absent or compressed basal cisterns on first CT scan: Ominous predictors of outcome in severe head injury. *J Neurosurg* 61(4):691–694.
6. Eisenberg, H.M. et al. 1990. Initial CT findings in 753 patients with severe head injury. A report from the NIH Traumatic Coma Data Bank. *J Neurosurg* 73(5):688–698.

7. Kishore, P.R. et al. 1981. Significance of CT in head injury: Correlation with intracranial pressure. *AJR Am J Roentgenol* 137(4):829–833.
8. Narayan, R.K. et al. 1982. Intracranial pressure: To monitor or not to monitor? A review of our experience with severe head injury. *J Neurosurg* 56(5):650–659.
9. O'Sullivan, M.G. et al. 1994. Role of intracranial pressure monitoring in severely head-injured patients without signs of intracranial hypertension on initial computerized tomography. *J Neurosurg* 80(1):46–50.
10. Stocchetti, N. et al. 2014. Clinical applications of intracranial pressure monitoring in traumatic brain injury: Report of the Milan consensus conference. *Acta Neurochir (Wien)* 156(8):1615–1622.
11. Peterson, E.C., Chesnut, R.M. 2011. Talk and die revisited: Bifrontal contusions and late deterioration. *J Trauma* 71(6):1588–1592.
12. Aarabi, B. et al. 2006. Outcome following decompressive craniectomy for malignant swelling due to severe head injury. *J Neurosurg* 104(4):469–479.
13. Wilberger, J.E., Jr., Harris, M., Diamond, D.L. 1991. Acute subdural hematoma: Morbidity, mortality, and operative timing. *J Neurosurg* 74(2):212–218.
14. Lobato, R.D. et al. 1988. Acute epidural hematoma: An analysis of factors influencing the outcome of patients undergoing surgery in coma. *J Neurosurg* 68(1):48–57.
15. Hutchinson, P.J. et al. 2011. Update on the RESCUEicp decompressive craniectomy trial. *Crit Care* 15(Suppl 1):P312.
16. Marion, D.W., Spiegel, T.P. 2000. Changes in the management of severe traumatic brain injury: 1991–1997. *Crit Care Med* 28:16–18.
17. Kirkman, M.A., Smith, M. 2014. Intracranial pressure monitoring, cerebral perfusion pressure estimation, and ICP/CPP-guided therapy: A standard of care or optional extra after brain injury? *Br J Anaesth* 112(1):35–46.
18. Alali, A.S. et al. 2013. Intracranial pressure monitoring in severe traumatic brain injury: Results from the American College of Surgeons Trauma Quality Improvement Program. *J Neurotrauma* 30(20):1737–1746.
19. Stein, S.C. et al. 2010. Relationship of aggressive monitoring and treatment to improved outcomes in severe traumatic brain injury. *J Neurosurg* 112:1105–1112.
20. Gerber, L.M. et al. 2013. Marked reduction in mortality in patients with severe traumatic brain injury. *J Neurosurg* 119(6):1583–1590.
21. Talving, P. et al. 2013. Intracranial pressure monitoring in severe head injury: Compliance with Brain Trauma Foundation guidelines and effect on outcomes: A prospective study. *J Neurosurg* 119(5):1248–1254.
22. Farahvar, A. et al. 2012. Increased mortality in patients with severe traumatic brain injury treated without intracranial pressure monitoring. *J Neurosurg* 117(4):729–734.
23. Chesnut, R.M. et al. 2012. A trial of intracranial-pressure monitoring in traumatic brain injury. *N Engl J Med* 367:2471–2481.
24. Whitmore, R.G. et al. 2012. Is aggressive treatment of traumatic brain injury cost-effective? *J Neurosurg* 116(5):1106–1113.
25. Sekhon, M.S. et al. 2014. Optic nerve sheath diameter on computed tomography is correlated with simultaneously measured intracranial pressure in patients with severe traumatic brain injury. *Intensive Care Med* 40(9):1267–1274.
26. Le Roux, P. et al. 2014. Consensus summary statement of the International Multidisciplinary Consensus Conference on Multimodality Monitoring in Neurocritical Care: A statement for healthcare professionals from the Neurocritical Care Society and the European Society of Intensive Care Medicine. *Neurocrit Care* 21(Suppl 2):S1–S26.
27. Mollman, H.D., Rockswold, G.L., Ford, S.E. 1988. A clinical comparison of subarachnoid catheters to ventriculostomy and subarachnoid bolts: A prospective study. *J Neurosurg* 68(5):737–741.

28. Sahuquillo, J. et al. 1999. Interhemispheric supratentorial intracranial pressure gradients in head-injured patients: Are they clinically important? *J Neurosurg* 90(1):16–26.
29. Chesnut, R. et al. 2014. Intracranial pressure monitoring: Fundamental considerations and rationale for monitoring. *Neurocrit Care* 21(Suppl 2):S64–S84.
30. Bekar, A. et al. 2009. Risk factors and complications of intracranial pressure monitoring with a fiberoptic device. *J Clin Neurosci* 16(2):236–240.
31. Kasotakis, G. et al. 2012. Intraparenchymal vs. extracranial ventricular drain intracranial pressure monitors in traumatic brain injury: Less is more? *J Am Coll Surg* 214(6):950–957.
32. Guyot, L.L. et al. 1998. Cerebral monitoring devices: Analysis of complications. *Acta Neurochir* 71:47–49.
33. North, B., Reilly, P. 1986. Comparison among three methods of intracranial pressure recording. *Neurosurgery* 18(6):730–732.
34. Wiesmann, M., Mayer, T.E. 2001. Intracranial bleeding rates associated with two methods of external ventricular drainage. *J Clin Neurosci* 8(2):126–128.
35. Anderson, R.C. et al. 2004. Complications of intracranial pressure monitoring in children with head trauma. *J Neurosurg* 101(Suppl 1):53–58.
36. Gelabert-Gonzalez, M. et al. 2006. The Camino intracranial pressure device in clinical practice. Assessment in a 1000 cases. *Acta Neurochir (Wien)* 148(4):435–441.
37. Matevosyan, K. et al. 2010. Coagulation factor levels in neurosurgical patients with mild prolongation of prothrombin time: Effect on plasma transfusion therapy. *J Neurosurg* 114:3–7.
38. Bauer, D.F. et al. 2010. The relationship between INR and development of hemorrhage with placement of ventriculostomy. *J Trauma* 70(5):111–127.
39. Martinez-Manas, R.M. et al. 2000. Camino intracranial pressure monitor: Prospective study of accuracy and complications. *J Neurol Neurosurg Psychiatry* 69:82–86.
40. Davis, J.W. et al. 2004. Placement of intracranial pressure monitors: Are "normal" coagulation parameters necessary? *J Trauma* 57(6):1173–1177.
41. Lozier, A.P. et al. 2002. Ventriculostomy-related infections: A critical review of the literature. *Neurosurgery* 51(1):170–181; discussion 181–182.
42. Camacho, E.F. et al. 2011. Infection rate and risk factors associated with infections related to external ventricular drain. *Infection* 39(1):47–51. Epub January 25, 2011.
43. Scheithauer, S. et al. 2009. Prospective surveillance of drain associated meningitis/ventriculitis in a neurosurgery and neurological intensive care unit. *J Neurol Neurosurg Psychiatry* 80(12):1381–1385.
44. Williams, T.A. et al. 2011. Decrease in proven ventriculitis by reducing the frequency of cerebrospinal fluid sampling from extraventricular drains. *J Neurosurg* 115(5):1040–1046.
45. Holloway, K.L. et al. 1996. Ventriculostomy infections: The effect of monitoring duration and catheter exchange in 584 patients. *J Neurosurg* 85(3):419–424.
46. Wong, G.K. et al. 2002. Failure of regular external ventricular drain exchange to reduce cerebrospinal fluid infection: Result of a randomised controlled trial. *J Neurol Neurosurg Psychiatry* 73(6):759–761.
47. Citerio, G. et al. 2008. Multicenter clinical assessment of the Raumedic Neurovent-P intracranial pressure sensor: A report by the BrainIT group. *Neurosurgery* 63:1152–1158.
48. Leverstein-van Hall, M.A. et al. 2010. A bundle approach to reduce the incidence of external ventricular and lumbar drain-related infections. *J Neurosurg* 112(2):345–353.
49. Honda, H. et al. 2010. Reducing the incidence of intraventricular catheter-related ventriculitis in the neurology-neurosurgical intensive care unit at a tertiary care center in St Louis, Missouri: An 8-year follow-up study. *Infect Control Hosp Epidemiol* 31(10):1078–1081.
50. Treggiari, M.M. et al. 2007. Role of intracranial pressure values and patterns in predicting outcome in traumatic brain injury: A systematic review. *Neurocrit Care* 6(2):104–112.

51. Andrews, P.J. et al. 2002. Predicting recovery in patients suffering from traumatic brain injury by using admission variables and physiological data: A comparison between decision tree analysis and logistic regression. *J Neurosurg* 97(2):326–336.
52. Badri, S. et al. 2012. Mortality and long-term functional outcome associated with intracranial pressure after traumatic brain injury. *Intensive Care Med* 38(11):1800–1809.
53. Balestreri, M. et al. 2006. Impact of intracranial pressure and cerebral perfusion pressure on severe disability and mortality after head injury. *Neurocrit Care* 4(1):8–13.
54. Saul, T.G., Ducker, T.B. 1982. Intracranial pressure monitoring in patients with severe head injury. *Am Surg* 48(9):477–480.
55. Stein, D.M. et al. 2011. Brief episodes of intracranial hypertension and cerebral hypoperfusion are associated with poor functional outcome after severe traumatic brain injury. *J Trauma* 71(2):364–373; discussion 73–74.
56. Majdan, M. et al. 2014. Timing and duration of intracranial hypertension versus outcomes after severe traumatic brain injury. *Minerva Anestesiol* 80(12):1261–1272.
57. Stein, D.M. et al. 2013. Timing of intracranial hypertension following severe traumatic brain injury. *Neurocrit Care* 18(3):332–340.
58. Vik, A. et al. 2008. Relationship of "dose" of intracranial hypertension to outcome in severe traumatic brain injury. *J Neurosurg* 109(4):678–684.
59. Stocchetti, N. et al. 2008. Refractory intracranial hypertension and "second-tier" therapies in traumatic brain injury. *Intensive Care Med* 34(3):461–467.
60. Sheth, K.N. et al. 2013. Intracranial pressure dose and outcome in traumatic brain injury. *Neurocrit Care* 18(1):26–32.
61. Colton, K. et al. 2014. Responsiveness to therapy for increased intracranial pressure in traumatic brain injury is associated with neurological outcome. *Injury* 45(12):2084–2088.
62. Miller, J.D. et al. 1977. Significance of intracranial hypertension in severe head injury. *J Neurosurg* 47(4):503–516.
63. Güiza, F. et al. 2013. Novel methods to predict increased intracranial pressure during intensive care and long-term neurologic outcome after traumatic brain injury: Development and validation in a multicenter data set. *Crit Care Med* 41(2):554–564.
64. Levin, H.S. et al. 1991. Intracranial hypertension in relation to memory functioning during the first year after severe head injury. *Neurosurgery* 28(2):196–199; discussion 200.
65. Bremmer, R. et al. 2010. The course of intracranial pressure in traumatic brain injury: Relation with outcome and CT-characteristics. *Neurocrit Care* 12(3):362–368.
66. Kahraman, S. et al. 2010. Heart rate and pulse pressure variability are associated with intractable intracranial hypertension after severe traumatic brain injury. *J Neurosurg Anesthesiol* 22(4):296–302.
67. Lazaridis, C. et al. 2014. Patient-specific thresholds of intracranial pressure in severe traumatic brain injury. *J Neurosurg* 120(4):893–900.
68. Marshall, L.F., Smith, R.W., Shapiro, H.M. 1979. The outcome with aggressive treatment in severe head injuries. Part I: The significance of intracranial pressure monitoring. *J Neurosurg* 50(1):20–25.
69. Marmarou, A. et al. 2005. Contribution of raised ICP and hypotension to CPP reduction in severe brain injury: Correlation to outcome. *Acta Neurochir* 95:277–280.
70. Ratanalert, S. et al. 2004. ICP threshold in CPP management of severe head injury patients. *Surg Neurol* 61(5):429–434; discussion 34–35.
71. Chambers, I.R., Treadwell, L., Mendelow, A.D. 2001. Determination of threshold levels of cerebral perfusion pressure and intracranial pressure in severe head injury by using receiver-operating characteristic curves: An observational study in 291 patients. *J Neurosurg* 94(3):412–416.
72. Menon, D.K. et al. 2004. Diffusion limited oxygen delivery following head injury. *Crit Care Med* 32:1384–1390.

73. Vespa, P. et al. 2005. Metabolic crisis without brain ischemia is common after traumatic brain injury: A combined microdialysis and positron emission tomography study. *J Cereb Blood Flow Metab* 25:763–774.
74. Stiefel, M.F. et al. 2006. Conventional neurocritical care and cerebral oxygenation after traumatic brain injury. *J Neurosurg* 105:568–575.
75. Chesnut, R.M. 2014. A conceptual approach to managing severe traumatic brain injury in a time of uncertainty. *Ann N Y Acad Sci* 1345(1):99–107.
76. Le Roux, P. 2014. Intracranial pressure after the BEST TRIP trial: A call for more monitoring. *Curr Opin Crit Care* 20(2):141–147.
77. Le Roux, P. et al. 2014. The international multidisciplinary consensus conference on multimodality monitoring in neurocritical care: A list of recommendations and additional conclusions. *Neurocrit Care* 21(Suppl 2):282–296.
78. Muzevic, D., Splavski, B. 2013. The Lund concept for severe traumatic brain injury. *Cochrane Database Syst Rev* 12:CD010193.
79. Stocchetti, N., Maas, A. 2014. Traumatic intracranial hypertension. *NEJM* 370: 2121–2130.
80. Bratton, S.L. et al. 2007. Guidelines for the management of severe traumatic brain injury. *J. Neurotrauma* 24:S1–S106.
81. Mendelson, A.A. et al. 2012. Intracranial pressure monitors in traumatic brain injury: A systematic review. *Can J Neurol Sci* 39(5):571–576.
82. Stevens, R.D. et al. 2012. Emergency neurological life support: Intracranial hypertension and herniation. *Neurocrit Care* 17(Suppl 1):S60–S65.
83. Swadron, S.P. et al. 2012. Emergency neurological life support: Traumatic brain injury. *Neurocrit Care* 17(Suppl 1):112–121.
84. Colton, K. et al. 2014. Intracranial pressure response after pharmacologic treatment of intracranial hypertension. *J Trauma Acute Care Surg* 77(1):47–53; discussion 53.
85. Kheirbek, T., Pascual, J.L. 2014. Hypertonic saline for the treatment of intracranial hypertension. *Curr Neurol Neurosci Rep* 14(9):482.
86. Alali, A.S. et al. 2014. Economic evaluation of decompressive craniectomy versus barbiturate coma for refractory intracranial hypertension following traumatic brain injury. *Crit Care Med* 42(10):2235–2243.
87. Dias, C. et al. 2014. Post-traumatic multimodal brain monitoring: Response to hypertonic saline. *J Neurotrauma* 31(22):1872–1880.
88. Wakai, A. et al. 2013. Mannitol for acute traumatic brain injury. *Cochrane Database Syst Rev* 8:CD001049.
89. Roberts, I., Sydenham, E. 2012. Barbiturates for acute traumatic brain injury. *Cochrane Database Syst Rev* 12:CD000033.
90. Urbano, L.A., Oddo, M. 2012. Therapeutic hypothermia for traumatic brain injury. *Curr Neurol Neurosci Rep* 12(5):580–591.
91. Sandsmark, D.K., Sheth, K.N. 2014. Management of increased intracranial pressure. *Curr Treat Options Neurol* 16(2):272.
92. Kolias, A.G., Kirkpatrick, P.J., Hutchinson, P.J. 2013. Decompressive craniectomy: Past, present and future. *Nat Rev Neurol* 9:405–415.
93. Hartings, J.A. et al. 2014. Surgical management of traumatic brain injury: A comparative-effectiveness study of 2 centers. *J Neurosurg* 120(2):434–446.
94. Ichai, C. et al. 2013. Half-molar sodium lactate infusion to prevent intracranial hypertensive episodes in severe traumatic brain injured patients: A randomized controlled trial. *Intensive Care Med* 39(8):1413–1422.
95. Eskandari, R. et al. 2013. Effective treatment of refractory intracranial hypertension after traumatic brain injury with repeated boluses of 14.6% hypertonic saline. *J Neurosurg* 119(2):338–346.

96. Rockswold, S.B. et al. 2013. A prospective, randomized phase II clinical trial to evaluate the effect of combined hyperbaric and normobaric hyperoxia on cerebral metabolism, intracranial pressure, oxygentoxicity, and clinical outcome in severe traumatic brain injury. *J Neurosurg* 118(6):1317–1328.
97. Clayton, T.J., Nelson, R.J., Manara, A.R. 2004. Reduction in mortality from severe head injury following introduction of a protocol for intensive care management. *Br J Anaesth* 93(6):761–767.
98. Fakhry, S.M. et al. 2004. Management of brain-injured patients by an evidence-based medicine protocol improves outcomes and decreases hospital charges. *J Trauma* 56(3):492–499; discussion 499–500.
99. Spain, D.A. et al. 1998. Effect of a clinical pathway for severe traumatic brain injury on resource utilization. *J Trauma* 45(1):101–104; discussion 4–5.
100. Arabi, Y.M. et al. 2010. Mortality reduction after implementing a clinical practice guidelines-based management protocol for severe traumatic brain injury. *J Crit Care* 25(2):190–195.
101. Haddad, S. et al. 2011. Relationship between intracranial pressure monitoring and outcomes in severe traumatic brain injury patients. *Anaesth Intensive Care* 39(6):1043–1050.
102. Bulger, E.M. et al. 2002. Management of severe head injury: Institutional variations in care and effect on outcome. *Crit Care Med* 30(8):1870–1876.
103. Barmparas, G. et al. 2012. Decreased intracranial pressure monitor use at level II trauma centers is associated within creased mortality. *Am Surg* 78(10):1166–1171.
104. Eisenberg, H.M. et al. 1988. High-dose barbiturate control of elevated intracranial pressure in patients with severe head injury. *J Neurosurg* 69(1):15–23.
105. Shiozaki, T. et al. 1993. Effect of mild hypothermia on uncontrollable intracranial hypertension after severe head injury. *J Neurosurg* 79(3):363–368.
106. Cooper, D.J. et al. 2011. Decompressive craniectomy in diffuse traumatic brain injury. *N Engl J Med* 364(16):1493–1502.
107. Cremer, O.L. et al. 2005. Effect of intracranial pressure monitoring and targeted intensive care on functional outcome after severe head injury. *Crit Care Med* 33(10):2207–2213.
108. Lane, P.L. et al. 2000. Intracranial pressure monitoring and outcomes after traumatic brain injury. *Can J Surg* 43(6):442–448.
109. Shafi, S. et al. 2008. Intracranial pressure monitoring in brain-injured patients is associated with worsening of survival. *J Trauma* 64(2):335–340.
110. Marmarou, A., Shulman, K., LaMorgese, J. 1975. Compartmental analysis of compliance and outflow resistance of the cerebrospinal fluid system. *J Neurosurg* 43(5):523–534.
111. Oddo, M., Le Roux, P. 2009. What is the etiology, pathogenesis and pathophysiology of elevated intracranial pressure? In *The Evidenced Based Practice of Critical Care*, eds. P. Neligan, C.S. Deutschman, 399–405. Philadelphia, PA: Elsevier Science.
112. Weiner, G.M. et al. 2010. Decompressive craniectomy for elevated intracranial pressure and its effect on the cumulative ischemic burden and therapeutic intensity levels after severe traumatic brain injury. *Neurosurgery* 66(6):1111–1118.
113. Czosnyka, M. et al. 1996. Monitoring of cerebral autoregulation in head-injured patients. *Stroke* 27(10):1829–1834.
114. Sorrentino, E. et al. 2012. Critical thresholds for cerebrovascular reactivity after traumatic brain injury. *Neurocrit Care* 16(2):258–266.
115. Czosnyka, M. et al. 1997. Continuous assessment of the cerebral vasomotor reactivity in head injury. *Neurosurgery* 41(1):11–17.
116. Steiner, L.A. et al. 2002. Continuous monitoring of cerebrovascular pressure reactivity allows determination of optimal cerebral perfusion pressure in patients with traumatic brain injury. *Crit Care Med* 30:733–738.

117. Robertson, C.S. et al. 1999. Prevention of secondary ischemic insults after severe head injury. *Crit Care Med* 27(10):2086–2095.
118. Aries, M.J. et al. 2012. Continuous determination of optimal cerebral perfusion pressure in traumatic brain injury. *Crit Care Med* 40(8):2456–2463.
119. Eriksson, E.A. et al. 2012. Cerebral perfusion pressure and intracranial pressure are not surrogates for brain tissue oxygenation in traumatic brain injury. *Clin Neurophysiol* 123:1255–1260.
120. Oddo, M. et al. 2011. Brain hypoxia is associated with short-term outcome after severe traumatic brain injury independently of intracranial hypertension and low cerebral perfusion pressure. *Neurosurgery* 69:1037–1045.
121. Schwarzmaier, S.M. et al. 2010. Temporal profile of thrombogenesis in the cerebral microcirculation after traumatic brain injury in mice. *J Neurotrauma* 27:121–130.
122. Singh, I.N. et al. 2006. Time course of post-traumatic mitochondrial oxidative damage and dysfunction in a mouse model of focal traumatic brain injury: Implications for neuroprotective therapy. *J Cereb Blood Flow Metab* 26:1407–1418.
123. Marmarou, A. et al. 2006. Predominance of cellular edema in traumatic brain swelling in patients with severe head injuries. *J Neurosurg* 104:720–730.
124. Castejón, O.J., de Castejón, H.V. 2004. Structural patterns of injured mitochondria in human oedematous cerebral cortex. *Brain Inj* 18(11):1107–1126.
125. Le Roux, P. et al. 1997. Cerebral arteriovenous difference of oxygen: A predictor of cerebral infarction and outcome in severe head injury. *J Neurosurg* 87:1–8.
126. Vespa, P.M. et al. 2007. Pericontusional brain tissue exhibits persistent elevation of lactate/pyruvate ratio independent of cerebral perfusion pressure. *Crit Care Med* 35(4):1153–1160.
127. Adamides, A.A. et al. 2009. Brain tissue lactate elevations predict episodes of intracranial hypertension in patients with traumatic brain injury. *J Am Coll Surg* 209:531–539.
128. Belli, A. et al. 2008. Metabolic failure precedes intracranial pressure rises in traumatic brain injury: A microdialysis study. *Acta Neurochir (Wien)* 150:461–469.
129. Le Roux, P. 2013. Invasive neurological and multimodality monitoring in the NeuroICU. In *Textbook of Neurocritical Care*, 2nd ed., eds. A.J. Layon, A. Gabrielli, W.A. Friedman, 127–145. Springer, London.
130. Le Roux, P., Lam, A. 2013. Monitoring blood flow and metabolism. In *Fundamentals of Neuroanesthesia: A Physiologic Approach to Clinical Practice*, eds. K.J. Ruskin, S.H. Rosenbaum, I.J. Rampil, 25–49. Oxford University Press, New York.
131. Johnston, A.J. et al. 2005. Effect of cerebral perfusion pressure augmentation on regional oxygenation and metabolism after head injury. *Crit Care Med* 33(1):189–195.
132. Tolias, C.M. et al. 2004. Normobaric hyperoxia—Induced improvement in cerebral metabolism and reduction in intracranial pressure in patients with severe head injury: A prospective historical cohort-matched study. *J Neurosurg* 101(3):435–444.
133. Oddo, M. et al. 2012. Anemia and brain oxygen after severe traumatic brain injury. *Intensive Care Med* 38:1497–1504.
134. Sala, N. et al. 2013. Cerebral extracellular lactate increase is predominantly nonischemic in patients with severe traumatic brain injury. *J Cereb Blood Flow Metab* 33(11):1815–1822.
135. Carre, E. et al. 2013. Metabolic crisis in severely head injured patients: Is ischemia just the tip of the iceberg? *Front Neurol* 4:146.
136. Bouzat, P. et al. 2014. Accuracy of brain multimodal monitoring to detect cerebral hypoperfusion after traumatic brain injury. *Crit Care Med* 43(2):445–452.
137. Figaji, A.A. et al. 2009. Brain tissue oxygen tension monitoring in pediatric severe traumatic brain injury. Part 1: Relationship with outcome. *Childs Nerv Syst* 25(10): 1325–1333.

138. Maloney-Wilensky, E. et al. 2009. Brain tissue oxygen and outcome after severe traumatic brain injury: A systematic review. *Crit Care Med* 37:2057–2063.
139. Timofeev, I. et al. 2011. Cerebral extracellular chemistry and outcome following traumatic brain injury: A microdialysis study of 223 patients. *Brain* 134:484–494.
140. Hu, X. et al. 2010. Forecasting ICP elevation based on prescient changes of intracranial pressure waveform morphology. *IEEE Trans Biomed Eng* 57(5):1070–1078.
141. Hemphill, J.C., Andrews, P., DeGeorgia, M. 2011. Medscape. Multimodal monitoring and neurocritical care bioinformatics. *Nat Rev Neurol* 7(8):451–460.
142. Sivaganesan, A., Manley, G.T., Huang, M.C. 2014. Informatics for neurocritical care: Challenges and opportunities. *Neurocrit Care* 20(1):132–141.
143. Bohman, L.E. et al. 2011. Medical management of compromised brain oxygen in patients with severe traumatic brain injury. *Neurocrit Care* 14(3):361–369.
144. Howells, T. et al. 2005. Pressure reactivity as a guide in the treatment of cerebral perfusion pressure in patients with brain trauma. *J Neurosurg* 102(2):311–317.
145. Nangunoori, R. et al. 2012. Brain tissue oxygen-based therapy and outcome after severe traumatic brain injury: A systematic literature review. *Neurocrit Care* 17(1):131–138.
146. Shutter, L. et al. 2014. BOOST-2 randomized trial results: ICP versus ICP+PbtO2 guided management of TBI. 12th Annual Neurocritical Care Society Meeting, Seattle, WA.
147. Ponce, L.L. et al. 2012. Position of probe determines prognostic information of brain tissue pO2 in severe traumatic brain injury. *Neurosurgery* 70(6):1492–1502.

16 Post-Traumatic Stress Disorder

Relationship to Traumatic Brain Injury and Approach to Treatment

Jonathon R. Howlett and Murray B. Stein

CONTENTS

INTRODUCTION

Post-traumatic stress disorder (PTSD) is characterized by a specific clinical syndrome including re-experiencing symptoms, avoidance, and alterations in arousal, cognition, and mood, resulting from exposure to severe traumatic events. PTSD was first officially recognized in the *Diagnostic and Statistical Manual of Mental Disorders, 3rd ed.* (*DSM III*) in 1980,[1] and since that time a great deal of knowledge has accumulated about the characteristics of post-traumatic symptomatology, the epidemiology of PTSD, and assessment and treatment of individuals suffering from this disorder. Traumatic events that can result in PTSD fall into a number of categories, including military combat, rape, physical assault, natural disaster, and witnessing violence.[2]

Since the beginning of the conflicts in Iraq and Afghanistan, it has become clear that PTSD has an important and complex relationship with traumatic brain injury (TBI). Both disorders are common in Iraq and Afghanistan veterans, and together they have been termed the "signature wounds" of these conflicts.[3] TBI has emerged as a clear and important risk factor for the development of PTSD, although the reasons for this association are not fully understood.[4,5] PTSD may also be an important mediator of the negative sequelae of TBI.[4] At the same time, the substantial overlap in symptoms of PTSD and postconcussive symptoms may result in considerable diagnostic confusion.[6] Finally, treatment of individuals with comorbid PTSD and TBI may present special challenges, and yet there is currently very little evidence base to guide pharmacologic and psychotherapeutic treatment in this population.

In this chapter, we begin with a brief overview of the epidemiology of PTSD, followed by a longer discussion of the epidemiology of comorbid PTSD and TBI, with special attention to the role of TBI as a risk factor for PTSD. We then consider diagnostic issues surrounding PTSD and TBI, including the substantial overlap in symptoms of PTSD and postconcussive syndrome and the resulting difficulty in attributing symptoms to either PTSD or TBI. Next, we briefly review evidence-based treatments for PTSD, including psychopharmacology and psychotherapy. Finally, we discuss special considerations in treating individuals with comorbid PTSD and TBI, as well as offer some preliminary recommendations in treating these patients while highlighting the need for more treatment studies in this population.

EPIDEMIOLOGY OF POST-TRAUMATIC STRESS DISORDER AND TRAUMATIC BRAIN INJURY

Research into the epidemiology of PTSD can be divided into studies of combat veterans and studies of civilian populations. Surveys of combat-related PTSD have focused largely on veterans of the Vietnam War and of the conflicts in Iraq and Afghanistan. It is estimated that 18.7% of male Vietnam veterans developed PTSD, and that 9.1% still suffered from PTSD in the late 1980s when surveys were conducted.[7] Given that PTSD was not yet clinically defined during the Vietnam War, research into the epidemiology of the disorder in Vietnam veterans did not begin until years after the end of the conflict, which may affect the accuracy of the results. In contrast, research into PTSD in Iraq and Afghanistan veterans has been ongoing since soon after these conflicts began.[2] However, estimates of the prevalence of PTSD in these groups have varied widely, likely due in part to variations in sample populations and in how PTSD was defined. Estimates of the prevalence of PTSD in non-treatment-seeking samples of Iraq and Afghanistan veterans have ranged from 5% to 20%.[8] Risk of PTSD is positively associated with level of combat exposure.[2,8]

As noted earlier, civilian populations can experience PTSD as a result of a variety of stressors including rape, assault, natural disasters, and motor vehicle accidents, among others. The lifetime prevalence of PTSD in the United States is estimated to be 6.8%.[9] Research has indicated that a large proportion of the general population has experienced at least one such stressor, but only a fairly small percentage go on to develop PTSD. For example, in a sample of adults from the Detroit metropolitan area, 89.2% of respondents had experienced at least one qualifying trauma based on *DSM* criteria,[10] yet only 9.2% of traumatic events resulted in PTSD. This rate varied depending on the type of traumatic event, with assaultive violence being most likely to lead to PTSD. Risk factors for the development of PTSD in individuals exposed to trauma include preexisting psychiatric disorders, family history of psychiatric disorders, poor social support, low IQ, and female gender.[2]

Understanding of the relationship between PTSD and TBI has evolved over time and has grown considerably on the basis of studies of veterans of the conflicts in Iraq and Afghanistan. Prior to these conflicts, comorbid PTSD and TBI received relatively little attention, and was considered by some authors to be a rare phenomenon. In particular, it was argued that impairment or loss of consciousness that occurs with severe TBI may prevent PTSD by interfering with the encoding of trauma-related memories.[11,12] An early literature review concluded that PTSD could, indeed, occur after TBI even in the absence of explicit memories of trauma, through nondeclarative memories, fear conditioning occurring outside of awareness, and reconstructed memories of traumatic events.[13]

Early epidemiological studies of veterans returning from Iraq and Afghanistan revealed high rates of comorbid PTSD and TBI, and that TBI was actually an important risk factor for PTSD. For example, a study of Iraq veterans found that 4.9% reported an injury with loss of consciousness, and an additional 10.3% reported an injury with altered mental status but without loss of consciousness.[4] The study found a strong association between TBI and PTSD: Of those who reported loss of consciousness, 43.9% developed PTSD, and of those who reported altered mental status, 27.3% developed PTSD. By contrast, in those with an injury without loss of consciousness or altered mental status, the rate of PTSD was 16.2%, and in those without an injury it was 9.1%. The relationship between TBI and PTSD remained significant after controlling for combat experiences. PTSD was also an important mediator of the relationship between TBI and poor physical health outcomes.[4] Another study of both Iraq and Afghanistan veterans found that 11% screened positive for PTSD and 12% reported a history consistent with mild TBI.[14] Mild TBI was associated with a 2.37-fold increase in the prevalence of PTSD. Of veterans with mild TBI, 34% also met criteria for PTSD. A survey of veterans performed by the RAND Corporation found that 19.5% of returned service members had sustained a probable TBI, and of these, about 34% were affected by probable PTSD (compared to an overall rate of 13.8% probable PTSD).[3]

Subsequent research has further clarified the relationship between TBI and PTSD. A study of U.S. Army Special Operations Command personnel found that 12.7% of subjects suffered at least one mild TBI.[15] Of these, 28% reported clinical levels of PTSD symptoms. Not only was TBI associated with the development of PTSD, but there was a dose–response relationship between exposure to blast or combination blast–blunt mild TBI and levels of PTSD symptoms. Another study using retrospective chart review specifically compared blast and nonblast TBI in service members deployed to Iraq and Afghanistan. Blast TBI resulted in a higher rate of re-experiencing symptoms, but PTSD symptoms were otherwise equivalent between the two groups.[16] Overall, 37.3% of those who suffered TBI met criteria for clinically significant PTSD symptoms. A prospective study of Marines who were assessed before and after deployment to Iraq and Afghanistan further probed the role of TBI as a risk factor for PTSD by accounting for combat intensity and predeployment characteristics. The study found that 56.8% of participants reported a history of predeployment TBI, while 19.8% sustained a deployment-related TBI.[5] A regression analysis was performed to predict postdeployment symptoms of PTSD on the basis of predeployment symptoms, predeployment TBI, combat intensity, and deployment-related TBI. Both predeployment symptoms and combat intensity predicted postdeployment PTSD symptoms. However, even when controlling for these factors, deployment-related TBI was a strong predictor of PTSD symptoms. A deployment-related TBI approximately doubled the rate of PTSD for participants with low predeployment symptoms.

Several studies have investigated the relationship between TBI, PTSD, and suicide in military populations. A study of military personnel referred for evaluation of TBI in Iraq found that the number of TBIs was associated with suicidal thoughts and behaviors, depression, PTSD, and TBI symptom severity.[17] The number of TBIs was associated with suicide risk after controlling for depression, PTSD, and symptom severity. It is important to note that this study assessed suicide based on questionnaire data and did not measure actual suicides. Another study of Iraq and Afghanistan veterans used psychological and social risk factors to estimate suicide risk, and was unable to find clear evidence that veterans with PTSD and TBI were at greater risk for suicide than those with PTSD alone.[18] A study of veterans deployed to Iraq and Afghanistan found that, after controlling for psychiatric comorbidities, TBI increased the risk for current suicidal ideation in male but not female veterans.[19] A case-control study of suicides in the U.S. Armed Forces from 2001 to 2009 found that TBI and PTSD did not increase the odds for suicide (although mood disorders did).[20]

Estimates of the connection between TBI and PTSD in civilian populations are variable; among those with TBI in cases of nonmilitary trauma, 14% to 56% have been found to have PTSD.[21] One study found no difference in rates of PTSD among those exposed to motor vehicle accidents with or without TBI.[22] However, a prospective cohort study of traumatically injured civilians found that those with traumatic brain injury were more likely to develop PTSD, with an odds ratio of 1.92.[23] Furthermore, patients with TBI were also more likely to develop other psychiatric illnesses including panic disorder, agoraphobia, and social anxiety compared to those with traumatic injuries but without TBI.

The aforementioned research has provided compelling evidence that TBI is a risk factor for PTSD. However, the causal mechanisms by which TBI increases the risk of PTSD remain unclear. Some authors have argued that damage to the brain may directly contribute to PTSD by compromising neural circuitry required to regulate fear.[24,25] Alternatively, TBI may increase the risk of PTSD by depleting the cognitive resources required to cope with stressors.[24] One possibility is that circumstances that surround TBI tend to be more emotionally traumatic than those that do not cause TBI. While studies have attempted to address this by controlling for combat severity or comparing TBI patients to those who have suffered other traumatic injuries, it is difficult to definitively control for all aspects of the traumatic situations.[5] It is also possible that in some cases, postconcussive symptoms (caused directly by trauma to the brain) are being mistaken for symptoms of PTSD, raising the probability of a PTSD diagnosis.[5] Overlap between postconcussive and PTSD symptoms will be discussed in the next section.

In summary, understanding of the epidemiologic relationship between TBI and PTSD has evolved rapidly over the past decade, and TBI has emerged as an important risk factor for the development of PTSD in both military and civilian populations. Several studies have shown that this relationship holds when controlling for preexisting psychiatric symptoms as well as when attempting to control for trauma severity. The role of TBI as a risk factor for psychiatric disorders appears not to be specific to PTSD but also to extend to other disorders. Given the high rates of comorbid TBI and PTSD, it has been difficult to disentangle the causal relationships between TBI and subsequent symptoms (i.e., which symptoms result from direct neurological insult and which are related to stress or emotional trauma). The diagnostic difficulties resulting from the overlap in symptoms of PTSD and postconcussive syndrome are discussed in the next section.

DIAGNOSIS OF PTSD AND TBI

PTSD as described in the American Psychiatric Association's *DSM-5* requires a history of exposure to a traumatic event meeting certain criteria, along with symptoms from each of four symptoms clusters (see Table 16.1).[26] The first cluster is intrusion, which includes intrusive memories, nightmares, or flashbacks. The second symptom cluster involves avoidance of trauma-related thoughts and feelings or external reminders. The third symptom cluster consists of negative alterations in cognitions and mood, including dissociative amnesia; negative emotions such as fear, horror, and guilt; loss of interest in activities; and detachment from others. The final symptom cluster involves alterations in arousal and reactivity, which include irritability, reckless behavior, hypervigilance, difficulty concentrating, and sleep difficulties.

There is significant overlap between PTSD and postconcussive symptoms. In particular, both syndromes can involve depressed mood, anxiety, insomnia, irritability, difficulty concentrating, fatigue, hyperarousal, and avoidance.[6] Emotional numbing, derealization, reduced awareness of one's surroundings, depersonalization, and amnesia can also be related to either PTSD or TBI.[24,27] Symptoms that are more specific to PTSD include re-experiencing symptoms, shame, and guilt. Headache, sensitivity to light and sound, and dizziness are more specific to postconcussive syndrome. Beyond the overlap in symptoms, there may also be more complex interactions between the symptomatology of PTSD and TBI. For example, pain related to a traumatic injury may serve as a trigger for the re-experiencing symptoms of PTSD.[27] Another difficulty in diagnosis is the possible presence of other comorbid conditions. For example, depression is often comorbid with both TBI and PTSD, and can result in concentration problems, memory problems, irritability, reduced motivation, and fatigue.[24] The overlap in symptoms makes differential diagnosis difficult in patients at risk for both TBI and PTSD, and careful attention must be paid to the presentation and history in each individual case. It has also been noted that several symptoms characteristic of postconcussive syndrome are fairly nonspecific in general, including headaches, sleep difficulty, irritability, and memory problems.[24] In some individuals, these symptoms may be unrelated to either TBI or PTSD.

Researchers have debated whether postconcussive symptoms in general may be caused by emotional and psychological stress, direct neurological injury, or both.[28] A cross-sectional study of Iraq and Afghanistan veterans found that PTSD was the strongest factor associated with postconcussive symptoms, even after removing overlapping symptoms from the PTSD score.[14] A prospective study of patients admitted to a hospital for trauma found that acute postconcussive symptoms were not

TABLE 16.1
Diagnostic Criteria (*DSM 5*) for Post-Traumatic Stress Disorder

Criterion	Symptom or Description
Criterion A: Stressor (one required)	The person was exposed to death, threatened death, actual or threatened serious injury, or actual or threatened sexual violence, as follows:
	1. Direct exposure
	2. Witnessing, in person
	3. Indirectly, by learning that a close relative or close friend was exposed to trauma
	4. Repeated or extreme indirect exposure to aversive details of the event(s), usually in the course of professional duties
Criterion B: Intrusion symptoms (one required)	1. Recurrent, involuntary, and intrusive memories
	2. Traumatic nightmares
	3. Dissociative reactions (e.g., flashbacks), which may occur on a continuum from brief episodes to complete loss of consciousness
	4. Intense or prolonged distress after exposure to traumatic reminders
	5. Marked physiologic reactivity after exposure to trauma-related stimuli
Criterion C: Avoidance (one required)	1. Trauma-related thoughts or feelings
	2. Trauma-related external reminders
Criterion D: Negative alterations in cognitions and mood (two required)	1. Inability to recall key features of the traumatic event
	2. Persistent (and often distorted) negative beliefs and expectations about oneself or the world
	3. Persistent distorted blame of self or others for causing the traumatic event or for resulting consequences
	4. Persistent negative trauma-related emotions
	5. Markedly diminished interest in (pre-traumatic) significant activities
	6. Feeling alienated from others
	7. Constricted affect: persistent inability to experience positive emotions
Criterion E: Alterations in arousal and reactivity (two required)	1. Irritable or aggressive behavior
	2. Self-destructive or reckless behavior
	3. Hypervigilance
	4. Exaggerated startle response
	5. Problems in concentration
	6. Sleep disturbance
Criterion F: Duration	Persistence of symptoms for more than one month
Criterion G: Functional significance	Significant symptom-related distress or functional impairment (e.g., social, occupational)

specific to mild traumatic brain injury but were predicted by prior affective or anxiety disorders.[29] A subsequent follow-up study found that mild TBI did not predict postconcussive symptoms at 3 months, but that PTSD was related to postconcussive symptoms.[30] A retrospective study of National Guard members using a self-report measure of postconcussive symptoms found that National Guard members with a TBI reported more postconcussive symptoms than those in a nonclinical group, but that those with PTSD reported higher postconcussive symptom severity than those with TBI.[31] Some studies have found an association between TBI and certain post-concussive symptoms after controlling for PTSD. For example, a large study found that TBI was associated with headaches after controlling for PTSD and depression.[4] Another study found that TBI with loss of consciousness was associated with headaches, memory problems, balance problems, and pain in the extremities after controlling for PTSD and depression.[32] One cross-sectional study found that PTSD and TBI were independently associated with postconcussive symptoms, and that those patients with both disorders were at greater risk of postconcussive symptoms than those with only PTSD, TBI, or neither.[33]

Several studies have found a role for PTSD as a mediator between TBI and poor health outcomes and functional impairment. One cross-sectional study of combat veterans found that PTSD is an important mediator between TBI and poor health outcomes.[4] Similarly, another study found that the association between TBI and functional impairment disappeared after controlling for PTSD.[34] A study examining neuropsychological outcomes in Iraq and Afghanistan veterans with PTSD, TBI, or both, found that PTSD contributed to objective cognitive deficits, but TBI in and of itself did not.[35] However, another study found that PTSD was associated with functional impairment, but that TBI with loss of consciousness was also associated with functional impairment even after controlling for PTSD and depression.[36] In considering the literature on PTSD, TBI, postconcussive symptoms, and poor outcomes, it is important to note that biomechanical injury to the brain may contribute to PTSD itself. The role of PTSD as a mediator between TBI and postconcussive symptoms, health outcomes, and functional impairment therefore does not rule out a role for neurological injury, as opposed to emotional trauma. For this reason, the cause of postconcussive symptoms and poor outcomes in TBI patients remain difficult to disentangle, despite ongoing research.

Some authors have expressed concern that in some cases PTSD symptoms caused by emotional stress may be incorrectly attributed to physical brain injury, and that this misattribution may actually cause harm to patients by reducing their expectations of recovery.[24] This process has been termed "diagnosis threat."[37] In order to avoid diagnosis threat, it may be advisable to be cautious when definitively attributing symptoms to neurological injuries rather than other causes, and also to emphasize that recovery from these symptoms is possible and that prognosis is not necessarily poor.

Because PTSD may contribute to postconcussive symptoms and poor outcomes, evaluation and treatment of comorbid PTSD is crucial in individuals with TBI. It should be noted that PTSD might be difficult to recognize in TBI patients, because there is evidence that some of these patients have a tendency to underreport symptoms.[28] TBI patients may have difficulties monitoring and summarizing their symptoms. It is important to be cognizant of the possibility of underreported symptoms when assessing PTSD symptoms in this population.

In summary, assessment of comorbid PTSD and TBI presents multiple diagnostic and conceptual challenges. There is significant overlap in diagnostic criteria between PTSD and postconcussive syndrome, although a few symptoms are more specific to each disorder. However, several symptoms of postconcussive syndrome are relatively nonspecific, and in some cases may be related to neither PTSD nor TBI. Despite ongoing research, there continues to be controversy over whether postconcussive symptoms are caused by biomechanical injury to the brain, by emotional and psychological stress, or both. PTSD also appears to be an important mediator between TBI and poor health outcomes and functional impairment in general. Some authors have argued that misattribution of PTSD symptoms to TBI may cause harm to patients by decreasing expectations for recovery. Given the probable role for PTSD in worsening postconcussive symptoms and more general outcomes in TBI patients, it is important to diagnose and treat PTSD in TBI patients. However, it should be noted that some TBI patients may be prone to underreporting symptoms of PTSD.

PTSD TREATMENT

Treatment for PTSD can be categorized as pharmacologic and nonpharmacologic. Among pharmacologic treatments, the strongest evidence exists for selective serotonin reuptake inhibitors (SSRIs). The only FDA approved medications for PTSD are sertraline and paroxetine, which are both SSRIs and have been shown effective in randomized controlled trials.[38–40] There is also strong evidence for fluoxetine, another SSRI, and venlafaxine, a serotonin norepinephrine reuptake inhibitor (SNRI).[41–44] Prazosin, an alpha-1 adrenergic antagonist, has shown efficacy in decreasing nightmares associated with PTSD and may also decrease other PTSD symptoms.[45–47]

Certain psychotherapies have been found effective for PTSD. Prolonged exposure (PE) is a type of trauma-focused cognitive behavioral therapy (CBT) that aims toward fear extinction through imaginal exposures (in which a patient repeatedly recounts memories of a trauma) and in vivo exposures (in which a patient is exposed to distressing situations in the present). PE has been shown to reduce symptoms of PTSD.[48] Another type of trauma-focused CBT, cognitive processing therapy (CPT) focuses on restructuring dysfunctional cognitions related to a traumatic event. It has also been found to be effective for PTSD.[49,50] Eye movement desensitization and reprocessing therapy (EMDR) is another evidence-based psychotherapy for PTSD that involves the use of bilateral eye movements in addition to other psychotherapy elements.[51] Within the Veterans Affairs system, both PE and CPT have been widely disseminated and are widely available.[52]

TREATMENT OF COMORBID PTSD AND TBI

Because PTSD is common in TBI patients and may be an important contributor to poor outcomes in these patients, it is important to identify and treat PTSD in this population. However, treatment of PTSD may present special challenges when comorbid with TBI. Unfortunately, there is currently little evidence to guide treatment in this population. Many PTSD treatment studies actually exclude individuals with TBI.[53] Therefore, treatment in this population must take account of the available evidence but also be adapted based on individualized considerations. Such considerations may include cognitive and other sequelae of TBI that may interfere in treatment, overlapping symptoms between PTSD and TBI, and possible tradeoffs when treatments are helpful for one condition but potentially harmful for the other.

Sequelae of TBI may require special attention in trauma-focused CBT for PTSD. For example, as discussed earlier, patients may sometimes attribute nonspecific symptoms to TBI and consequently assume that prognosis for recovery is poor. It may therefore be helpful to address these attributions and normalize the symptoms, reducing distress and increasing expectations for recovery.[24] It is also possible that cognitive dysfunction related to TBI may impair patients' ability to understand, remember, and engage with material in the context of cognitive therapy. This may be addressed by simplifying the delivery of therapy, providing written instructions for exercises to compensate for memory problems, and using other strategies to minimize reliance on attentional focus.[27,28] Impulse control problems related to TBI may also make exposure exercises more difficult to tolerate. These patients may therefore benefit from greater emphasis on the rationale for these exercises in order to encourage engagement with treatment.[27] Amnesia related to TBI may interfere in imaginal exposures in PE, in which the patient repeatedly recounts a traumatic experience in order to facilitate fear extinction. In patients who cannot remember the trauma, imaginal exposures may focus on reconstructions of the traumatic experience rather than direct memories. Alternatively, therapy may emphasize in vivo exposure, in which patients are exposed to distressing situations in the present, rather than to distressing memories.[24,28] To date, no controlled trials have been performed on psychotherapy for PTSD with comorbid TBI. Two early case reports described treatment of individuals with PTSD and TBI using CBT with modifications, such as the use of personal digital assistants and cognitive breaks,[21,54,55] with improvements in mood and anxiety. More recently, a retrospective study of patients undergoing CPT for PTSD found similar treatment adherence in patients with or without mild TBI.[56] A case series of slightly modified PE treatment for individuals with comorbid PTSD and TBI found that the subjects significantly improved in symptoms of PTSD and depression from pre- to posttreatment.[57] Studies of a residential treatment program for comorbid PTSD and TBI using a modified version of CPT found a reduction in both PTSD and postconcussive symptoms over the course of treatment.[58,59] A controlled trial of CBT for acute stress disorder after TBI showed efficacy in preventing PTSD, providing further evidence that TBI patients are able to benefit from CBT.[60] Overall, more research is still needed to determine optimal approaches to psychotherapy in this population.

Given the overlapping domains of dysfunction in both TBI and PTSD, treatments that target symptoms common to both conditions may be especially helpful in this population. For example, SSRIs may be beneficial for both PTSD and TBI.[53] Additionally, stimulants may be prescribed for attentional dysfunctions, hypnotics for sleep difficulties, and anticonvulsants for affect dysregulation.[6] The risks of such treatments must also be taken into account (e.g., substance use disorder).

Some pharmacologic treatments for PTSD may carry special risk in TBI patients. For example, individuals with a history of TBI have a higher rate of seizures, gait and balance problems, and deficits in sensory processing. Certain psychotropic medications may worsen sensory or balance problems. Other medications, including buproprion, maprotiline, and amoxapine, can increase the risk for seizures.[53] These risks should be considered before starting psychotropic medications, and patients should be monitored for worsening of symptoms. Several types of medications, including antipsychotics, anticonvulsants, anxiolytics, and any anticholinergic medication, may contribute to cognitive deficits associated with TBI.[53] Medications often used for TBI may also worsen PTSD symptoms. For example, stimulants to combat fatigue and attentional problems related to TBI may theoretically worsen hyperarousal in PTSD,[53] although empirical evidence that this is commonly encountered as an adverse event is lacking. Finally, many psychotropic medications have complex effects on sleep, which is often impaired in both TBI and PTSD.[53] Considering the potential risks of psychotropic medications in these populations, it may be advisable to start at low doses and titrate slowly with careful monitoring for adverse effects.

The prevalence of comorbid TBI and PTSD, and the complexities of clinical care in this population, require special approaches in the delivery of care. The diversity of symptoms in this population, including both medical and mental health problems, means that coordination of care between specialists is crucial.[61] The need for coordination also extends to research, as better design and interpretation of studies in this population will benefit from expertise in psychiatry, neuropsychology, neurology, neurosurgery, and physical medicine and rehabilitation, among others.[6]

In summary, there is currently very little evidence base for the treatment of comorbid PTSD and TBI. Treatment must be individualized based on the presentation of each patient. Cognitive sequelae of TBI may interfere with trauma-focused psychotherapy (though preliminary evidence suggests that many patients with mild-to-moderate cognitive deficits do benefit from such therapies), and therapy may need to be modified to account for cognitive impairment. Certain medications may carry benefit for both disorders, such as antidepressants, stimulants, anticonvulsants, and hypnotics. However, careful attention must be paid to the risks of each medication. In general, it is advisable to start at a low dose, titrate slowly, and monitor closely. This population also requires attention in terms of delivery of care, including special screening procedures and coordination between specialties. More research is clearly needed to determine the best treatments for comorbid PTSD and TBI.

CONCLUSIONS

Comorbid PTSD and TBI is a common and challenging clinical situation that requires attention to develop better strategies for assessment and treatment. Accumulating evidence has identified TBI as an important risk factor for PTSD. The causal relationship between TBI and PTSD remains unclear, but may involve both neurological and psychological factors. Further, there is significant overlap in symptoms of PTSD and TBI, making differential diagnosis difficult. Multiple studies have shown that PTSD is associated with greater reporting of postconcussive symptoms, creating debate over whether these symptoms are related to neurological injury, psychological stress, or both. When assessing this evidence, it is important to note that PTSD itself may be related to neurological injury, and that the association between PTSD and postconcussive symptoms cannot definitively implicate a psychological cause for these symptoms. There are a number of effective pharmacologic and nonpharmacologic treatments for PTSD, but the treatment of comorbid PTSD and TBI may be complicated by several considerations. These include overlapping symptoms, cognitive and other sequelae of TBI that may interfere in treatment, and trade-offs in which a treatment for one condition may adversely affect the other condition. Unfortunately, there is currently very little research to guide treatment in this population, and there is a pressing need for treatment studies in this population.

REFERENCES

1. American Psychiatric Association. 1980. *Diagnostic and Statistical Manual of Mental Disorders*, 3rd ed. Washington, DC: American Psychiatric Association.
2. Breslau, N. 2012. Epidemiology of posttraumatic stress disorder in adults. In *The Oxford Handbook of Traumatic Stress Disorders*, eds. J.G. Beck, D.M. Sloan, 84–97. New York: Oxford University Press.
3. Tanielian, T.L., Jaycox, L. 2008. *Invisible Wounds of War: Psychological and Cognitive Injuries, Their Consequences, and Services to Assist Recovery*, vol. 720. Santa Monica, CA: Rand Corporation.
4. Hoge, C.W. et al. 2008. Mild traumatic brain injury in U.S. soldiers returning from Iraq. *N Engl J Med* 358(5):453–463.
5. Yurgil, K.A. et al. 2014. Association between traumatic brain injury and risk of posttraumatic stress disorder in active-duty Marines. *JAMA Psychiatry* 71(2):149–157.
6. Stein, M., McAllister, T. 2009. Exploring the convergence of posttraumatic stress disorder and mild traumatic brain injury. *Am J Psychiatry* 166(7):768–776.
7. Dohrenwend, B.P. et al. 2006. The psychological risks of Vietnam for U.S. veterans: A revisit with new data and methods. *Science* 313(5789):979–982.
8. Ramchand, R. et al. 2010. Disparate prevalence estimates of PTSD among service members who served in Iraq and Afghanistan: Possible explanations. *J Trauma Stress* 23(1):59–68.
9. Kessler, R.C. et al. 2005. Lifetime prevalence and age-of-onset distributions of DSM-IV disorders in the National Comorbidity Survey Replication. *Arch Gen Psychiatry* 62(6):593–602.
10. Breslau, N. et al. 2004. Estimating post-traumatic stress disorder in the community: Lifetime perspective and the impact of typical traumatic events. *Psychol Med* 34(5):889–898.
11. Mayou, R., Bryant, B., Duthie, R. 1993. Psychiatric consequences of road traffic accidents. *BMJ* 307(6905):647.
12. Sbordone, R.J., Liter, J.C. 1995. Mild traumatic brain injury does not produce posttraumatic stress disorder. *Brain Inj* 9(4):405–412.
13. Bryant, R.A. 2001. Posttraumatic stress disorder and traumatic brain injury: Can they co-exist? *Clin Psychol Rev* 21(6):931–948.
14. Schneiderman, A.I., Braver, E.R., Kang, H.K. 2008. Understanding sequelae of injury mechanisms and mild traumatic brain injury incurred during the conflicts in Iraq and Afghanistan: Persistent postconcussive symptoms and posttraumatic stress disorder. *Am J Epidemiol* 167(12):1446–1452.
15. Kontos, A.P. et al. 2013. Residual effects of combat-related mild traumatic brain injury. *J Neurotrauma* 30(8):680–686.
16. Kennedy, J.E. et al. 2010. Posttraumatic stress symptoms in OIF/OEF service members with blast-related and non-blast-related mild TBI. *NeuroRehabilitation* 26(3):223–231.
17. Bryan, C.J., Clemans, T.A. 2013. Repetitive traumatic brain injury, psychological symptoms, and suicide risk in a clinical sample of deployed military personnel. *JAMA Psychiatry* 70(7):686–691.
18. Barnes, S.M., Walter, K.H., Chard, K.M. 2012. Does a history of mild traumatic brain injury increase suicide risk in veterans with PTSD? *Rehabil Psychol* 57(1):18.
19. Wisco, B.E. et al. 2014. Traumatic brain injury, PTSD, and current suicidal ideation among Iraq and Afghanistan U.S. veterans. *J Trauma Stress*. 27(2):244–248.

20. Skopp, N.A. et al. 2012. Relations between suicide and traumatic brain injury, psychiatric diagnoses, and relationship problems, active component, US Armed Forces, 2001–2009. *MSMR* 19(2):7–11.
21. Carlson, K.F. et al. 2011. Prevalence, assessment, and treatment of mild traumatic brain injury and posttraumatic stress disorder: A systematic review of the evidence. *J Head Trauma Rehabil* 26(2):103–115.
22. Jones, C., Harvey, A.G., Brewin, C.R. 2005. Traumatic brain injury, dissociation, and posttraumatic stress disorder in road traffic accident survivors. *J Trauma Stress* 18(3):181–191.
23. Bryant, R.A. et al. 2010. The psychiatric sequelae of traumatic injury. *Am J Psychiatry* 167(3):312–320.
24. Bryant, R. 2011. Post-traumatic stress disorder vs. traumatic brain injury. *Dialogues Clin Neurosci* 13(3):251.
25. Vasterling, J.J., Verfaellie, M., Sullivan, K.D. 2009. Mild traumatic brain injury and posttraumatic stress disorder in returning veterans: Perspectives from cognitive neuroscience. *Clin Psychol Rev* 29(8):674–684.
26. American Psychiatric Association. 2013. *Diagnostic and Statistical Manual of Mental Disorders*, 5th ed. Washington, DC: American Psychiatric Association.
27. Bryant, R.A., Hopwood, S. 2006. Commentary on "Trauma to the psyche and soma." *Cogn Behav Pract* 13(1):17–23.
28. Bryant, R.A. 2001. Posttraumatic stress disorder and mild brain injury: Controversies, causes and consequences. *J Clin Exp Neuropsychol* 23(6):718–728.
29. Meares, S. et al. 2008. Mild traumatic brain injury does not predict acute postconcussion syndrome. *J Neurol Neurosurg Psychiatry* 79(3):300–306.
30. Meares, S. et al. 2011. The prospective course of postconcussion syndrome: The role of mild traumatic brain injury. *Neuropsychology* 25(4):454.
31. Soble, J.R. et al. 2014. Normative Data for the Neurobehavioral Symptom Inventory (NSI) and post-concussion symptom profiles among TBI, PTSD, and nonclinical samples. *Clin Neuropsychol.* 28(4):614–632.
32. Wilk, J.E. et al. 2012. Mild traumatic brain injury (concussion), posttraumatic stress disorder, and depression in U.S. soldiers involved in combat deployments: Association with postdeployment symptoms. *Psychosom Med* 74(3):249–257.
33. Brenner, L.A. et al. 2010. Traumatic brain injury, posttraumatic stress disorder, and postconcussive symptom reporting among troops returning from Iraq. *J Head Trauma Rehabil* 25(5):307–312.
34. Ragsdale, K.A. et al. 2013. Posttraumatic stress disorder in OEF/OIF veterans with and without traumatic brain injury. *J Anxiety Disord* 27(4):420–426.
35. Shandera-Ochsner, A.L. et al. 2013. Neuropsychological effects of self-reported deployment-related mild TBI and current PTSD in OIF/OEF veterans. *Clin Neuropsychol* 27(6):881–907.
36. Verfaellie, M. et al. 2012. Chronic postconcussion symptoms and functional outcomes in OEF/OIF veterans with self-report of blast exposure. *J Int Neuropsychol Soc* 19(1):1.
37. Howe, L.L.S. 2009. Giving context to post-deployment post-concussive-like symptoms: Blast-related potential mild traumatic brain injury and comorbidities. *Clin Neuropsychol* 23(8):1315–1337.
38. Brady, K. et al. 2000. Efficacy and safety of sertraline treatment of posttraumatic stress disorder: A randomized controlled trial. *JAMA* 283(14):1837–1844.
39. Davidson, J.R.T. et al. 2001. Multicenter, double-blind comparison of sertraline and placebo in the treatment of posttraumatic stress disorder. *Arch Gen Psychiatry* 58(5):485–492.
40. Marshall, R.D. et al. 2001. Efficacy and safety of paroxetine treatment for chronic PTSD: A fixed-dose, placebo-controlled study. *Am J Psychiatry* 158(12):1982–1988.

41. Connor, K.M. et al. 1999. Fluoxetine in post-traumatic stress disorder. Randomised, double-blind study. *Br J Psychiatry* 175(1):17–22.
42. Davidson, J. et al. 2006. Treatment of posttraumatic stress disorder with venlafaxine extended release: A 6-month randomized controlled trial. *Arch Gen Psychiatry* 63(10):1158–1165.
43. Martenyi, F. et al. 2002. Fluoxetine versus placebo in posttraumatic stress disorder. *J Clin Psychiatry* 63(3):199–206.
44. Tucker, P. et al. 2001. Paroxetine in the treatment of chronic posttraumatic stress disorder: Results of a placebo-controlled, flexible-dosage trial. *J Clin Psychiatry* 62(11):860–868.
45. Raskind, M.A. et al. 2007. A parallel group placebo controlled study of prazosin for trauma nightmares and sleep disturbance in combat veterans with post-traumatic stress disorder. *Biol Psychiatry* 61(8):928–934.
46. Raskind, M.A. et al. 2003. Reduction of nightmares and other PTSD symptoms in combat veterans by prazosin: A placebo-controlled study. *Am J Psychiatry* 160(2):371–373.
47. Raskind, M.A. et al. 2013. A trial of prazosin for combat trauma PTSD with nightmares in active-duty soldiers returned from Iraq and Afghanistan. *Am J Psychiatry* 170(9):1003–1010.
48. Foa, E.B. et al. 2005. Randomized trial of prolonged exposure for posttraumatic stress disorder with and without cognitive restructuring: Outcome at academic and community clinics. *J Consult Clin Psychol* 73(5):953.
49. Monson, C.M. et al. 2006. Cognitive processing therapy for veterans with military-related posttraumatic stress disorder. *J Consult Clin Psychol* 74(5):898.
50. Resick, P.A. et al. 2008. A randomized clinical trial to dismantle components of cognitive processing therapy for posttraumatic stress disorder in female victims of interpersonal violence. *J Consult Clin Psychol* 76(2):243.
51. Rothbaum, B.O. 1997. A controlled study of eye movement desensitization and reprocessing in the treatment of posttraumatic stress disordered sexual assault victims. *Bull Menninger Clin.* 61(3):317–334.
52. Capehart, B., Bass, D. 2012. Review: Managing posttraumatic stress disorder in combat veterans with comorbid traumatic brain injury. *J Rehabil Res Dev* 49(5):789–812.
53. McAllister, T.W. 2009. Psychopharmacological issues in the treatment of TBI and PTSD. *Clin Neuropsychol* 23(8):1338–1367.
54. Batten, S.V., Pollack, S.J. 2008. Integrative outpatient treatment for returning service members. *J Clin Psychol* 64(8):928–939.
55. McGrath, J. 1997. Cognitive impairment associated with post-traumatic stress disorder and minor head injury: A case report. *Neuropsychol Rehabil* 7(3):231–239.
56. Davis, J.J. et al. 2013. Treatment adherence in cognitive processing therapy for combat-related PTSD with history of mild TBI. *Rehabil Psychol* 58(1):36.
57. Wolf, G.K. et al. 2012. A preliminary examination of prolonged exposure therapy with Iraq and Afghanistan veterans with a diagnosis of posttraumatic stress disorder and mild to moderate traumatic brain injury. *J Head Trauma Rehabil* 27(1):26–32.
58. Chard, K.M. et al. 2011. Exploring the efficacy of a residential treatment program incorporating cognitive processing therapy-cognitive for veterans with PTSD and traumatic brain injury. *J Trauma Stress* 24(3):347–351.
59. Walter, K.H., Kiefer, S.L., Chard, K.M. 2012. Relationship between posttraumatic stress disorder and postconcussive symptom improvement after completion of a posttraumatic stress disorder/traumatic brain injury residential treatment program. *Rehabil Psychol* 57(1):13.
60. Bryant, R.A. et al. 2003. Treating acute stress disorder following mild traumatic brain injury. *Am J Psychiatry* 160(3):585–587.
61. Lew, H.L. et al. 2008. Overlap of mild TBI and mental health conditions in returning OIF/OEF service members and veterans. *J Rehabil Res Dev* 45(3):xi.

17 Traumatic Brain Injury and Potential for Neuromodulation

Shervin Rahimpour and Shivanand P. Lad

CONTENTS

INTRODUCTION

Traumatic brain injury (TBI) is a major health problem with devastating consequences and an enormous socioeconomic burden on individuals and their families.[1] While significant progress has been made in understanding TBI pathophysiology, current interventions are still insufficient to prevent long-term disabilities.[2]

Recent insights on the brain's innate neuroplasticity provide an array of therapeutic possibilities (yet to be fully understood) that may allow for neuromodulation following a TBI to limit damage and promote recovery. Neuromodulation strategies using various forms of electrical stimulation have recently been applied to treat functional deficits in animal models and clinical stroke trials. The results from these studies suggest that neuromodulation may augment both motor and cognitive deficits after brain injury. As imaging, targeting, and technologies continue to rapidly improve, future studies will validate the feasibility of these techniques and clinical applications for TBI.

CURRENT LANDSCAPE

There are multiple therapeutic strategies that are currently employed for tackling the oftentimes challenging neurologic disabilities of TBI patients. Despite continued progress and efforts in occupational, physical, and cognitive rehabilitation, there are many significant unmet clinical needs for more specific and effective therapies. After decades of research, there has yet to be a successful phase III clinical trial intervention for TBI. This may be in part because TBI is a heterogeneous condition. There has been recent emphasis on a clear definition of the condition and its subtypes to provide increased homogeneity for the purposes of clinical trials. The pathophysiology of TBI is complex and includes inflammation, oxidative stress, apoptosis, excitotoxicity, and mitochondrial dysfunction. Recent advances in our understanding of neural circuitry, ability to image brain connectivity, and technical advances have all rekindled the excitement for innovative neuromodulation approaches to enhance the functional recovery in patients with acquired brain disorders such as TBI.

TRANSCRANIAL MAGNETIC STIMULATION

Transcranial magnetic stimulation (TMS), introduced in 1985, is a technique based on Faraday's principle of electromagnetic induction.[3] By passing a very brief and large electric current though a coil held over the scalp, a rapidly changing magnetic field is created that then penetrates through the skull and induces electric currents in particular brain regions. When a single TMS pulse is applied, the resulting electric current can be of sufficient magnitude to depolarize cortical neurons, either through a direct effect on their axon hillock or indirectly by depolarizing interneurons. The induced electrical stimulus can simultaneously activate many different populations of neurons, mainly in the neocortex, some of which project axons to or from the site of stimulation. Therefore, varying effects can be induced on the cerebral cortex and in functionally related brain regions via synaptic spread. Factors such as coil orientation and baseline cortical activity influence the effects of TMS, and patient-specific factors such as anatomical or neurophysiologic derangements can have similar consequences.[4]

SHORT-TERM EFFECTS OF REPETITIVE TMS

When applied in repetitive trains, pulses are referred to as repetitive TMS (rTMS). This can be further classified as "conventional," when all pulses are delivered at a constant interval, or as "patterned" when brief rTMS bursts at a high inner frequency are interleaved by short pauses of no stimulation. Examples of patterned rTMS include theta burst (TBS) and quadripulse stimulation. Different neuronal populations show different thresholds to electrical stimulation, allowing for some selectivity in their activation depending on the rTMS parameters used. In most cases, conventional rTMS frequencies between 0.2 and 1 Hz (i.e., low-frequency rTMS) tend to cause a decrease in cortical excitability, most likely due to preferential stimulation of GABAergic neurons, while frequencies greater than 5 Hz (i.e., high-frequency rTMS) have the opposite effect. The inhibitory effects of rTMS can be enhanced when low-frequency pulses are preceded by a brief high-frequency priming and vice versa. Moreover, rTMS has been shown to induce changes in regional CBF in a stimulation-dependent manner, with low-frequency rTMS decreasing CBF and high-frequency rTMS increasing CBF, likely a response to changes in cortical excitability.

LONG-TERM EFFECTS OF rTMS

After a period of stimulation that usually ranges between 10 to 15 min, the facilitatory or inhibitory effects of rTMS can be observed, lasting for approximately 10 minutes after the end of the intervention. However, such effects are of greater magnitude following repeated sessions and may persist for several weeks. These long-lasting effects seen after repeated stimulation can prove to be crucial for the potential therapeutic role of rTMS by inducing an environment supportive of neural plasticity. Even though the mechanisms responsible for long-term effects of TMS are yet to be fully understood, evidence from pharmacological studies suggests that they could be related to the modulation of NMDA glutamatergic receptors, a process comparable to the induction of LTP and LTD. Hence, high- or low-frequency rTMS can be used to induce LTP or LTD, respectively. This effect can be demonstrated in the actual target site for stimulation, as well as in functionally or anatomically remote brain regions.[4,5]

CORTICAL STIMULATION

Unlike TMS techniques, direct stimulation of the cortex after surgical exposure enables direct access to the target structures and may allow for more precise stimulation of cortical areas. Direct cortical stimulation can induce plasticity in somatosensory areas. Stimulation during rehabilitation following motor cortex injury improves motor function in both rats and primates.[6,7] In animal experiments using intracortical microstimulation, cortical surface electrodes stimulating the motor cortex at subthreshold levels during rehabilitation after focal motor cortex ischemia can increase forelimb function.[8,9] This functional improvement following motor cortex infarction is induced by restoration of movement representation in peri-infarct areas, as well as the emergence of novel representations. Areas of movement representation shifted several microns after repeated stimuli and increased in size when rats received intracortical microstimulation in the forelimb area of the motor cortex. Histological examination revealed an increase in spine density in pyramidal cell layers III and V.

Similarly, cortical electrical stimulus parameters that improved motor performance in rats following focal ischemic injury to the motor cortex were also associated with increased dendritic processes in perilesional areas. The improvement in motor function following electrical stimulation correlated with increased markers of dendritic plasticity and decreased markers of astrogliosis by immunohistochemistry in both the perilesional cortex and the contralesional anterior horn of the cervical spinal cord. Thus, cortical electrical stimulation may lead to motor function improvements due to increased neuronal structural plasticity. Likewise, direct cortical stimulation has also been used to drive plasticity in rodents subjected to a lesioning injury. Unilateral corticospinal tract injury followed by 10 days of electrical stimulation of the motor cortex resulted in forelimb function recovery and increased axon outgrowth. A follow-up study by the same group demonstrated that motor cortex stimulation after pyramidotomy can increase the length of axons from the primary motor cortex to the spinal cord, as well as to the red nucleus and cuneate nucleus.[10] Increased outgrowth with stimulation may be due to neurotrophin release, such as brain-derived neurotrophic factor, and increased activity. The pyramidotomy is a mechanism of injury that is different from TBI, but one can cautiously extrapolate that electrical stimulation may enhance neurotrophin release in brain tissue damaged by diffuse axonal injury or cortical contusion. These preclinical studies suggest that direct cortical stimulation may enhance functional plasticity after cortical injury.[11]

HUMAN TRIALS OF DIRECT CORTICAL STIMULATION

Motor cortex stimulation has been clinically tested in patients with trigeminal neuralgia and central pain from stroke with mixed results. A meta-analysis of motor cortex stimulation to reduce pain showed efficacy in approximately 50% of patients (with 40%–50% improvement). However, most studies are retrospective series without control groups, and long-term loss of pain relief following stimulation was common. Extradural motor cortex stimulation has also been used in patients with Parkinson's disease (PD) when DBS is either contraindicated or proven ineffective.

Based on anecdotal reports showing improved motor function when stroke patients with chronic pain were treated by motor cortex stimulation, an initial study using subthreshold epidural motor cortex stimulation on a hemiparetic stroke patient was performed. Stimulation for 3 weeks during stroke rehabilitation led to improved pincer movement of the previously paretic hand, improved scores on the upper-extremity Fugl–Meyer motor scale and Stroke Impact Scale, and reduced flexor posture. This single-patient study suggesting the efficacy of epidural motor cortex stimulation prompted several prospective trials using the same techniques on stroke patients with motor deficits. In a small prospective randomized trial of 12 stroke patients assigned to either rehabilitation and motor cortical stimulation or rehabilitation alone, the group receiving electrical stimulation showed Fugl–Meyer motor scale improvements in upper-extremity function compared with the control group during 6 months of follow-up. Similarly, prospective multicenter studies using motor cortex stimulation during 3 weeks or 6 weeks of the rehabilitation phase have reported improvements in measures of upper-extremity function.[12–14]

NONINVASIVE STIMULATION FOR POST-TRAUMATIC COGNITIVE SEQUELAE

In addition to its putative effects on motor function, TMS has been shown to enhance various cognitive functions.[15] In healthy subjects, TMS has been suggested to enhance performance in analogical reasoning, recognition memory, and choice-reaction time.

Working memory may also be enhanced in a frequency-dependent manner when the midline parietal cortex is stimulated by TMS. However, whether TMS results in functional enhancement or impairment may depend on the stimulation site and parameters as working memory impairment has been reported with TMS of the cerebellum or dorsolateral prefrontal cortex (DLPFC).

There are some data demonstrating the benefit of TMS for acquired brain injury. Among patients with unilateral stroke, TMS of the unaffected parietal cortex reduced symptoms of contralesional visuospatial neglect. In a pilot study of two patients with right hemispheric infarction who had unilateral spatial neglect, 2 weeks of TMS to the unaffected parietal cortex led to significant improvements in skills relevant to visual neglect. In addition, TMS may also be effective in improving recovery from poststroke aphasias. Aphasic patients show increased metabolic activation of the nondominant inferior frontal and superior temporal areas as evaluated by PET. Previous studies used low-frequency TMS to suppress the hypothetical overactivity of the contralesional inferior frontal gyrus in patients with chronic aphasia that resulted in a significant increase in discourse productivity and improved picture naming.[5,16]

DEEP BRAIN STIMULATION (DBS)

DBS is an effective therapy for treatment of movement disorder such as Parkinson's disease, essential tremor, and dystonia. The mechanism by which DBS is thought to improve symptoms in these diseases is disruption of abnormal neural synchrony between affected brain regions. Rather than inhibiting neural activity, DBS alters activity patterns to moderate abnormal brain function related to disease symptoms. For example, DBS of the subthalamic nucleus in patients with PD induces widespread normalization of activity by increasing activation of motor areas during movement execution and decreased overactivity at rest, as well as modulating metabolic changes in the globus pallidus internus (GPi), caudal midbrain, and posterior parietal lobe. Some of the most common sequelae of TBI are movement disorders, including tremor, dystonia, and parkinsonism, and as a result the utility of DBS in a medically refractory post-traumatic movement disorder is a topic of significant interest.

DBS FOR POST-TRAUMATIC TREMOR

Post-traumatic tremor has been reported as early as a few weeks and as late as 2 years following injury. Limited data exist for the effectiveness of DBS on post-traumatic tremors, with limited follow-up intervals. Successful stimulation of the ventralis intermedius (VIM) nucleus of the thalamus, which is the accepted stimulation target for essential tremor, has been reported in several single case reports. Cerebellar projection neurons are concentrated in the VIM, and high-frequency stimulation can modulate the activity of these projections. Foote et al.[17] employed a strategy to target both the cerebellothalamic and pallidothalamic circuits in post-traumatic tremor by implantation of two ipsilateral DBS electrodes in parallel in the VIM and ventralis oralis anterior/posterior (VOA/VOP, receives inputs from the basal ganglia, in particular from the GPi) with good tremor response, however percent reduction (38%–67%) did not differ considerably between VIM plus VOA/VOP stimulation and VIM stimulation alone. Issar et al.[18] recently published a small case series in which post-traumatic tremor patients received unilateral VIM DBS (n = 3), bilateral VIM DBS (n = 1), or bilateral GPi DBS (n = 1). Among the four patients who underwent VIM stimulation, all reported reduction of their tremor which improved their ability to perform daily activities. However, patients also experienced some side effects, including dystonic movements of the upper extremities, gait and balance problems, paresthesia, and slurred speech. Similarly, of nine post-traumatic tremor patients reported in a broad cases series, three patients who received bilateral thalamic implants developed speech impairments. Lim et al. reported the case of a patient with a poststroke Holmes' tremor who underwent simultaneous implantation in the unilateral VIM, VOA, and GPi. This patient experienced moderate tremor reduction with GPi stimulation alone, but neither VIM nor VOA stimulation provided additional benefit, demonstrating the need for patient-specific target selection.[19–22]

DBS FOR POST-TRAUMATIC DYSTONIA AND PARKINSONISM

Another common sequelae of TBI is dystonia, occurring at a wide range of time points after injury from 1 day up to 6 years. Post-traumatic dystonia is most commonly found in patients with injuries in the basal ganglia and thalamus. Reduced inhibitory control in basal ganglia-thalamocortical motor loops leading to hyperactivity in frontal motor areas has been suggested to result in dystonia, but the mechanisms of post-traumatic dystonia are complex, as different injury constellations will produce different dystonic symptoms.

DBS for post-traumatic dystonia has been reported in a few case reports and one small case series. In a post-traumatic patient with right hemidystonia, DBS of the ventroposterolateral nucleus of the thalamus resulted in a drastic reduction of symptoms. Similarly, a patient with a left-sided low-frequency tremor and hemidystonia 9 years post-TBI received contralateral GPi stimulation, resulting in reductions of arm dystonic movements and abnormal posture for at least 4 years. In another case study, a patient with chronic dystonic posturing and tremor who received DBS of the GPi showed a significant reduction of symptoms up to 2 years after stimulation. Recently, Kim et al.[23] reported a 38%–94% improvement in Burke–Fahn–Marsden Dystonia Rating Scale movement and disability scores in four patients who underwent unilateral GPi DBS for the treatment of secondary hemidystonia associated with TBI. Improvement in dystonia secondary to anoxic brain injury has also been reported following stimulation of either the GPi or the VOA nucleus of the thalamus.[19–22]

Post-traumatic parkinsonism, which can involve tremor, hypokinesia, rigidity, and postural instability, is another potential sequelae of TBI that may be amenable to DBS. Although the mechanism of these symptoms after TBI is unclear and more direct comparisons are needed between patients with idiopathic PD and those with post-traumatic parkinsonism, there may be common pathophysiological elements that could suggest the potential for a therapeutic response to DBS.[19–22]

DBS FOR RECOVERY OF COGNITIVE FUNCTION

Cognitive sequelae of TBI range from a state of minimal consciousness, to problems with memory and executive functioning, to the long-term development of neuropsychological disorders. Noninvasive techniques have been clinically applied to treat the cognitive sequelae of stroke, while in TBI, DBS has been studied as a way to arouse patients from a minimally conscious state (MCS). In addition, DBS is currently under investigation for the treatment of depression, a common sequelae following TBI. In fact, TBI patients commonly show symptoms of depression as measured by multiple depression scales and major depressive disorder is reported as the most common psychiatric illness resulting from TBI. As DBS has shown encouraging results in depression in both humans and animal models, it may be a useful treatment option for TBI survivors suffering from depression. Other potential sequelae of TBI, such as deficits in memory and arousal, may also be amenable to DBS.[24]

DBS IN THE MINIMALLY CONSCIOUS STATE

There are only a few studies reporting the use of DBS in TBI patients despite numerous applications of DBS in other neurological diseases. This is due in part to the wide heterogeneity of functional deficits and anatomical damage in patients with TBI, and equally to the ethical considerations for initiating DBS studies in this patient population. Identifying a specific subset of TBI patients with a functional deficit that would be expected to respond to DBS targeting a specific location is a central challenge for the design of clinical trials in TBI patients.

It has been proposed that DBS in patients who are in a chronic MCS after injury may facilitate the recovery of consciousness. A clinical trial involving vegetative or minimally conscious patients, studied DBS of the midbrain reticular formation or central thalamus in vegetative or minimally conscious patients, beginning 4–8 months following injury. The patients were continuously stimulated and observed for 10 years, and 8 of the 21 patients emerged from a vegetative state and were able to follow verbal instructions. In a follow-up study, Yamamoto et al.[25] created a set of electrophysiological criteria for DBS implantation in this same cohort by analyzing electroencephalography, auditory brainstem responses, somatosensory evoked potentials, and pain-related evoked potentials. Among the 16 patients who met these criteria and received DBS, there was a significantly higher proportion of patients emerging from the vegetative state after DBS (8 out of 10) compared with the patients not treated by DBS (2 out of 6). Caution, however, must be applied in interpreting the effectiveness of DBS within an early time frame following injury, since patients are likely to exhibit some spontaneous recovery. Recovery from MCS has been documented up to a year following injury. These challenges in interpreting the efficacy of DBS in treating consciousness disorders may be overcome through the use of double-blind and/or placebo-controlled prospective studies.[24]

In a more recent case report by Schiff et al.,[26] DBS of the central thalamus was shown to lead to behavioral improvements in an MCS patient. The patient was minimally conscious prior to DBS for 6.5 years following injury but still showed improvements, specifically with regard to arousal state, limb control, and verbalization. As explained by Schiff et al.,[26] this impressive change in behavioral function may have been due to general preservation of cerebral networks in this patient and widespread innervation of frontal cortex and basal ganglia by thalamic association nuclei. By activating the association nuclei of the thalamus with DBS, general arousal and awareness may have been enhanced.[24]

DBS FOR POST-TRAUMATIC DEPRESSION

Depression is very common after TBI, with prevalence rates as high as 77%. Although a history of TBI has traditionally been considered an exclusion criterion for repetitive TMS trials due to the increased risk of seizure, there is one case report of a patient with TBI-related depression who was successfully treated with repetitive TMS to the right DLPFC. The DLPFC was likely chosen as the TMS target due to the fact that PET studies have shown that a pattern of increased metabolism in the subgenual cingulate cortex (SCC) and decreased metabolism in the DLPFC is correlated with active depression, and reversal of this pattern is associated with symptom resolution. On this basis, several trials of SCC DBS for treatment-resistant depression (TRD) have been undertaken.

Stimulation of the SCC was originally described by Mayberg et al.[27] in 6 patients, with follow-up in a larger group of 20 patients. At 12 months, 55% of patients were responders (50% or greater reduction in the 17-item Hamilton Rating Scale for Depression [HRSD-17]), and 35% achieved or were within 1 point of remission (scoring 8 or less on the HRSD-17). More recently, a group at Emory University reported a study of 17 patients with treatment-resistant unipolar and bipolar depression undergoing DBS of the SCC. Patients underwent a single-blind lead-in sham stimulation phase for 4 weeks followed by open-label active stimulation for 24 weeks. Patients then entered a single-blind discontinuation phase that was discontinued after the first 3 patients experienced complete relapse of symptoms when stimulation was stopped. At 2 years after the onset of active stimulation, 92% of patients were responders and 58% were in remission (n = 12). A multicenter study of SCC DBS yielded more modest results, with 29% of 21 patients achieving responder status at 12 months. Multiple clinical trials in the United States, Canada, and Europe are underway to continue the evaluation of SCC DBS in TRD. The anterior limb of the internal capsule and adjacent ventral striatum/nucleus accumbens, structures that have been shown to have significant overlap with the SCC in connections to the medial temporal lobes, frontal limbic cortices, and autonomic subcortical structures, have also been the focus of several investigations as a DBS target for TRD. In general, open-label trials in this target area have reported a 40%–70% response rate, but supportive randomized controlled data are lacking. Although promising, DBS as a treatment for depression is still at an early stage of application, and these results must be interpreted with caution.[24,27–30]

DBS FOR POST-TRAUMATIC EPILEPSY

Epilepsy is a significant delayed comorbidity occurring in TBI patients, with 30-year cumulative incidences of 2.1%, 4.2%, and 16.7% for mild, moderate, and severe injuries, respectively. The clinical study of DBS in pharmacoresistant epilepsy has included the targeting of the anterior nucleus of the thalamus (ANT), centromedian nucleus of the thalamus, hippocampus, caudate nucleus, subthalamic nucleus, and cerebellum. DBS targeting the ANT was approved for the treatment of epilepsy in the European Union in 2010, but still awaits approval in the United States from the Food and Drug Administration (FDA). Human studies have shown that ANT stimulation results in prominent activation in the ipsilateral cingulate gyrus, insular cortex, and lateral neocortical temporal structures and suggest that thalamic DBS may drive cortical inhibitory circuits. Greater benefit from ANT stimulation has been observed in patients with temporal seizure foci, potentially reflecting participation of the mesial temporal lobe and the ANT in the limbic circuit of Papez.

Another aspect of DBS that may be critical for its application in the setting of TBI is the concept of closed-loop stimulation. Clinical DBS experiences to date have mostly involved open-loop stimulation, where the device is programmed for the constant delivery of stimulation within selected, fixed parameters. In closed-loop stimulation, the system is capable of performing real-time seizure detection and delivering responsive stimulation, and the stimulation target depends on each patient's seizure focus or foci. The responsive neurostimulation system (RNS) (NeuroPace, Inc.) was recently approved by the FDA for use in the United States and employs separate electrodes for seizure detection and stimulation delivery. Alternating current systems that allow detection and stimulation on the same electrode lead are currently under investigation. As electrophysiological biomarkers of other TBI comorbidities are discovered, closed-loop stimulation may have the potential for application in TBI beyond the treatment of epilepsy.[31–34]

CONCLUSIONS

Deep brain stimulation has been used to treat movement disorders for the past 15 years and more recently to treat neuropsychiatric disorders. The use of neurostimulation in movement and psychiatric disorders does not necessarily justify its use in TBI, but as our understanding of the therapeutic mechanisms of DBS and the pathophysiological mechanisms of TBI continue to develop, it is likely that DBS will be applied to counteract the network abnormalities responsible for some TBI symptoms. The studies outlined in this review suggest potential applications of DBS to improve motor function, mood, memory, and arousal in TBI patients. Major challenges in developing these approaches include the identification of specific phenotypes of TBI patients whose symptoms are amenable to therapeutic DBS. TBI is not one disorder; rather, it is a heterogeneous mix of brain injuries and associated deficits. There is increasing recognition of the importance in recognizing brain network abnormalities related to altered functional and structural abnormalities in TBI patients. The expanding potential to modulate functional brain networks with spatially precise neuromodulation therapies gives hope to millions of patients to target functional deficits in a patient-specific fashion.

REFERENCES

1. Lu, J. et al. 2013. Traumatic brain injury: Methodological approaches to estimate health and economic outcomes. *J Neurotrauma* 30:1925–1933.

2. Bazarian, J.J. et al. 2009. Long-term neurologic outcomes after traumatic brain injury. *J Head Trauma Rehabil* 24:439–451.
3. George, M.S. et al. 1995. Daily repetitive transcranial magnetic stimulation (rTMS) improves mood in depression. *Neuroreport* 6:1853–1856.
4. Fitzgerald, P.B., Fountain, S., Daskalakis, Z.J. 2006. A comprehensive review of the effects of rTMS on motor cortical excitability and inhibition. *Clin Neurophysiol* 117:2584–2596.
5. Kwon, T.G. et al. 2014. Effective method of combining rTMS and motor training in stroke patients. *Restor Neurol Neurosci* 32:223–232.
6. Kleim, J.A. et al. 2003. Motor cortex stimulation enhances motor recovery and reduces peri-infarct dysfunction following ischemic insult. *Neurolog Res* 25:789–793.
7. Plautz, E.J. et al. 2003. Post-infarct cortical plasticity and behavioral recovery using concurrent cortical stimulation and rehabilitative training: A feasibility study in primates. *Neurolog Res* 25:801–810.
8. Carmel, J.B. et al. 2013. Motor cortex electrical stimulation promotes axon outgrowth to brain stem and spinal targets that control the forelimb impaired by unilateral corticospinal injury. *European J Neurosci* 37:1090–1102.
9. Brus-Ramer, M. et al. 2007. Electrical stimulation of spared corticospinal axons augments connections with ipsilateral spinal motor circuits after injury. *J Neurosci: Official J Soc Neurosci* 27:13793–13801.
10. Carmel, J.B., Martin, J.H. 2014. Motor cortex electrical stimulation augments sprouting of the corticospinal tract and promotes recovery of motor function. *Frontiers Integrative Neurosci* 8, 51.
11. Suzuki, K. et al. 2012. Comparison of the after-effects of transcranial direct current stimulation over the motor cortex in patients with stroke and healthy volunteers. *Int J Neurosci* 122:675–681.
12. Allison, T. et al. 1991. Potentials-evoked in human and monkey cerebral-cortex by stimulation of the median nerve—A review of scalp and intracranial recordings. *Brain* 114:2465–2503.
13. Kujirai, T. et al. 1993. Corticocortical inhibition in human motor cortex. *J Physiol-Lond* 471:501–519.
14. Bolognini, N., Maravita, A. 2007. Proprioceptive alignment of visual and somatosensory maps in the posterior parietal cortex. *Curr Biol* 17:1890–1895.
15. Luber, B., Lisanby, S.H. 2014. Enhancement of human cognitive performance using transcranial magnetic stimulation (TMS) *Neuroimage* 2014; 85 Pt 3:961–970.
16. Gonzalez-Garcia, N. et al. 2011. Effects of rTMS on Parkinson's disease: A longitudinal fMRI study. *J Neurol* 258:1268–1280.
17. Foote, K.D., Okun, M.S. 2005. Ventralis intermedius plus ventralis oralis anterior and posterior deep brain stimulation for posttraumatic Holmes tremor: Two leads may be better than one: Technical note. *Neurosurgery* 56(2 Suppl):E445.
18. Issar, N.M., Hedera, P., Phibbs, F.T., Konrad, P.E., Neimat, J.S. 2013. Treating posttraumatic tremor with deep brain stimulation: Report of five cases. *Parkinsonism Relat Disord.* 19(12):1100–1105.
19. Nayernouri, T. 1985. Posttraumatic parkinsonism. *Surg Neurol* 24:263–264.
20. Krauss, J.K. et al. 1994. The treatment of posttraumatic tremor by stereotaxic surgery—Symptomatic and functional outcome in a series of 35 patients. *J Neurosurg* 80:810–819.
21. Loher, T.J. et al. 2000. Long-term follow-up study of chronic globus pallidus internus stimulation for posttraumatic hemidystonia—Case report. *J Neurosurg* 92:457–460.
22. Krauss, J.K., Jankovic, J. 2002. Head injury and posttraumatic movement disorders. *Neurosurgery* 50:927–939.

23. Kim, J.P., Chang, W.S., Chang, J.W. 2012. The long-term surgical outcomes of secondary hemidystonia associated with post-traumatic brain injury. *Acta Neurochir (Wien)* 154(5):823–830.
24. Giacino, J. et al. 2012. Central thalamic deep brain stimulation to promote recovery from chronic posttraumatic minimally conscious state: Challenges and opportunities. *Neuromodulation* 15:339–349.
25. Yamamoto, T., Kobayashi, K., Kasai, M., Oshima, H., Fukaya, C., Katayama, Y. 2005. DBS therapy for the vegetative state and minimally conscious state. *Acta Neurochir Suppl* 93:101–104.
26. Schiff, N.D., Giacino, J.T., Kalmar, K. et al. 2007. Behavioural improvements with thalamic stimulation after severe traumatic brain injury. *Nature* 448(7153):600–603.
27. Mayberg, H.S. et al. 2005. Deep brain stimulation for treatment-resistant depression. *Neuron* 45:651–660.
28. Kumar, R. et al. 1998. Double-blind evaluation of subthalamic nucleus deep brain stimulation in advanced Parkinson's disease. *Neurology* 51:850–855.
29. Rodriguez-Oroz, M.C. et al. 2005. Bilateral deep brain stimulation in Parkinson's disease: A multicentre study with 4 years follow-up. *Brain* 128:2240–2249.
30. Weaver, F.M. et al. 2009. Bilateral deep brain stimulation vs. best medical therapy for patients with advanced Parkinson disease: A randomized controlled trial. *JAMA* 301:63–73.
31. Bowman, E.S. 1993. Etiology and clinical course of pseudoseizures—Relationship to trauma, depression, and dissociation. *Psychosomatics* 34:333–342.
32. Temkin, N.R. et al. 1999. Valproate therapy for prevention of posttraumatic seizures: A randomized trial. *J Neurosurg* 91:593–600.
33. Golarai, G. et al. 2001. Physiological and structural evidence for hippocampal involvement in persistent seizure susceptibility after traumatic brain injury. *J Neurosci* 21:8523–8537.
34. Temkin, N.R. 2001. Antiepileptogenesis and seizure prevention trials with antiepileptic drugs: Meta-analysis of controlled trials. *Epilepsia* 42:515–524.

18 Enhanced Functional Outcome from Traumatic Brain Injury with Brain–Machine Interface Neuromodulation

Neuroprosthetic Scaling in Relation to Injury Severity

Dennis A. Turner

CONTENTS

INTRODUCTION

There are a number of residual issues patients face after experiencing even mild closed head injury, as well as moderate and severe injuries, including changes in memory, attention, cognitive processing, and consciousness.[1–3] It is believed that neuroprosthetic devices may both facilitate recovery from the basic head injury (i.e., to help the brain heal)[4–6] as well as synchronize and activate circuits that are deficient or impaired from damage caused by the head injury.[7–9] We will discuss clinical problems associated with traumatic brain injury, depending on severity, and how brain–machine interface (BMI) approaches may help recovery or function, through a combination of brain recording and stimulation systems. The core concept of a brain–machine interface would be to measure the "intent" of the brain to perform an action, such as memory retrieval, motion, or perhaps focusing on an activity.[10,11] Neuroprosthetic devices could then link the measured "intent" to an internal or external cue or device to facilitate the action. The last link in effective brain control systems and brain–machine interface is an automatic feedback to enhance and fine-tune performance on the task. Similar to intrinsic motor function, for example, where "intent" is defined, then a motor plan has to be internally created (i.e., by the basal ganglia and thalamus), then smoothly executed (by motor cortex, cerebellum and associated motor circuitry), and refinement via external vision or sensory input is critical for improving the function.[11] We discuss several possible treatment routes relative to each level of severity of head injury, and how a feedback, control circuit might be implemented, and critical approaches now available as well as under development.

As a parallel to traumatic brain injury, movement disorders treatment has evolved to include a number of regions within the brain, focusing on pathways involved with both motor pattern generation (i.e., basal ganglia in concert with cortical areas involved with generating intent), motor output (i.e., motor cortex), and sensory feedback.[12–15] As these areas have been further studied over time, additional subregions have been included as treatment possibilities, for example, the subthalamic nucleus, a new target derived from motor systems analysis in nonhuman primates, but rapidly translated into clinical usefulness.[13,15,16] There are now at least three interconnected regions that appear to be involved in treatment of abnormal motion and for application of conventional deep brain stimulation (DBS). Interestingly, DBS applied to the three interconnected regions (i.e., globus pallidus, subthalamic nucleus, and ventral intermediate thalamus) provide different types of symptom relief with some overlap. DBS provides essentially a point source of electrical input into the brain (i.e., a 1.5 mm contact point; Figure 18.1) but can affect widespread regions through both direct local neuronal changes but particularly direct stimulation of afferent and efferent axons coursing through each area.[12–15] In effect, the point source stimulation that DBS provides can reset nearly the entire brain within a short period of time after activation, showing that dynamic stimulation can have widespread effects (as shown in Figure 18.1). There are multiple efforts ongoing to optimize DBS treatment effects, including improving stimulation patterns, measuring widespread EEG changes (i.e., beta oscillations) and using these as a feedback system to monitor treatment efficacy, and to transform DBS systems into "smart" neuroprosthetic systems, along the lines of brain–machine interface approaches.

A natural extension of DBS for movement disorders would then be to optimize both the number of stimulation sites as well as coordinate feedback control, to extend DBS toward a full implementation of brain–machine interface. For example, detecting surrogate signals within the brain (either locally or globally) that might be linked to the movement disorder symptoms (i.e., tremor, rigidity, bradykinesia, etc.), analyzing these signals in real-time, and then altering the DBS output to dynamically route stimulation to both multiple sites and in various patterns to improve the disease symptoms. It is not clear if single or multiple sites of DBS stimulation might be more

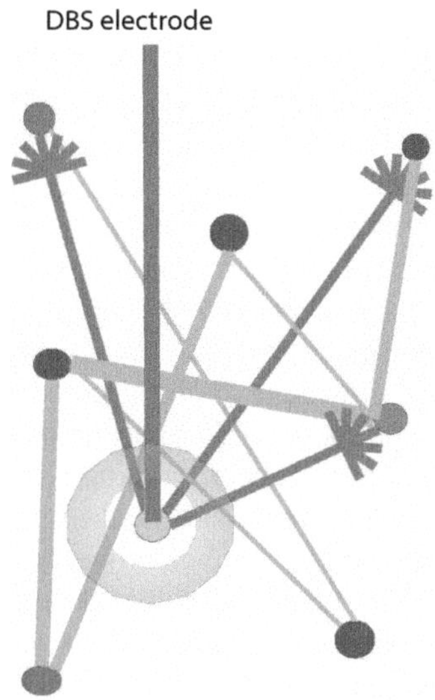

FIGURE 18.1 This diagram shows a simplified brain network, exemplified by a small number of "nodes" with interconnections of various strengths. A deep brain stimulating (DBS) electrode has been inserted close to a node, with stimulation effects extending outward from the region of stimulation. Since the DBS electrode is at a particularly defined location, then network elements projecting from that location are activated, indicating how location can introduce a high degree of specificity to possible treatment routes.

effective, but more advanced tools are needed for individual patients to assess (in a predictive sense) where to place electrodes and how to stimulate. Similarly, a revolution in epilepsy treatment is ongoing to dynamically record and stimulate seizures, with now a first-generation Neuropace device available (with four recording and four stimulating channels).[17–19] The location and predictive efficacy of sites and types of stimulation for both movement disorders and epilepsy is a current strong need, with sufficient brain detail and knowledge of the underlying circuitry to be useful for placing electrodes.

GRAPHICAL NETWORK ANALYSIS OF BRAIN FUNCTION

Understanding of functional networks within the brain requires integration of all available brain anatomical and physiological studies, including magnetic resonance imaging (MRI), functional MRI (fMRI, typically based on venous blood oxygen levels), diffusion tensor imaging MRI (DTI MRI), and dynamic information, such as scalp electroencephalogram (EEG), electrocorticogram from direct recordings on the brain surface (ECoG), and depth electrode recordings (from specific short-term electrodes or DBS long-term electrodes). One scheme to integrate these various data hinges on identifying structural regions or "nodes" on MRI + fMRI, then adding further physiological data as relevant to each node or small region, for a dynamic approach, termed graphical network analysis (Figure 18.1).[18,20–23]

For short periods of time, MRI or EEG analysis of the brain can demonstrate coordination and a low level of synchronized activity between various areas of the brain, likely due to close interconnections anatomically and physiologically, even when the patient is intentionally unfocused on any specific task, "resting state analysis."[21,24] This background activity has been termed resting or default mode network but is actually complex as to how the brain maintains a functional background, on which tasks may then build and alter the network. In contrast, performance of a specific cognitive, memory, sensory, or movement task can be defined by subtracting out the default mode network, outlining specific activation regions.[25]

For many years there have presumed to be "circuits" within the brain underlying specific functions. For example, a detailed analysis of motor pathways in patients has been performed using repetitive real or imaginary motion (i.e., of the finger).[26,27] These detailed motor circuits have been analyzed to include premotor and parietal cortices (for intent to act or move), the basal ganglia, thalamus, and finally motor cortex, for actual movement initiation and performance. If a movement actually occurs (i.e., a real rather than an "imaginary" motion), then the cerebellum performs feedback monitoring of the actual motor pattern through dense proprioceptive input into the cerebellar cortex. In some other cases, detailed circuits or regions of the brain are typically invoked in a specific sequence, to provide a "signature" of that particular function, which occurs via enhanced connectivity and timing between brain regions. To parcellate brain regions on imaging studies, Brodman's areas are often highlighted, which are based on detailed morphological analysis of human brain and cell pattern differences between regions. Since there are many such areas, then an MRI scan can be dissected into these regions, and the timing between regions ascertained by assessing functional blood flow changes; anatomical connectivity can be evaluated via DTI to obtain preferential direction of axonal connectivity.[18,21,28]

As in the motor circuits, these specific circuits and regions of the brain involved in a functional network have been determined usually from a bottom-up approach, where areas known to be involved with movement generation may be lesioned or apparent in various disease states (i.e., Parkinson's disease), or involved with stroke, and then the pattern of abnormal motion noted.[12,15,16] However, a top-down approach allows every brain region on the MRI to be independent, then the regions involved or synchronized during detailed brain MRI scans may be highlighted solely on the base of activity. Large-scale top-down network models can be independent of underlying known brain circuits, hence they can be very revealing in terms of unexpected areas of the brain involved in task generation, highlighting that multiple regions are coactive even in limited tasks, forming highly dynamic networks underlying function. This "neutral" top-down approach to brain function has, in many instances, confirmed known bottom-up networks known from disease or lesion studies, but has added a new dimension in terms of widespread interactions outside of known areas of involvement.[20,29–31]

A large goal of graphical network analysis has been to distill specific regions dynamically involved with brain functions by focusing on the limited nodes involved and the connectivity between nodes, based on MRI/fMRI/DTI MRI and further physiological data as available. However, a large limitation of these studies is their short duration, usually only during a 30–60 min MRI session, and in many cases provocative symptoms are too complex or short-lived to be effectively imaged or evoked during this short time. Thus, the addition of high-density physiological information will be critical to add to the MRI network framework to allow a dynamic approach over time. Physiological information could include widespread scalp EEG, ECoG from cortical recordings, and/or depth electrodes.[21,30,32,33] The further critical extension of these complex data sets will be to analyze, in a predictive sense, how treatment inputs could affect the network circuitry. For example, there are now attempts to build large-scale models of how DBS influences motor pathways, including time-dependent properties, but recordings of sufficient density to validate these models have not been feasible, nor have they been positioned within the MRI framework to allow realistic determination of location.

Thus, what is needed is an anatomically accurate MRI model that can accommodate both physiological data in multiple relevant locations and modeling predictions, possibly from a bottom-up data set. The modeling predictions can then include the effect of single or multiple inputs placed into specific locations, such as DBS or Neuropace electrodes, and how the network can be influenced. To be realistic, stimulation parameters will also need to be included as well as location. This type of modeling could then be used to optimize the number of electrodes placed, for example, to enhance connectivity between various regions, with cooperativity between stimulation sites defined (i.e., alternating vs. intermittent stimulation), as well as appropriate recording sites for determination of the specific effect the stimulation provides to the network. In this way the model could include modulation of the stimulation pattern, location, and interactions to fine-tune a network into a desired target range, presumably correlated with symptom relief for the specific disease. This type of model remains highly visionary but critically needed for extending DBS or other techniques of activating/modulating brain to a more effective level, and to further diseases beyond movement disorders.

TBI: OPPORTUNITIES TO INTERVENE

The severity of TBI ranges from mild (i.e., no permanent structural sequelae) to severe, typically as estimated from the Glasgow Coma Score (GCS), which is useful for large studies on outcomes and treatment of TBI.[1,34,35] The symptoms referable to each level of severity vary, hence treatment aimed at symptom relief and/or enhancement of recovery may vary in terms of how invasive the treatment may need to be, how many areas of brain activity should be modulated, and how constant the treatment should be active. We briefly review the common problems associated with the different forms of head injury and how specific issues for each form may be resolved via a neuroprosthetics and/or brain–machine interface approach. Fortunately, MRI and graphical network analysis is becoming increasingly used in TBI analysis, facilitating the understanding of how treatment could potentially be implemented.[23,32,36,37] In some cases of mild TBI, graphical network analysis (or a scalp EEG version) may be very helpful simply to detect and alert the presence of a head injury, for example, in a military sense, since even a mild head injury may transiently affect judgment and performance on complex tasks for a period. Thus, markers of TBI have been sought to improve detection and analyze severity of TBI.

MILD TBI

Mild closed head injury is usually defined as GCS 13–15 with only transient loss of neurological function and no lesions noted on imaging studies (i.e., brain CT).[3] In this situation there is apparent complete recovery from an initial mild injury. However, in many cases there can be persistent cognitive changes, including frontal lobe function, attention, and a heavy memory load, as in studying for courses and tests.[3] This is particularly true if the patient is exposed to multiple concussions with a critical window of time, so that there are now guidelines on when a patient can return to sports.[35,38,39] Even if the patient is intact neurologically and there is no obvious evidence of persistent injury,[38,40,41] symptoms such as depression and attention are accompanied by changes even in the default mode resting network.[21,37,42] Many of these symptoms may improve over time, though issues such as memory may linger, particularly if there are multiple TBI occurrences, with less than adequate recovery between each.[38,40,41] For example, this is now a large issue in sports concussions, where insufficient recovery after even mild injuries may lead to prominent symptoms.

These patients in some cases have persisting symptoms that are bothersome but are not likely to lead to sufficient issues for the patient to consider an invasive system, such as DBS, to even partially treat their symptoms. Further, symptoms may be episodic and associated with specific tasks, such as studying, which require more intense concentration and memory function. In these specific instances it may be helpful to have some form of enhancement to overcome network issues, but perhaps a temporary, extracranial device may be most useful, though applied in a BMI context (Figure 18.2). For example, there are now many extracranial scalp EEG devices available, though the recording fidelity is typically very suspect since electrodes are often applied over hair in an inconsistent fashion. However, newer devices may overcome some of the limitation of scalp EEG through improved electrode design. For example, the Emotiv device offers a headset with multiple electrodes and excellent connectivity (http://www.emotiv.com/apps/epoc/299/), but does not offer any control over brain circuitry. However, there has been considerable development insight into small applied constant and alternating currents injected across the skull and how these currents may influence brain function in a subtle sense.[8,9,43–45] What is missing from these studies is an appreciation of how subtle interference across the skull may actually influence function, and having continuous scalp EEG may help fine-tune such studies. This type of external system may be advantageous (and not too cumbersome) for specific activities such as studying, where incorporation and recall of new material may be improved to a degree (perhaps mild). Though this is possible with scalp stimulation/recording (as shown in Figure 18.2), this could also be achieved in a more long-term and stable sense with direct cortical electrodes (i.e., ECoG) and direct cortical stimulation. The advantage of an implant is that the function is constant from day to day (since no reapplication is needed) and once trained, the functional improvement may be consistent also.

Thus, for mild TBI the optimal form of BMI implementation may be intermittent, external headgear, but with both a stimulation component and a recording component, with an intelligent processing system to assess how overall brain networks are

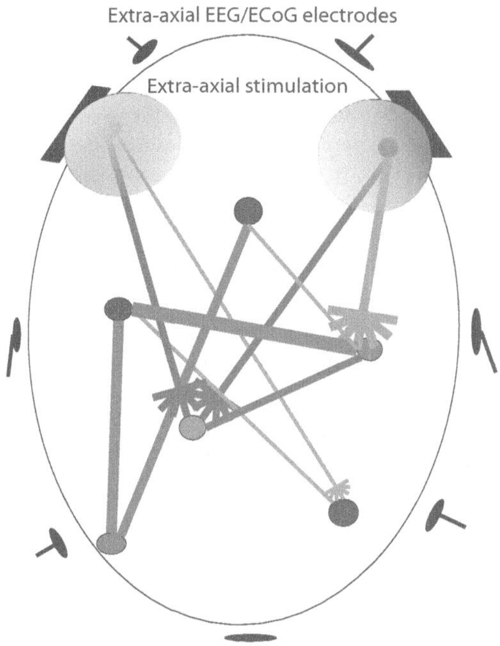

FIGURE 18.2 Extra-axial recording electrodes (scalp electroencephalogram [EEG], epidural, or subdural electrocorticogram [ECoG], outside of the brain parenchyma) together with extra-axial brain stimulation (scalp transcranial stimulation, epidural or direct cortical stimulation). Both recording and stimulation systems have been implemented but not in a closed-loop or feedback fashion together to manipulate the intracranial network. This type of approach is much less specific than a direct brain stimulation approach, as in Figure 18.1.

being modulated, and to dynamically adjust the transcranial stimulation to reach a desired endpoint or network status. This type of system should be studied in a rigorous sense for its ability to enhance everyday activities. In contrast, DBS systems implanted near the fornix may enhance memory through direct septal/hippocampal circuit stimulation, as a control for enhancing memory.[46]

MODERATE TBI

Moderate TBI (GCS 9–12) may demonstrate a mixture of both white matter and gray matter injury to the point of a surgical lesion sometimes present and usually an abnormal MRI scan (i.e., extra-axial hematoma, contusions, etc.).[33,34,47] After this type of injury the patient may have some permanent sequelae, including seizures (from the cerebral contusion component), memory loss, depression, difficulties with attention, and frontal lobe deficits.[34] Analysis of brain networks may show more severe disruption of connectivity between various regions, affecting a variety of functions.[21,48] For example, memory formation and retrieval is a complex function of the brain, involving attention to the underlying incorporation of memory, storage of a temporary form and then eventual more permanent storage, as well as repetition and emotional coloring, which can enhance memory storage (or at least reduce its disappearance).[25] Memory alterations in TBI may vary as a function of cortical damage, white matter connectivity between regions, and hippocampal integrity.[33] Generally, in moderate TBI the patient is alert (i.e., no coma initially) though perhaps confused from the multiple regions of brain injury, so the connectivity relating to functional activities is the major issue.

In many of these patients there could be a two-phase investigation parallel to epilepsy evaluation. In severe refractory epilepsy, an initial evaluation with scalp EEG, video analysis of events in comparison to the EEG, and perhaps invasive (but temporary) recordings are performed. The goal of these recordings is to capture patient-specific events and to analyze the locations of the abnormal networks that are involved in the generation of these events. The equivalent in head injury would be MRI and then video EEG (or more invasive depth/ECoG recordings) to identify deficits with provocation, such as memory abnormalities, to localize where in the networks the connectivity is altered. Instead of positive phenomena such as seizures, TBI patients may show predominantly negative events, such as lack of memory formation, but still due to network pathology. In epilepsy, the second phase would then be an intervention, such as a surgical procedure to remove (i.e., temporal lobectomy), disconnect (i.e., corpus callosotomy), or stimulate to abort a preseizure state (i.e., Neuropace focal demand stimulation), in general reducing synchrony between various parts of the brain, dampening connectivity. In TBI the opposite may be needed, i.e., multisite stimulation to enhance coordination and synchrony that has been lost or reduced by the injury, particularly with white matter lesions.[4]

The type of system that would be needed would be a focused, multisite recording and stimulation system. For enhanced memory, for example, a number of approaches could be used. A generalized approach to improving memory could be simple one-site DBS with septal/fornix stimulation to improve overall cholinergic tone in the memory acquisition systems; this type of DBS is now being tested, for example, in Alzheimer's patients with focal memory deficits.[46] However, in abnormal connectivity associated with moderate to severe TBI this may be insufficient since other important pathways are deficient. Thus, a multisite recording and stimulating system could be developed, providing a cooperative connectivity between brain regions and substituting natural connections with electrical connections, for example, if a frontal lobe site is disconnected, but to facilitate attention and timing this could be stimulating in a specific pattern during incorporation of a memory, triggered by a hippocampal site[25] with the goal of improving overall network connectivity. The location of the multisite recording-stimulation system would be set for the patient's specific symptoms, similar to an epilepsy case. A simple immediate solution would be the Neuropace device, which is currently set up for demand stimulation based on EEG input, with a pattern of enhanced synchrony identified, then stimulation triggered. The stimulation could be triggered with a specific pattern of activity in the hippocampus, for example, via a recording electrode, with a frontal lobe electrode then activated to enhance the memory. A longer term solution may require multiple additional recording and stimulation sites, such as 50–100, with much more complex processing, as recently proposed in the DARPA SUBNETs program. Such types of systems do not currently exist but could be developed, for example, as a second- or third-generation Neuropace device.[45]

Thus, in moderate TBI there may be sufficient long-lasting symptoms (as in epilepsy) that an initial evaluation session may be fruitful, followed by some form of patient-specific intervention to enhance connectivity for specific symptom reduction. This intervention should incorporate the principles of a complex BMI, with multiple recording sites, using depth or electrocorticogram electrodes for long term stability, a processing unit to evaluate demand through processing complex signals in a network sense, then multiple stimulating sites, coordinated and cooperatively activated, to achieve a long-term closed loop solution based on the dynamic recordings.[10]

SEVERE TBI

In these injuries the patient is severely affected by the TBI, with GCS in the 3–8 range, and usually either comatose or with minimal interaction with the environment.[22,34,49] In this case the primary symptom to be treated may simply be arousal or sufficient improvement in consciousness to allow more self care, rather than improving function in society as with mild to moderate TBI.[50,51] Consciousness requires an intact reticular formation (i.e., midbrain and thalamic) as well as hemispheric function.[49,52,53] Schiff et al.[54] have further refined which structures are intact and which may be stimulated (i.e., with DBS) to enhance disorders of consciousness.[51,54–56] Likewise, neuromodulation approaches to enhance consciousness have been tried in human patients with variable effects, particularly medial thalamic stimulation to enhance arousal[4,7,57] but also cortical stimulation[58] and transcranial stimulation.[59] However, this approach has not generalized since it has proven very difficult to identify appropriate candidates for DBS treatment, and even in TBI patients to know their level of consciousness.[49] Though a few cases have shown benefit from medial thalamic DBS to enhance awareness,[54] further trials will be needed, such as an ongoing long-term trial (Rezai trial, NCT01377952 on clinical trials.gov, no results posted).

For example, use of MRI and scalp EEG may be helpful in identifying modulation of underlying cerebral networks in response to external stimuli to identify levels of consciousness.[51–53,56] If a patient with severe TBI recovers sufficiently over time, then the approach described in moderate TBI with a focused, patient-symptom-specific intervention may be tailored. However, if the patient with severe TBI remains with a lowered level of arousal or unconsciousness, then neuromodulation approaches to activate arousal networks may be important. Theoretically, this could be approached with depth electrode monitoring of multiple network sites together with MRI, particularly in thalamic and midbrain reticular locations, but this has not yet been performed. Likewise, external provocation to define alertness may be helpful. The graphical network approach and then interventions based upon critical nodes may be very helpful in identifying both the underlying pathological circuitry as well as how to stimulate in specific patients.[49,56,60] This multifaceted approach may help define critical, abnormal nodes in consciousness (beyond medial thalamus) that might be stimulated in a cooperative approach using internal electrodes.

BRAIN–MACHINE INTERFACE TO ENHANCE RECOVERY

Most of the current approaches to treatment of TBI assume that the patient is at a static level of functioning, and intervention may be focused to treat specific, residual symptoms arising from an injury (i.e., memory loss). However, chronic brain–machine interfaces show that the underlying network can be modified over time, particularly in relation to feedback, including both neuronal functional changes as well as network alterations.[11] As in stroke, TBI involves a gradual rewiring and significant changes in brain function as compensatory circuits take over.[61–64] The entire recovery process is in itself dynamic and can be significantly altered by external events. For example, many of the known anticonvulsants affecting action potentials (for example, phenytoin, carbamazepine, etc.) also significantly retard stroke and TBI recovery, due to overlap with developing axonal connections, which require Na+-channel bursting for pathfinding.[65] In stroke an occasional seizure may actually enhance recovery, due to the known effects of seizures on inducing axonal sprouting, a critical process in brain rewiring. Further, studies in stroke recovery have shown that even after stable recovery, stroke-induced loss of function can still be modified, as demonstrated in the Northstar cortical stimulation trial.[62] This trial paired cortical stimulation of residual, functioning motor cortex (or associated areas) with rehabilitation approaches to enhance functional recovery in a paretic limb. Thus, functional use combined with stimulation may improve late outcome in TBI as well as stroke.

However, stimulation early after TBI may also accelerate recovery, possibly through stimulation of axonal sprouting and enhanced connectivity, once the initial injury is resolving (i.e., after 7–10 days).[2,5,6,8,9,36,66] Because brain–machine interfaces involve constant interplay between outputs from the brain and inputs with brain activation, these approaches may enhance circuitry modification over time, taking advantage of inherent plasticity.[26,45] Thus, in addition to treating specific symptoms referable to TBI, an early BMI approach to specific areas of pathology may enhance connectivity and rewiring through direct axon stimulation, a highly novel approach, since axonal recovery depends in large part upon activity-dependent mechanisms.

A large area of rehabilitation has built up around this core concept of activity-induced recovery, but it has been very difficult to show that external rehabilitation in particular can facilitate recovery of function, over and beyond that which the brain can inherently demonstrate from innate mechanisms.[36] In contrast, direct brain stimulation may show a much higher efficacy for functional recovery over time through direct enhancement of axonal recovery in terms of brain rewiring.[2,5,9] Though TBI and stroke recovery may differ in some ways, particularly the heterogeneity of TBI mechanisms of injury and specific areas of damage in comparison to stroke, many of the principles of rewiring the brain may be relevant for both areas.

SUMMARY

There may be multiple roles for deployment of BMI technology in TBI, depending on the specific purpose needed and particularly symptoms to be treated. These may vary from simply detecting and alerting the presence of even mild TBI, to treating residual mild TBI symptoms, to more complex moderate and severe TBI issues. Since there are currently no invasive approaches to TBI investigation or treatment, it is unclear what may be ethically appropriate and tolerated.[51] Thus, the invasiveness of the approach (i.e., noninvasive scalp EEG and stimulation vs. invasive depth/cortical recordings/stimulation) should be scaled for the severity of the symptom. This has proven problematic for treatment of minimally conscious states, for example, where it remains very murky as to who may benefit from such stimulation and whether it should be more widely disseminated.[50] However, we now have available a continuum of noninvasive and invasive devices for potential treatment of TBI, beginning with scalp EEG-based BMI paired with noninvasive brain stimulation[8,59] for potentially contingent stimulation, more advanced epilepsy-type approaches and possible demand stimulation to enhance connectivity (Neuropace), and the application of both current DBS systems or smarter new DBS systems (with recording capability) for more severe TBI states, particularly if disorders of consciousness can be improved. Further, BMI and neuromodulation approaches may enhance recovery from TBI and stroke, by directly enhancing axonal sprouting and other mechanisms of neural compensation, as well as invoking and modulating inherent neural recovery schemes.

The main limitations to the application of these approaches is defining the symptoms to be treated, how to optimize predictive network function to surgically localize where to place electrodes, and dynamic analysis of circuitry to extend graphical network analysis to a larger range of states, including provocative states where clear dysfunction is present (i.e., memory tasks). All of these comprehensive approaches may likewise parallel new, advanced epilepsy approaches and may eventually feed back to a multitude of other, less defined CNS diseases where the type of pathological network alteration is unknown. For example, psychiatric diseases may be primarily abnormal network phenomena, which could be characterized and treated with similar approaches.

REFERENCES

1. Maas, A.I. et al. 2013. Advancing care for traumatic brain injury: Findings from the IMPACT studies and perspectives on future research. *Lancet Neurol* 12(12):1200–1210.
2. Nudo, R.J. 2013. Recovery after brain injury: Mechanisms and principles. *Front Hum Neurosci* 7:887.
3. Kristman, V.L. et al. 2014. Methodological issues and research recommendations for prognosis after mild traumatic brain injury: Results of the international collaboration on mild traumatic brain injury prognosis. *Arch Phys Med Rehabil* 95(3 Suppl):S265–S277.
4. Schiff, N.D., Plum, F., Rezai, A.R. 2002. Developing prosthetics to treat cognitive disabilities resulting from acquired brain injuries. *Neurol Res* 24(2):116–124.
5. De, F.P. et al. 2009. The new neuroscience frontier: Promoting neuroplasticity and brain repair in traumatic brain injury. *Clin Neuropsychol* 23(8):1391–1399.
6. Dancause, N., Nudo, R.J. 2011. Shaping plasticity to enhance recovery after injury. *Prog Brain Res* 192:273–295.
7. Schiff, N.D., Rezai, A.R., Plum, F. 2000. A neuromodulation strategy for rational therapy of complex brain injury states. *Neurol Res* 22(3):267–272.
8. Villamar, M.F. et al. 2012. Noninvasive brain stimulation to modulate neuroplasticity in traumatic brain injury. *Neuromodulation* 15(4):326–338.
9. Guggenmos, D.J. et al. 2013. Restoration of function after brain damage using a neural prosthesis. *Proc Natl Acad Sci U S A* 110(52):21177–21182.

10. Patil, P.G., Turner, D.A. 2008. The development of brain–machine interface neuroprosthetic devices. *Neurotherapeutics* 5(1):137–146.
11. Nicolelis, M.A., Lebedev, M.A. 2009. Principles of neural ensemble physiology underlying the operation of brain-machine interfaces. *Nat Rev Neurosci* 10(7):530–540.
12. Schroeder, U. et al. 2003. Subthalamic nucleus stimulation affects a frontotemporal network: A PET study. *Ann Neurol* 54(4):445–450.
13. Grafton, S.T. et al. 2006. Normalizing motor-related brain activity: Subthalamic nucleus stimulation in Parkinson disease. *Neurology* 66(8):1192–1199.
14. Geday, J. et al. 2009. STN-stimulation in Parkinson's disease restores striatal inhibition of thalamocortical projection. *Hum Brain Mapp* 30(1):112–121.
15. Albaugh, D.L., Shih, Y.Y. 2014. Neural circuit modulation during deep brain stimulation at the subthalamic nucleus for Parkinson's disease: What have we learned from neuroimaging studies? *Brain Connect* 4(1):1–14.
16. Bergman, H., Wichmann, T., DeLong, M.R. 1990. Reversal of experimental parkinsonism by lesions of the subthalamic nucleus. *Science* 249(4975):1436–1438.
17. Morrell, M.J., RNS System in Epilepsy Study Group. 2011. Responsive cortical stimulation for the treatment of medically intractable partial epilepsy. *Neurology* 77(13):1295–1304.
18. Caeyenberghs, K. et al. 2013. Topological correlations of structural and functional networks in patients with traumatic brain injury. *Front Hum Neurosci* 7:726.
19. Irimia, A., Van Horn, J.D. 2014. Systematic network lesioning reveals the core white matter scaffold of the human brain. *Front Hum Neurosci* 8:51.
20. Sporns, O. 2012. From simple graphs to the connectome: Networks in neuroimaging. *NeuroImage* 62(2):881–886.
21. Stevens, M.C. et al. 2012. Multiple resting state network functional connectivity abnormalities in mild traumatic brain injury. *Brain Imaging Behav* 6(2):293–318.
22. Thibaut, A. et al. 2012. Metabolic activity in external and internal awareness networks in severely brain-damaged patients. *J Rehabil Med* 44(6):487–494.
23. Sharp, D.J., Scott, G., Leech, R. 2014. Network dysfunction after traumatic brain injury. *Nat Rev Neurol* 10(3):156–166.
24. Palacios, E.M. et al. 2013. Resting-state functional magnetic resonance imaging activity and connectivity and cognitive outcome in traumatic brain injury. *JAMA Neurol* 70(7):845–851.
25. Watrous, A.J. et al. 2013. Frequency-specific network connectivity increases underlie accurate spatiotemporal memory retrieval. *Nat Neurosci* 16(3):349–356.
26. Rebesco, J.M., Miller, L.E. 2011. Stimulus-driven changes in sensorimotor behavior and neuronal functional connectivity application to brain-machine interfaces and neurorehabilitation. *Prog Brain Res* 192:83–102.
27. Zhou, Y. et al. 2013. Characterization of thalamo-cortical association using amplitude and connectivity of functional MRI in mild traumatic brain injury. *J Magn Reson Imaging* 39(6):1558–1568.
28. Pandit, A.S. et al. 2013. Traumatic brain injury impairs small-world topology. *Neurology* 80(20):1826–1833.
29. Sporns, O. 2013. Network attributes for segregation and integration in the human brain. *Curr Opin Neurobiol* 23(2):162–171.
30. van den Heuvel, M.P., Sporns, O. 2013. Network hubs in the human brain. *Trends Cogn Sci* 17(12):683–696.
31. Sporns, O. 2014. Contributions and challenges for network models in cognitive neuroscience. *Nat Neurosci* 17:652–660.
32. Karmonik, C. et al. 2013. Comparison of functional network integrity in TBI and orthopedic control patients using graph-theoretical analysis. *Conf Proc IEEE Eng Med Biol Soc* 2013:1366–1369.

33. Palacios, E.M. et al. 2013. Long-term declarative memory deficits in diffuse TBI: Correlations with cortical thickness, white matter integrity and hippocampal volume. *Cortex* 49(3):646–657.
34. Frattalone, A.R., Ling, G.S. 2013. Moderate and severe traumatic brain injury: Pathophysiology and management. *Neurosurg Clin N Am* 24(3):309–319.
35. Harmon, K.G. et al. 2013. American Medical Society for Sports Medicine position statement: Concussion in sport. *Br J Sports Med* 47(1):15–26.
36. Ham, T.E., Sharp, D.J. 2012. How can investigation of network function inform rehabilitation after traumatic brain injury? *Curr Opin Neurol* 25(6):662–669.
37. Zhou, Y. et al. 2012. Default-mode network disruption in mild traumatic brain injury. *Radiology* 265(3):882–892.
38. Meehan, W.P., 3rd et al. 2012. Increasing recovery time between injuries improves cognitive outcome after repetitive mild concussive brain injuries in mice. *Neurosurgery* 71(4):885–891.
39. Bryan, C.J. 2013. Repetitive traumatic brain injury (or concussion) increases severity of sleep disturbance among deployed military personnel. *Sleep* 36(6):941–946.
40. Signoretti, S. et al. 2011. The pathophysiology of concussion. *PM R* 3(10 Suppl 2):S359–S368.
41. Hylin, M.J. et al. 2013. Repeated mild closed head injury impairs short-term visuospatial memory and complex learning. *J Neurotrauma* 30(9):716–726.
42. Tang, L. et al. 2011. Thalamic resting-state functional networks: Disruption in patients with mild traumatic brain injury. *Radiology* 260(3):831–840.
43. Paulus, W. 2011. Transcranial electrical stimulation (tES - tDCS; tRNS, tACS) methods. *Neuropsychol Rehabil* 21(5):602–617.
44. Demirtas-Tatlidede, A. et al. 2012. Noninvasive brain stimulation in traumatic brain injury. *J Head Trauma Rehabil* 27(4):274–292.
45. Gharabaghi, A. et al. 2014. Coupling brain-machine interfaces with cortical stimulation for brain-state dependent stimulation: Enhancing motor cortex excitability for neurorehabilitation. *Front Hum Neurosci* 8:122.
46. Laxton, A.W., Lipsman, N., Lozano, A.M. 2013. Deep brain stimulation for cognitive disorders. *Handb Clin Neurol* 116:307–311.
47. Tavazzi, B. et al. 2005. Cerebral oxidative stress and depression of energy metabolism correlate with severity of diffuse brain injury in rats. *Neurosurgery* 56(3):582–589; discussion 582–589.
48. Olsen, A. et al. 2015. Altered cognitive control activations after moderate-to-severe traumatic brain injury and their relationship to injury severity and everyday-life function. *Cereb Cortex* 25(8):2170–2180.
49. Di Perri, C. et al. 2014. Functional neuroanatomy of disorders of consciousness. *Epilepsy Behav* 30:28–32.
50. Glannon, W. 2008. Neurostimulation and the minimally conscious state. *Bioethics* 22(6):337–345.
51. Giacino, J.T. et al. 2014. Disorders of consciousness after acquired brain injury: The state of the science. *Nat Rev Neurol* 10(2):99–114.
52. Fernandez-Espejo, D. et al. 2008. Cerebral response to speech in vegetative and minimally conscious states after traumatic brain injury. *Brain Inj* 22(11):882–890.
53. Fernandez-Espejo, D. et al. 2012. A role for the default mode network in the bases of disorders of consciousness. *Ann Neurol* 72(3):335–343.
54. Schiff, N.D. et al. 2007. Behavioural improvements with thalamic stimulation after severe traumatic brain injury. *Nature* 448(7153):600–603.
55. Schiff, N.D. 2012. Moving toward a generalizable application of central thalamic deep brain stimulation for support of forebrain arousal regulation in the severely injured brain. *Ann N Y Acad Sci* 1265:56–68.

56. Schiff, N.D., Nauvel, T., Victor, J.D. 2014. Large-scale brain dynamics in disorders of consciousness. *Curr Opin Neurobiol* 25C:7–14.
57. Schiff, N.D. 2013. Central thalamic deep brain stimulation for support of forebrain arousal regulation in the minimally conscious state. *Handb Clin Neurol* 116:295–306.
58. Canavero, S. et al. 2009. Bifocal extradural cortical stimulation-induced recovery of consciousness in the permanent post-traumatic vegetative state. *J Neurol* 256(5):834–836.
59. Thibaut, A. et al. 2014. tDCS in patients with disorders of consciousness: Sham-controlled randomized double-blind study. *Neurology* 82(13):1112–1118.
60. Goldfine, A.M., Schiff, N.D. 2011. Consciousness: Its neurobiology and the major classes of impairment. *Neurol Clin* 29(4):723–737.
61. Hummel, F.C., Cohen, L.G. 2006. Non-invasive brain stimulation: A new strategy to improve neurorehabilitation after stroke? *Lancet Neurol* 5(8):708–712.
62. Levy, R. et al. 2008. Cortical stimulation for the rehabilitation of patients with hemiparetic stroke: A multicenter feasibility study of safety and efficacy. *J Neurosurg* 108(4):707–714.
63 Benowitz, L.I., Carmichael, S.T. 2010. Promoting axonal rewiring to improve outcome after stroke. *Neurobiol Dis* 37(2):259–266.
64. Edwardson, M.A. et al. 2013. New modalities of brain stimulation for stroke rehabilitation. *Exp Brain Res* 224(3):335–358.
65. Bhullar, I.S. et al. 2014. More harm than good: Antiseizure prophylaxis after traumatic brain injury does not decrease seizure rates but may inhibit functional recovery. *J Trauma Acute Care Surg* 76(1):54–60; discussion 60–61.
66. Chen, H.I. et al. 2014. Harnessing plasticity for the treatment of neurosurgical disorders: An overview. *World Neurosurg* 82(5):648–659.
67. Rivara, F.P., Graham, R. 2014. Sports-related concussions in youth: Report from the Institute of Medicine and National Research Council. *JAMA* 311(3):239–240.

19 Neurocognitive and Psychiatric Symptoms following Mild Traumatic Brain Injury

Wesley R. Cole and Jason M. Bailie

CONTENTS

Traumatic brain injury (TBI) is associated with a wide variety of symptoms that impact cognition and emotional health. The vast majority of TBI incurred in the United States and across the world are of the mild severity (MTBI). Symptoms after MTBI are generally regarded as transient. Common symptoms include sensory and somatic symptoms (headache, blurred vision, dizziness, sleep problems), neurobehavioral and cognitive complaints (subjective memory problems, poor concentration, slowed information processing, and other cognitive difficulties), and neuropsychiatric complaints related to emotional difficulties (irritability, depression, anxiety).[1,2]

NEUROCOGNITIVE SYMPTOMS AND EXPECTED RECOVERY

Studies exploring self-reported neurobehavioral complaints after MTBI indicate that patients experience the most severe symptoms in the acute stage of injury (<72 hours after injury) and generally start improving over the next several days.[3] The majority of patients experience full recovery within a couple of weeks to

3 months.[1,3,4] However, well-designed prospective studies indicate a small percentage of patients continue to complain of symptoms 12 months after injury.[5]

Large-scale meta-analytical studies suggest that individuals with MTBI have good neurobehavioral and cognitive recovery.[4,6–8] Specifically, recovery in cognitive functioning is found to have a similar trajectory to self-reported postconcussive symptoms. That is, after MTBI, disruptions in cognitive functioning may be observed within the first 6 days of injury with rapid improvement observed and patients becoming statistically indistinguishable from controls on cognitive tests at 1 month postinjury.[9] In a meta-analysis, Belanger and colleagues[6] investigated the long-term impact concussion has on specific cognitive domains. Within 90 days of injury, their analysis revealed a large impact on memory and verbal fluency with moderate impact on attention, language, and visuospatial ability. In patients more than 90 days from injury, the extent of cognitive impairment was dependent on their involvement in litigation as well as the sampling context (e.g., clinical setting versus prospectively studied individuals). The most methodologically sound studies (i.e., prospective) supported the conclusion that MTBI has no long-term impact on global cognitive functioning ($d = 0.04$) after 90 days. There was, however, evidence of an isolated impairment in verbal fluency ($d = 0.98$), though this finding was based on a single study.[10] In contrast, studies that included litigants and studies that recruited participants from clinical settings both found moderate to large impacts in areas of attention, executive functioning, learning, and delayed memory.

In addition to neurocognitive symptoms, individuals sustaining a TBI of any level of severity are at higher risk of developing psychological distress, especially depression,[11,12] anxiety, and post-traumatic stress disorder (PTSD).[13] The prevalence rates of psychological distress after TBI have been reported as low as 4%–5%[14] and as high as 49%–63%.[15] The nature of the TBI as well as pre- and postinjury factors impact the presentation of subsequent psychological distress.

PERSISTENT POSTCONCUSSIVE SYMPTOMS (PPCS)

Though large-scale studies suggest individuals can expect complete recovery following MTBI, there is a subsample of individuals who appear to have persistent problems.[16–18] There has been criticism that the mainstream approaches to investigating MTBI substantiating the high recovery rate for the *majority* of patients (i.e., large group studies, meta analyses, etc.) obscure the *minority* of patients with persistent symptoms.[19–22] For example, Iverson[19] demonstrated that statistical techniques relying on group analyses will not detect a hypothetical subgroup of individuals who are experiencing persistent symptoms. That is, there very well could be a subgroup of patients with persistent postconcussive symptoms (PPCS) who are veiled in traditional methodological and statistical approaches. Though the results of large-scale meta-analyses are extremely valuable in understanding MTBI, Iverson cautioned that the results should not be used to *unequivocally* state that MTBI cannot cause long-term problems in some individuals.

Estimates suggest 10% to 15% of patients who sustain an MTBI do not recover at 1 year postinjury,[18] though it is important to mention this estimate has been criticized.[23] The criticism is based primarily on methodological issues. Longitudinal studies indicate that presence or absence of clinically significant PPCS may vary over time. For example, Lange and colleagues[24] assessed a group of service members with MTBI at 2 months and 12 months postinjury. They observed that more than half of participants had persistent symptoms at both time points. However, the individuals who had persistent symptoms at 2 months were not entirely the same group of participants who had persistent symptoms at 12 months. Specifically, approximately 20% of the sample with persistent symptoms at 2 months improved at 12 months, while almost 17% of the sample who did not have significant symptoms at 2 months reported increased and clinically significant PPCS at 12 months. Similar variability in symptom complaints following an MTBI has also been observed in civilian samples in which the specific type of symptoms reported changed over the course of a year.[23,25,26]

Studies on this topic continue to overcome methodological issues[22,27] and account for confounding situational (e.g., secondary gain), social, psychological, and medical comorbidities[28] that have limited previous work in this area. The prevalence of postconcussive symptoms, as well as its existence as a nosology, remains controversial.

Despite the ongoing debate, MTBI is known to have a substantial impact on disability in the United States, which indirectly speaks to the persistence of postconcussive symptoms. Employment issues represent a noteworthy indicator of overall disability and a sizeable number of patients (5%–20%) with MTBI, experience functional disability manifested by employment problems.[29–33] Given the extremely high incidence of MTBI, believed to be in excess of 100 to 600 per 100,000 people,[34] even a relatively small percentage of patients with persistent disability represent a large societal and healthcare issue.

PSYCHOLOGICAL DISTRESS AND MILD TRAUMATIC BRAIN INJURY (MTBI)

A traumatic brain injury of any severity is by definition "traumatic," and therefore indicates the individual sustaining the injury has experienced trauma. Thus, if there is emotional distress in response to the trauma sustained, and that distress persists, the individual may meet criteria for PTSD or some other psychiatric condition.[35] In the military, as many as one-third of the service members who sustained an MTBI while on combat deployment reported post-traumatic stress symptoms.[36] In those service members who had an MTBI with a loss of consciousness, 44% met criteria for PTSD, compared to only 9% of soldiers who did not sustain a concussion.

Post-traumatic symptoms (PTS) and PTSD after MTBI are also observed in civilians.[37] Between 5% and 14% of civilians with MTBI will have "subthreshold" symptoms of acute stress disorder before being discharged from the hospital[38,39] and 80% of these individuals will go on to develop PTSD.[40] Overall, rates of PTS in civilians who suffered an MTBI is estimated to be between 13% and 40%.[41–45] Oddly, rates of PTSD appear to be higher after MTBI than moderate or severe TBI.[46] There is evidence that individuals who experience longer post-traumatic amnesia (PTA) report less intrusive memories and are less likely to have subsequent PTSD or psychological distress.[47] This suggests that loss of consciousness (LOC) or PTA surrounding injury may be a protective factor and could explain why PTSD rates are higher after

MTBI than moderate or severe TBI, as more severe injuries are typically associated with longer LOC or PTA.[46]

Causal Links between TBI and Psychological Distress

The causal link between brain injury and psychological distress remains unclear. It has traditionally been thought that psychiatric responses after TBI are a reaction to persistent symptoms. That is, loss of functional abilities and independence, necessary lifestyle changes, or a loss of sense of self due to personality, mood, or cognitive changes resulting from the TBI contribute to feelings of depression, anxiety, or even PTSD. Other theories suggest PTSD is a learned response to trauma. That is, PTSD is hypothetically a classically conditioned response to the traumatic injury event.[48] Therefore, it has been posited that, absent of LOC or PTA, the disorientation, confusion, or other alterations of consciousness that occurs after an injury may result in inaccurate or disorganized encoding of memories of the event, subsequently increasing the risk of developing PTS.[35,49]

Though a TBI is a "physical event" and PTSD or other psychiatric problems are a "psychological reaction" to an event, there is growing evidence supporting a link between biological and psychological processes and the subsequent physical and psychological recovery.[37] As such, the injury could cause the psychological distress via physiological mechanisms. When directly linking biology and psychology, research has focused on the limbic regions and inferior frontal and medial temporal lobe structures.[50,51] Langeland and Olff[52] suggested PTS are related to an individual's dysregulated biological response to stress, with direct implications for the hypothalamus–pituitary–adrenal (HPA) axis. Similarly, a limbic-frontal model has been implicated in the development of depression after TBI.[53] Conversion disorder (i.e., psychological manifestation of physical symptoms) after TBI may also be related to injuries to the limbic, frontal, and temporal lobe structures.

There is an alternative viewpoint to the notion that psychological distress is caused by MTBI or persistent symptoms, with significant evidence building that psychological distress is independent of injury and even some evidence that PPCS may be better predicted by psychological distress rather than MTBI. Postconcussive symptoms are not specific to TBI, as many uninjured individuals report experiencing PPCS.[54] Brenner and colleagues[55] demonstrated that self-reported PPCS in service members returning from deployment were independently linked with either MTBI or PTSD.

As early as 1947, there were suggestions that a physical injury is not necessary for persistent injury symptoms.[51] Miller, in 1961, described "accident neurosis" as a psychogenic manifestation of postinjury symptoms rather than being organically caused.[56] Over 25 years ago, Lishman proposed that organic factors were more relevant in the acute injury phase, whereas psychological factors were more relevant during chronic stages of postinjury recovery.[57] In 1996, McClelland suggested that symptoms in the first 3 months after injury were due to physical/biological reasons, and symptoms after this period were due to psychological reasons.[58] More recently, Iverson and Lange[13] state that if symptoms are actually due to TBI, they should be present in the first week after injury.

Growing evidence suggests that the causal link between injury, PPCS, and psychological distress may be in the opposite direction than expected. That is, psychological distress may be a better predictor of the presence of PPCS than an actual injury. Hoge and colleagues[59] investigated the links between injury and subsequent symptoms, including psychological distress and PTSD, in service members returning from deployment. Their findings were somewhat controversial when first published, as they concluded that health problems reported by soldiers 3 to 4 months after sustaining MTBI were largely mediated by PTSD and not the actual injury. Since this publication, numerous other studies have demonstrated a similar effect. That is, non-TBI factors such as psychological distress contribute to the presence or worsening of symptoms in both military and civilian populations.[45,60–63] Specifically, psychological distress, notably depression and anxiety, present within the first 2 weeks postinjury, significantly increases the odds of displaying PPCS at 6 months postinjury.[64]

Premorbid Psychological Distress

Though psychological distress experienced in the acute stage of injury impacts recovery, this variable may not be as important as preinjury mental health. The presence of psychiatric issues, especially high levels of stress, prior to sustaining an injury is an important predictor for both psychological distress and PPCS after injury.[65–69] Ponsford and colleagues[63] found that preinjury factors, notably physical or psychiatric problems, but not the actual MTBI, predicted PPCS. In a review of related civilian literature, McCauley and colleagues[45] concluded that preinjury estimates of depressed mood and estimates of resilience were significant predictors of postinjury anxiety and PPCS, even after accounting for other potentially important factors such as age, gender, and education. However, they reported other preinjury factors predicted PPCS as well, including intelligence, demographics, poor social support, and personality factors.[45]

OTHER VARIABLES IMPACTING NEUROCOGNITIVE AND PSYCHIATRIC OUTCOMES

Demographic Variables

Other researchers in addition to McCauley and colleagues[45] have indicated a variety of premorbid factors may impact an individual's clinical presentation after injury. Several studies have indicated that women are more likely than men to experience PPCS,[70–72] though at least one study[73] reported the opposite. With regard to age, some studies suggest that increased age is a risk factor for PPCS,[71,73] while others have found little or no correlation between PPCS reporting and demographic factors such as age.[74,75] These studies also found no relationship between level of education or socioeconomic status and PPCS.[74,75] Overall, the research in this area can be contradictory and is still emerging, although it suggests that premorbid factors are important in considering an individual's recovery after MTBI.

Blast-Related MTBI

Blast-related MTBI has unique pathophysiological mechanisms compared to blunt-force MTBI.[76] MTBI, especially those related to blast overpressure exposure, has been a cardinal injury sustained by service members deployed to Operation Iraqi Freedom, Operation Enduring Freedom, and Operation New Dawn.[59] Beyond the battlefield, blast-exposure to civilians through acts of terrorism (e.g., Boston Marathon bombing) and accidents (e.g., industrial plant explosions) presents a challenge to our medical community both in terms of response to mass causality incidents[77,78] and proper treatment.[79]

Though mechanically and pathophysiologically unique, blast-related MTBI results in a similar behavioral presentation to nonblast-related MTBI during the acute stage. Luethcke and colleagues[80] evaluated 39 individuals who sustained a blast-related MTBI and 38 individuals with a nonblast but combat-related MTBI within 72 hours of injury. The two groups were similar on measures of cognitive functioning, sleep, global mental health, and post-traumatic stress. The nonblast MTBI group reported more problems with balance, nausea, and vomiting, whereas the blast-related MTBI group reported more problems with hearing. In general, acute blast-related MTBI appears to result in similar clinical presentation to other causes of MTBI. Similarly, multiple studies have indicated that the long-term psychological and cognitive outcome of blast-related MTBI is no different than MTBI from other blunt force mechanisms.[81,82]

Though the clinical presentation of blast-related MTBI and blunt force trauma may be similar, Taber and colleagues[83] recently observed that veterans with a prior history of blast-exposure had poorer white matter integrity compared to healthy controls. Interestingly, this relationship was observed in participants who had notable blast-exposure but who did not have any clinically significant symptoms following the event that would warrant a diagnosis of MTBI. However, the white-matter abnormalities in this blast-exposed group were related to poorer performance on cognitive tasks. The level of exposure to blast overpressure required to result in injury to the brain is not well delineated, and the impact of blast-exposure on neurological functioning may not be fully accounted for by current clinical diagnostic criteria for MTBI.

Repetitive Injuries

Repeated head injuries also impact recovery. Generally speaking, a history of repeated concussions (e.g., three or more) can be associated with greater vulnerability to subsequent injuries[84] and greater memory impairment in the acute to subacute phase of injury.[85] In a study of concussions in the National Hockey League, Benson[86] observed that the time out of play (days) increased 2.25 times for every subsequent concussion. In a systematic review, Belanger and colleagues[87] reported that the overall impact of repeated head injuries on neuropsychological outcome was statistically significant but not clinically meaningful ($d = 0.06$); this same study revealed that memory ($d = 0.16$) and executive functioning ($d = 0.24$) may be particularly vulnerable to multiple concussions.

There is early evidence to suggest repetitive "subconcussive" blows to the head can have neurological consequences that may have long-term neurobehavioral sequelae.[88] These subconcussive events that do not result in a clinically diagnosable concussion are extremely common in contact sports such as rugby and American football. In the last 2 years advanced neuroimaging techniques have been used in investigations of repeated nonconcussive head injuries. These studies have typically evaluated individuals before and after periods of repetitive injury, such as an American football season. Abnormalities on neuroimaging using functional magnetic resonance imaging[89,90] and diffusion tensor imaging[83,91] in individuals exposed to frequent subconcussive events have both been reported. These changes on neuroimaging are correlated with head impact frequency and associated with cognitive and behavioral changes, as assessed by neuropsychological tests. This line of research on repeated subconcussive events dovetails with concerns about the long-term impact of multiple concussions on later cognitive and mental health problems as well as later development of neurodegenerative disease such as chronic traumatic encephalopathy (CTE).[92,93]

Medical Comorbidities

Polytrauma is common in individuals sustaining MTBI, and therefore concurrent physical injuries will play a role in the recovery from MTBI. However, research in the military indicates an inverse relationship between PPCS and other physical injuries sustained in the event. That is, the lowest levels of self-reported postconcussive symptoms are observed in those individuals who have the highest amount of physical injuries. Previous studies investigating combat-related TBI revealed that individuals with combat-related MTBI and bodily injuries tend to endorse fewer neurobehavioral and psychiatric symptoms than patients with only an MTBI.[24,94,95] French and colleagues[94] purported several explanations to account for this relationship. Primarily, the reduction of self-reported psychiatric and somatic symptoms may be a positive reaction to the tangible recovery that follows a physical injury. Recovery from MTBI, or at least perception of symptom severity, may be bolstered by the consistent and predictable recovery that often occurs with orthopedic injuries and wounds. In contrast, the recovery pattern and treatment response rate associated with "invisible wounds" such as MTBI and PTSD can be nebulous to patients and treatment providers. Those with overt injuries may receive more external sympathy from society that reacts strongly to overt combat injuries such as amputation.

METHODOLOGICAL ISSUES IN ASSESSING PPCS

The current literature suggests a complicated relationship between MTBI, PPCS, and psychological distress. Most studies are correlational in nature, and therefore the nature of causality between brain injury, PPCS, and psychological distress is not certain. Methodological issues in the assessment of PPCS and psychological distress further complicate the issue. Many assessments of PPCS and psychological distress rely on self-report of symptoms, which can be inherently flawed. Self-report can vary both between and within individuals, reducing test reliability and subsequently reducing test validity. Though clinicians would like to assume everyone is accurate, thorough, and objective in how they report symptoms, many individuals can be vague, emotional, and incomplete.[28] Additionally, there are numerous biases, which have also been referred to as psychological responses to MTBI or a predisposition to PPCS,[96] that can influence an individual's accuracy in reporting symptoms after injury. These biases have been thoroughly reviewed by Iverson and colleagues,[97] as well as several others, and are briefly reviewed here.

Individuals may be subject to overreporting of symptoms, and Iverson and colleagues liken this to "cognitive hypochondriasis and preoccupation."[97] This can manifest as reporting a previously resolved symptom as unresolved, exaggerating the extent and severity of current symptoms, or fabricating symptoms. Conversely, some individuals may underreport symptoms, which Iverson describes as "stoic." This may occur in athletes or military personnel who do not want to be withheld from play or duty, respectively, or individuals who are minimizing ongoing symptoms and problems for other reasons.

Expectation as etiology or the "nocebo effect" may also impact the reliability of self-report assessment tools.[98–100] That is, the actual injury is not responsible for the development of symptoms. Rather the individual's expectation that they should develop symptoms as a result of a negative event causes misperceptions of postinjury symptoms. In reality, individuals may be attributing benign, everyday events that are common to all people to a recent injury. This is similar to what others have described as the "good-old-days" bias. This results from people overestimating their preinjury levels of functioning, leading to a larger discrepancy between their perceptions of how they were before the injury and how they are after the injury. In studies looking at the accuracy of reporting preinjury functioning, researchers have found individuals with a TBI report fewer preinjury symptoms (i.e., higher levels of functioning) than populations of healthy, uninjured adults.[99–105]

The final bias is referred to as *diagnosis threat*. This was first described by Suhr and Gunstad[106] and results from the attention given to a diagnosis of head injury and its symptoms by medical professionals, family members, friends, or others. The message "you have a head injury" conveyed to the patient can lead to increased symptom reporting or worse performance on objective assessments, such as neurocognitive testing. This bias raises the concern for *iatrogenic disorder*, literally referring to disorder brought on by medical treatment. This is especially a concern when treating MTBI. Studies have demonstrated that the language used by medical professionals when treating individuals who recently sustained an MTBI can impact expected outcomes.[107–109] Terms that sound more medically serious such as "mild *traumatic* brain injury" can result in perceptions of a more serious injury and/or expectations of worse outcomes as compared to terms such as "concussion" or "mild head injury."

Studies investigating outcome following MTBI are also impacted by the presence or absence of the potential for secondary gain by the individual with the injury. Individuals who have something to gain by being "disabled" or otherwise functionally impacted by the injury may exaggerate or fabricate symptoms or deliberately perform poorly on objective testing, such as neurocognitive assessment. Secondary gain can include financial incentives (e.g., disability payments, settlements from litigation), avoiding obligations (e.g., school, work, duty, punishment), or attention from others. In individuals involved in litigation or disability claims, as many as 40% are likely to be exaggerating symptoms on self-report measures and cognitive testing.[110–112] A growing body of literature suggests the participant's effort on neurocognitive testing accounts for a significant portion of the variance in predicting testing outcomes.[110,113–116] Individuals with MTBI may even misrepresent their symptoms to the degree where they clinically resemble individuals with more serious injuries. In a study demonstrating this, Lange and colleagues[117] found that individuals with MTBI who failed performance validity tests scored worse than individuals with severe TBI who had passed performance validity tests. Given such findings, it is a critical part of the clinical evaluation to determine the credibility of the patient's symptom presentation before concluding that cognitive impairment or psychological distress is present.

BIOPSYCHOSOCIAL MODEL

The existent literature paints a complicated picture of the relationship between preinjury factors, the injury, physical symptoms and medical comborbidities, psychological distress, and PPCS. Trying to link PPCS to only physiological or only psychological reasons is an overly simplistic view. Rather, understanding PPCS requires a multifactorial conceptualization, taking into account biological, psychological, and social factors, as well as both pre- and postinjury variables, to adequately understand poor outcome following injury.[28] Therefore, PPCS, including neurocognitive problems and psychological distress, are best understood within the context of a "biopsychosocial" model (Figure 19.1).

Conceptualizing post-traumatic symptoms from a biopsychosocial perspective was first described by Symonds in 1937,[118] and more recently by McLean and colleagues in 2005[119] in the context of post-traumatic chronic pain, and Wood and colleagues[120] as a "diathesis stress model" to describe the link between conversion and PPCS. Within the context of TBI, and especially MTBI, this model has been extensively expanded on by Iverson and colleagues.[97,121] Approaching MTBI and PPCS from a biopsychosocial model provides a framework for assessment that has been described as the only "reasonable way to have a comprehensive, unbiased understanding of this disorder."[121]

The biopsychosocial framework necessitates a multisystems, multimethod approach to assessment that extends well beyond neurocognitive and psychiatric evaluation. Iverson and colleagues[121] describe PPCS as a "diagnosis of exclusion," where clinicians should systematically rule out alternative explanations or diagnoses that may account for persistent symptoms. This requires an extensive history, careful documentation of the nature and course of symptoms, objective testing across multiple domains (e.g., neurological, cognitive, vestibular, vision), internal and external factors potentially influencing symptom presentation (e.g., secondary gain, malingering, iatrogenesis), consideration of preexisting conditions, and rule outs of coexisting medical or mental health conditions that may account for symptom presentation. The biopsychosocial approach to assessment is consistent with current recommended clinical guidelines and will more accurately guide clinicians to areas that should be

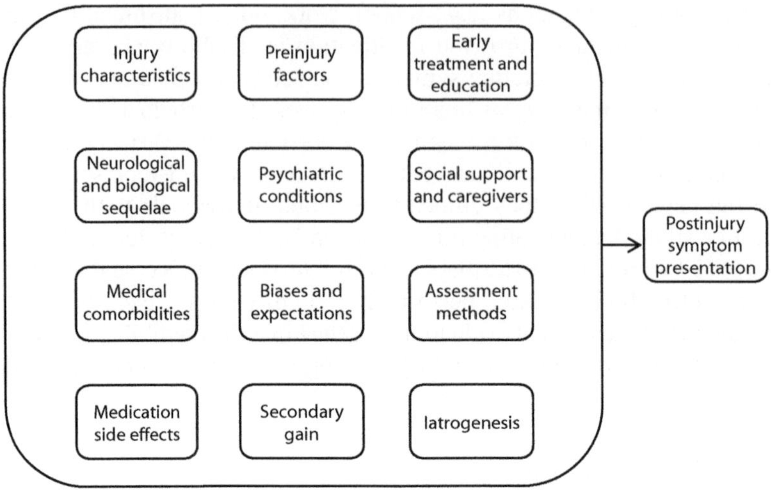

FIGURE 19.1 Biopsychosocial factors that influence symptom presentation after MTBI. (With kind permission from Springer Science+Business Media: *The Little Black Book of Neuropsychology: A Syndrome Based Approach*, M.R. Schoenberg and J.G. Scott, Editors, Post-concussion syndrome, pp. 745–764, Iverson, G.L. and R.T. Lange, 2011; Iverson, G.L., A biopsychosocial conceptualization of poor outcome from mild traumatic brain injury, in *PTSD and Mild Traumatic Brain Injury*, J.J. Vasterling, R.A. Bryant, and T.M. Keane, Editors, 2012, The Guilford Press: New York; Iverson, G.L., N.D. Zasler, and R.T. Lange, Post-concussive disorder, in *Brain Injury Medicine: Principles and Practice*, N.D. Zasler, D. Katz, and R.D. Zafonte, Editors. 2007, Demos: New York.)

the focus of treatment, aid in reducing the risk of iatrogenensis, decrease unnecessary burden on medical systems, and ideally improve long-term neurocognitive and psychiatric outcomes after MTBI.

REFERENCES

1. Carroll, L.J. et al. 2004. Prognosis for mild traumatic brain injury: Results of the WHO Collaborating Centre Task Force on Mild Traumatic Brain Injury. *J Rehabil Med* 43 (Suppl):84–105.
2. Cicerone, K.D., Kalmar, K. 1995. Persistent postconcussion syndrome: The structure of subjective complaints after mild traumatic brain injury. *J Head Trauma Rehabil* 10(3):1–17.
3. McCrea, M. et al. 2003. Acute effects and recovery time following concussion in collegiate football players: The NCAA Concussion Study. *JAMA* 290(19):2556–2563.
4. Belanger, H.G., Vanderploeg, R.D. 2005. The neuropsychological impact of sports-related concussion: A meta-analysis. *J Int Neuropsychol Soc* 11(4):345–357.
5. Iverson, G.L. 2005. Outcome from mild traumatic brain injury. *Curr Opin Psychiatry* 18(3):301–317.
6. Belanger, H.G. et al. 2005. Factors moderating neuropsychological outcomes following mild traumatic brain injury: A meta-analysis. *J Int Neuropsychol Soc* 11(3): 215–227.

7. Rohling, M.L. et al. 2011. A meta-analysis of neuropsychological outcome after mild traumatic brain injury: Re-analysis and reconsiderations of Binder et al. (1997), Frencham et al. (2005), and Pertab et al. (2009). *Neuropsychology* 25:608–623.
8. Frenchman, K., Fox, A., Maybery, M. 2005. Neuropsychological studies of mild traumatic brain injury: A meta-analytic review of research since 1995. *J Clin Exp Neuropsychol* 27:334–351.
9. Schretlen, D.J., Shapiro, A.M. 2003. A quantitative review of the effects of traumatic brain injury on cognitive functioning. *Int Rev Psychiatry* 15(4):341–349.
10. Barrow, I.M. et al. 2003. Can within-category naming identify subtle cognitive deficits in the mild traumatic brain-injured patient? *J Trauma* 54(5):888–895; discussion 895–897.
11. Seel, R.T. et al. 2003. Depression after traumatic brain injury: A National Institute on Disability and Rehabilitation Research Model Systems multicenter investigation. *Arch Phys Med Rehabil* 84(2):177–184.
12. Kreutzer, J.S., Seel, R.T., Gourley, E. 2001. The prevalence and symptom rates of depression after traumatic brain injury: A comprehensive examination. *Brain Inj* 15(7):563–576.
13. Iverson, G.L., Lange, R.T. 2011. Post-concussion syndrome. In *The Little Black Book of Neuropsychology: A Syndrome Based Approach*, eds. M.R. Schoenberg, J.G. Scott, 745–764. New York: Springer.
14. Yang, C.-C. et al. 2009. Early clinical characteristics of patients with persistent post-concussion symptoms: A prospective study. *Brain Inj* 23(4):299–306.
15. Dischinger, P.C. et al. 2009. Early predictors of postconcussive syndrome in a population of trauma patients with mild traumatic brain injury. *J Trauma* 66(2):289–296; discussion 296–297.
16. Ruff, R.M., Camenzuli, L., Mueller, J. 1996. Miserable minority: Emotional risk factors that influence the outcome of a mild traumatic brain injury. *Brain Inj* 10(8):551–565.
17. Ruff, R.M. et al. 1994. Selected cases of poor outcome following a minor brain trauma: Comparing neuropsychological and positron emission tomography assessment. *Brain Inj* 8(4):297–308.
18. Alexander, M.P. 1995. Mild traumatic brain injury: Pathophysiology, natural history, and clinical management. *Neurology* 45(7):1253–1260.
19. Iverson, G.L. 2010. Mild traumatic brain injury meta-analyses can obscure individual differences. *Brain Inj* 24(10):1246–1255.
20. Pertab, J.L., James, K.M., Bigler, E.D. 2009. Limitations of mild traumatic brain injury meta-analyses. *Brain Inj* 23(6):498–508.
21. Rohling, M.L., Larrabee, G.J., Millis, S.R. 2012. The "Miserable Minority" following mild traumatic brain injury: Who are they and do meta-analyses hide them? *Clin Neuropsychol* 26(2):197–213.
22. Bigler, E.D. et al. 2013. Reaffirmed limitations of meta-analytic methods in the study of mild traumatic brain injury: A response to Rohling et al. *Clin Neuropsychol* 27(2):176–214.
23. Greiffenstein, M.F. 2009. Clinical myths of forensic neuropsychology. *Clin Neuropsychol* 23(2):286–296.
24. Lange, R.T. et al. 2011. 12-month outcome of mild traumatic brain injury and polytrauma in U.S. military service members. In *Coping with Blast-Related Traumatic Brain Injury in Returning Troops*, ed. B.K. Wiederhold. Amsterdam: IOS Press.
25. Rutherford, W.H., Merrett, J.D., McDonald, J.R. 1979. Symptoms at one year following concussion from minor head injuries. *Injury* 10(3):225–230.
26. Roe, C. et al. 2009. Post-concussion symptoms after mild traumatic brain injury: Influence of demographic factors and injury severity in a 1-year cohort study. *Disabil Rehabil* 31(15):1235–1243.

27. Larrabee, G.J. et al. 2013. Meta-analytic methods and the importance of non-TBI factors related to outcome in mild traumatic brain injury: Response to Bigler et al. (2013). *Clin Neuropsychol* 27(2):215–237.
28. Iverson, G.L. 2012. A biopsychosocial conceptualization of poor outcome from mild traumatic brain injury. In *PTSD and Mild Traumatic Brain Injury*, eds. J.J. Vasterling, R.A. Bryant, T.M. Keane. New York: The Guilford Press.
29. Kay, T. et al. 1992. Toward a neuropsychological model of functional disability after mild traumatic brain injury. *Neuropsychology* 6(4):371–384.
30. Guerin, F. et al. 2006. Vocational outcome indicators in atypically recovering mild TBI: A post-intervention study. *NeuroRehabilitation* 21(4):295–303.
31. Shames, J. et al. 2007. Return to work following traumatic brain injury: Trends and challenges. *Disabil Rehabil* 29(17):1387–1395.
32. Drake, A.I. et al. 2000. Factors predicting return to work following mild traumatic brain injury: A discriminant analysis. *J Head Trauma Rehabil* 15(5):1103–1112.
33. Cancelliere, C. et al. 2014. Systematic review of return to work after mild traumatic brain injury: Results of the International Collaboration on Mild Traumatic Brain Injury Prognosis. *Arch Phys Med Rehabil* 95(3 Suppl):S201–S209.
34. Cassidy, J.D. et al. 2004. Incidence, risk factors and prevention of mild traumatic brain injury: Results of the WHO Collaborating Centre Task Force on Mild Traumatic Brain Injury. *J Rehabil Med* 43(Suppl):28–60.
35. Brenner, L.A. 2011. Neuropsychological and neuroimaging findings in traumatic brain injury and post-traumatic stress disorder. *Dialogues Clin Neurosci* 13(3):311.
36. Tanielian, T.L., Jaycox, L. 2008. *Invisible Wounds of War: Psychological and Cognitive Injuries, Their Consequences, and Services to Assist Recovery*, vol. 1. Santa Monica, CA: Rand Corporation.
37. Vasterling, J.J., Dikmen, S. 2012. Mild traumatic brain injury and posttraumatic stress disorder: Clinical and conceptual complexities. *J Int Neuropsychol Soc* 18(3):390–393.
38. Broomhall, L.G. et al. 2009. Early stage assessment and course of acute stress disorder after mild traumatic brain injury. *J Nerv Ment Dis* 197(3):178–181.
39. Harvey, A.G., Bryant, R.A. 1998. Predictors of acute stress following mild traumatic brain injury. *Brain Inj* 12(2):147–154.
40. Harvey, A.G., Bryant, R.A. 2000. Two-year prospective evaluation of the relationship between acute stress disorder and posttraumatic stress disorder following mild traumatic brain injury. *Am J Psychiatry* 157(4):626–628.
41. Bryant, R.A., Harvey, A.G. 1995. Acute stress response: A comparison of head injured and non-head injured patients. *Psychol Med* 25(4):869–873.
42. Bryant, R.A. Harvey, A.G. 1998. Relationship between acute stress disorder and posttraumatic stress disorder following mild traumatic brain injury. *Am J Psychiatry* 155(5):625–629.
43. Ohry, A., Rattok, J., Solomon, Z. 1996. Post-traumatic stress disorder in brain injury patients. *Brain Inj* 10(9):687–695.
44. Hickling, E.J. et al. 1998. Traumatic brain injury and posttraumatic stress disorder: A preliminary investigation of neuropsychological test results in PTSD secondary to motor vehicle accidents. *Brain Inj* 12(4):265–274.
45. McCauley, S.R. et al. 2013. Preinjury resilience and mood as predictors of early outcome following mild traumatic brain injury. *J Neurotrauma* 30(8):642–652.
46. Bombardier, C.H. et al. 2006. Posttraumatic stress disorder symptoms during the first six months after traumatic brain injury. *J Neuropsychiatry Clin Neurosci* 18(4):501–508.
47. Bryant, R.A. et al. 2009. Post-traumatic amnesia and the nature of post-traumatic stress disorder after mild traumatic brain injury. *J Int Neuropsychol Soc* 15(6):862–867.
48. Bryant, R.A. 2001. Posttraumatic stress disorder and traumatic brain injury: Can they co-exist? *Clin Psychol Rev* 21(6):931–948.

49. Halligan, S.L. et al. 2003. Posttraumatic stress disorder following assault: The role of cognitive processing, trauma memory, and appraisals. *J Consult Clin Psychol* 71(3):419–431.
50. Mooney, G., Speed, J. 2001. The association between mild traumatic brain injury and psychiatric conditions. *Brain Injury* 15(10):865–877.
51. Bigler, E.D. 2008. Neuropsychology and clinical neuroscience of persistent post-concussive syndrome. *J Int Neuropsychol Soc* 14(1):1–22.
52. Langeland, W., Olff, M. 2008. Psychobiology of posttraumatic stress disorder in pediatric injury patients: A review of the literature. *Neurosci Biobehav Rev* 32(1): 161–174.
53. McCrory, P. et al. 2009. Consensus statement on concussion in sport—The Third International Conference on Concussion in Sport held in Zurich, November 2008. *Phys Sportsmed* 37(2):141–159.
54. Lange, R.T., Iverson, G.L., Rose, A. 2011. Depression strongly influences postconcussion symptom reporting following mild traumatic brain injury. *J Head Trauma Rehabil* 26(2):127–137.
55. Brenner, L.A. et al. 2010. Traumatic brain injury, posttraumatic stress disorder, and postconcussive symptom reporting among troops returning from Iraq. *J Head Trauma Rehabil* 25(5):307–312.
56. Miller, H. 1961. Accident neurosis. *Br Med J* 1(5231):992–998.
57. Lishman, W. 1988. Physiogenesis and psychogenesis in the "post-concussional syndrome." *Br J Psychiatry* 153(4):460–469.
58. McClelland, R. 1996. The postconcussional syndrome: A rose by any other name. *J Psychosom Res* 40(6):563–568.
59. Hoge, C.W. et al. 2008. Mild traumatic brain injury in U.S. Soldiers returning from Iraq. *N Engl J Med* 358(5):453–463.
60. Meares, S. et al. 2011. The prospective course of postconcussion syndrome: The role of mild traumatic brain injury. *Neuropsychology* 25(4):454–465.
61. Silva, M.A. et al. 2012. Abnormal neurological exam findings in individuals with mild traumatic brain injury (mTBI) versus psychiatric and healthy controls. *Clin Neuropsychol* 26(7):1102–1116.
62. Stulemeijer, M. et al. 2007. Cognitive complaints after mild traumatic brain injury: Things are not always what they seem. *J Psychosom Res* 63(6):637–645.
63. Ponsford, J. et al. 2012. Predictors of postconcussive symptoms 3 months after mild traumatic brain injury. *Neuropsychology* 26(3):304–313.
64. Hou, R. et al. 2012. When a minor head injury results in enduring symptoms: A prospective investigation of risk factors for postconcussional syndrome after mild traumatic brain injury. *J Neurol Neurosurg Psychiatry* 83(2):217–223.
65. Silverberg, N.D., Iverson, G.L. 2011. Etiology of the post-concussion syndrome: Physiogenesis and psychogenesis revisited. *NeuroRehabilitation* 29(4):317–329.
66. Bay, E., Bergman, K. 2006. Symptom experience and emotional distress after traumatic brain injury. *Care Manag J* 7(1):3–9.
67. Fenton, G. et al. 1993. The postconcussional syndrome: Social antecedents and psychological sequelae. *Br J Psychiatry* 162:493–497.
68. Mooney, G., Speed, J., Sheppard, S. 2005. Factors related to recovery after mild traumatic brain injury. *Brain Inj* 19(12):975–987.
69. Ponsford, J. et al. 2000. Factors influencing outcome following mild traumatic brain injury in adults. *J Int Neuropsychol Soc* 6(5):568–579.
70. Preiss-Farzanegan, S.J. et al. 2009. The relationship between gender and postconcussion symptoms after sport-related mild traumatic brain injury. *PM R* 1(3):245–253.
71. King, N.S., Kirwilliam, S. 2011. Permanent post-concussion symptoms after mild head injury. *Brain Inj* 25(5):462–470.

72. Meares, S. et al. 2008. Mild traumatic brain injury does not predict acute postconcussion syndrome. *J Neurol Neurosurg Psychiatry* 79(3):300–306.
73. Garden, N., Sullivan, K.A. 2010. An examination of the base rates of post-concussion symptoms: The influence of demographics and depression. *Appl Neuropsychol* 17(1):1–7.
74. Kashluba, S., Paniak, C., Casey, J.E. 2008. Persistent symptoms associated with factors identified by the WHO Task Force on Mild Traumatic Brain Injury. *Clin Neuropsychol* 22(2):195–208.
75. Sigurdardottir, S. et al. 2009. Post-concussion symptoms after traumatic brain injury at 3 and 12 months post-injury: A prospective study. *Brain Inj* 23(6):489–497.
76. Duckworth, J.L., Grimes, J.B., Ling, G. 2013. Pathophysiology of battlefield associated traumatic brain injury. *Pathophysiology* 20:23–30.
77. Noji, E.K. et al. 2005. Investigation of Federal Bureau of Investigation bomb-related death and injury data in the United States between 1988 and 1997. *Mil Med* 170(7):595–598.
78. Kapur, G.B. et al. 2005. The United States twenty-year experience with bombing incidents: Implications for terrorism preparedness and medical response. *J Trauma* 59(6):1436–1444.
79. Bochicchio, G.V. et al. 2008. Blast injury in a civilian trauma setting is associated with a delay in diagnosis of traumatic brain injury. *Am Surg* 74(3):267–270.
80. Luethcke, C.A. et al. 2011. Comparisson of concussive symptoms, cognitive performance, and psycholgoical symptoms between acute blast-versus non-blast-induced mild traumatic brain injury. *J Int Neuropsychol Soc* 17(1):36–45.
81. Belanger, H.G. et al. 2009. Cognitive sequelae of blast-related versus other mechanisms of brain trauma. *J Int Neuropsychol Soc* 15(1):1–8.
82. Lange, R.T. et al. 2012. Neuropsychological outcome from blast versus non-blast: Mild traumatic brain injury in U.S. military service members. *J Int Neuropsychol Soc* 18(3):595–605.
83. Taber, K.H. et al. 2015. White matter compromise in veterans exposed to primary blast forces. *J Head Trauma Rehabil* 30(1): E15–E25.
84. Collins, M.W. et al. 2002. Cumulative effects of concussion in high school athletes. *Neurosurgery* 51(5):1175–1179; discussion 1180–1181.
85. Iverson, G.L. et al. 2004. Cumulative effects of concussion in amateur athletes. *Brain Inj* 18(5):433–443.
86. Benson, B.W. et al. 2011. A prospective study of concussions among National Hockey League players during regular season games: The NHL-NHLPA Concussion Program. *CMAJ* 183(8):905–911.
87. Belanger, H.G., Spiegel, E., Vanderploeg, R.D. 2010. Neuropsychological performance following a history of multiple self-reported concussions: A meta-analysis. *J Int Neuropsychol Soc* 16(2):262–267.
88. Bailes, J.E. et al. 2013. Role of subconcussion in repetitive mild traumatic brain injury. *J Neurosurg* 119(5):1235–1245.
89. Breedlove, E.L. et al. 2012. Biomechanical correlates of symptomatic and asymptomatic neurophysiological impairment in high school football. *J Biomech* 45(7): 1265–1272.
90. Talavage, T.M. et al. 2014. Functionally-detected cognitive impairment in high school football players without clinically-diagnosed concussion. *J Neurotrauma* 31(4):327–338.
91. Bazarian, J.J. et al. 2014. Persistent, long-term cerebral white matter changes after sports-related repetitive head impacts. *PLoS One* 9(4):e94734.
92. Baugh, C.M. et al. 2012. Chronic traumatic encephalopathy: Neurodegeneration following repetitive concussive and subconcussive brain trauma. *Brain Imaging Behav* 6(2):244–254.

93. Turner, R.C. et al. 2012. Repetitive traumatic brain injury and development of chronic traumatic encephalopathy: A potential role for biomarkers in diagnosis, prognosis, and treatment? *Front Neurol* 3:186.
94. French, L.M. et al. 2012. Influence of bodily injuries on symptom reporting following uncomplicated mild traumatic brain injury in US military service members. *J Head Trauma Rehabil* 27(1):63–74.
95. Kennedy, J.E. et al. 2010. Symptoms in military service members after blast mTBI with and without associated injuries. *NeuroRehabilitation* 26(3):191–197.
96. McCrae, M.A. 2008. *Mild Traumatic Brain Injury and Postconcussion Syndrome: The New Evidence Base for Diagnosis and Treatment*. New York: Oxford University Press.
97. Iverson, G.L. et al. 2012. Conceptualizing outcome from mild traumatic brain injury. In *Brain Injury Medicine: Principles and Practice*, eds. N.D. Zasler, D. Katz, R.D. Zafonte, 470–497. New York: Demos Medical Publishing.
98. Ferguson, R.J. et al. 1999. Postconcussion syndrome following sports-related head injury: Expectation as etiology. *Neuropsychology* 13(4):582–589.
99. Gunstad, J., Suhr, J.A. 2001. "Expectation as etiology" versus "the good old days": Postconcussion syndrome symptom reporting in athletes, headache sufferers, and depressed individuals. *J Int Neuropsychol Soc* 7(3):323–333.
100. Mittenberg, W. et al. 1992. Symptoms following mild head injury: Expectation as aetiology. *J Neurol Neurosurg Psychiatry* 55(3):200–204.
101. Davis, C.H. 2002. Self-perception in mild traumatic brain injury. *Am J Phys Med Rehabil* 81(8):609–621.
102. Gunstad, J., Suhr, J.A. 2004. Cognitive factors in postconcussion syndrome symptom report. *Arch Clin Neuropsychol* 19(3):391–405.
103. Hilsabeck, R.C., Gouvier, W.D., Bolter, J.F. 1998. Reconstructive memory bias in recall of neuropsychological symptomatology. *J Clin Exp Neuropsychol* 20(3): 328–338.
104. Iverson, G.L. et al. 2010. "Good old days" bias following mild traumatic brain injury. *Clin Neuropsychol* 24(1):17–37.
105. Lange, R.T., Iverson, G.L., Rose, A. 2010. Post-concussion symptom reporting and the "good-old-days" bias following mild traumatic brain injury. *Arch Clin Neuropsychol* 25(5):442–450.
106. Suhr, J.A., Gunstad, J. 2002. "Diagnosis threat": The effect of negative expectations on cognitive performance in head injury. *J Clin Exp Neuropsychol* 24(4):448–457.
107. Kempe, C.B., Sullivan, K.A., Edmed, S.L. 2013. CE the effect of varying diagnostic terminology within patient discharge information on expected mild traumatic brain injury outcome. *Clin Neuropsychol* 27(5):762–778.
108. Sullivan, K.A., Edmed, S.L., Kempe, C. 2014. The effect of injury diagnosis on illness perceptions and expected postconcussion syndrome and posttraumatic stress disorder symptoms. *J Head Trauma Rehabil* 29(1):54–64.
109. Weber, M., Edwards, M.G. 2010. The effect of brain injury terminology on university athletes' expected outcome from injury, familiarity and actual symptom report. *Brain Inj* 24(11):1364–1371.
110. Green, P. et al. 2001. Effort has a greater effect on test scores than severe brain injury in compensation claimants. *Brain Inj* 15(12):1045–1060.
111. Larrabee, G.J. 2003. Detection of malingering using atypical performance patterns on standard neuropsychological tests. *Clin Neuropsychol* 17(3):410–425.
112. Larrabee, G.J., Millis, S.R., Meyers, J.E. 2009. 40 plus or minus 10, a new magical number: Reply to Russell. *Clin Neuropsychol* 23(5):841–849.
113. Armistead-Jehle, P., Buican, B. 2012. Evaluation context and Symptom Validity Test performances in a U.S. military sample. *Arch Clin Neuropsychol* 27(8):828–839.

114. Green, P. 2007. The pervasive influence of effort on neuropsychological tests. *Phys Med Rehabil Clin N Am* 18(1):43–68, vi.
115. Fox, D.D. 2011. Symptom validity test failure indicates invalidity of neuropsychological tests. *Clin Neuropsychol* 25(3):488–495.
116. Lange, R.T. et al. 2010. Influence of poor effort on self-reported symptoms and neurocognitive test performance following mild traumatic brain injury. *J Clin Exp Neuropsychol* 32(9):961–972.
117. Lange, R.T. et al. 2012. Influence of poor effort on neuropsychological test performance in U.S. military personnel following mild traumatic brain injury. *J Clin Exp Neuropsychol* 34(5):453–466.
118. Symonds, C. 1937. The assessment of symptoms following head injury. *Guys Hospital Gaz* 51:461–468.
119. McLean, S.A. et al. 2005. The development of persistent pain and psychological morbidity after motor vehicle collision: Integrating the potential role of stress response systems into a biopsychosocial model. *Psychosom Med* 67(5):783–790.
120. Wood, R.L. 2004. Understanding the "miserable minority": A diasthesis-stress paradigm for post-concussional syndrome. *Brain Inj* 18(11):1135–1153.
121. Iverson, G.L., Zasler, N.D., Lange, R.T. 2007. Post-concussive disorder. In *Brain Injury Medicine: Principles and Practice*, eds. N.D. Zasler, D. Katz, R.D. Zafonte. New York: Demos.

Index

Page numbers followed by f and t indicate figures and tables, respectively.

B

C

D

G

O

Q

R

U

V

W

X

An environmentally friendly book printed and bound in England by www.printondemand-worldwide.com

PEFC Certified
This product is from sustainably managed forests and controlled sources
www.pefc.org
PEFC/16-33-415

This book is made of chain-of-custody materials; FSC materials for the cover and PEFC materials for the text pages.